URBAN STORMWATER
MANAGEMENT
IN THE UNITED STATES

Committee on Reducing Stormwater Discharge
Contributions to Water Pollution

Water Science and Technology Board

Division on Earth and Life Studies

NATIONAL RESEARCH COUNCIL
OF THE NATIONAL ACADEMIES

THE NATIONAL ACADEMIES PRESS
Washington, D.C.
www.nap.edu

THE NATIONAL ACADEMIES PRESS 500 Fifth Street, N.W. Washington, DC 20001

NOTICE: The project that is the subject of this report was approved by the Governing Board of the National Research Council, whose members are drawn from the councils of the National Academy of Sciences, the National Academy of Engineering, and the Institute of Medicine. The members of the committee responsible for the report were chosen for their special competences and with regard for appropriate balance.

Support for this project was provided by the U.S. Environmental Protection Agency under Award No. 68-C-03-081. Any opinions, findings, conclusions, or recommendations expressed in this publication are those of the author(s) and do not necessarily reflect the views of the organizations or agencies that provided support for the project.

The views and conclusions contained in this document are those of the authors and should not be interpreted as necessarily representing the official policies, either expressed or implied of the U.S. Government.

International Standard Book Number-13: 978-0-309-12539-0 (Book)
International Standard Book Number-10: 0-309-12539-1 (Book)
International Standard Book Number-13: 978-0-309-12540-6 (PDF)
International Standard Book Number-10: 0-309-12540-5 (PDF)
Library of Congress Control Number: 2008940395

Urban Stormwater Management in the United States is available from the National Academies Press, 500 Fifth Street, N.W., Lockbox 285, Washington, DC 20055; (800) 624-6242 or (202) 334-3313 (in the Washington metropolitan area); Internet, http://www.nap.edu. Cover photo courtesy of Roger Bannermann.

Printed in the United States of America.

THE NATIONAL ACADEMIES
Advisers to the Nation on Science, Engineering, and Medicine

The **National Academy of Sciences** is a private, nonprofit, self-perpetuating society of distinguished scholars engaged in scientific and engineering research, dedicated to the furtherance of science and technology and to their use for the general welfare. Upon the authority of the charter granted to it by the Congress in 1863, the Academy has a mandate that requires it to advise the federal government on scientific and technical matters. Dr. Ralph J. Cicerone is president of the National Academy of Sciences.

The **National Academy of Engineering** was established in 1964, under the charter of the National Academy of Sciences, as a parallel organization of outstanding engineers. It is autonomous in its administration and in the selection of its members, sharing with the National Academy of Sciences the responsibility for advising the federal government. The National Academy of Engineering also sponsors engineering programs aimed at meeting national needs, encourages education and research, and recognizes the superior achievement of engineers. Dr. Charles M. Vest is president of the National Academy of Engineering.

The **Institute of Medicine** was established in 1970 by the National Academy of Sciences to secure the services of eminent members of appropriate professions in the examination of policy matters pertaining to the health of the public. The Institute acts under the responsibility given to the National Academy of Sciences by its congressional charter to be an adviser to the federal government and, upon its own initiative, to identify issues of medical care, research, and education. Dr. Harvey V. Fineberg is president of the Institute of Medicine.

The **National Research Council** was organized by the National Academy of Sciences in 1916 to associate the broad community of science and technology with the Academy's purposes of furthering knowledge and advising the federal government. Functioning in accordance with general policies determined by the Academy, the Council has become the principal operating agency of both the National Academy of Sciences and the National Academy of Engineering in providing services to the government, the public, and the scientific and engineering communities. The Council is administered jointly by both Academies and the Institute of Medicine. Dr. Ralph, J. Cicerone and Dr. Charles M. Vest are chair and vice-chair, respectively, of the National Research Council.

www.national-academies.org

COMMITTEE ON REDUCING STORMWATER DISCHARGE CONTRIBUTIONS TO WATER POLLUTION

CLAIRE WELTY, *Chair*, University of Maryland, Baltimore County
LAWRENCE E. BAND, University of North Carolina, Chapel Hill
ROGER T. BANNERMAN, Wisconsin Department of Natural Resources, Madison
DEREK B. BOOTH, Stillwater Sciences, Inc., Santa Barbara, California
RICHARD R. HORNER, University of Washington, Seattle
CHARLES R. O'MELIA, Johns Hopkins University, Baltimore, Maryland
ROBERT E. PITT, University of Alabama, Tuscaloosa
EDWARD T. RANKIN, Institute for Local Government Administration and Rural Development, Ohio University, Athens
THOMAS R. SCHUELER, Chesapeake Stormwater Network, Baltimore, Maryland
KURT STEPHENSON, Virginia Polytechnic Institute and State University, Blacksburg
XAVIER SWAMIKANNU, California EPA, Los Angeles Regional Water Board
ROBERT G. TRAVER, Villanova University, Philadelphia, Pennsylvania
WENDY E. WAGNER, University of Texas School of Law, Austin
WILLIAM E. WENK, Wenk Associates, Inc., Denver, Colorado

National Research Council Staff

LAURA J. EHLERS, Study Director
ELLEN A. DE GUZMAN, Research Associate

Preface

Stormwater runoff from the built environment remains one of the great challenges of modern water pollution control, as this source of contamination is a principal contributor to water quality impairment of waterbodies nationwide. In addition to entrainment of chemical and microbial contaminants as stormwater runs over roads, rooftops, and compacted land, stormwater discharge poses a physical hazard to aquatic habitats and stream function, owing to the increase in water velocity and volume that inevitably result on a watershed scale as many individually managed sources are combined. Given the shift of the world's population to urban settings, and that this trend is expected to be accompanied by continued wholesale landscape alteration to accommodate population increases, the magnitude of the stormwater problem is only expected to grow.

In recognition of the need for improved control measures, in 1987 the U.S. Congress mandated the U.S. Environmental Protection Agency (EPA), under amendments to the Clean Water Act, to control certain stormwater discharges under the National Pollutant Discharge Elimination System. In response to this federal legislation, a permitting program was put in place by EPA as the Phase I (1990) and Phase II (1999) stormwater regulations, which together set forth requirements for municipal separate storm sewer systems and industrial activities including construction. The result of the regulatory program has been identification of hundreds of thousands of sources needing to be permitted, which has put a strain on EPA and state administrative systems for implementation and management. At the same time, achievement of water quality improvement as a result of the permit requirements has remained an elusive goal.

To address the seeming intractability of this problem, the EPA requested that the National Research Council (NRC) review its current permitting program for stormwater discharge under the Clean Water Act and provide suggestions for improvement. The broad goals of the study were to better understand the links between stormwater pollutant discharges and ambient water quality, to assess the state of the science of stormwater management, and to make associated policy recommendations. More specifically, the study was asked to:

(1) Clarify the mechanisms by which pollutants in stormwater discharges affect ambient water quality criteria and define the elements of a "protocol" to link pollutants in stormwater discharges to ambient water quality criteria.

(2) Consider how useful monitoring is for both determining the potential of a discharge to contribute to a water quality standards violation and for determining the adequacy of stormwater pollution prevention plans. What specific parameters should be monitored and when and where? What effluent

limits and benchmarks are needed to ensure that the discharge does not cause or contribute to a water quality standards violation?

(3) Assess and evaluate the relationship between different levels of stormwater pollution prevention plan implementation and in-stream water quality, considering a broad suite of best management practices (BMPs).

(4) Make recommendations for how to best stipulate provisions in stormwater permits to ensure that discharges will not cause or contribute to exceedances of water quality standards. This should be done in the context of general permits. As a part of this task, the committee will consider currently available information on permit and program compliance.

(5) Assess the design of the stormwater permitting program implemented under the Clean Water Act.

There are a number of related topics that one might expect to find in this report that are excluded, because EPA requested that the study be limited to problems addressed by the agency's stormwater regulatory program. Specifically, nonpoint source pollution from agricultural runoff, septic systems, combined sewer overflows, sanitary sewer overflows, and concentrated animal feeding operations are not addressed in this report. In addition, alteration of the urban base-flow hydrograph from a number of causes that are not directly related to storm events (e.g., interbasin transfers of water, leakage from water supply pipes, lawn irrigation, and groundwater withdrawals) is a topic outside the scope of the report and therefore not included in any depth.

In developing this report, the committee benefited greatly from the advice and input of EPA representatives, including Jenny Molloy, Linda Boornazian, and Mike Borst; representatives from the City of Austin; representatives from King County, Washington, and the City of Seattle; and representatives from the Irvine Ranch Water District. The committee heard presentations by many of these individuals in addition to Chris Crockett, City of Philadelphia Water Department; Pete LaFlamme and Mary Borg, Vermont Department of Environmental Conservation; Michael Barrett, University of Texas at Austin; Roger Glick, City of Austin; Michael Piehler, UNC Institute of Marine Sciences, Keith Stolzenbach, UCLA; Steve Burges, University of Washington; Wayne Huber, Oregon State University; Don Theiler, King County; Charlie Logue, Clean Water Services, Hillsboro, Oregon; Don Duke, Florida Gulf Coast University; Mike Stenstrom, UCLA; Gary Wolff, California Water Board; Paula Daniels, City of Los Angeles Public Works; Mark Gold, Heal the Bay; Geoff Brosseau, California Stormwater Quality Association; Steve Weisberg, Southern California Coastal Water Research Project; Chris Crompton, Southern California Stormwater Monitoring Coalition; David Beckman, NRDC; and Eric Strecker, Geosyntec. We also thank all those stakeholders who took time to share with us their perspectives and wisdom about the various issues affecting stormwater.

The committee was fortunate to have taken several field trips in conjunction with committee meetings. The following individuals are thanked for their participation in organizing and guiding these trips: Austin (Kathy Shay, Mike Kelly, Matt Hollon, Pat Hartigan, Mateo Scoggins, David Johns, and Nancy McClintock); Seattle (Darla Inglis, Chris May, Dan Powers, Scott Bawden, Nat Scholz, John Incardona, Kate McNeil, Bob Duffner, and Curt Crawford); and Los Angeles (Peter Postlmayr, Matthew Keces, Alan Bay, and Sat Tamaribuchi).

Completion of this report would not have been possible without the Herculean efforts of project study director Laura Ehlers. Her powers to organize, probe, synthesize, and keep the committee on track with completing its task were simply remarkable. Meeting logistics and travel arrangements were ably assisted by Ellen De Guzman and Jeanne Aquilino.

This report has been reviewed in draft form by individuals chosen for their diverse perspectives and technical expertise, in accordance with procedures approved by the NRC's Report Review Committee. The purpose of this independent review is to provide candid and critical comments that will assist the institution in making its published report as sound as possible and to ensure that the report meets institutional standards for objectivity, evidence, and responsiveness to the study charge. The review comments and draft manuscript remain confidential to protect the integrity of the deliberative process. We wish to thank the following individuals for their review of this report: Michael Barrett, University of Texas; Bruce Ferguson, University of Georgia; James Heaney, University of Florida; Daniel Medina, CH2MHILL; Margaret Palmer, University of Maryland Chesapeake Biological Laboratory; Kenneth Potter, University of Wisconsin; Joan Rose, Michigan State University; Eric Strecker, Geosyntec; and Bruce Wilson, Minnesota Pollution Control Agency.

Although the reviewers listed above have provided many constructive comments and suggestions, they were not asked to endorse the conclusions and recommendations nor did they see the final draft of the report before its release. The review of this report was overseen by Michael Kavanaugh, Malcolm Pirnie, Inc., and Richard Conway, Union Carbide Corporation, retired. Appointed by the NRC, they were responsible for making certain that an independent examination of this report was carried out in accordance with institutional procedures and that all review comments were carefully considered. Responsibility for the final content of this report rests entirely with the authoring committee and institution.

Claire Welty,
Committee Chair

Contents

Summary

Urbanization is the changing of land use from forest or agricultural uses to suburban and urban areas. This conversion is proceeding in the United States at an unprecedented pace, and the majority of the country's population now lives in suburban and urban areas. The creation of impervious surfaces that accompanies urbanization profoundly affects how water moves both above and below ground during and following storm events, the quality of that stormwater, and the ultimate condition of nearby rivers, lakes, and estuaries.

The National Pollutant Discharge Elimination System (NPDES) program under the Clean Water Act (CWA) is the primary federal vehicle to regulate the quality of the nation's waterbodies. This program was initially developed to reduce pollutants from industrial process wastewater and municipal sewage discharges. These point sources were known to be responsible for poor, often drastically degraded conditions in receiving waterbodies. They were easily regulated because they emanated from identifiable locations, such as pipe outfalls. To address the role of stormwater in causing or contributing to water quality impairments, in 1987 Congress wrote Section 402(p) of the CWA, bringing stormwater control into the NPDES program, and in 1990 the U.S. Environmental Protection Agency (EPA) issued the Phase I Stormwater Rules. These rules require NPDES permits for operators of municipal separate storm sewer systems (MS4s) serving populations over 100,000 and for runoff associated with industry, including construction sites five acres and larger. In 1999 EPA issued the Phase II Stormwater Rule to expand the requirements to small MS4s and construction sites between one and five acres in size.

With the addition of these regulated entities, the overall NPDES program has grown by almost an order of magnitude. EPA estimates that the total number of permittees under the stormwater program at any time exceeds half a million. For comparison, there are fewer than 100,000 non-stormwater (meaning wastewater) permittees covered by the NPDES program. To manage the large number of permittees, the stormwater program relies heavily on the use of general permits to control industrial, construction, and Phase II MS4 discharges. These are usually statewide, one-size-fits-all permits in which general provisions are stipulated.

To comply with the CWA regulations, industrial and construction permittees must create and implement a stormwater pollution prevention plan, and MS4 permittees must implement a stormwater management plan. These plans document the stormwater control measures (SCMs) (sometimes known as best management practices or BMPs) that will be used to prevent stormwater emanating from these sources from degrading nearby waterbodies. These SCMs range from structural methods such as detention ponds and bioswales to nonstructural methods such as designing new development to reduce the percentage of impervious surfaces.

A number of problems with the stormwater program as it is currently implemented have been recognized. First, there is limited information available on the effectiveness and longevity of many SCMs, thereby contributing to uncertainty in their performance. Second, the requirements for monitoring vary depending on the regulating entity and the type of activity. For example, a subset of industrial facilities must conduct "benchmark monitoring" and the results often exceed the values established by EPA or the states, but it is unclear whether these exceedances provide useful indicators of potential water quality problems. Finally, state and local stormwater programs are plagued by a lack of resources to review stormwater pollution prevention plans and conduct regular compliance inspections. For all these reasons, the stormwater program has suffered from poor accountability and uncertain effectiveness at improving the quality of the nation's waters.

In light of these challenges, EPA requested the advice of the National Research Council's Water Science and Technology Board on the federal stormwater program, considering all entities regulated under the program (i.e., municipal, industrial, and construction). The following statement of task guided the work of the committee:

(1) Clarify the mechanisms by which pollutants in stormwater discharges affect ambient water quality criteria and define the elements of a "protocol" to link pollutants in stormwater discharges to ambient water quality criteria.

(2) Consider how useful monitoring is for both determining the potential of a discharge to contribute to a water quality standards violation and for determining the adequacy of stormwater pollution prevention plans. What specific parameters should be monitored and when and where? What effluent limits and benchmarks are needed to ensure that the discharge does not cause or contribute to a water quality standards violation?

(3) Assess and evaluate the relationship between different levels of stormwater pollution prevention plan implementation and in-stream water quality, considering a broad suite of SCMs.

(4) Make recommendations for how to best stipulate provisions in stormwater permits to ensure that discharges will not cause or contribute to exceedances of water quality standards. This should be done in the context of general permits. As a part of this task, the committee will consider currently available information on permit and program compliance.

(5) Assess the design of the stormwater permitting program implemented under the CWA.

Chapter 2 of this report presents the regulatory history of stormwater control in the United States, focusing on relevant portions of the CWA and the federal and state regulations that have been created to implement the Act. Chapter 3 reviews the scientific aspects of stormwater, including sources of pollutants in stormwater, how stormwater moves across the land surface, and its impacts on receiving waters. Chapter 4 evaluates the current industrial and MS4 monitoring requirements, and it considers the multitude of models available for linking stormwater discharges to ambient water quality. Chapter 5 considers the vast suite of both structural and nonstructural measures designed to control stormwater and reduce its pollutant loading to waterbodies. In Chapter 6, the limitations and possibilities associated with a new regulatory approach are explored, as are those of a more traditional but enhanced scheme. This new approach, which rests on the broad foundation of correlative studies demonstrating the effects of urbanization on aquatic ecosystems, would reduce the impact of stormwater on receiving waters beyond any efforts currently in widespread practice.

THE CHALLENGE OF REGULATING STORMWATER

Although stormwater has been long recognized as contributing to water quality impairment, the creation of federal regulations to deal with stormwater quality has occurred only in the last 20 years. Because this longstanding environmental problem is being addressed so late in the development and management of urban areas, the laws that mandate better stormwater control are generally incomplete and are often in conflict with state and local rules that have primarily stressed the flood control aspects of stormwater management (i.e., moving water away from structures and cities as fast as possible). Many prior investigators have observed that stormwater discharges would ideally be regulated through direct controls on land use, strict limits on both the quantity and quality of stormwater runoff into surface waters, and rigorous monitoring of adjacent waterbodies to ensure that they are not degraded by stormwater discharges. Future land-use development would be controlled to minimize stormwater discharges, and impervious cover and volumetric restrictions would serve as proxies for stormwater loading from many of these developments. Products that contribute pollutants through stormwater—like de-icing materials, fertilizers, and vehicular exhaust—would be regulated at a national level to ensure that the most environmentally benign materials are used.

Presently, however, the regulation of stormwater is hampered by its association with a statute that focuses primarily on specific pollutants and ignores the volume of discharges. Also, most stormwater discharges are regulated on an individualized basis without accounting for the cumulative contributions from multiple sources in the same watershed. Perhaps most problematic is that the requirements governing stormwater dischargers leave a great deal of discretion to the dischargers themselves in developing stormwater pollution prevention plans and self-monitoring to ensure compliance. These problems are exacerbated by the fact that the dual responsibilities of land-use planning and stormwater management within local governments are frequently decoupled.

EPA's current approach to regulating stormwater is unlikely to produce an accurate or complete picture of the extent of the problem, nor is it likely to adequately control stormwater's contribution to waterbody impairment. The lack of rigorous end-of-pipe monitoring, coupled with EPA's failure to use flow or alternative measures for regulating stormwater, make it difficult for EPA to develop enforceable requirements for stormwater discharg-ers. Instead, the stormwater permits leave a great deal of discretion to the regu-lated community to set their own standards and to self-monitor. Current statis-tics on the states' implementation of the stormwater program, discharger com-pliance with stormwater requirements, and the ability of states and EPA to in-corporate stormwater permits with Total Maximum Daily Loads are uniformly discouraging. Radical changes to the current regulatory program (see Chapter 6) appear necessary to provide meaningful regulation of stormwater dischargers in the future.

Flow and related parameters like impervious cover should be consid-ered for use as proxies for stormwater pollutant loading. These analogs for the traditional focus on the "discharge" of "pollutants" have great potential as a federal stormwater management tool because they provide specific and measur-able targets, while at the same time they focus regulators on water degradation resulting from the increased volume as well as increased pollutant loadings in stormwater runoff. Without these more easily measured parameters for evaluat-ing the contribution of various stormwater sources, regulators will continue to struggle with enormously expensive and potentially technically impossible at-tempts to determine the pollutant loading from individual dischargers or will rely too heavily on unaudited and largely ineffective self-reporting, self-policing, and paperwork enforcement.

EPA should engage in much more vigilant regulatory oversight in the national licensing of products that contribute significantly to stormwater pollution. De-icing chemicals, materials used in brake linings, motor fuels, asphalt sealants, fertilizers, and a variety of other products should be examined for their potential contamination of stormwater. Currently, EPA does not appar-ently utilize its existing licensing authority to regulate these products in a way that minimizes their contribution to stormwater contamination. States can also enact restrictions on or tax the application of pesticides or other particularly toxic products. Even local efforts could ultimately help motivate broader scale, federal restrictions on particular products.

The federal government should provide more financial support to state and local efforts to regulate stormwater. State and local governments do not have adequate financial support to implement the stormwater program in a rig-orous way. At the very least, Congress should provide states with financial sup-port for engaging in more meaningful regulation of stormwater discharges. EPA should also reassess its allocation of funds within the NPDES program. The agency has traditionally directed funds to focus on the reissuance of NPDES wastewater permits, while the present need is to advance the NPDES stormwater program because NPDES stormwater permittees outnumber wastewater permit-tees more than five fold, and the contribution of diffuse sources of pollution to degradation of the nation's waterbodies continues to increase.

EFFECTS OF URBANIZATION ON WATERSHEDS

Urbanization causes change to natural systems that tends to occur in the following sequence. First, land use and land cover are altered as vegetation and topsoil are removed to make way for agriculture, or subsequently buildings, roads, and other urban infrastructure. These changes, and the introduction of a constructed drainage network, alter the hydrology of the local area, such that receiving waters in the affected watershed experience radically different flow regimes than prior to urbanization. Nearly all of the associated problems result from one underlying cause: loss of the water-retaining and evapotranspirating functions of the soil and vegetation in the urban landscape. In an undeveloped area, rainfall typically infiltrates into the ground surface or is evapotranspirated by vegetation. In the urban landscape, these processes of evapotranspiration and water retention in the soil are diminished, such that stormwater flows rapidly across the land surface and arrives at the stream channel in short, concentrated bursts of high discharge. This transformation of the hydrologic regime is a wholesale reorganization of the processes of runoff generation, and it occurs throughout the developed landscape. When combined with the introduction of pollutant sources that accompany urbanization (such as lawns, motor vehicles, domesticated animals, and industries), these changes in hydrology have led to water quality and habitat degradation in virtually all urban streams.

The current state of the science has documented the characteristics of stormwater runoff, including its quantity and quality from many different land covers, as well as the characteristics of dry weather runoff. In addition, many correlative studies show how parameters co-vary in important but complex and poorly understood ways (e.g., changes in macroinvertebrate or fish communities associated with watershed road density or the percentage of impervious cover). Nonetheless, efforts to create mechanistic links between population growth, land-use change, hydrologic alteration, geomorphic adjustments, chemical contamination in stormwater, disrupted energy flows and biotic interactions, and changes in ecological communities are still in development. Despite this assessment, there are a number of overarching truths that remain poorly integrated into stormwater management decision-making, although they have been robustly characterized for more than a decade and have a strong scientific basis that reaches even farther back through the history of published investigations.

There is a direct relationship between land cover and the biological condition of downstream receiving waters. The possibility for the highest levels of aquatic biological condition exists only with very light urban transformation of the landscape. Conversely, the lowest levels of biological condition are inevitable with extensive urban transformation of the landscape, commonly seen after conversion of about one-third to one-half of a contributing watershed into impervious area. Although not every degraded waterbody is a product of intense urban development, all highly urban watersheds produce severely degraded receiving waters.

The protection of aquatic life in urban streams requires an approach that incorporates all stressors. Urban Stream Syndrome reflects a multitude of effects caused by altered hydrology in urban streams, altered habitat, and polluted runoff. Focusing on only one of these factors is not an effective management strategy. For example, even without noticeably elevated pollutant concentrations in receiving waters, alterations in their hydrologic regimes are associated with impaired biological condition. More comprehensive biological monitoring of waterbodies will be critical to better understanding the cumulative impacts of urbanization on stream condition.

The full distribution and sequence of flows (i.e., the flow regime) should be taken into consideration when assessing the impacts of stormwater on streams. Permanently increased stormwater volume is only one aspect of an urban-altered storm hydrograph. It contributes to high in-stream velocities, which in turn increase streambank erosion and accompanying sediment pollution of surface water. Other hydrologic changes, however, include changes in the sequence and frequency of high flows, the rate of rise and fall of the hydrograph, and the season of the year in which high flows can occur. These all can affect both the physical and biological conditions of streams, lakes, and wetlands. Thus, effective hydrologic mitigation for urban development cannot just aim to reduce post-development peak flows to predevelopment peak flows.

Roads and parking lots can be the most significant type of land cover with respect to stormwater. They constitute as much as 70 percent of total impervious cover in ultra-urban landscapes, and as much as 80 percent of the directly connected impervious cover. Roads tend to capture and export more stormwater pollutants than other land covers in these highly impervious areas, especially in regions of the country having mostly small rainfall events. As rainfall amounts become larger, pervious areas in most residential land uses become more significant sources of runoff, sediment, nutrients, and landscaping chemicals. In all cases, directly connected impervious surfaces (roads, parking lots, and roofs that are directly connected to the drainage system) produce the first runoff observed at a storm-drain inlet and outfall because their travel times are the quickest.

MONITORING AND MODELING

The stormwater monitoring requirements under the EPA Stormwater Program are variable and generally sparse, which has led to considerable skepticism about their usefulness. This report considers the amount and value of the data collected over the years by municipalities (which are substantial on a nationwide basis) and by industries, and it makes suggestions for improvement. The MS4 and particularly the industrial stormwater monitoring programs suffer from a paucity of data, from inconsistent sampling techniques, and from requirements that are difficult to relate to the compliance of individual dischargers. For these reasons, conclusions about stormwater management are usually made with incomplete information. Stormwater management would benefit most substantially from a well-balanced monitoring program that encompasses chemical, biological, and physical parameters from outfalls to receiving waters.

Many processes connect sources of pollution to an effect observed in a downstream receiving water—processes that can be represented in watershed models, which are the key to linking stormwater dischargers to impaired receiving waters. The report explores the current capability of models to make such links, including simple models and more involved mechanistic models. At the present time, stormwater modeling has not evolved enough to consistently say whether a particular discharger can be linked to a specific waterbody impairment. Some quantitative predictions can be made, particularly those that are based on well-supported causal relationships of a variable that responds to changes in a relatively simple driver (e.g., modeling how a runoff hydrograph or pollutant loading change in response to increased impervious land cover). However, in almost all cases, the uncertainty in the modeling and the data (including its general unavailability), the scale of the problems, and the presence of multiple stressors in a watershed make it difficult to assign to any given source a specific contribution to water quality impairment.

Because of a 10-year effort to collect and analyze monitoring data from MS4s nationwide, the quality of stormwater from urbanized areas is well characterized. These results come from many thousands of storm events, systematically compiled and widely accessible; they form a robust dataset of utility to theoreticians and practitioners alike. These data make it possible to accurately estimate stormwater pollutant concentrations from various land uses. Additional data are available from other stormwater permit holders that were not originally included in the database and from ongoing projects, and these should be acquired to augment the database and improve its value in stormwater management decision-making.

Industry should monitor the quality of stormwater discharges from certain critical industrial sectors in a more sophisticated manner, so that permitting authorities can better establish benchmarks and technology-based effluent guidelines. Many of the benchmark monitoring requirements and effluent guidelines for certain industrial subsectors are based on inaccurate and old information. Furthermore, there has been no nationwide compilation and analysis of industrial benchmark data, as has occurred for MS4 monitoring data, to better understand typical stormwater concentrations of pollutants from various industries.

Continuous, flow-weighted sampling methods should replace the traditional collection of stormwater data using grab samples. Data obtained from too few grab samples are highly variable, particularly for industrial monitoring programs, and subject to greater uncertainly because of experimenter error and poor data-collection practices. In order to use stormwater data for decision making in a scientifically defensible fashion, grab sampling should be abandoned as a credible stormwater sampling approach for virtually all applications. It should be replaced by more accurate and frequent continuous sampling methods that are flow weighted. Flow-weighted composite monitoring should continue for the duration of the rain event. Emerging sensor systems that provide high temporal resolution and real-time estimates for specific pollutants should be further investigated, with the aim of providing lower costs and more extensive monitoring systems to sample both streamflow and constituent loads.

Watershed models are useful tools for predicting downstream impacts from urbanization and designing mitigation to reduce those impacts, but they are incomplete in scope and do not offer definitive causal links between polluted discharges and downstream degradation. Every model simulates only a subset of the multiple interconnections between physical, chemical, and biological processes found in any watershed, and they all use a grossly simplified representation of the true spatial and temporal variability of a watershed. To speak of a "comprehensive watershed model" is thus an oxymoron, because the science of stormwater is not sufficiently far advanced to determine causality between all sources, resulting stressors, and their physical, chemical, and biological responses. Thus, it is not yet possible to create a protocol that mechanistically links stormwater dischargers to the quality of receiving waters. The utility of models with more modest goals, however, can still be high—as long as the questions being addressed by the model are in fact relevant and important to the functioning of the watershed to which that model is being applied, and sufficient data are available to calibrate the model for the processes included therein.

STORMWATER MANAGEMENT APPROACHES

A fundamental component of EPA's stormwater program is the creation of stormwater pollution prevention plans that document the SCMs that will be used to prevent the permittee's stormwater discharges from degrading local waterbodies. Thus, a consideration of these measures—their effectiveness in meeting different goals, their cost, and how they are coordinated with one another—is central to any evaluation of the stormwater program. The statement of task asks for an evaluation of the relationship between different levels of stormwater pollution prevention plan implementation and in-stream water quality. Although the state of knowledge has yet to reveal the mechanistic links that would allow for a full assessment of that relationship, enough is known to design systems of SCMs, on a site-scale or local watershed scale, that can substantially reduce the effects of urbanization.

The characteristics, applicability, goals, effectiveness, and cost of nearly 20 different broad categories of SCMs to treat the quality and quantity of stormwater runoff are discussed in Chapter 5, organized as they might be applied from the rooftop to the stream. SCMs, when designed, constructed, and maintained correctly, have demonstrated the ability to reduce runoff volume and peak flows and to remove pollutants. A multitude of case studies illustrates the use of SCMs in specific settings and demonstrates that a particular SCM can have a measurable positive effect on water quality or a biological metric. However, the implementation of SCMs at the watershed scale has been too inconsistent and too recent to be able to definitively link their performance to the prolonged sustainment—at the watershed level—of receiving water quality, in-stream habitat, or stream geomorphology.

Individual controls on stormwater discharges are inadequate as the sole solution to stormwater in urban watersheds. SCM implementation needs to be designed as a system, integrating structural and nonstructural SCMs and incorporating watershed goals, site characteristics, development land use, construction erosion and sedimentation controls, aesthetics, monitoring, and maintenance. Stormwater cannot be adequately managed on a piecemeal basis due to the complexity of both the hydrologic and pollutant processes and their effect on habitat and stream quality. Past practices of designing detention basins on a site-by-site basis have been ineffective at protecting water quality in receiving waters and only partially effective in meeting flood control requirements.

Nonstructural SCMs such as product substitution, better site design, downspout disconnection, conservation of natural areas, and watershed and land-use planning can dramatically reduce the volume of runoff and pollutant load from a new development. Such SCMs should be considered first before structural practices. For example, lead concentrations in stormwater have been reduced by at least a factor of 4 after the removal of lead from gasoline. Not creating impervious surfaces or removing a contaminant from the runoff stream simplifies and reduces the reliance on structural SCMs.

SCMs that harvest, infiltrate, and evapotranspirate stormwater are critical to reducing the volume and pollutant loading of small storms. Urban municipal separate stormwater conveyance systems have been designed for flood control to protect life and property from extreme rainfall events, but they have generally failed to address the more frequent rain events (<2.5 cm) that are key to recharge and baseflow in most areas. These small storms may only generate runoff from paved areas and transport the "first flush" of contaminants. SCMs designed to remove this class of storms from surface runoff (runoff-volume-reduction SCMs—rainwater harvesting, vegetated, and subsurface) can also help address larger watershed flooding issues.

Performance characteristics are starting to be established for most structural and some nonstructural SCMs, but additional research is needed on the relevant hydrologic and water quality processes within SCMs across

different climates and soil conditions. Typical data such as long-term load reduction efficiencies and pollutant effluent concentrations can be found in the International Stormwater BMP Database. However, understanding the processes involved in each SCM is in its infancy, making modeling of these SCMs difficult. Seasonal differences, the time between storms, and other factors all affect pollutant loadings emanating from SCMs. Research is needed that moves away from the use of percent removal and toward better simulation of SCM performance. Research is particularly important for nonstructural SCMs, which in many cases are more effective, have longer life spans, and require less maintenance than structural SCMs. EPA should be a leader in SCM research, both directly by improving its internal modeling efforts and by funding state efforts to monitor and report back on the success of SCMs in the field.

The retrofitting of urban areas presents both unique opportunities and challenges. Promoting growth in these areas is desirable because it takes pressure off the suburban fringes, thereby preventing sprawl, and it minimizes the creation of new impervious surfaces. However, it is more complex than Greenfields development because of the need to upgrade existing infrastructure, the limited availability and affordability of land, and the complications caused by rezoning. These sites may be contaminated, requiring cleanup before redevelopment can occur. Both innovative zoning and development incentives, along with the careful selection SCMs, are needed to achieve fair and effective storm-water management in these areas. For example, incentive or performance zoning could be used to allow for greater densities on a site, freeing other portions of the site for SCMs. Publicly owned, consolidated SCMs should be strongly considered as there may be insufficient land to have small, on-site systems. The performance and maintenance of the former can be overseen more effectively by a local government entity. The types of SCMs that are used in consolidated facilities—particularly detention basins, wet/dry ponds, and stormwater wetlands—perform multiple functions, such as prevention of streambank erosion, flood control, and large-scale habitat provision.

INNOVATIVE STORMWATER MANAGEMENT
AND REGULATORY PERMITTING

There are numerous innovative regulatory strategies that could be used to improve the EPA's stormwater program. The course of action most likely to check and reverse degradation of the nation's aquatic resources would be to **base all stormwater and other wastewater discharge permits on watershed boundaries instead of political boundaries.** Watershed-based permitting is the regulated allowance of discharges of water and wastes borne by those discharges to waters of the United States, with due consideration of: (1) the implications of those discharges for preservation or improvement of prevailing ecological conditions in the watershed's aquatic systems, (2) cooperation among political jurisdictions sharing a watershed, and (3) coordinated regulation and management of all discharges having the potential to modify the hydrology and water quality of the watershed's receiving waters.

Responsibility and authority for implementation of watershed-based permits would be centralized with a municipal lead permittee working in partnership with other municipalities in the watershed as co-permittees. Permitting authorities (designated states or, otherwise, EPA) would adopt a minimum goal in every watershed to avoid any further loss or degradation of designated beneficial uses in the watershed's component waterbodies and additional goals in some cases aimed at recovering lost beneficial uses. Permittees, with support by the states or EPA, would then move to comprehensive impact source analysis as a foundation for targeting solutions. The most effective solutions are expected to lie in isolating, to the extent possible, receiving waterbodies from exposure to those impact sources. In particular, low-impact design methods, termed Aquatic Resources Conservation Design in this report, should be employed to the fullest extent feasible and backed by conventional SCMs when necessary.

The approach gives municipal co-permittees more responsibility, with commensurately greater authority and funding, to manage all of the sources discharging, directly or through municipally owned conveyances, to the waterbodies comprising the watershed. This report also outlines a new monitoring program structured to assess progress toward meeting objectives and the overlying goals, diagnosing reasons for any lack of progress, and determining compliance by dischargers. The proposal further includes market-based trading of credits among dischargers to achieve overall compliance in the most efficient manner and adaptive management to determine additional actions if monitoring demonstrates failure to achieve objectives.

As a first step to taking the proposed program nationwide, a pilot program is recommended that will allow EPA to work through some of the more predictable impediments to watershed-based permitting, such as the inevitable limits of an urban municipality's authority within a larger watershed.

Short of adopting watershed-based permitting, other smaller-scale changes to the EPA stormwater program are possible. These recommendations do not preclude watershed-based permitting at some future date, and indeed they lay the groundwork in the near term for an eventual shift to watershed-based permitting.

Integration of the three permitting types is necessary, such that construction and industrial sites come under the jurisdiction of their associated municipalities. Federal and state NPDES permitting authorities do not presently have, and can never reasonably expect to have, sufficient personnel to inspect and enforce stormwater regulations on more than 100,000 discrete point source facilities discharging stormwater. A better structure would be one where the NPDES permitting authority empowers the MS4 permittees to act as the first tier of entities exercising control on stormwater discharges to the MS4 to protect water quality. The National Pretreatment Program, EPA's successful treatment program for municipal and industrial wastewater sources, could serve as a model for integration.

To improve the industrial, construction, and MS4 permitting programs in their current configuration, EPA should (1) issue guidance for MS4, industrial, and construction permittees on what constitutes a design storm for water quality purposes; (2) issue guidance for MS4 permittees on methods to identify high-risk industrial facilities for program prioritization such as inspections; (3) support the compilation and collection of quality industrial stormwater effluent data and SCM effluent quality data in a national database; and (4) develop numerical expressions of the MS4 standard of "maximum extent practicable." Each of these issues is discussed in greater detail in Chapter 6.

***

Watershed-based permitting will require additional resources and regulatory program support. Such an approach shifts more attention to ambient outcomes as well as expanded permitting coverage. Additional resources for program implementation could come from shifting existing programmatic resources. For example, some state permitting resources may be shifted away from existing point source programs toward stormwater permitting. Strategic planning and prioritization could shift the distribution of federal and state grant and loan programs to encourage and support more watershed-based stormwater permitting programs. However, securing new levels of public funds will likely be required. All levels of government must recognize that additional resources may be required from citizens and businesses (in the form of taxes, fees, etc.) in order to operate a more comprehensive and effective stormwater permitting program.

1
Introduction

URBANIZATION AND ITS IMPACTS

The influence of humans on the physical and biological systems of the Earth's surface is not a recent manifestation of modern societies; instead, it is ubiquitous throughout our history. As human populations have grown, so has their footprint, such that between 30 and 50 percent of the Earth's surface has now been transformed (Vitousek et al., 1997). Most of this land area is not covered with pavement; indeed, less than 10 percent of this transformed surface is truly "urban" (Grübler, 1994). However, urbanization causes extensive changes to the land surface beyond its immediate borders, particularly in ostensibly rural regions, through alterations by agriculture and forestry that support the urban population (Lambin et al., 2001). Within the immediate boundaries of cities and suburbs, the changes to natural conditions and processes wrought by urbanization are among the most radical of any human activity.

In the United States, population is growing at an annual rate of 0.9 percent (U.S. Census Bureau, *http://www.census.gov/compendia/statab/2007edition.html*); the majority of the population of the United States now lives in suburban and urban areas (Figure 1-1). Because the area appropriated for urban land uses is growing even faster, these patterns of growth all but guarantee that the influences of urban land uses will continue to expand over time. Cities and suburbia obviously provide the homes and livelihood for most of the nation's population. But, as this report makes clear, these benefits have been accompanied by significant environmental change. Urbanization of the landscape profoundly affects how water moves both above and below ground during and following storm events; the quality of that stormwater (defined in Box 1-1); and the ultimate condition of nearby rivers, lakes, and estuaries. Unlike agriculture, which can display significant interchange with forest cover over time scales of a century (e.g., Hart, 1968), there is no indication that once-urbanized land ever returns to a less intensive state. Urban land, however, does continue to change over time; by one estimate, 42 percent of land currently considered "urban" in the United States will be redeveloped by 2030 (Brookings Institute, 2004). In their words, "nearly half of what will be the built environment in 2030 doesn't even exist yet" (p. vi). This truth belies the common belief that efforts to improve management of stormwater are doomed to irrelevancy because so much of the landscape is already built. Opportunities for improvement have indeed been lost, but many more still await an improved management approach.

Measures of urbanization are varied, and the disparate methods of quantifying the presence and influence of human activity tend to confound analyses of environmental effects. Population density is a direct metric of human presence, but it is not the most relevant measure of the influence of those people on their surrounding landscape. Expressions of the built environment, most commonly

13

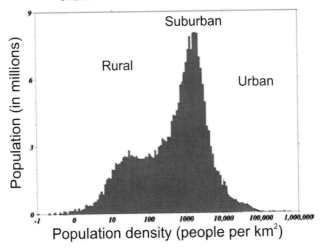

FIGURE 1-1 Histogram of population for the United States, based on 2000 census data. The median population density is about 1,000 people/km^2. SOURCE: Modified from Pozzi and Small (2005), who place the rural–suburban boundary at 100 people/km^2. Reprinted, with permission, from ASPRS (2005). Copyright 2005 by the American Society for Photogrammetry and Remote Sensing.

BOX 1-1
What Is "Stormwater"?

"Stormwater" is a term that is used widely in both scientific literature and regulatory documents. It is also used frequently throughout this report. Although all of these usages share much in common, there are important differences that benefit from an explicit discussion.

Most broadly, stormwater runoff is the water associated with a rain or snow storm that can be measured in a downstream river, stream, ditch, gutter, or pipe shortly after the precipitation has reached the ground. What constitutes "shortly" depends on the size of the watershed and the efficiency of the drainage system, and a number of techniques exist to precisely separate stormwater runoff from its more languid counterpart, "baseflow." For small and highly urban watersheds, the interval between rainfall and measured stormwater discharges may be only a few minutes. For watersheds of many tens or hundreds of square miles, the lag between these two components of storm response may be hours or even a day.

From a regulatory perspective, stormwater must pass through some sort of engineered conveyance, be it a gutter, a pipe, or a concrete canal. If it simply runs over the ground surface, or soaks into the soil and soon reemerges as seeps into a nearby stream, it may be water generated by the storm but it is not regulated stormwater.

This report emphasizes the first, more hydrologically oriented definition. However, attention is focused mainly on that component of stormwater that emanates from those parts of a landscape that have been affected in some fashion by human activities ("urban stormwater"). Mostly this includes water that flows over the ground surface and is subsequently collected by natural channels or artificial conveyance systems, but it can also include water that has infiltrated into the ground but nonetheless reaches a stream channel relatively rapidly and that contributes to the increased stream discharge that commonly accompanies almost any rainfall event in a human-disturbed watershed.

road density or pavement coverage as a percentage of gross land area, are more likely to determine stormwater runoff-related consequences. An inverse metric, the percentage of mature vegetation or forest across a landscape, expresses the magnitude of related, but not identical, impacts to downstream systems. Alternatively, these measures of land cover can be replaced by measures of land use, wherein the types of human activity (e.g., residential, industrial, commercial) are used as proxies for the suite of hydrologic, chemical, and biological changes imposed on the surrounding landscape.

All of these metrics of urbanization are strongly correlated, although none can directly substitute for another. They also are measured differently, which renders one or another more suitable for a given application. Land use is a common measure in the realm of urban planning, wherein current and future conditions for a city or an entire region are characterized using equivalent categories across parcels, blocks, or broad regions. Road density can be reliably and rapidly measured, either manually or in a Geographic Information System environment, and it commonly displays a very good correlation with other measures of human activity. "Land cover," however, and particularly the percentage of impervious cover, is the metric most commonly used in studying the effects of urban development on stormwater, because it clearly expresses the hydrologic influence and watershed scale of urbanization. Box 1-2 describes the ways in which the percent of impervious cover in a watershed is measured.

There is no universally accepted terminology to describe land-cover or land-use conditions along the rural-to-urban gradient. Pozzi and Small (2005), for example, identified "rural," "suburban," and "urban" land uses on the basis of population density and vegetation cover, but they did not observe abrupt transitions that suggested natural boundaries (see Figure 1-1). In contrast, the Center for Watershed Protection (2005) defined the same terms but used impervious area percentage as the criterion, with such labels as "rural" (0 to 10 percent imperviousness), "suburban" (10 to 25 percent imperviousness), "urban" (25 to 60 percent imperviousness) and "ultra-urban" (greater than 60 percent imperviousness).

Beyond the problems posed by precise yet inconsistent definitions for commonly used words, none of the boundaries specified by these definitions are reflected in either hydrologic or ecosystem responses. Hydrologic response is strongly dependent on both land cover and drainage connectivity (e.g., Leopold, 1968); ecological responses in urbanizing watersheds do not show marked thresholds along an urban gradient (e.g., Figure 1-2) and they are dependent on not only the sheer magnitude of urban development but also the spatial configuration of that development across the watershed (Alberti et al., 2006). This report, therefore, uses such terms as "urban" and "suburban" under their common usage, without implying or advocating for a more precise (but ultimately limited and discipline-specific) definition.

Changing land cover and land use influence the physical, chemical, and biological conditions of downstream waterways. The specific mechanisms by which this influence occurs vary from place to place, and even a cursory review

BOX 1-2
Measures of Impervious Cover

The percentage of impervious surface or cover in a landscape is the most frequently used measure of urbanization. Yet this parameter has its limitations, in part because it has not been consistently used or defined. Most significant is the distinction between *total* impervious area (TIA) and *effective* impervious area (EIA). TIA is the "intuitive" definition of imperviousness: that fraction of the watershed covered by constructed, non-infiltrating surfaces such as concrete, asphalt, and buildings. Hydrologically, however, this definition is incomplete for two reasons. First, it ignores nominally "pervious" surfaces that are sufficiently compacted or otherwise so low in permeability that the rate of runoff from them is similar or indistinguishable from pavement. For example, Burges and others (1998) found that the impervious unit-area runoff was only 20 percent greater than that from pervious areas—primarily thin sodded lawns over glacial till—in a western Washington residential subdivision. Clearly, this hydrologic contribution cannot be ignored entirely.

The second limitation of TIA is that it includes some paved surfaces that may contribute nothing to the stormwater-runoff response of the downstream channel. A gazebo in the middle of parkland, for example, probably will impose no hydrologic changes into the catchment except for a very localized elevation of soil moisture at the edge of its roof. Less obvious, but still relevant, would be the different downstream consequences of rooftops that drain alternatively into a piped storm-drain system with direct discharge into a natural stream or onto splash blocks that disperse the runoff onto the garden or lawn at each corner of the building. This metric therefore cannot recognize any stormwater mitigation that may result from alternative runoff-management strategies, for example, pervious pavements or rainwater harvesting.

The first of these TIA limitations, the production of significant runoff from nominally pervious surfaces, is typically ignored in the characterization of urban development. The reason for such an approach lies in the difficulty in identifying such areas and estimating their contribution, and because of the credible belief that the degree to which pervious areas shed water as overland flow should be related, albeit imperfectly, with the amount of *impervious* area: where construction and development are more intense and cover progressively greater fractions of the

of the literature demonstrates that many different factors can be important, such as changes to flow regime, physical and chemical constituents in the water column, or the physical form of the stream channel itself (Paul and Meyer, 2001). Not all of these changes are present in any given system—lakes, wetlands, and streams can be altered by human activity in many different ways, each unique to the activity and the setting in which it occurs. Nonetheless, direct influences of land-use change on freshwater systems commonly include the following (Naiman and Turner, 2000):

- Altering the composition and structure of the natural flora and fauna,
- Changing disturbance regimes,
- Fragmenting the land into smaller and more diverse parcels, and
- Changing the juxtaposition between parcel types.

Historically, human-induced alteration was not universally seen as a problem. In particular, dams and other stream-channel "improvements" were a watershed, it is more likely that the intervening green spaces have been stripped and compacted during construction and only imperfectly rehabilitated for their hydrologic functions during subsequent "landscaping."

The second of these TIA limitations, inclusion of non-contributing impervious areas, is formally addressed through the concept of EIA, defined as the impervious surfaces with direct hydraulic connection to the downstream drainage (or stream) system. Thus, any part of the TIA that drains onto pervious (i.e., "green") ground is excluded from the measurement of EIA. This parameter, at least conceptually, captures the hydrologic significance of imperviousness. EIA is the parameter normally used to characterize urban development in hydrologic models.

The direct measurement of EIA is complicated. Studies designed specifically to quantify this parameter must make direct, independent measurements of both TIA and EIA (Alley and Veenhuis, 1983; Laenen, 1983; Prysch and Ebbert, 1986). The results can then be generalized either as a correlation between the two parameters or as a "typical" value for a given land use. Sutherland (1995) developed an equation that describes the relationship between EIA and TIA. Its general form is:

$$EIA = A\,(TIA)^B$$

where A and B are a unique combination of numbers that satisfy the following criteria:

TIA = 1 then EIA = 0%
TIA = 100 then EIA = 100%

A commonly used version of this equation (EIA = $0.15\,TIA^{1.41}$) was based on samples from highly urbanized land uses in Denver, Colorado (Alley and Veenhuis, 1983; Gregory et al., 2005). These results, however, are almost certainly region- and even neighborhood-specific, and, although highly relevant to watershed studies, they can be quite laborious to develop.

common activity of municipal and federal engineering works of the mid-20th century (Williams and Wolman, 1984). "Flood control" implied a betterment of conditions, at least for streamside residents (Chang, 1992). And fisheries "enhancements," commonly reflected by massive infrastructure for hatcheries or artificial spawning channels, were once seen as unequivocal benefits for fish populations (White, 1996; Levin et al., 2001).

By almost any currently applied metric, however, the net result of human alteration of the landscape to date has resulted in a degradation of the conditions in downstream watercourses. Many prior researchers, particularly when considering ecological conditions and metrics, have recognized a crude but monotonically declining relationship between human-induced landscape alteration and downstream conditions (e.g., Figure 1-2; Horner et al., 1997; Davies and Jackson, 2006). These include metrics of physical stream-channel conditions (e.g., Bledsoe and Watson, 2001), chemical constituents (e.g., Figure 1-3; House et al., 1993), and biological communities (e.g., Figure 1-4; Steedman, 1988; Wang et al., 1997).

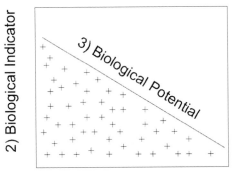

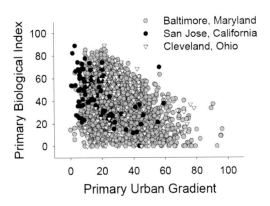

FIGURE 1-2 Conceptual model (top) and actual response (bottom) of a biological system's response to stress. The "Urban Gradient of Stressors" might be a single metric of urbanization, such as percent watershed impervious or road density; the "Biological Indicator" may be single-metric or multi-metric measures of the level of disturbance in an aquatic community. The right-declining line traces the limits of a "factor-ceiling distribution" (Thomson et al., 1986), wherein individual sites (i.e., data points) have a wide range of potential values for a given position along the urban gradient but are not observed above a maximum possible limit of the biological index. The bottom graph illustrates actual biological responses, using a biotic index developed to show responses to urban impacts plotted against a standardized urban gradient comprising urban land use, road density, and population. SOURCE: Top figure reprinted, with permission, from Davies and Jackson (2006). Copyright by the Ecological Society of America. Bottom figure reprinted, with permission, from Barbour et al. (2006). Copyright by the Water Environment Research Foundation.

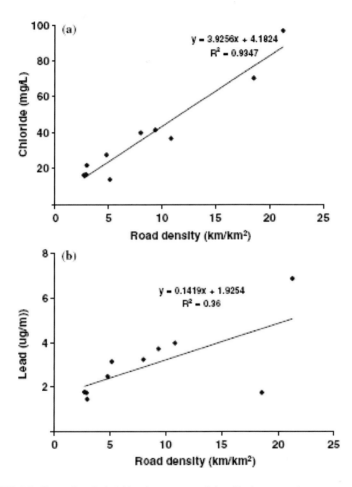

FIGURE 1-3 Example relationships between road density (a surrogate measure of urban development) and common water quality constituents. Direct causality is not necessarily implied by such relationships, but the monotonic increase in concentrations with increasing "urbanization," however measured, is near-universal. SOURCE: Reprinted, with permission, from Chang and Carlson (2005). Copyright 2005 by Springer.

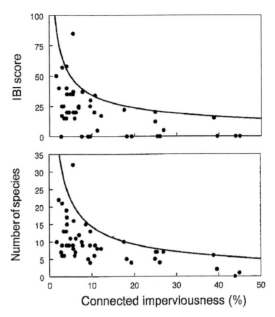

FIGURE 1-4   Plots of Effective Impervious Area (EIA, or "connected imperviousness") against metrics of biologic response in fish populations. SOURCE: Reprinted, with permission, from Wang et al. (2001). Copyright 2001 by Springer.

The association between watercourse degradation and landscape alteration in general, and urban development in particular, seems inexorable. The scientific and regulatory challenge of the last three decades has been to decouple this relationship, in some cases to reverse its trend and in others to manage where these impacts are to occur.

WHAT'S WRONG WITH THE NATION'S WATERS?

Since passage of the Water Quality Act of 1948 and the Clean Water Act (CWA) of 1972, 1977, and 1987, water quality in the United States has measurably improved in the major streams and rivers and in the Great Lakes. However, substantial challenges and problems remain. Major reporting efforts that have examined state and national indicators of condition, such as CWA 305(b) reports (EPA, 2002) and the Heinz State of the Nation's Ecosystem report (Heinz Center, 2002), or environmental monitoring that was designed to provide statistically valid estimates of condition (e.g., National Wadeable Stream Assessment; EPA, 2006), have confirmed widespread impairments related to diffuse sources of pollution and stressors.

The National Water Quality Inventory (derived from Section 305b of the CWA) compiles data in relation to use designations and water quality standards.

As discussed in greater detail in Chapter 2, such standards include both (1) a description of the use that a waterbody is supposed to achieve (such as a source of drinking water or a cold water fishery) and (2) narrative or numeric criteria for physical, chemical, and biological parameters that allow the designated use to be achieved. As of 2002, 45 percent of assessed streams and rivers, 47 percent of assessed lakes, 32 percent of assessed estuarine areas, 17 percent of assessed shoreline miles, 87 percent of near-coastal ocean areas, 51 percent of assessed wetlands, 91 percent of assessed Great Lakes shoreline miles, and 99 percent of assessed Great Lakes open water areas were not meeting water quality standards set by the states (2002 EPA Report to Congress).[1]

The U.S. Environmental Protection Agency (EPA) has also embarked on a five-year statistically valid survey of the nation's waters (http://www.epa.gov/owow/monitoring/guide.pdf). To date, two waterbody types—coastal areas and wadeable streams—have been assessed. The most recent data indicate that 42 percent of wadeable streams are in poor biological condition and 25 percent are in fair condition (EPA, 2006). The overall condition of the nation's estuaries is generally fair, with Puerto Rico and Northeast Coast regions rated poor, the Gulf Coast and West Coast regions rated fair, and the Southeast Coast region rated good to fair (EPA, 2007). These condition ratings for the National Estuary Program are based on a water quality index, a sediment quality index, a benthic index, and a fish tissue contaminants index.

The impairment of waterbodies is manifested in a multitude of ways. Indeed, EPA's primary process for reporting waterbody condition (Section 303(d) of the CWA—see Chapter 2) identifies over 200 distinct types of impairments. As shown in Table 1-1, these have been categorized into 15 broad categories, encompassing about 94 percent of all impairments. 59,515 waterbodies fall into one of the top 15 categories, while the total reported number of waterbodies impaired from all causes is 63,599 (which is an underestimate of the actual total because not all waterbodies are assessed). Mercury, microbial pathogens, sediments, other metals, and nutrients are the major pollutants associated with impaired waterbodies nationwide. These constituents have direct impacts on aquatic ecosystems and public health, which form the basis of the water quality standards set for these compounds. Sediments can harm fish and macroinvertebrate communities by introducing sorbed contaminants, decreasing available light in streams, and smothering fish eggs. Microbial pathogens can cause disease to humans via both ingestion and dermal contact and are frequently cited as the cause of beach closures and other recreational water hazards in lakes and estuaries. Nutrient over-enrichment can promote a cascade of events in waterbodies from algal blooms to decreases in dissolved oxygen and associated fish kills. Metals like mercury, pesticides, and other organic compounds that enter

---

[1] EPA does not yet have the 2004 assessment findings compiled in a consistent format from all the states. EPA is also working on processing the states 2006 Integrated Reports as the 303(d) portions are approved and the states submit their final assessment findings. Susan Holdsworth, EPA, personal communication, September 2007.

waterways can be taken up by fish species, accumulating in their tissues and presenting a health risk to organisms (including humans) that consume the fish.

However, Table 1-1 can be misleading if it implies that degraded *water quality* is the primary metric of impairment. In fact, many of the nation's streams, lakes, and estuaries also suffer from fundamental changes in their flow regime and energy inputs, alteration of aquatic habitats, and resulting disruption of biotic interactions that are not easily measured via pollutant concentrations. Such waters may not be listed on State 303(d) lists because of the absence of a corresponding water quality standard that would directly indicate such conditions (like a biocriterion). Figure 1-5A, B, and C show examples of such impacted waterbodies.

TABLE 1-1 Top 15 Categories of Impairment Requiring CWA Section 303(d) Action

Cause of Impairment	Number of Waterbodies	Percent of the Total
Mercury	8,555	14%
Pathogens	8,526	14%
Sediment	6,689	11%
Metals (other than mercury)	6,389	11%
Nutrients	5,654	10%
Oxygen depletion	4,568	8%
pH	3,389	6%
Cause unknown - biological integrity	2,866	5%
Temperature	2,854	5%
Habitat alteration	2,220	4%
PCBs	2,081	3%
Turbidity	2,050	3%
Cause unknown	1,356	2%
Pesticides	1,322	2%
Salinity/TDS/chlorides	996	2%

Note: "Waterbodies" refers to individual river segments, lakes, and reservoirs. A single waterbody can have multiple impairments. Because most waters are not assessed, however, there is no estimate of the number of unimpaired waters in the United States. SOURCE: EPA, National Section 303(d) List Fact Sheet (http://iaspub.epa.gov/waters/national_rept.control). The data are based on three-fourths of states reporting from 2004 lists, with the remaining from earlier lists and one state from a 2006 list.

FIGURE 1-5A Headwater tributary in Philadelphia suffering from Urban Stream Syndrome. SOURCE: Courtesy of Chris Crockett, Philadelphia Water Department.

FIGURE 1-5B A destabilized stream in Vermont. SOURCE: Courtesy of Pete LaFlamme, Vermont Department of Environmental Conservation.

FIGURE 1-5C An urban stream, the Lower Oso Creek in Orange County, California, following a storm event. Oso Creek was formerly an ephemeral stream, but heavy development in the contributing watershed has created perennial flow—stormwater flow during wet weather and minor wastewater discharges and authorized non-stormwater discharges such as landscape irrigation runoff during dry weather. Courtesy of Eric Stein, Southern California Coastal Research Water Project.

Over the years, the greatest successes in improving the nation's waters have been in abating the often severe impairments caused by municipal and industrial point source discharges. The pollutant load reductions required of these facilities have been driven by the National Pollutant Discharge Elimination System (NPDES) permit requirements of the CWA (see Chapter 2). Although the majority of these sources are now controlled, further declines in water quality remain likely if the land-use changes that typify more diffuse sources of pollution are not addressed (Palmer and Allan, 2006). These include land-disturbing agricultural, silvicultural, urban, industrial, and construction activities from which hard-to-monitor pollutants emerge during wet-weather events. Pollution from these landscapes has been almost universally acknowledged as the most pressing challenge to the restoration of waterbodies and aquatic ecosystems nationwide. All population and development forecasts indicate a continued worsening of the environmental conditions caused by diffuse sources of pollution under the nation's current growth and land-use trajectories.

Recognition of urban stormwater's role in the degradation of the nation's waters is but the latest stage in the history of this byproduct of the human environment. Runoff conveyance systems have been part of cities for centuries, but they reflected only the desire to remove water from roads and walkways as rapidly and efficiently as possible. In some arid environments, rainwater has always been collected for irrigation or drinking; elsewhere it has been treated as an unmetered, and largely benign, waste product of cities. Minimal (unengineered) ditches or pipes drained developed areas to the nearest natural watercourse. Where more convenient, stormwater shared conveyance with wastewater, eliminating the cost of a separate pipe system but commonly resulting in sewage overflows during rainstorms. Recognition of downstream flooding that commonly resulted from upstream development led to construction of stormwater storage ponds or vaults in many municipalities in the 1960s, but their performance has typically fallen far short of design objectives (Booth and Jackson, 1997; Maxted and Shaver, 1999; Nehrke and Roesner, 2004). Water-quality treatment has been a relatively recent addition to the management of stormwater, and although a significant fraction of pollutants can be removed through such efforts (e.g., Strecker et al., 2004; see http://www.bmpdatabase.org), the constituents remaining even in "treated" stormwater represent a substantial, but largely unappreciated, impact to downstream watercourses.

Of the waterbodies that have been assessed in the United States, impairments from urban runoff are responsible for about 38,114 miles of impaired rivers and streams, 948,420 acres of impaired lakes, 2,742 square miles of impaired bays and estuaries, and 79,582 acres of impaired wetlands (2002 305(b) report). These numbers must be considered an underestimate, since the urban runoff category does not include stormwater discharges from municipal separate storm sewer systems (MS4s) and permitted industries, including construction. Urban stormwater is listed as the "primary" source of impairment for 13 percent of all rivers, 18 percent of all lakes, and 32 percent of all estuaries (2000 305(b) report). Although these numbers may seem low, urban areas cover just 3 percent of the land mass of the United States (Loveland and Auch, 2004), and so their influence is disproportionately large. Indeed, developed and developing areas that are a primary focus of stormwater regulations contain some of the most degraded waters in the country. For example, in Ohio few sites with greater than 27 percent imperviousness can meet interim CWA goals in nearby waterbodies, and biological degradation is observed with much less urban development (Miltner et al., 2004). Numerous authors have found similar patterns (see Meyer et al., 2005).

Although no water quality inventory data have been made available from the EPA since 2002, the dimensions of the stormwater problem can be further gleaned from several past regional and national water quality inventories. Many of these assessments are somewhat dated and are subject to the normal data and assessment limitations of national assessment methods, but they indicate that stormwater runoff has a deleterious impact on nearly all of the nation's waters. For example:

- Harvesting of shellfish is prohibited, restricted, or conditional in nearly 40 percent of all shellfish beds nationally due to high bacterial levels, and urban runoff and failing septic systems are cited as the prime causes. Reopening of shellfish beds due to improved wastewater treatment has been more than offset by bed closures due to rapid coastal development (NOAA, 1992; EPA, 1998).

- In 2006 there were over 15,000 beach closings or swimming advisories due to bacterial levels exceeding health and safety standards, with polluted runoff and stormwater cited as the cause of the impairment 40 percent of the time (NRDC, 2007).

- Pesticides were detected in 97 percent of urban stream water samples across the United States, and exceeded human health and aquatic life benchmarks 6.7 and 83 percent of the time, respectively (USGS, 2006). In 94 percent of fish tissues sampled in urban areas nationwide, organochlorine compounds were detected.

- Urban development was responsible for almost 39 percent of freshwater wetland loss (88,960 acres) nationally between 1998 and 2004 (Dahl, 2006), and the direct impact of stormwater runoff in degrading wetland quality is predicted to affect an even greater acreage (Wright et al., 2006).

- Eastern brook trout are present in intact populations in only 5 percent of more than 12,000 subwatersheds in their historical range in eastern North America, and urbanization is cited as a primary threat in 25 percent of the remaining subwatersheds with reduced populations (Trout Unlimited, 2006).

- Increased flooding is common throughout urban and suburban areas, sometimes as a consequence of improperly sited development (Figure 1-6A) but more commonly as a result of increasing discharges over time resulting from progressive urbanization farther upstream (Figure 1-6B). According to FEMA (undated), property damage from all types of flooding, from flash floods to large river floods, averages $2 billion a year.

- The chemical effects of stormwater runoff are pervasive and severe throughout the nation's urban waterways, and they can extend far downstream of the urban source. Stormwater discharges from urban areas to marine and estuarine waters cause greater water column toxicity than similar discharges from less urban areas (Bay et al., 2003).

- A variety of studies have shown that stormwater runoff is a vector of pathogens with potential human health implications in both freshwater (Calderon et al., 1991) and marine waters (Dwight et al., 2004; Colford et al., 2007).

FIGURE 1-6 (A) New residential construction in the path of episodic stream discharge (Issaquah, Washington); (B) recent flooding of an 18th-century tavern in Collegeville, Pennsylvania following a storm event in an upstream developing watershed. SOURCES: Top, Derek Booth, Stillwater Sciences, Inc., and bottom, Robert Traver, Villanova University.

WHY IS IT SO HARD TO REDUCE
THE IMPACTS OF STORMWATER?

"Urban stormwater" is the runoff from a landscape that has been affected in some fashion by human activities, during and immediately after rain. Most visibly, it is the water flow over the ground surface, which is collected by natural channels and artificial conveyance systems (pipes, gutters, and ditches) and ultimately routed to a stream, river, lake, wetland, or ocean. It also includes water that has percolated into the ground but nonetheless reaches a stream channel relatively rapidly (typically within a day or so of the rainfall), contributing to the high discharge in a stream that commonly accompanies rainfall. The subsurface flow paths that contribute to this stormflow response are typically quite shallow, in the upper layers of the soil, and are sometimes termed "interflow." They stand in contrast to deeper groundwater paths, where water moves at much lower velocities by longer paths and so reaches the stream slowly, over periods of days, weeks, or months. This deeper flow sustains streamflow during rainless periods and is usually called baseflow, as distinct from "stormwater." A formal distinction between these types of runoff is sometimes needed for certain computational procedures, but for most purposes a qualitative understanding is sufficient.

These runoff paths can be identified in virtually all modified landscapes, such as agriculture, forestry, and mining. However, this report focuses on those settings with the particular combination of activities that constitute "urbanization," by which we mean to include the commonly understood conversion (whether incremental or total) of a vegetated landscape to one with roads, houses, and other structures.

Although the role of urban stormwater in degrading the nation's waters has been recognized for decades (e.g., Klein, 1979), reducing that role has been notoriously difficult. This difficulty arises from three basic attributes of what is commonly termed "stormwater":

1.   It is produced from literally everywhere in a *developed* landscape;

2.   Its production and delivery are episodic, and these fluctuations are difficult to attenuate; and

3.   It accumulates and transports much of the collective waste of the urban environment.

Wherever grasslands and forest are replaced by urban development in general, and impervious surfaces in particular, the movement of water across the landscape is radically altered (see Figure 1-7). Nearly all of the associated problems result from one underlying cause: loss of the water-retaining function of the soil and vegetation in the urban landscape. In an undeveloped, vegetated landscape, soil structure and hydrologic behavior are strongly influenced by biological activities that increase soil porosity (the ratio of void space to total soil volume) and the number and size of macropores, and thus the storage and conductivity of water as it moves through the soil. Leaf litter on the soil surface dissipates raindrop energy; the soil's organic content reduces detachment of small soil particles and maintains high surface infiltration rates. As a consequence, rainfall typically infiltrates into the ground surface or is evapotranspired by vegetation, except during particularly intense rainfall events (Dunne and Leopold, 1978).

In the urban landscape, these processes of evapotranspiration and water retention in the soil may be lost for the simple reason that the loose upper layers of the soil and vegetation are gone—stripped away to provide a better foundation for roads and buildings. Even if the soil still exists, it no longer functions if precipitation is denied access because of paving or rooftops. In either case, a stormwater

Local Hydrologic Cycle

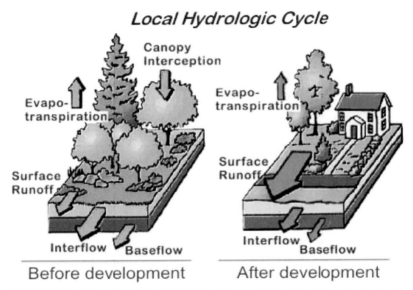

FIGURE 1-7 Schematic of the hydrologic pathways in humid-region watersheds, before and after urban development. The sizes of the arrows suggest relative magnitudes of the different elements of the hydrologic cycle, but conditions can vary greatly between individual catchments and only the increase in surface runoff in the post-development condition is ubiquitous. SOURCE: Adapted from Schueler (1987) and Maryland Department of the Environment; http://www.mde.state.md.us/Programs/WaterPrograms.

runoff reservoir of tremendous volume is removed from the stormwater runoff system; water that may have lingered in this reservoir for a few days or many weeks, or been returned directly to the atmosphere by evaporation or transpiration by plants, now flows rapidly across the land surface and arrives at the stream channel in short, concentrated bursts of high discharge.

This transformation of the hydrologic regime from one where subsurface flow once dominated to one where overland flow now dominates is not simply a readjustment of runoff flow paths, and it does not just result in a modest increase in flow volumes. It is a wholesale reorganization of the processes of runoff generation, and it occurs throughout the developed landscape. As such, it can affect every aspect of that runoff (Leopold, 1968)—not only its rate of production, its volume, and its chemistry, but also what it indirectly affects farther downstream (Walsh et al., 2005a). This includes erosion of mobile channel boundaries, mobilization of once-static channel elements (e.g., large logs), scavenging of contaminants from the surface of the urban landscape, and efficient transfer of heat from warmed surfaces to receiving waterbodies. These changes have commonly inspired human reactions—typically with narrow objectives but carrying additional, far-ranging consequences—such as the piping of once-exposed channels, bank armoring, and construction of large open-water detention ponds (e.g., Lieb and Carline, 2000).

This change in runoff regime is also commonly accompanied by certain land-use activities that have the potential to generate particularly harmful or toxic discharges, notably those commercial activities that are the particular focus of the industrial NPDES permits. These include manufacturing facilities, transport of freight or passengers, salvage yards, and a more generally defined category of "sites where industrial materials, equipment, or activities are exposed to stormwater" (e.g., EPA, 1992).

Other human actions are associated with urban landscapes that do not affect stormwater directly, but which can further amplify the negative consequences of altered flow. These actions include clearing of riparian vegetation around streams and wetlands, introduction of atmospheric pollutants that are subsequently deposited, inadvertent release of exotic chemicals into the environment, and channel crossings by roads and utilities. Each of these additional actions further degrades downstream waterbodies and increases the challenge of finding effective methods to reverse these changes (Boulton, 1999). There is little doubt as to why the problem of urban stormwater has not yet been "solved"—because every functional element of an aquatic ecosystem is affected. Urban stormwater has resulted in such widespread impacts, both physical and biological, in aquatic systems across the world that this phenomenon has been termed the "Urban Stream Syndrome" (see Figure 1-5; Walsh et al., 2005b).

Of the many possible ways to consider these conditions, Karr (1991) has recommended a simple yet comprehensive grouping of the major stressors arising from urbanization that influence aquatic assemblages (Figure 1-8). These include chemical pollutants (water quality and toxicity); changes to flow magnitude, frequency, and seasonality of various discharges; the physical aspects of stream, lake, or wetland habitats; the energy dynamics of food webs, sunlight, and temperature; and biotic interactions between native and exotic species. Stormwater and stormwater-related impacts encompass all of these categories, some directly (e.g., water chemistry) and some indirectly (e.g., habitat, energy dynamics). Because of the wide-ranging effects of stormwater, programs to abate stormwater impacts on aquatic systems must deal with a broad range of impairments far beyond any single altered feature, whether traditional water-chemistry parameters or flow rates and volumes.

The broad spatial scale of where and how these impacts are generated suggests that solutions, if effective, should be executed at an equivalent scale. Although the "problem" of stormwater runoff is manifested most directly as an altered hydrograph or elevated concentrations of pollutants, it is ultimately an expression of land-use change at a landscape scale. Symptomatic solutions, applied only at the end of a stormwater collection pipe, are not likely to prove fully effective because they are not functioning at the scale of the original disturbance (Kloss and Calarusse, 2006).

The landscape-scale generation of stormwater has a number of consequences for any attempt to reduce its effects on receiving waters, as described below.

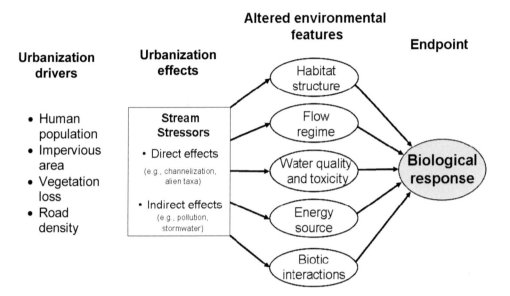

FIGURE 1-8 Five features that are affected by urban development and, in turn, affect biological conditions in urban streams. SOURCES: Modified from Karr (1991), Karr and Yoder (2004), and Booth (2005). Reprinted, with permission, from Karr (1991). Copyright 2001 by Ecological Society of America. Reprinted, with permission, from Karr and Yoder (2004). Copyright 2004 by American Society of Civil Engineers. Reprinted, with permission, from Booth (2005). Copyright 2005 by the North American Benthological Society.

Sources and Volumes

The "source" of stormwater runoff is dispersed, making collection and centralized treatment challenging. To the extent that collection is successful, however, the flip side of this condition—very large volumes—becomes manifest. Either an extensive infrastructure brings stormwater to centralized facilities, whose operation and maintenance may be relatively straightforward (e.g., Anderson et al., 2002) but of modest effectiveness, or stormwater remains dispersed for management, treatment, or both across the landscape (e.g., Konrad and Burges, 2001; Holman-Dodds et al., 2003; Puget Sound Action Team, 2005; Walsh et al., 2005a; Bloom, 2006; van Roon, 2007), better mimicking the natural processes of runoff generation but requiring a potentially unlimited number of "facilities" that may have their own particular needs for space, cost, and maintenance.

Treatment Challenges

Regardless of the scale at which treatment is attempted, technological difficulties are significant because of the variety of "pollutants" that must be addressed. These include physical objects, from large debris to microscopic particles; chemical constituents, both dissolved and immiscible; and less easily categorized properties such as temperature. Wastewater treatment plants manage a similarly broad range of pollutants, but stormwater flows have highly unsteady inflows and, when present, typically much greater volumes to treat.

Industrial sources of stormwater pose a particularly challenging problem because potential generators of polluted or toxic runoff are widespread and are regulated under NPDES permitting by their *activities*, not by the specific category of industrial activity under which they fall. This complicates any systematic effort to identify those entities that should be regulated (Duke et al., 1999). Even for the limited number of regulated generators, pollution prevention measures are of uncertain effectiveness.

Soil erosion from construction sites is another pollution source that has proven difficult to effectively control. Although most bare sites are relatively small and only short-lived, at any given time there can be many sites under construction, each of which can deliver sediment loads to downstream waterbodies at rates that exceed background levels by many orders of magnitude (e.g., Wolman and Schick, 1967). Relatively effective approaches and technologies exist to dramatically reduce the magnitude of these sediment discharges (e.g., Raskin et al., 2005), but they depend on conscientious installation and regular maintenance. Enforcement of such requirements, normally a low-priority activity of local departments of building or public works, is commonly lacking.

Another difference between the stormwater and wastewater streams is that stormwater treatment must address not only "pollutants" but also physically and ecologically deleterious changes in flow rate and total runoff volume. Treating these changes constitutes a particularly difficult task for two reasons. First, there is simply more runoff, as a rule, and so replicating the predevelopment hydrograph is not an option—the increased volume of runoff guarantees that some discharges, some of the time, must be allowed to increase. Second, there is little agreement on what constitutes "adequate" or "effective" treatment for the various attributes of flow. Even the most basic metrics, such as the magnitude of peak flow, can require extensive infrastructure to achieve (e.g., Booth and Jackson, 1997); other flow metrics that correlate more directly with undesired effects on physical and biological systems can require even greater efforts to match. In many cases, the urban-induced transformation of the flow regime makes true "mitigation" virtually impossible.

Widespread Cause and Effects

The spatial scale of stormwater generation and its impacts is wide-ranging. "Generators" are literally landscape-wide, and impacts can occur at every location in the path followed by urban runoff, from source to receiving waterbody (Hamilton et al., 2004). There are few ways to demonstrate causal connections between distributed landscape sources and cumulative downstream effects (Allan, 2004), and so site-specific mitigation typically provides little lasting improvement in the watershed as a whole (Maxted and Shaver, 1997).

Stormwater Measurements

The desired attributes of stormwater runoff are normally expressed through a combination of physical and chemical parameters. These parameters are commonly presumed to have direct correlation to attributes of human or ecological concern, such as the condition of human or fish communities, or the stability of a stream channel, even though these parameters do not directly measure those effects. The most commonly measured physical parameters are hydrologic and simply measure the rate of flow past a specified location. Both the absolute, instantaneous magnitude of that flow rate (i.e., the discharge) and the variations in that rate over multiple time scales (i.e., how rapidly the discharge varies over an hour, a day, a season, etc.) can be captured by analysis of a continuous time series of a flow. Obviously, however, a nearly unlimited number of possible metrics, capturing a multitude of temporal scales, could be defined (Poff et al., 1997, 2006; Cassin et al., 2004; Konrad et al., 2005; Roy et al., 2005; Chang, 2007). Commonly only a single parameter—the peak storm discharge for a given return period (Hollis, 1975)—has been emphasized in the past. Mitigation of urban-induced flow increases have followed this narrow approach, typically by endeavoring to reduce peak discharge by use of detention ponds but leaving the underlying increase in runoff volumes—and the associated augmentation of both frequency and duration of high discharges—untouched. This partly explains why evaluation of downstream conditions commonly document little improvement resulting from traditional flow-mitigation measures (e.g., Maxted and Shaver, 1997; Roesner et al., 2001; May and Horner, 2002).

Other physical parameters, less commonly measured or articulated, can also express the conditions of downstream watercourses. Measures of size or complexity, particularly for stream channels, are particularly responsive to the changes in flow regime and discharge. Booth (1990) suggested that discriminating between *channel expansion*, the proportional increase in channel cross-sectional area with increasing discharge, and *channel incision*, the catastrophic vertical downcutting that sometimes accompanies urban-induced flow increases, captures important end-members of the physical response to hydrologic change. The former (proportional expansion) is more thoroughly documented (Hammer, 1972; Hollis and Luckett, 1976; Morisawa and LaFlure, 1982; Neller, 1988;

Whitlow and Gregory, 1989; Booth and Jackson, 1997; Moscrip and Montgomery, 1997; Booth and Henshaw, 2001); the latter (catastrophic incision) is more difficult to quantify but has been recognized in both urban and agricultural settings (e.g., Simon, 1989). Both types of changes result not only in a larger channel but also in substantial simplification and loss of features normally associated with high-quality habitat for fish and other in-stream biota. The sediment released by these "growing channels" also can be the largest component of the overall sediment load delivered to downstream waterbodies (Trimble, 1997; Nelson and Booth, 2002).

Chemical parameters (or, historically, "water-quality parameters"; see Dinius, 1987; Gergel et al., 2002) cover a host of naturally and anthropogenically occurring constituents in water. In flowing water these are normally expressed as instantaneous measurements of concentration. In waterbodies with long residence times, such as lakes, these may be expressed as either concentrations or as loads (total accumulated amounts, or total amounts integrated over an extended time interval). The CWA defined a list of priority pollutants, of which a subset is regularly measured in many urban streams (e.g., Field and Pitt, 1990). Parameters that are not measured may or may not be present, but without assessment they are rarely recognized for their potential (or actual) contribution to waterbody impairment.

Other attributes of stormwater do not fit as neatly into the categories of water quantity or water quality. Temperature is commonly measured and is normally treated as a water quality parameter, although it is obviously not a chemical property of the water (LeBlanc et al., 1997; Wang et al., 2003). Similarly, direct or indirect measures of suspended matter in the water column (e.g., concentration of total suspended solids, or secchi disk depths in a lake) are primarily physical parameters but are normally included in water quality metrics. Flow velocity is rarely measured in either context, even though it too correlates directly to stream-channel conditions. Even more direct expressions of a flow's ability to transport sediment or other debris, such as shear stress or unit stream power, are rarely reported and virtually never regulated.

***

Urban runoff degrades aquatic systems in multiple ways, which confounds our attempts to define causality or to demonstrate clear linkages between mitigation and ecosystem improvement. It is generally recognized from the conceptual models that seek to describe this system that no single element holds the key to ecosystem condition. All elements must be functional, and yet every element can be affected by urban runoff in different ways. These impacts occur at virtually all spatial scales, from the site-specific to the landscape; this breadth and diversity challenges our efforts to find effective solutions.

This complexity and the continued growth of the built environment also present fundamental social choices and management challenges. Stormwater control measures entail substantial costs for their long-term maintenance, monitoring to determine their performance, and enforcement of their use—all of which must be weighed against their (sometimes unproven) benefits. Furthermore, the overarching importance of impervious surfaces inextricably links stormwater management to land-use decisions and policy. For example, where a reversal of the effects of urbanization cannot be realized, more intensive land-use development in certain areas may be a paradoxically appropriate response to reduce the overall impacts of stormwater. That is, increasing population density and impervious cover in designated urban areas may reduce the creation of impervious surface and the associated ecological impacts in areas that will remain undeveloped as a result. In these highly urban areas (with very high percentages of impervious surface), aquatic conditions in local streams will be irreversibly changed and the Urban Stream Syndrome may be unavoidable to some extent. Where these impacts occur and what effort and cost will be used to avoid these impacts are both fundamental issues confronting the nation as it attempts to address stormwater.

IMPETUS FOR THE STUDY AND REPORT ROADMAP

In 1972 Congress amended the Federal Water Pollution Control Act (subsequently referred to as the Clean Water Act) to require control of discharges of pollutants to waters of the United States from point sources. Initial efforts to improve water quality using NPDES permits focused primarily on reducing pollutants from industrial process wastewater and municipal sewage discharges. These point source discharges were clearly and easily shown to be responsible for poor, often drastically degraded conditions in receiving waterbodies because they tended to emanate from identifiable and easily monitored locations, such as pipe outfalls.

As pollution control measures for industrial process wastewater and municipal sewage were implemented and refined during the 1970s and 1980s, more diffuse sources of water pollution have become the predominant causes of water quality impairment, including stormwater runoff. To address the role of stormwater in causing water quality impairments, Congress included Section 402(p) in the CWA; this section established a comprehensive, two-phase approach to stormwater control using the NPDES program. In 1990 EPA issued the Phase I Stormwater Rule (55 Fed. Reg. 47990; November 16, 1990) requiring NPDES permits for operators of municipal separate storm sewer systems (MS4s) serving populations over 100,000 and for runoff associated with industrial activity, including runoff from construction sites five acres and larger. In 1999 EPA issued the Phase II Stormwater Rule (64 Fed. Reg. 68722; December 8, 1999), which expanded the requirements to small MS4s in urban areas and to construction sites between one and five acres in size.

Since EPA's stormwater program came into being, several problems inherent in its design and implementation have become apparent. As discussed in more detail in Chapter 2, problems stem to a large extent from the diffuse nature of stormwater discharges combined with a regulatory process that was created for point sources (the NPDES permitting approach). These problems are compounded by the shear number of entities requiring oversight. Although exact numbers are not available, EPA estimates that the number of regulated MS4s is about 7,000, including 1,000 Phase I municipalities and 6,000 from Phase II. The number of industrial permittees is thought to be around 100,000. Each year, the construction permit covers around 200,000 permittees each for both Phase I (five acres or greater) and Phase II (one to five acres) projects. Thus, the total number of permittees under the stormwater program at any time numbers greater than half a million. There are fewer than 100,000 non-stormwater (meaning wastewater) permittees covered by the NPDES program, such that stormwater permittees account for approximately 80 percent of NPDES-regulated entities. To manage this large number of permittees, the stormwater program relies heavily on the use of general permits to control industrial, construction, and Phase II MS4 discharges, which are usually statewide, one-size-fits-all permits in which general provisions are stipulated.

An example of the burden felt by a single state is provided by Michigan (David Drullinger, Michigan Department of Environmental Quality Water Bureau, personal communication, September 2007). The Phase I Stormwater regulations that became effective in 1990 regulate 3,400 industrial sites, 765 construction sites per year, and five large cities in Michigan. The Phase II regulations, effective since 1999, have extended the requirements to 7,000 construction sites per year and 550 new jurisdictions, which are comprised of about 350 "primary jurisdictions" (cities, villages, and townships) and 200 "nested jurisdictions" (county drains, road agencies, and public schools). Often, only a handful of state employees are allocated to administer the entire program (see the survey in Appendix C).

In order to comply with the CWA regulations, permittees must fulfill a number of requirements, including the creation and implementation of a stormwater pollution prevention plan, and in some cases, monitoring of stormwater discharges. Stormwater pollution prevention plans document the stormwater control measures (SCMs; sometimes known as best management practices or BMPs) that will be used to prevent or slow stormwater from quickly reaching nearby waterbodies and degrading their quality. These include structural methods such as detention ponds and nonstructural methods such as designing new development to reduce the percentage of impervious surfaces. Unfortunately, data on the degree of pollutant reduction that can be assigned to a particular SCM are only now becoming available (see Chapter 5).

Other sources of variability in EPA's stormwater program are that (1) there are three permit types (municipal, industrial, and construction), (2) some states and local governments have assumed primacy for the program from EPA while others have not, and state effluent limits or benchmarks for stormwater discharges may differ from the federal requirements, and (3) whether there are monitoring requirements varies depending on the regulating entity and the type of activity. For industrial stormwater there are 29 sectors of industrial activity covered by the general permit, each of which is characterized by a different suite of possible contaminants and SCMs.

Because of the industry-, site-, and community-specific nature of stormwater pollution prevention plans, and because of the lack of resources of most NPDES permitting authorities to review these plans and conduct regular compliance inspections, water quality-related accountability in the stormwater program is poor. Monitoring data are minimal for most permittees, despite the fact that they are often the only indicators of whether an adequate stormwater program is being implemented. At the present time, available monitoring data indicate that many industrial facilities routinely exceed "benchmark values" established by EPA or the states, although it is not clear whether these exceedances provide useful indicators of stormwater pollution prevention plan inadequacies or potential water quality problems. These uncertainties have led to mounting and contradictory pressure from permittees to eliminate monitoring requirements entirely as well as from those hoping for greater monitoring requirements to better understand the true nature of stormwater discharges and their impact.

To improve the accountability of it Stormwater Program, EPA requested advice on stormwater issues from the National Research Council's (NRC's) Water Science and Technology Board as the next round of general permits is being prepared. Although the drivers for this study have been in the industrial stormwater arena, this study considered all entities regulated under the NPDES program (municipal, industrial, and construction). The following statement of task guided the work of the committee:

(1) Clarify the mechanisms by which pollutants in stormwater discharges affect ambient water quality criteria and define the elements of a "protocol" to link pollutants in stormwater discharges to ambient water quality criteria.

(2) Consider how useful monitoring is for both determining the potential of a discharge to contribute to a water quality standards violation and for determining the adequacy of stormwater pollution prevention plans. What specific parameters should be monitored and when and where? What effluent limits and benchmarks are needed to ensure that the discharge does not cause or contribute to a water quality standards violation?

(3) Assess and evaluate the relationship between different levels of stormwater pollution prevention plan implementation and in-stream water quality, considering a broad suite of SCMs.

(4) Make recommendations for how to best stipulate provisions in stormwater permits to ensure that discharges will not cause or contribute to exceedances of water quality standards. This should be done in the context of general permits. As a part of this task, the committee will consider currently available information on permit and program compliance.

(5) Assess the design of the stormwater permitting program implemented under the CWA.

The report is intended to inform decision makers within EPA, affected industries, public stormwater utilities, other government agencies and the private sector about potential options for managing stormwater.

EPA requested that the study be limited to those issues that fall under the agency's current regulatory scheme for stormwater, which excludes nonpoint sources of pollution such as agricultural runoff and septic systems. Thus, these sources are not extensively covered in this report. The reader is referred to NRC (2000, 2005) for more detailed information on the contribution of agricultural runoff and septic systems to waterbody impairment and on innovative technologies for treating these sources. Also at the request of EPA, concentrated animal feeding operations and combined sewer overflows were not a primary focus. However, the committee felt that in order to be most useful it should opine on certain critical effects of regulated stormwater beyond the delivery of traditional pollutants. Thus, changes in stream flow, streambank erosion, and habitat alterations caused by stormwater are considered, despite the relative inattention given to them in current regulations.

Chapter 2 presents the regulatory history of stormwater control in the United States, focusing on relevant portions of the CWA and the regulations that have been created to implement the Act. Federal, state, and local programs for or affecting stormwater management are described and critiqued. Chapter 3 deals with the first item in the statement of task. It reviews the scientific aspects of stormwater, including sources of pollutants in stormwater, how stormwater moves across the land surface, and its impacts on receiving waters. It reflects the best of currently available science, and addresses biological endpoints that go far beyond ambient water quality criteria. Methods for monitoring and modeling stormwater (the subject of the second item in the statement of task) are described in Chapter 4. The material evaluates the usefulness of current benchmark and MS4 monitoring requirements, and suggestions for improvement are made. The latter half of the chapter considers the multitude of models available for linking stormwater discharges to ambient water quality. This analysis makes it clear that stormwater pollution cannot yet be treated as a deterministic system (in which the contribution of individual dischargers to a waterbody impairment can be identified) without significantly greater investment in model development. Addressing primarily the third item in the statement of task, Chapter 5 considers the vast suite of both structural and nonstructural measures designed to control stormwater and reduce its pollutant loading to waterbodies. It also takes on relevant larger-scale concepts, such as the benefit of stormwater management within a watershed framework. In Chapter 6, the limitations and possibilities associated with a new regulatory approach are explored, as are those of an enhanced but more traditional scheme. Numerous suggestions for improving the stormwater permitting process for municipalities, industrial sites, and construction are made. Along with Chapter 2, this chapter addresses the final two items in the committee's statement of task.

REFERENCES

Alberti, M., D. B. Booth, K. Hill, B. Coburn, C. Avolio, S. Coe, and D. Spirandelli. 2006. The impact of urban patterns on aquatic ecosystems: an empirical analysis in Puget lowland sub-basins. Landscape Urban Planning, doi:10.1016/j.landurbplan.2006.08.001.

Allan, J. D. 2004. Landscapes and riverscapes: the influence of land use on stream ecosystems. Annual Review of Ecology, Evolution, and Systematics 35:257–284.

Alley, W. A., and J. E. Veenhuis. 1983. Effective impervious area in urban runoff modeling. Journal of Hydrological Engineering ASCE 109(2):313–319.

Anderson, B .C., W. E. Watt, and J. Marsalek. 2002. Critical issues for stormwater ponds: learning from a decade of research. Water Science and Technology 45(9):277–283.

Barbour, M. T., M. J. Paul, D. W. Bressler, A. H. Purcell, V. H. Resh, and E. T. Rankin. 2006. Bioassessment: a tool for managing aquatic life uses for urban streams. Water Environment Research Foundation Research Digest 01-WSM-3.

Bay, S., B. H. Jones, K. Schiff, and L. Washburn. 2003. Water quality impacts of stormwater discharges to Santa Monica Bay . Marine Environmental Research 56:205–223.

Bledsoe, B. P., and C. C. Watson. 2001. Effects of urbanization on channel instability. Journal of the American Water Resources Association 37(2):255–270.

Bloom, M. F. 2006. Low Impact Development approach slows down drainage, reduces pollution. Water and Wastewater International 21(4):59.

Booth, D. B. 1990. Stream channel incision in response following drainage basin urbanization. Water Resources Bulletin 26:407–417.

Booth, D. B. 2005. Challenges and prospects for restoring urban streams: a perspective from the Pacific Northwest of North America. Journal of the North American Benthological Society 24(3):724–737.

Booth, D. B., and C. R. Jackson. 1997. Urbanization of aquatic systems—degradation thresholds, stormwater detention, and the limits of mitigation. Water Resources Bulletin 33:1077–1090.

Booth, D. B., and P. C. Henshaw. 2001. Rates of channel erosion in small urban streams. Pp. 17–38 *In*: Land Use and Watersheds: Human Influence on Hydrology and Geomorphology in Urban and Forest Areas. M. Wigmosta and S. Burges (eds.). AGU Monograph Series, Water Science and Application, Volume 2.

Boulton, A. J. 1999. An overview of river health assessment: philosophies, practice, problems and prognosis. Freshwater Biology 41(2):469–479.

Brookings Institute. 2004. Toward a new metropolis: the opportunity to rebuild America. Arthur C. Nelson, Virginia Polytechnic Institute and State University. Discussion paper prepared for The Brookings Institution Metropolitan Policy Program.

Burges, S. J., M. S. Wigmosta, and J. M. Meena. 1998. Hydrological effects of land-use change in a zero-order catchment. Journal of Hydrological Engineering 3:86–97.

Calderon, R., E. Mood, and A. Dufour. 1991. Health effects of swimmers and nonpoint sources of contaminated water. International Journal of Environmental Health Research 1:21–31.

Cassin, J., R. Fuerstenberg, F. Kristanovich, L. Tear, and K. Whiting. 2004. Application of normative flow on small streams in Washington State—hydrologic perspective. Pp. 4281–*4299 In:* Proceedings of the 2004 World Water and Environmental Resources Congress: Critical Transitions in Water and Environmental Resources Management.

Center for Watershed Protection (CWP). 2005. An integrated framework to restore small urban watersheds. Ellicott City, MD: CWP. Available at http://www.cwp.org/Store/usrm.htm. Last accessed September 23, 2008.

Chang, H. 2007. Comparative streamflow characteristics in urbanizing basins in the Portland Metropolitan Area, Oregon, USA. Hydrological Processes 21(2):211–222.

Chang, H., and T. N. Carlson. 2005. Water quality during winter storm events in Spring Creek, Pennsylvania USA. Hydrobiologia 544(1):321–332.

Chang, H. H. 1992. Fluvial Processes in River Engineering. Malabar, FL: Krieger Publishing.

Colford, J. M., Jr, T. J. Wade, K. C. Schiff, C. C. Wright, J. F. Griffith, S. K. Sandhu, S. Burns, J. Hayes, M. Sobsey, G. Lovelace, and S. Weisberg. 2007. Water quality indicators and the risk of illness at non-point source beaches in Mission Bay, California. Epidemiology 18(1):27–35.

Dahl, T. 2006. Status and trends of wetlands in the conterminous United States: 1998–2004. Washington, DC: U.S. Department of the Interior Fish and Wildlife Service.

Davies, S. P., and S. K. Jackson. 2006. The biological condition gradient: a descriptive model for interpreting change in aquatic ecosystems. Ecological Applications 16(4):1251–1266.

Dinius, S. H. 1987. Design of an index of water quality. Water Resources Bulletin 23(5):833–843.

Duke, L. D., K. P. Coleman, and B. Masek. 1999. Widespread failure to comply with U.S. storm water regulations for industry—Part I: Publicly available data to estimate number of potentially regulated facilities. Environmental Engineering Science 16(4):229–247.

Dunne, T., and L. B. Leopold. 1978. Water in Environmental Planning. New York: W. H. Freeman.

Dwight, R. H., D. B. Baker, J. C. Semenza, and B. H. Olson. 2004. Health effects associated with recreational coastal water use: urban vs. rural California. American Journal of Public Health 94(4):565–567.

EPA (U.S. Environmental Protection Agency). 1992. Storm Water Management for Industrial Activities, Developing Pollution Prevention Plans and Best Management Practices. Available at http://www.ntis.gov.

EPA. 1998. EPA Project Beach. Washington, DC: EPA Office of Water.

EPA. 2000. National Water Quality Inventory. 305(b) List. Washington, DC: EPA Office of Water.

EPA. 2002. 2000 National Water Quality Inventory. EPA-841-R-02-001. Washington, DC: EPA Office of Water.

EPA. 2006. Wadeable Streams Assessment: A Collaborative Survey of the Nation's Streams. EPA 841-B-06-002. Washington, DC: EPA Office of Water.

EPA. 2007. National Estuary Program Coastal Condition Report. EPA-842-B-06-001. Washington, DC: EPA Office of Water and Office of Research and Development.

FEMA (Federal Emergency Management Agency). No date. Flood. A report of the Subcommittee on Disaster Reduction. Available at http://www.sdr.gov. Last accessed September 23, 3008.

Field, R., and R. E. Pitt. 1990. Urban storm-induced discharge impacts: U.S. Environmental Protection Agency research program review. Water Science and Technology 22(10–11):1–7.

Gergel, S. E., M. G. Turner, J. R. Miller, J. M. Melack, and E. H. Stanley. 2002. Landscape indicators of human impacts to riverine systems. Aquatic Sciences 64(2):118–128.

Gregory, M., J. Aldrich, A. Holtshouse, and K. Dreyfuss-Wells. 2005. Evaluation of imperviousness impacts in large, developing watersheds. Pp. 115–150 *In:* Efficient Modeling for Urban Water Systems, Monograph 14. W. James, E. A. McBean, R. E. Pitt, and S. J. Wright (eds.). Guelph, Ontario, Canada: CHI.

Grübler, A. 1994. Technology. Pp. 287–*328 In:* Changes in Land Use and Land Cover: A Global Perspective. W. B. Meyer and B. L. Turner II (eds.). Cambridge: Cambridge University Press.

Hamilton, P. A., T. L. Miller, and D. N. Myers. 2004. Water Quality in the Nation's Streams and Aquifers—Overview of Selected Findings, 1991–2001. U.S. Geological Survey Circular 1265. Available at http://pubs.usgs.gov/circ/2004/1265/pdf/circular1265.pdf. Last accessed September 23, 2008.

Hammer, T. R. 1972. Stream and channel enlargement due to urbanization. Water Resources Research 8:1530–1540.

Hart, J. F. 1968. Loss and abandonment of cleared farm land in the Eastern United States. Annals of the Association of American Geographers 58(5):417–440.

Heinz Center. 2002. The State of the Nation's Ecosystems. Measuring the Lands, Waters, and Living Resources of the United States. Cambridge: Cambridge University Press.

Hollis, G. E. 1975. The effect of urbanization on floods of different recurrence interval. Water Resources Research 11:431–435.

Hollis, G. E., and J. K. Luckett. 1976. The response of natural river channels to urbanization: two case studies from southeast England. Journal of Hydrology 30:351–363.

Holman-Dodds, J. K., A. A. Bradley, and K. W. Potter. 2003. Evaluation of hydrologic benefits of infiltration based urban storm water management. Journal of the American Water Resources Association 39(1):205–215.

Horner, R. R., D. B. Booth, A. A. Azous, and C. W. May. 1997. Watershed determinants of ecosystem functioning. Pp. 251-274 *In*: Effects of Watershed Development and Management on Aquatic Ecosystems. L. A. Roesner (ed.). Proceedings of the Engineering Foundation Conference, Snowbird, UT, August 4–9, 1996.

House, M. A., J. B. Ellis, E. E. Herricks, T. Hvitved-Jacobsen, J. Seager, L. Lijklema, H. Aalderink, and I. T. Clifforde. 1993. Urban drainage: Impacts on receiving water quality. Water Science and Technology 27(12):117–158.

Karr, J. R. 1991. Biological integrity: a long-neglected aspect of water resource management. Ecological Applications 1:66–84.

Karr, J. R., and C. O. Yoder. 2004. Biological assessment and criteria improve TMDL planning and decision making. Journal of Environmental Engineering 130:594–604.

Klein, R. D. 1979. Urbanization and stream quality impairment. Water Resources Bulletin 15:948–969.

Kloss, C., and C. Calarusse. 2006. Rooftops to rivers—green strategies for controlling stormwater and combined sewer overflows. New York: National Resources Defense Council. Available at http://www.nrdc.org/water/pollution/rooftops/rooftops.pdf. Last accessed September 23, 2008.

Konrad, C. P., and S. J. Burges. 2001. Hydrologic mitigation using on-site residential storm-water detention. Journal of Water Resources Planning and Management 127:99–107.

Konrad, C. P., D. B. Booth, and S. J. Burges. 2005. Effects of urban development in the Puget Lowland, Washington, on interannual streamflow patterns: consequences for channel form and streambed disturbance. Water Resources Research 41(7):1–15.

Laenen, A. 1983. Storm runoff as related to urbanization based on data collected in Salem and Portland, and generalized for the Willamette Valley, Oregon. U.S. Geological Survey Water-Resources Investigations Report 83-4238, 9 pp.

Lambin, E. F., B. L. Turner, H. J. Geist, S. B. Agbola, A. Angelsen, J. W. Bruce, O. T. Coomes, R. Dirzo, G. Fischer, C. Folke, P. S. George, K. Homewood, J. Imbernon, R. Leemans, X. Li, E. F. Moran, M. Mortimore, P. S. Ramakrishnan, J. F. Richards, H. Skånes, W. Steffen, G. D. Stone, U. Svedin, T. A. Veldkamp, C. Vogel, and J. Xu. 2001. The causes of land-use and land-cover change: moving beyond the myths. Global Environmental Change 11(4):261–269.

LeBlanc, R. T., R. D. Brown, and J. E. FitzGibbon. 1997. Modeling the effects of land use change on the water temperature in unregulated urban streams. Journal of Environmental Management 49(4):445–469.

Leopold, L. B. 1968. Hydrology for urban land planning: a guidebook on the hydrologic effects of urban land use. U.S. Geological Survey Circular 554. Washington, DC: USGS.

Levin, P. S., R. W. Zabel, and J. G. Williams. 2001. The road to extinction is paved with good intentions: negative association of fish hatcheries with threatened salmon. Proceedings of the Royal Society—Biological Sciences (Series B) 268(1472):1153–1158.

Lieb, D. A., and R. F. Carline. 2000. Effects of urban runoff from a detention pond on water quality, temperature and caged gammarus minus (say) (amphipoda) in a headwater stream. Hydrobiologia 441:107–116.

Loveland, T., and R. Auch. 2004. The changing landscape of the eastern United States. Washington, DC: U.S. Geological Survey. Available at http://www.usgs.gov/125/articles/eastern_us.html. Last accessed November 25, 2007.

Maxted, J. R., and E. Shaver. 1997. The use of retention basins to mitigate stormwater impacts on aquatic life. Pp. 494-512 *In:* Effects of Watershed Development and Management on Aquatic Ecosystems. L. A. Roesner (Ed.). New York: American Society of Civil Engineers.

Maxted, J. R., and E. Shaver. 1999. The use of detention basins to mitigate stormwater impacts to aquatic life. Pp. *6–15 In:* National Conference on Retrofit Opportunities for Water Resource Protection in Urban Environments, Chicago, February 9–12, 1998. EPA/625/R-99/002. Washington, DC: EPA Office of Research and Development.

May, C. W., and R. R. Horner. 2002. The limitations of mitigation-based stormwater management in the pacific northwest and the potential of a conservation strategy based on low-impact development principles. Pp. 1-16 *In:* Global Solutions for Urban Drainage. Proceedings of the Ninth International Conference on Urban Drainage.

Meyer, J. L., M. J. Paul, and W. K. Taulbee. 2005. Stream ecosystem function in urbanizing landscapes. Journal of the North American Benthological Society 24:602–612.

Miltner, R. J., White, D., and C. O. Yoder. 2004. The biotic integrity of streams in urban and suburbanizing landscapes. Landscape and Urban Planning 69:87–100.

Morisawa, M., and E. LaFlure. 1982. Hydraulic geometry, stream equilibrium and urbanization. Pp. 333–350 *In:* Adjustments of the Fluvial System. D. D. Rhodes and G. P. Williams (eds.). London: Allen and Unwin.

Moscrip, A. L., and D. R. Montgomery. 1997. Urbanization, flood frequency, and salmon abundance in Puget Lowland streams. Journal of the American Water Resources Association 33:1289–1297.

Naiman, R. J., and M. G. Turner. 2000. A future perspective on North America's freshwater ecosystems. Ecological Applications 10(4):958–970.

Neller, R. J. 1988. A comparison of channel erosion in small urban and rural catchments, Armidale, New South Wales. Earth Surface Processes and Landforms 13:1–7.

Nelson, E. J., and D. B. Booth. 2002. Sediment budget of a mixed-land use, urbanizing watershed. Journal of Hydrology 264:51–68.

Nehrke, S. M., and L. A. Roesner. 2004. Effects of design practice for flood control and best management practices on the flow-frequency curve. Journal of Water Resources Planning and Management 130(2):131-139.

NOAA (National Oceanic and Atmospheric Administration). 1992. 1990 Shellfish Register of Classified Estuarine Waters. Data supplement. Rockville, MD: National Ocean Service.

NRC (National Research Council). 2000. Watershed Management for Potable Water Supply: Assessing the New York City Strategy. Washington, DC: National Academies Press.

NRC. 2005. Regional Cooperation for Water Quality Improvement in Southwestern Pennsylvania. Washington, DC: National Academies Press.

NRDC (Natural Resources Defense Council). 2007. Testing the Waters: A Guide to Water Quality at Vacation Beaches (17th ed.). New York: NRDC.

Palmer, M. A., and J. D. Allan. 2006. Restoring Rivers. Issues in Science & Technology. Washington, DC: National Academies Press.

Paul, M. J., and J. L. Meyer. 2001. Streams in the urban landscape. Annual Review of Ecology and Systematics 32:333–365.

Poff, N. L., J. D. Allan, M. B. Bain, J. R. Karr, K. L. Prestegaard, B. D. Richter, R. E. Sparks, and J. C. Stromberg. 1997. The natural flow regime: a paradigm for river conservation and restoration. BioScience 47(11):769–784.

Poff, N. L., B. P. Bledsoe, and C. O. Cuhaciyan. 2006. Hydrologic variation with land use across the contiguous United States: geomorphic and ecological consequences for stream ecosystems. Geomorphology 79 (3–4):264–285.

Pozzi, F., and C. Small. 2005. Analysis of urban land cover and population density in the United States. Photogrammetric Engineering and Remote Sensing 71(6):719–726.

Prysch, E. A., and J. C. Ebbert. 1986. Quantity and quality of storm runoff from three urban catchments in Bellevue, Washington. USGS Water-Resources Investigations Report 86-4000, 85 pp.

Puget Sound Action Team. 2005. Low Impact Development: Technical Guidance Manual for Puget Sound. Available at http://www.psat.wa.gov/Programs/LID.htm. Last accessed September 23, 2008.

Raskin, L., A. DePaoli, and M. J. Singer. 2005. Erosion control materials used on construction sites in California. Journal of Soil and Water Conservation 60(4):187–192.

Roesner, L. A., B. P. Bledsoe, and R. W. Brashear. 2001. Are best-management-practice criteria really environmentally friendly? Journal of Water Resources Planning and Management 127(3):150-154.

Roy, A. H., M. C. Freeman, B. J. Freeman, S. J. Wenger, W. E. Ensign, and J. L. Meyer. 2005. Investigating hydrological alteration as a mechanism of fish assemblage shifts in urbanizing streams. Journal of the North American Benthological Society 24:656–678.

Schueler, T. 1987. Controlling Urban Runoff: A Practical Manual for Planning and Designing Urban Best Management Practices. Washington, DC: Metropolitan Washington Council of Governments.

Simon, A. 1989. A model of channel response in disturbed alluvial channels. Earth Surface Processes and Landforms 14:11–26.

Steedman, R. J. 1988. Modification and assessment of an index of biotic integrity to quantify stream quality in Southern Ontario. Canadian Journal of Fisheries and Aquatic Sciences 45:492–501.

Strecker, E. W., M. M. Quigley, B. Urbonas, and J. Jones. 2004. Analyses of the expanded EPA/ASCE International BMP Database and potential implications for BMP design. In: Proceedings of the World Water and Environmental Congress 2004, June 27–July 1, 2004, Salt Lake City, UT. G. Sehlke, D. F. Hayes, and D. K. Stevens (eds.). Reston, VA: ASCE.

Sutherland, R. 1995. Methods for estimating the effective impervious area of urban watersheds. Watershed Protection Techniques 2(1):282-284. Ellicott City, MD: Center for Watershed Protection.

Thomson, J. D., G. Weiblen, B. A. Thomson, S. Alfaro, and P. Legendre. 1986. Untangling multiple factors in spatial distributions: lilies, gophers, and rocks. Ecology 77:1698–1715.

Trimble, S. W. 1997. Contribution of stream channel erosion to sediment yield from an urbanizing watershed. Science 278:1442–1444.

Trout Unlimited. 2006. Eastern Brook Trout: Status and Threats. Eastern Brook Trout Joint Venture. Arlington, VA: Trout Unlimited.

USGS (U.S. Geological Survey). 2006. The quality of our nation's waters: pesticides in the nation's streams and ground water: 1992–2001. National Water Quality Assessment Program. USGS Circular 1291. Reston, VA: USGS.

van Roon, M. 2007. Water localisation and reclamation: steps towards low impact urban design and development. Journal of Environmental Management 83(4):437–447.

Vitousek, P. M., H. A. Mooney, J. Lubchenco, and J. M. Melillo. 1997. Human domination of Earth's ecosystems. Science 277(5325):494–499.

Walsh, C. J., T. D. Fletcher, and A. R. Ladson. 2005a. Stream restoration in urban catchments through redesigning stormwater systems: looking to the catchment to save the stream. Journal of the North American Benthological Society 24:690–705.

Walsh, C. J., A. H. Roy, J. W. Feminella, P. D. Cottingham, P. M. Groffman, and R. P. Morgan. 2005b. The urban stream syndrome: current knowledge and the search for a cure. Journal of the North American Benthological Society 24(3):706–723.

Wang, L., J. Lyons, P. Kanehl, and R. Gatti. 1997. Influences of watershed land use on habitat quality and biotic integrity in Wisconsin streams. Fisheries 22(6):6–12.

Wang, L., J. Lyons, P. Kanehl, and R. Bannerman. 2001. Impacts of urbanization on stream habitat and fish across multiple spatial scales. Environmental Management 28(2):255–266.

Wang, L., J. Lyons, and P. Kanehl. 2003. Impacts of urban land cover on trout streams in Wisconsin and Minnesota. Transactions of the American Fisheries Society 132(5):825–839.

White, R. J. 1996. Growth and development of North American stream habitat management for fish. Canadian Journal of Fisheries and Aquatic Sciences 53(Suppl 1):342–363.

Whitlow, J. R., and K. J. Gregory. 1989. Changes in urban stream channels in Zimbabwe. Regulated Rivers: Research and Management 4:27–42.

Williams, G. P., and M. G. Wolman. 1984. Downstream Effects of Dams on Alluvial Rivers. U.S. Geological Survey Professional Paper 1286.

Wolman, M. G., and Schick, A. 1967. Effects of construction on fluvial sediment, urban and suburban areas of Maryland. Water Resources Research 3:451–464.

Wright, T., J. Tomlinson, T. Schueler, and K. Cappiella. 2006. Direct and indirect impacts of urbanization on wetland quality. Wetlands and Watersheds Article 1. Ellicott City, MD: Center for Watershed Protection.

2
The Challenge of Regulating Stormwater

Although stormwater has long been regarded as a major culprit in urban flooding, only in the past 30 years have policymakers appreciated the significant role stormwater plays in the impairment of urban watersheds. This recent rise to fame has led to a cacophony of federal, state, and local regulations to deal with stormwater, including the federal Clean Water Act (CWA) implemented by the U.S. Environmental Protection Agency (EPA). Perhaps because this longstanding environmental problem is being addressed so late in the development and management of urban watersheds, the laws that mandate better stormwater control are generally incomplete and were often passed for other purposes, like industrial waste control.

This chapter discusses the regulatory programs that govern stormwater, particularly the federal program, explaining how these programs manage stormwater only impartially and often inadequately. While progress has been made in the regulation of urban stormwater—from the initial emphasis on simply moving it away from structures and cities as fast as possible to its role in degrading neighboring waterbodies—a significant number of gaps remain in the existing system. Chapter 6 returns to these gaps and considers the ways that at least some of them may be addressed.

FEDERAL REGULATORY FRAMEWORK FOR STORMWATER

The Clean Water Act

The CWA is a comprehensive piece of U.S. legislation that has a goal of restoring and maintaining the chemical, physical, and biological integrity of the nation's waters. Its long-term goal is the elimination of polluted discharges to surface waters (originally by 1985), although much of its current effort focuses on the interim goal of attaining swimmable and fishable waters. Initially enacted as the Federal Water Pollution Control Act in 1948, it was revised by amendments in 1972 that gave it a stronger regulatory, water chemistry-focused basis to deal with acute industrial and municipal effluents that existed in the 1970s. Amendments in 1987 broadened its focus to deal with more diffuse sources of impairments, including stormwater. Improved monitoring over the past two decades has documented that although discharges have not been eliminated, there has been a widespread lessening of the effects of direct municipal and industrial wastewater discharges.

A timeline of federal regulatory events over the past 125 years relevant to stormwater, which includes regulatory precursors to the 1972 CWA, is shown in Table 2-1. The table reveals that while there was a flourish of regulatory activity related to stormwater during the mid-1980s to 1990s, there has been much less regulatory activity since that time.

TABLE 2-1 Legal and Regulatory Milestones for the Stormwater Program

1886	**Rivers and Harbors Act.** A navigation-oriented statute that was used in the 1960s and 1970s to challenge unpermitted pollutant discharges from industry.
1948 *1952* *1955*	**Federal Water Pollution Control Act.** Provided matching funds for wastewater treatment facilities, grants for state water pollution control programs, and limited federal authority to act against interstate pollution.
1965	**Water Quality Act.** Required states to adopt water quality standards for interstate waters subject to federal approval. It also required states to adopt state implementation plans, although failure to do so would not result in a federally implemented plan. As a result, enforceable requirements against polluting industries, even in interstate waters, was limited.
1972	**Federal Water Pollution Control Act.** First rigorous national law prohibiting the discharge of pollutants into surface waters without a permit. • Goal is to restore and maintain health of U.S. waters • Protection of aquatic life and human contact recreation by 1983 • Eliminate discharge of pollutants by 1985 • Wastewater treatment plant financing
	Clean Water Act Section 303(d) • Contains a water quality-based strategy for waters that remain polluted after the implementation of technology-based standards. • Requires states to identify waters that remain polluted, to determine the total maximum daily loads that would reverse the impairments, and then to allocate loads to sources. If states do not perform these actions, EPA must.
	Clean Water Act Section 208 • Designated and funded the development of regional water quality management plans to assess regional water quality, propose stream standards, identify water quality problem areas, and identify wastewater treatment plan long-term needs. These plans also include policy statements which provide a common consistent basis for decision making.
1977 *1981*	**Clean Water Act Sections 301 and 402** • Control release of toxic pollutants to U.S. waters • Technology treatment standards for conventional pollutants and priority toxic pollutants. • Recognition of technology limitations for some processes.
1977	*NRDC vs. Costle.* Required EPA to include stormwater discharges in the National Pollution Discharge Elimination System (NPDES) program.
1987	**Clean Water Act Amended Sections 301 and 402** • Control toxic pollutants discharged to U.S. waters. • Manage urban stormwater pollution. • Numerical criteria for all toxic pollutants. • Integrated control strategies for impaired waters. • Stormwater permit programs for urban areas and industry. • Stronger enforcement penalties. • Anti-backsliding provisions.

Table continues next page

TABLE 2-1 continued

1990	**EPA's Phase I Stormwater Permit Rules are Promulgated** • Application and permit requirements for large and medium municipalities • Application and permit requirements for light and heavy industrial facilities based on Standard Industrial Classification (SIC) Codes, and construction activity ≥ 5 acres
1999	**EPA's Phase II Stormwater Permit Rules are Promulgated** • Permit requirements for census-defined urbanized areas • Permit requirements for construction sites 1 to 5 acres
1997-2001	**Total Maximum Daily Load (TMDL) Program Litigation** • Courts order EPA to establish TMDLs in a number of states if the states fail to do so. The TMDLs assign Waste Load Allocations for stormwater discharges which must be incorporated as effluent limitations in stormwater permits.
2006-2008	**Section 323 of the Energy Policy Act of 2005** • EPA promulgates rule (2006) to exempt stormwater discharges from oil and gas exploration, production, processing, treatment operations, or transmission facilities from NPDES stormwater permit program. • In 2008, courts order EPA to reverse the rule which exempted certain activities in the oil and gas exploration industry from storm water regulations. In *Natural Resources Defense Council vs. EPA* (9th Cir. 2008), the court held that it was "arbitrary and capricious" to exempt from the Clean Water Act stormwater discharges containing sediment contamination that contribute to a violation of water quality standards.
2007	**Energy Independence and Security Act of 2007** • Requires all federal development and redevelopment projects with a footprint above 5,000 square feet to achieve predevelopment hydrology to the "maximum extent technically feasible."

The Basic NPDES Program: Regulating Pollutant Discharges

The centerpiece of the CWA is its mandate "that all *discharges* into the nation's waters are unlawful, unless specifically authorized by a permit" [42 U.S.C. §1342(a)]. Discharges do not include all types of pollutant flows, however. Instead, "discharges" are defined more narrowly as "point sources" of pollution, which in turn include only sources that flow through a discrete conveyance, like a pipe or ditch, into a lake or stream [33 U.S.C. §§ 1362(12) and (14)]. Much of the focus of the CWA program, then, is on limiting pollutants emanating from these discrete, point sources directly into waters of the United States. Authority to control nonpoint sources of pollution, like agricultural runoff (even when drained via pipes or ditches), is generally left to the states with more limited federal oversight and direction.

All point sources of pollutants are required to obtain a National Pollutant Discharge Elimination System (NPDES) permit and ensure that their pollutant discharges do not exceed specified effluent standards. Congress also commanded that rather than tie effluent standards to the needs of the receiving waterbody—an exercise that was far too scientifically uncertain and time-consuming—the effluent standards should first be based on the best available pollution technology or the equivalent. In response to a very ambitious mandate, EPA has promulgated very specific, quantitative discharge limits for the wastewater produced by over 30 industrial categories of sources based on what the best pollution control technology could accomplish, and it requires at least secondary treatment for the effluent produced by most sewage treatment plants. Under the terms of their permits, these large sources are also required to self-monitor their effluent at regular intervals and submit compliance reports to state or federal regulators.

EPA quickly realized after passage of the CWA in 1972 that if it were required to develop pollution limits for all point sources, it would need to regulate hundreds of thousands and perhaps even millions of small stormwater ditches and thousands of small municipal stormwater outfalls, all of which met the technical definition of "point source". It attempted to exempt all these sources, only to have the D.C. Circuit Court read the CWA to permit no exemptions [*NRDC vs. Costle*, 568 F.2d 1369 (D.C. Cir. 1977)]. In response, EPA developed a "general" permit system (an "umbrella" permit that covers multiple permittees) for smaller outfalls of municipal stormwater and similar sources, but it generally did not require these sources to meet effluent limitations or monitor their effluent.

It should be noted that, while the purpose of the CWA is to ensure protection of the physical, biological, and chemical integrity of the nation's waters, the enforceable reach of the Act extends only to the discharges of "pollutants" into waters of the United States [33 U.S.C. § 1311(a); cf. PUD No. 1 of Jefferson County v. Washington Department of Ecology, 511 U.S. 700 (1994) (providing states with broad authority under section 401 of the CWA to protect designated uses, not simply limit the discharge of pollutants)]. Even though "pollutant" is defined broadly in the Act to include virtually every imaginable substance added to surface waters, including heat, it has not traditionally been read to include water volume [33 U.S.C. § 1362(6)]. Thus, the focus of the CWA with respect to its application to stormwater has traditionally been on the water quality of stormwater and not on its quantity, timing, or other hydrologic properties. Nonetheless, because the statutory definition of "pollutant" includes "industrial, municipal, and agricultural waste discharged into water," using transient and substantial increases in flow in urban watersheds as a proxy for pollutant loading seems a reasonable interpretation of the statute. EPA Regions 1 and 3 have considered flow control as a particularly effective way to track sediment loading, and they have used flow in TMDLs as a surrogate for pollutant loading (EPA Region 3, 2003). State trial courts have thus far ruled that municipal separate storm sewer system (MS4) permits issued under delegated federal authority can impose restrictions on flow where changes in flow impair the beneficial uses of surface waters (Beckman, 2007). EPA should consider more formally clarifying that significant, transient increases in flow in urban watersheds serve as a legally valid proxy for the loading of pollutants. This clarification will allow regulators to address the problems of stormwater in more diverse ways that include attention to water volume as well as to the concentration of individual pollutants.

Stormwater Discharge Program

By 1987, Congress became concerned about the significant role that stormwater played in contributing to water pollution, and it commanded EPA to regulate a number of enumerated stormwater discharges more rigorously. Specifically, Section 402(p), introduced in the 1987 Amendments to the CWA, directs EPA to regulate some of the largest stormwater discharges—those that occur at industrial facilities and municipal storm sewers from larger cities and other significant sources (like large construction sites)—by requiring permits and promulgating discharge standards that require the equivalent of the best available technology [42 U.S.C. § 1342(p)(3)]. Effectively, then, Congress grafted larger stormwater discharges onto the existing NPDES program that was governing discharges from manufacturing and sewage treatment plants.

Upon passage of Section 402(p), EPA divided the promulgation of its stormwater program into two phases that encompass increasingly smaller discharges. The first phase, finalized in 1990, regulates stormwater discharges from ten types of industrial operations (this includes the entire manufacturing sector), construction occurring on five or more acres, and medium or large storm sewers in areas that serve 100,000 or more people [40 C.F.R. § 122.26(a)(3) (1990); 40 C.F.R. § 122.26 (b)(14) (1990)]. The second phase, finalized in 1995, includes smaller municipal storm sewer systems and smaller construction sites (down to one acre) [60 Fed. Reg. 40,230 (Aug. 7, 1995) (codified at 40 C.F.R. Parts 122, 124 (1995)]. If these covered sources fail to apply for a permit, they are in violation of the CWA.

Because stormwater is more variable and site specific with regard to its quality and quantity than wastewater, EPA found it necessary to diverge in two important ways from the existing NPDES program governing discharges from industries and sewage treatment plants. First, stormwater discharge limits are not federally specified in advance as they are with discharges from manufacturing plants. Even though Congress directed EPA to require stormwater sources to install the equivalent of the best available technology or "best management practices," EPA concluded that the choice of these best management practices (referred to in this report as stormwater control measures or SCMs) would need to be source specific. As a result, although EPA provides constraints on the choices available, it generally leaves stormwater sources with responsibility for developing a stormwater pollution prevention plan and the state with the authority to approve, amend, or reject these plans (EPA, 2006, p. 15).

Second, because of the great variability in the nature of stormwater flow, some sources are not required to monitor the pollutants in their stormwater discharges. Even when monitoring is required, there is generally a great deal of flexibility for regulated parties to self-monitor as compared with the monitoring requirements applied to industrial waste effluent (not stormwater from industries). More specifically, for a small subset of stormwater sources such as Phase I MS4s, some monitoring of effluent during a select number of storms at a select number of outfalls is required (EPA, 1996a, p. VIII-1). A slightly larger number of identified stormwater dischargers, primarily industrial, are only required to collect grab samples four times during the year and visually sample and report on them (so-called benchmark monitoring). The remaining stormwater sources are not required to monitor their effluent at all (EPA, 1996a). States and localities may still demand more stringent controls and rigorous stormwater monitoring, particularly in areas undergoing a Total Maximum Daily Load (TMDL) assessment, as discussed below. Yet, even for degraded waters subject to TMDLs, any added monitoring that might be required will be limited only to the pollutants that cause the degraded condition [40 C.F.R. §§ 420.32-420.36 (2004)].

Water Quality Management

Since technology-based regulatory requirements imposed on both stormwater and more traditional types of discharges are not tied to the conditions of the receiving water—that is, they require sources only to do their technological best to eliminate pollution—basic federal effluent limits are not always adequate to protect water quality. In response to this gap in protection, Congress has developed a number of programs to ensure that waters are not degraded below minimal federal and state goals [e.g., 33 U.S.C. §§ 1288, 1313(e), 1329, 1314(l)]. Among these, the TMDL program involves the most rigorous effort to control both point and nonpoint sources to ensure that water quality goals are met [33 U.S.C. § 1313(d)].

Under the TMDL program, states are required to list waterbodies not meeting water quality standards and to determine, for each degraded waterbody, the "total maximum daily load" of the problematic pollutant that can be allowed without violating the applicable water quality standard. The state then determines what types of additional pollutant loading reductions are needed, considering not only point sources but also nonpoint sources. It then promulgates controls on these sources to ensure further reductions to achieve applicable water quality goals.

The TMDL process has four separate components. The first two components are already required of the states through other sections of the CWA: (1) identify beneficial uses for all waters in the state and (2) set water quality standards that correlate with these various uses. The TMDL program adds two components by requiring that states then (3) identify segments where water quality goals have not been met for one or more pollutants and (4) develop a plan that will ensure added reductions are made by point and/or nonpoint sources to meet water quality goals in the future. Each of these is discussed below.

Beneficial Uses. States are required to conduct the equivalent of "zoning" by identifying, for each water segment in the state, a beneficial use, which consists of ensuring that the waters are fit for either recreation, drinking water, aquatic life, or agricultural, industrial, and other purposes [33 U.S.C. § 1313(c)(2)(A)]. All states have derived "narrative definitions" to define the beneficial uses of waterbodies that are components of all water quality standard programs. Many of these narrative criteria are conceptual in nature and tend to define general aspects of the beneficial uses. For categories such as *aquatic life uses,* most states have a single metric for differentiating uses by type of stream (e.g., coldwater vs. warmwater fisheries). In general, the desired biological characteristics of the waterbody are not well defined in the description of the beneficial use. Some states, such as Ohio, have added important details to their beneficial uses by developing tiered aquatic life uses that recognize a strong gradient of anthropogenic background disturbance that controls whether a waterbody can attain a certain water quality and biological functioning (see Box 2-1; Yoder and Rankin, 1998). Any aquatic life use tier less stringent than the CWA interim goal of "swimmable–fishable" requires a Use Attainability Analysis to support a finding that restoration is not currently feasible and recovery is not likely in a reasonable period of time. This analysis and proposed designation must undergo public comment and review and are always considered temporary in nature. More importantly, typically one or more tiers above the operative interim goal of "swimmable–fishable" are provided. This method typically will protect the highest attainable uses in a state more effectively than having only single uses.

The concept of tiered beneficial uses and use attainability is especially important with regard to urban stormwater because of the potential irreversibility of anthropogenic development and the substantial costs that might be incurred in attempting to repair degraded urban watersheds to "swimmable–fishable" or higher status. Indeed, it is important to consider what public benefits and costs might occur for different designated uses. For example, large public benefits (in terms of aesthetics and safety) might be gained from initial improvements in an urban stream (e.g., restoring base flow) that achieve modest aquatic use and protect secondary human contact. However, achieving designated uses associated with primary human contact or exceptional aquatic habitat may be much more costly, such that the perceived incremental public gains may be much lower than the costs that must be expended to achieve that more ambitious designation.

BOX 2-1
Ohio's Tiered Aquatic Life Uses

"Designated" or "beneficial" uses for waterbodies are an important aspect of the CWA because they are the explicit water quality goals or endpoints set for each water or class of waters. Ohio was one of the first states to implement tiered aquatic life uses (TALUs) in 1978 as part of its water quality standards (WQS). Most states have a single aquatic life use for a class of waters based on narrative biological criteria (e.g., warmwater or cold-water fisheries) although many states now collect data that would allow identification of multiple tiers of condition. EPA has recognized the management advantages inherent to tiered aquatic life uses and has developed a technical document on how to develop the scientific basis that would allow States to implement tiered uses (EPA, 2005a; Davies and Jackson, 2006).

Ohio's TALUs reflect the mosaic of natural features across Ohio and over 200 years of human changes to the natural landscape. Widespread information on Ohio's natural history (e.g., Trautman's 1957 *Fishes of Ohio*) provided strong evidence that the potential fauna of streams was not uniform, but varied geographically. Based on this knowledge, Ohio developed a more protective aquatic life use tier to protect streams of high biological diversity that harbored unique assemblages of rare or sensitive aquatic species (e.g., fish, mussels, invertebrates). In its WQS in 1978, Ohio established a narrative Exceptional Warmwater Habitat (EWH) aquatic life use to supplement its more widespread general or "Warmwater Habitat" aquatic life use (WWH) (Yoder and Rankin, 1995).

The CWA permits states to assign aquatic life uses that do not meet the baseline swimmable-fishable goals of the CWA under specific circumstances after conducting a Use Attainability Analysis (UAA), which documents that higher CWA aquatic life use goals (e.g., WWH and EWH in Ohio) are not feasibly attainable. These alternate aquatic life uses are always considered temporary in case land use changes or technology changes to make restoration feasible. The accrual of more than ten years of biological assessment data by the late 1980s and extensive habitat and stressor data provided a key link between the stressors that limited attainment of a higher aquatic life use in certain areas and reaches of Ohio streams. This assessment formed the basis for several "modified" (physical) warm-water uses for Ohio waters and a "limited" use (limited resource water, LRW) for mostly small ephemeral or highly artificial waters (Yoder and Rankin, 1995). Table 2-2 summa-rizes the biological and physical characteristics of Ohio TALUs and the management con-sequences of these uses. Channelization typically maintained by county or municipal drainage and flood control efforts, particularly where such changes have been extensive, are the predominant cause of Modified and Limited aquatic life uses. Extensive channel modification in urban watersheds has led to some modified warmwater habitat (MWH) and LRW uses in urban areas. There has been discussion of developing specific "urban" aquatic life uses; however the complexity of multiple stressors and the need to find a clear link between the sources limiting aquatic life and feasible remediation is just now being addressed in urban settings (Barbour et al., 2006).

The TALUs in Ohio (EWH→LRW) reflect a gradient of landscape and direct physical changes, largely related to changes to instream habitat and associated hydrological fea-tures. Aquatic life uses and the classification strata based on ecoregion and stream size (headwater, wadeable, and boatable streams) provide the template for the biocriteria ex-pectations for Ohio streams (see Box 2-2). Identification of the appropriate tiers for streams and UAA are a routine part of watershed monitoring in Ohio and are based on biological, habitat, and other supporting data. Any recommendations for changes in aquatic life uses are subject to public comment when the Ohio WQS are changed.

Ohio's water quality standards contain specific listings by stream or stream reach with notations about the appropriate aquatic life use as well as other applicable uses (e.g., rec-reation). Much of the impact of tiered uses on regulated entities or watershed management

TABLE 2-2 Key features associated with tiered aquatic life uses in the Ohio WQS.
SOURCE: EPA (2005a), Appendix B.

Aquatic Life Use	Key Attributes	Why a Waterbody Would Be Designated	Practical Impacts (compared to a baseline of WWH)
Warmwater Habitat (WWH)	Balanced assemblages of fish/invertebrates comparable to least impacted *regional* reference condition	Either supports biota consistent with numeric biocriteria for that ecoregion or exhibits the habitat potential to support recovery of the aquatic fauna	Baseline regulatory requirements consistent with the CWA "fishable" and "protection & propagation" goals; criteria consistent with U.S. EPA guidance with State/regional modifications as appropriate
Exceptional Warmwater Habitat (EWH)	Unique and/or diverse assemblages; comparable to upper quartile of *statewide* reference condition	Attainment of the EWH biocriteria demonstrated by both organism groups	More stringent criteria for D.O., temperature, ammonia, and nutrient targets; more stringent restrictions on dissolved metals translators; restrictions on nationwide dredge & fill permits; may result in more stringent wastewater treatment requirements
Coldwater Habitat (CWH)	Sustained presence of Salmonid or non-salmonid coldwater aquatic organisms; bonafide trout fishery	Bioassessment reveals coldwater species as defined by Ohio EPA (1987); put-and-take trout fishery managed by Ohio DNR	Same as above except that common metals criteria are more stringent; may result in more stringent wastewater treatment requirements
Modified Warmwater Habitat (MWH)	Warmwater assemblage dominated by species tolerant of low D.O., excessive nutrients, siltation, and/or habitat modifications	Impairment of the WWH biocriteria; existence and/or maintenance of hydrological modifications that cannot be reversed or abated to attain the WWH biocriteria; a use attainability analysis is required	Less stringent criteria for D.O., ammonia, and nutrient targets; less restrictive applications of dissolved metals translators; Nationwide permits apply without restrictions or exception; may result in less restrictive wastewater treatment requirements
Limited Resource Waters (LRW)	Highly degraded assemblages dominated exclusively by tolerant species; *should not* reflect acutely toxic conditions	Extensive physical and hydrological modifications that cannot be reversed and which preclude attainment of higher uses; a use attainability analysis is required	Chemical criteria are based on the prevention of acutely lethal conditions; may result in less restrictive wastewater treatment requirements

efforts arises from the tiered chemical and stressor criteria associated with each TALU. Criteria for compounds such as ammonia and dissolved oxygen vary with aquatic life use (see Table 2-2). Furthermore, application of management actions in Ohio, ranging from assigning antidegradation tiers, awarding funding for wastewater infrastructure and other projects, to issuing CWA Section 401/404 permits, are influence by the TALU and the biological assemblages present.

Ohio has been expanding its use of tiered uses by proposing tiered uses for wetlands (*http://www.epa.state.oh.us/dsw/rules/draft_1-53_feb06.pdf*) and developing new aquatic life uses for very small (primary headwater, PHW) streams. Both of these water types have a strong intersection with urban construction and stormwater practices. In Ohio this is especially so because the proposed mitigation standards for steams and wetlands are linked to TALUs (Ohio EPA, 2007).

Davies and Jackson (2006) present a good summary of the Maine rationale for TALUs: "(1) identifying and preserving the highest quality resources, (2) more accurately depicting existing conditions, (3) setting realistic and attainable management goals, (4) preserving incremental improvements, and (5) triggering management action when conditions decline" (Davies et al., 1999). Appendices A and B of EPA (2005a) provide more detailed information about the TALUs in Maine and Ohio, respectively.

Water Quality Criteria. Once a state has created a list of beneficial uses for its waters, water quality criteria are then determined that correspond with these uses. These criteria can target chemical, biological, or physical parameters, and they can be either numeric or narrative.

In response to the acute chemical water pollution that existed when the CWA was written, the primary focus of water quality criteria was the control of toxic and conventional pollutants from wastewater treatment plants. EPA developed water quality criteria for a wide range of conventional pollutants and began working on criteria for a list of priority pollutants. These were generally in the form of numeric criteria that are then used by states to set their standards for the range of waterbody types that exist in that state. While states do not have to adopt EPA water quality criteria, they must have a scientific basis for setting their own criteria. In practice, however, states have promulgated numerical water quality standards that can vary by as much as 1,000-fold for the same contaminant but are still considered justified by the available science [e.g., the water quality criteria for dioxin—*Natural Resources Defense Council, Inc. vs. EPA*, 16 F.3d 1395, 1398, 1403-05 (4th Cir. 1993)].

The gradual abatement of point source impairments and increased focus on ambient monitoring and nonpoint source pollutants has led to a gradual, albeit inconsistent, shift by states toward (1) biological and intensive watershed monitoring and (2) consideration of stressors that are not typical point source pollutants including nutrients, bedded sediments, and habitat loss. For these parameters, many states have developed narrative criteria (e.g., "nutrients levels that will not result in noxious algal populations"), but these can be subjective and hard to enforce.

The use of biological criteria (biocriteria) has gained in popularity because traditional water quality monitoring is now perceived as insufficient to answer questions about the wide range of impairments caused by activities other than wastewater point sources, including stormwater (GAO, 2000). As described in Box 2-2, Ohio has defined biocriteria in its water quality standards based on multimetric indices from reference sites that quantify the baseline expectations for each tier of aquatic life use.

Antidegradation. The antidegradation provision of the water quality standards deals with waters that already achieve or exceed baseline water quality criteria for a given designated use. Antidegradation provisions must be considered before any regulated activity can be authorized that may result in a lowering of water quality which includes biological criteria. These provisions protect the existing beneficial uses of a water and only allow a lowering of water quality (but never lower than the baseline criteria associated with the beneficial use) where necessary to support important social and economic development. It essentially asks the question: is the discharge or activity necessary? States with refined designated uses and biological criteria have used these programs to their advantage to craft scientifically sound, protective, yet flexible antidegradation rules (see Ohio and Maine). Antidegradation is not a replacement for tiered

BOX 2-2
Ohio's Biocriteria

After it implemented tiered aquatic life uses in 1978, Ohio developed numeric biocriteria in 1990 (Ohio WQS; Ohio Administrative Code 3745-1) as part of its WQS. Since designated uses were formulated and described in ecological terms, Ohio felt that it was natural that the criteria should be assessed on an ecological basis (Yoder, 1978). Subsequent to the establishment of the EWH tier in its WQS, Ohio expanded its biological monitoring efforts to include both macroinvertebrates and fish (Yoder and Rankin, 1995) and established consistent and robust monitoring methodologies that have been maintained to the present. This core of consistently collected data has allowed the application of analytical tools, including multimetric indices such as the Index of Biotic Integrity (IBI), the Invertebrate Community Index (ICI), and other multivariate tools. The development of aquatic ecoregions (Omernik, 1987, 1995; Gallant et al., 1989), a practical definition of biological integrity (Karr and Dudley, 1981), multimetric assessment tools (Karr, 1981; Karr et al., 1986), and reference site concepts (Hughes et al., 1986) provided the basis for developing Ohio's ecoregion-based numeric criteria.

Successful application of biocriteria in Ohio was dependent on the ability to accurately classify aquatic ecosystem changes based on primarily natural abiotic features of the environment. Ohio's reference sites, on which the biocriteria are based, reflect spatial differences that were partially explained by aquatic ecoregions and stream size. Biological indices were calibrated and stratified on this basis to arrive at biological criteria that present minimally acceptable baseline ecological index scores (e.g., IBI, ICI). Ohio biocriteria stratified by ecoregion aquatic life use and stream size are depicted in Figure 2-1.

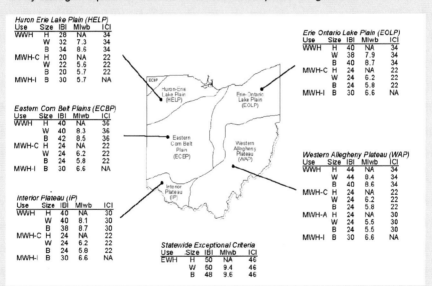

FIGURE 2-1 Numeric biological criteria adopted by Ohio EPA in 1990, using three biological indices [IBI, ICI, and the Modified Index of well-being (MIwb), which is used to assessed fish assemblages] and showing stratification by stream size, ecoregion, and designated use (warmwater habitat, WWH; modified warmwater habitat-channelized, MWH-C; modified warmwater habitat-impounded, MWH-I; and exceptional warmwater habitat, EWH). SOURCE: EPA (2006, Appendix B). The basis for the Ohio biocriteria and sampling methods is found in Ohio EPA (1987, 1989a,b), DeShon (1995), and Yoder and Rankin (1995).

uses, which provide a permanent floor against lowering water quality protection. Tiered beneficial uses and refined antidegradation rules can have substantial influence on stormwater programs because they influence the goals and levels of protection assigned to each waterbody.

Monitoring Programs to Identify Degraded Segments. Monitoring strategies by the states generally follow the regulatory efforts of EPA and seek to identify those waterbodies where water quality standards are not being met. Much of the initial ambient monitoring (i.e., monitoring of receiving waterbodies) was chemical based and focused on documenting changes in pollutant concentrations and exceedances of water quality criteria. Biological monitoring techniques have a long history of use as indicators of water quality impacts. However, it was not until such tools became more widespread—initially in states like Maine, North Carolina, and Ohio—that the extent of stormwater and other stressor effects on waterbodies became better understood. The biological response to common nonpoint stressors has driven the consideration of new water quality criteria (e.g., for nutrients, bedded sediments) that were not major considerations under an effluent-dominated paradigm of water management.

In parallel with the increase in biocriteria has been the development of biological monitoring to measure beneficial use attainment. Integrated biological surveys have revealed impairments of waterbodies that go beyond those caused by typical point sources (EPA, 1996b; Barbour et al., 1999a). The substantial increase in biological assemblage monitoring during the 1980s was enhanced by the development of more standard methods (Davis, 1995; Barbour et al., 1999a,b; Klemm et al., 2003) along with conceptual advances in the development of assessment tools (Karr, 1981; Karr and Chu, 1999). Development of improved classification tools (e.g., ecoregions, stream types), the reference site concept (Stoddard et al., 2006), and analytical approaches including multivariate (e.g., discriminant analysis) and multimetric indices such as IBI and ICI (see Box 2-3; Karr et al., 1986; DeShon, 1995) resulted in biological criteria being developed for several states. Biological monitoring approaches are becoming a widespread tool for assessing attainment of aquatic life use designation goals inherent to state water quality standards. Development of biocriteria represents a maturation of the use of biological data and provides institutional advantages for states in addressing pollutants without numeric criteria (e.g., nutrients) and non-chemical stressors such as habitat (Yoder and Rankin, 1998).

Setting Loads and Restricting Loading. Section 303d of the CWA requires that states compare existing water quality data with water quality standards set by the states, territories, and tribes. For those waters found to be in violation of their water quality standards, Section 303d requires that the state develop a TMDL. Currently, approximately 20,000 of monitored U.S. waters are in non-attainment of water quality standards, as evidenced by not meeting at least one specific narrative or numeric physical, chemical, or biological criterion, and thus require the development of a TMDL.

BOX 2-3
Commonly Used Biological Assessment Indices

Much of the initial work using biological data to assess the effects of pollution on inland streams and rivers was a response to Chicago's routing of sewage effluents into the Illinois River in the late 1800s. Early research focused on the use of indicator species, singly or in aggregate, and how they changed along gradients of effluent concentrations (Davis, 1990, 1995). In the 1950s Ruth Patrick used biological data to assess rivers by observing longitudinal changes in taxonomic groups, and later in the 1950s and 1960s "diversity indices" (e.g., Shannon-Wiener index, Shannon and Weaver, 1949) were used to assess aquatic communities (Washington, 1984; Davis 1990, 1995). These indices were various mathematical constructs that measured attributes such as richness and evenness of species abundance in samples and are still widely used today in ecological studies. Similarity indices are another approach that is used to compare biological assemblages between sites. There are a wide multitude of such indices (e.g., Bray-Curtis, Jaccard) and all use various mathematical constructs to examine species in common and absent between samples.

Biotic indices are generally of more recent origin (1970s to the present). Hilsenhoff (1987, 1988) assigned organic pollution tolerances to macroinvertebrate taxa and then combined these ratings in a biotic index that is still widely used for macroinvertebrates. Karr (1981) developed the Index of Biotic Integrity (IBI), a "multimetric" index that is composed of a series of 12 metrics of a Midwest stream fish community. This approach has been widely adopted and adapted to many types of waterbodies (streams, lakes, rivers, estuaries, wetlands, the Great Lakes, etc.) and organism groups and is probably the most widely used biotic index approach in the United States. Examples include the periphyton IBI (PIBI; Hill et al., 2000) for algal communities, the Invertebrate Community Index (ICI; DeShon, 1995) and benthic IBI (B-IBI, Kerans and Karr, 1994) for macroinvertebrates, a benthic IBI for estuaries (B-IBI; Weisberg et al., 1997), and a vegetative IBI for wetlands (VIBI-E; Mack, 2007).

Various multivariate statistical approaches have also been used to assess aquatic assemblages, often concurrently with multimetric indices. Maine, for example, uses a discriminant analysis that assesses stream stations by comparison to reference sites (Davies and Tsomides, 1997). Predictive modeling approaches, incorporating both biotic and environmental variables, have been widely used in Great Britain and Europe (River Invertebrate Prediction and Classification System, RIVPACS; Wright et al., 1993), Australia (AUS-RIVAS; Simpson and Norris, 2000), and more recently in the United States by Hawkins et al. (2000).

All of these approaches now have a wide scientific literature supporting their use and application. EPA (2002a) reports that most states have a biomonitoring program with at least one organism group to assess key waters in their states, although the level of implementation and sophistication varies by state. For example, only four states have numeric biocriteria in their state water quality standards, although 11 more are developing such biocriteria based on one or more of the above monitoring approaches (EPA, 2002a). The key to implementation of any of these approaches is to set appropriate goals for waters that can be accurately measured and then to use this type of information to identify limiting stressors (e.g., EPA Stressor Identification Process; EPA, 2000a).

The TMDL process includes an enforceable pollution control plan for degraded waters based on a quantification of the loading of pollutants and an understanding of problem sources within the watershed [33 U.S.C. § 1313(d)(1)(C)]. Both point and nonpoint sources of the problematic pollutants, including runoff from agriculture, are typically considered and their contributions to the problem are assessed. A plan is then developed that may require these sources to reduce their loading to a level (the TMDL) that ensures that the water will ultimately meet its designated use. Most of the TMDL requirements have been developed through regulation. Additional effluent limits for point sources discharging into segments subject to TMDLs are incorporated into the NPDES permit.

Total Maximum Daily Load Program and Stormwater

The new emphasis on TMDLs and the revelation that impacts are primarily from diffuse sources has increased the attention given to stormwater. If a TMDL assigns waste load allocations to stormwater discharges, these must be incorporated as effluent limitations into stormwater permits. In addition, the TMDL program provides a new opportunity for states to regulate stormwater sources more vigorously. In degraded waterbodies, effluent reductions for point sources are not limited by what is economically feasible but instead include requirements that will ensure that the continued degradation of the receiving water is abated. If a permitted stormwater source is contributing pollutants to a degraded waterbody and the state believes that further reductions in pollution from that source are needed, then more stringent discharge limitations are required. For example, in *City of Arcadia vs. State Water Resources Control Board* [135 Cal. App. 4th 1392 (Ca. Ct. App. 2006)], the court held in part that California's zero trash requirements for municipal storm drains, resulting from state TMDLs, were not inconsistent with TMDL requirements or the CWA. Thus, the maximum-extent-practicable standard for MS4s, as well as other technology-based requirements for other stormwater permittees, are a floor, not a ceiling, for permit requirements when receiving waters are impaired (Beckman, 2007). Finally, since the TMDL program expects the states to regulate any source—point or nonpoint—that it considers problematic, any source of stormwater is fair game, regardless of whether it is listed in Section 402p, and regardless of whether it is a "point source." Nonpoint source runoff from agricultural and silvicultural operations is in fact a common target for TMDL-driven restrictions [see, e.g., *Pronsolino vs. Nastri*, 291 F.3d 1123, 1130 (9th Cir. 2002), upholding restrictions on nonpoint sources, such as logging, compelled by State's TMDLs)].

Despite the potential for positive interaction between stormwater regulation and the TMDL program, there appears to be little activity occurring at the stormwater–TMDL interface. This is partly because the TMDL program itself has been slow in developing. In 2000, the National Wildlife Federation applied 36 criteria to the 50 states' water quality programs and concluded that 75 percent of the states had failed to develop meaningful TMDL programs (National Wildlife Federation, 2000, pp. 1–2). The General Accounting Office (GAO, 1989) identified the lack of implementation of TMDLs as a major impediment to attaining the goals of the CWA, which led to a spate of lawsuits filed by environmental groups to reverse this pattern. The result was numerous settlements with ambitious deadlines for issuing TMDLs.

Commentators blame the delays in these TMDL programs on inadequate ambient monitoring data and on the technical and political challenges of causally linking individual sources to problems of impairment. In a 2001 report, for example, the National Research Council (NRC) noted that unjustified and poorly supported water quality standards, a lack of monitoring, uncertainty in the relevant models, and a failure to use biocriteria to assess beneficial uses directly all contributed to the delays in states' abilities to bring their waters into attainment through the TMDL program (NRC, 2001). Each of these facets is not only technically complicated but also expensive. The cost of undertaking a rigorous TMDL program in a single state has been estimated to be about $4 billion per state, assuming that each state has 100 watersheds in need of TMDLs (Houck, 1999, p. 10476).

As a result, the technical demands of the TMDL program make for a particularly bad fit with the technical impediments already present in monitoring and managing stormwater. As mentioned earlier, the pollutant loadings in stormwater effluent vary dramatically over time and stormwater is notoriously difficult to monitor for pollutants. It is thus difficult to understand how much of a pollutant a stormwater point source contributes to a degraded waterbody, much less determine how best to reduce that loading so that the waterbody will meet its TMDL. As long as the focus in these TMDLs remains on pollutants rather than flow (a point raised earlier that will be considered again), the technical challenges of incorporating stormwater sources in a water quality-based regulatory program are substantial. Without considerable resources for modeling and monitoring, the regulator has insufficient tools to link stormwater contributions to water quality impairments.

These substantial challenges in linking stormwater sources back to TMDLs are reflected by the limited number of reports and guidance documents on the subject. In one recent report, for example, EPA provides 17 case studies in which states and EPA regions incorporated stormwater control measures into TMDL plans, but it is not at all clear from this report that these efforts are widespread or indicative of greater statewide activity (EPA, 2007a). Indeed, it almost appears that these case studies represent the universe of efforts to link TMDLs and stormwater management together. The committee's statement of task also appears to underscore, albeit implicitly, EPA's difficulty in making scientific connections between the TMDL and stormwater programs. This challenge is returned to in Chapter 6, which suggests some ways that the two can be joined together more creatively.

Other Statutory Authorities that Control Stormwater

Although the CWA is by far the most direct statutory authority regulating stormwater discharges, there are other federal regulatory authorities that could lead to added regulation of at least some stormwater sources of pollution.

Critical Resources

If there is evidence that stormwater flows or pollutants are adversely impacting either endangered species habitat or sensitive drinking water sources, federal law may impose more stringent regulatory restrictions on these activities. Under the Endangered Species Act, stormwater that jeopardizes the continued existence of endangered species may need to be reduced to the point that it no longer threatens the endangered or threatened populations in measurable ways, especially if the stormwater discharge results from the activity of a federal agency [16 U.S.C. §§ 1536(a), 1538(a)].

Under the Safe Drinking Water Act, a surface water supply of drinking water must conduct periodic "sanitary surveys" to ensure the quality of the supply (see 40 C.F.R. § 142.16). During the course of these surveys, significant stormwater contributions to pollution may be discovered that are out of compliance or not regulated under the Clean Water Act because they are outside of an MS4 area. Such a discovery could lead to more rigorous regulation of stormwater discharges. For a groundwater source that supplies 50 percent or more of the drinking water for an area and for which there is no reasonably available alternative source, the aquifer can be designated as a "Sole Source Aquifer" and receive greater protection under the Safe Drinking Water Act [42 U.S.C. § 300(h)-3(e)]. Stormwater sources that result from federally funded projects are also more closely monitored to ensure they do not cause significant contamination to these sole source aquifers.

Some particularly sensitive water supplies are covered by both programs. The Edwards Aquifer underlying parts of Austin and San Antonio, Texas, for example, is identified as a "Sole Source Aquifer." There are also several endangered species of fish and salamander in that same area. As a result, both the Safe Drinking Water Act and the Endangered Species Act demand more rigorous stormwater management programs to protect this delicate watershed.

Stormwater is also regulated indirectly by floodplain control requirements promulgated by the Federal Emergency Management Agency (FEMA). In order for a community to participate in the FEMA National Flood Insurance Program, it must fulfill a number of requirements, including ensuring that projects will not increase flood heights, including flood levels adjacent to the project site [see, e.g., 44 C.F.R. § 60.3(d)].

Contaminated Sites

Continuous discharges of contaminated stormwater and other urban pollutants (particularly through combined sewer overflows) have led to highly contaminated submerged sediments in many urban bays and rivers throughout the United States. In several cases where the sediment contamination was perceived as presenting a risk to human health or has led to substantial natural resource damages, claims have been filed under the federal hazardous waste cleanup statute commonly known as Superfund (42 U.S.C. § 9601 et seq.). This liability under the Comprehensive Environmental Response, Compensation, and Liability Act (CERCLA) technically applies to any area—whether submerged or not—as long as there is a "release or a threat of release of a hazardous substance" and the hazardous substances have accumulated in such a way as to lead to the "incurrence of response [cleanup] costs" or to "natural resource damages" [42 U.S.C. §9607(a)]. Although only a few municipalities and sewer systems have been sued, Superfund liability is theoretically of concern for possibly a much larger number of cities or even industries whose stormwater contains hazardous substances and when at least some of the discharges were either in violation of a permit or unpermitted. The National Oceanic and Atmospheric Administration brought suit against the City of Seattle and the Municipality of Metropolitan Seattle alleging natural resource damages to Elliott Bay resulting from pollution in stormwater and combined sewer overflows; the case was settled in 1991 (*United States vs. City of Seattle*, No. C90-395WD, *http://www.gc.noaa.gov/natural-office1.html*). While some of the elements for liability remain unresolved by the courts, such as whether some or all of the discharges are exempted under the "federally permitted release" defense of CERCLA [42 U.S.C. § 9601(10)(H)], which exempts surface water discharges that are covered by a general or NPDES permit from liability, the prospect of potential liability is still present.

Diversion of Stormwater Underground or into Wetlands

In some areas, stormwater is eliminated by discharging it into wetlands. If done through pipes or other types of point sources, these activities require a permit under the CWA. Localities or other sources that attempt to dispense with their stormwater discharges in this fashion must thus first acquire an NPDES permit.

Even without a direct discharge into wetlands, stormwater can indirectly enter wetland systems and substantially impair their functioning. In a review of more than 50 studies, the Center for Watershed Protection found that increased urbanization and development increased the amount of stormwater to wetlands, which in turn "led to increased ponding, greater water level fluctuation and/or hydrologic drought in urban wetlands" (Wright et al., 2006). They found that, in some cases, the ability of the wetlands to naturally remove pollutants became overwhelmed by pollutant loadings from stormwater.

An even more common method of controlling stormwater is to discharge it underground. Technically, these subsurface discharges of stormwater, including dry wells, bored wells, and infiltration galleries, are considered by EPA to be infiltration or "Class V" wells, which require a permit under the CWA as long as they are in proximity to an underground source of drinking water (40 C.F.R. Parts 144, 146). While EPA's definition excludes surface impoundments and excavated trenches lined with stone (provided they do not include subsurface fluid distribution systems or amount to "improved sinkholes" that involve the man-made modification of a naturally occurring karst depression for the purpose of stormwater control), most other types of subsurface drainage systems are covered regardless of the volume discharged (40 C.F.R. § 144.81(4)).

Given EPA's recent description of SCMs considered to be Class V injection wells (EPA, 2008), most SCMs that rely on infiltration are exempted. For example, if an infiltration trench is wider than it is deep, it is exempted from the Class V well regulations. Residential septic systems are also exempted [see 40 C.F.R. §§ 144.1(g)(1)(ii) and (2)(iii)]. However, those that involve deeper dry wells or infiltration galleries appear to require Class V well permits under the Safe Drinking Water Act. Because the use of these SCMs is likely to involve expensive compliance requirements, dischargers may steer away from them.

Air Contaminants

Air pollutants from vehicular exhaust and industrial sources that precipitate on roads and parking lots can also be collected in stormwater and increase pollutant loading (see Chapter 3 discussion of atmospheric deposition). While the Clean Air Act regulates these sources of air contamination, it does not eliminate them. Stormwater that is contaminated with air pollutants may consist of both "legal" releases of air pollutants, as well as "illegal" releases emitted in violation of a permit, although the distinction between the two groups of pollutants is effectively impossible to make in practice.

Pesticides and Other Chemical Products Applied to Land and Road Surfaces

EPA regulates the licensing of pesticides as well as chemicals and chemical mixtures, although its actual authority to take action, such as restricting product use or requiring labeling, varies according to the statute and whether the product is new or existing. Although EPA technically is allowed to consider the extent to which a chemical is accumulating in stormwater in determining whether additional restrictions of the chemical are needed, EPA is not aware of any instances in its Toxic Substances Control Act (TSCA) chemical regulatory decision-making in which it actually used this authority to advance water quality protection (Jenny Molloy, EPA, personal communication, March 13, 2008).

In its pesticide registration program, EPA does routinely consider a pesticide's potential for adverse aquatic effects from stormwater runoff in determining whether the pesticide constitutes an unreasonable risk (Bill Jordan, EPA, personal communication, March 14, 2008). EPA has imposed use restrictions on a number of individual pesticides, such as prohibiting aerial applications, requiring buffer strips, or reducing application amounts. Presumably states and localities are tasked with primary enforcement responsibility for most of these use restrictions. EPA has also required a surface water monitoring program as a condition of the re-registration for atrazine and continues to evaluate available surface water and groundwater data to assess pesticide risks (Bill Jordan, EPA, personal communication, March 14, 2008).

EPA STORMWATER PROGRAM

Stormwater is defined in federal regulations as "storm water runoff, snow melt runoff, and surface runoff and drainage" [40 CFR §122.26(b)(13)]. EPA intended that the term describe runoff from precipitation-related events and not include any type of non-stormwater discharge (55 Fed. Reg. 47995). A brief discussion of the evolution of the EPA's stormwater program is followed by an explanation of the permitting mechanisms and the various ways in which the program has been implemented by the states. As shown in Figure 2-2, the entire NPDES program has grown by almost an order of magnitude over the past 35 years in terms of the number of regulated entities, which explains the reliance of the program on general rather than individual permits. Both phases of the stormwater program have brought a large number of new entities under regulation.

Historical Background

States like Florida, Washington, Maryland, Wisconsin, and Vermont and some local municipalities such as Austin, Texas, Portland, Oregon, and Bellevue, Washington, preceded the EPA in implementing programs to mitigate the adverse impacts of stormwater quality and quantity on surface waters. The State of Florida, after a period of experimentation in the late 1970s, adopted a rule that required a state permit for all new stormwater discharges and for modifications to existing discharges if flows or pollutants increased (Florida Administrative Code, Chapter 17-25, 1982). The City of Bellevue, WA, established a municipal utility in 1974 to manage stormwater for water quality, hydrologic balance, and flood management purposes using an interconnected system of natural areas and existing drainage features.

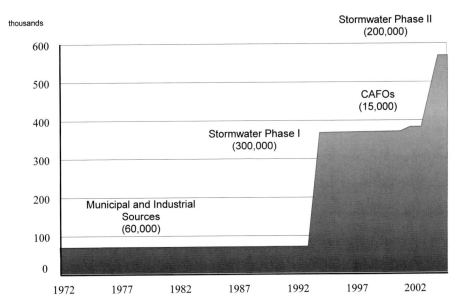

FIGURE 2-2 The number of permittees under the NPDES program of the Clean Water Act from 1972 to the present. Note that concentrated Animal Feeding Operations (CAFOs) are not considered in this report. SOURCE: Courtesy of Linda Boornazian, EPA.

EPA first considered regulating stormwater in 1973. At that time, it exempted from NPDES permit coverage conveyances carrying stormwater runoff not contaminated by industrial or commercial activity, unless the discharge was determined by the Administrator to be a significant contributor of pollutants to surface waters (38 Fed. Reg. 13530, May 22, 1973). EPA reasoned that while these stormwater conveyances were point sources, they were not suitable for end-of-pipe, technology-based controls because of the intermittent, variable, and less predictable nature of stormwater discharges. Stormwater pollution would be better managed at the local agency level through nonpoint source controls such as practices that prevent pollutants from entering the runoff. Further, EPA justified its decision by noting that the enormous numbers of individual permits that the Agency would have to issue would be administratively burdensome and divert resources from addressing industrial process wastewater and municipal sewage discharges, which presented more identifiable problems.

The Natural Resources Defense Council (NRDC) successfully challenged the EPA's selective exemption of stormwater point sources from the NPDES regulatory permitting scheme in federal court [*NRDC vs. Train*, 396 F.Supp. 1393 (D.D.C. 1975), *aff'd NRDC vs. Costle* 568 F.2d. 1369 (D.C. Cir. 1977)]. The court ruled that EPA did not have the authority to exempt point source discharges from the NPDES permit program, but recognized the Agency's discretion to use reasonable procedures to manage the administrative burden and to define what constitutes a stormwater point source. Consequently, EPA issued a rule establishing a comprehensive permit program for all stormwater discharges (except rural runoff) including municipal separate storm sewer systems (MS4s), which were to be issued "general" or area permits after a period of study (41 Fed. Reg. 11307, March 18, 1976). Individual permits were required for stormwater discharges from industrial or commercial activity, or where the stormwater discharge was designated by the permitting authority to be a significant contributor of pollutants. Comprehensive revisions to the NPDES regulations were published next, retaining the broad definition of stormwater discharges subject to the NPDES permit program and requiring permit application requirements similar to those for industrial wastewater discharges, including testing for an extended list of pollutants (44 Fed. Reg. 32854, June 7, 1979; 45 Fed. Reg. 33290, May 19, 1980).

The new NPDES regulations resulted in lawsuits filed in federal courts by a number of major trade associations, member companies, and environmental groups challenging several aspects of the NPDES program, including the stormwater provisions. The cases were consolidated in the D.C. Circuit Court of Appeals, and EPA reached a settlement with the industry petitioners on July 7, 1982, agreeing to propose changes to the stormwater regulations to balance environmental concerns with the practical limitations of issuing individual NPDES permits and limited resources. The Agency significantly narrowed the definition of stormwater point sources to conveyances contaminated by process wastes, raw materials, toxics, hazardous pollutants, or oil and grease, and it reduced application requirements by dividing stormwater discharges into two groups based on their potential for significant pollution problems (47 Fed. Reg. 52073, November 18, 1982). EPA issued a final rule retaining the broad coverage of stormwater point sources, and a two-tiered classification to administratively regulate these stormwater discharges (49 Fed. Reg. 37998, September 26, 1984).

The rule generated considerably controversy; trade associations and industry contended that application deadlines would be impossible to meet and that the sampling requirements were excessive, while the environmental community expressed a concern that additional changes or delays would exacerbate the Agency's failure to regulate sources of stormwater pollution. On the basis of the post-promulgation comments received, EPA determined that it was necessary to obtain additional data on stormwater discharges to assess their significance, and it conducted meetings with industry groups, who indicated an interest in providing representative data on the quality of stormwater discharges of their membership. The Agency determined that the submission of representative data was the most practical and efficient means of determining appropriate permit terms and conditions, as well as priorities for the multitude of stormwater point source discharges that needed to be permitted (50 Fed. Reg. 32548, August 12, 1985).

In the mean time, the U.S. House of Representatives and the Senate both passed bills to amend the CWA in mid-1985. The separate bills were reconciled in Conference Committee, and on February 4, 1987, Congress passed the Water

Quality Act (WQA), which specifically addressed stormwater discharges. The WQA added Section 402(p) to the CWA, which requires stormwater permits to be issued prior to October 1992 for (i) municipal stormwater discharges from large and medium municipalities based on the 1990 census; (ii) discharges associated with industrial activity; and (iii) a stormwater discharge that the Administrator determines contributes to the violation of a water quality standard or is a significant contributor of pollutants to waters of the United States. MS4s were required to reduce pollutants in stormwater discharges to the "maximum extent practicable" (MEP). Industrial and construction stormwater discharges must meet the best conventional technology (BCT) standard for conventional pollutants and the best available technology economically achievable (BAT) standard for toxic pollutants. EPA and the NPDES-delegated states were given the flexibility to issue municipal stormwater permits on a system-wide or jurisdiction-wide basis. In addition, the WQA amended Section 402(l)(2) of the CWA to not require a permit for stormwater discharges from mining and oil and gas operations if the stormwater discharge is not contaminated by contact, and it amended Section 502(14) of the CWA to exclude agricultural stormwater discharges from the definition of point source.

These regulations had been informed by the National Urban Runoff Program, conducted from 1978 to 1983 to characterize the water quality of stormwater runoff from light industrial, commercial, and residential areas (Athayde et al., 1983). The majority of samples collected were analyzed for eight conventional pollutants and three heavy metals, and a subset was analyzed for 120 priority pollutants. The study indicated that on an annual loading basis, some of the conventional pollutants were greater than the pollutant loadings resulting from municipal wastewater treatment plants. In addition, the study found that a significant number of samples exceeded EPA's water quality criteria for freshwater.

The Federal Highway Administration conducted studies over a ten-year period ending in 1990 to characterize the water quality of stormwater runoff from roadways (Driscoll et al., 1990). A total of 993 individual stormwater events at 31 highway sites in 11 states were monitored for eight conventional pollutants and three heavy metals. In addition, a subset of samples was analyzed for certain other conventional pollutant parameters. The studies found that urban highways had significantly higher pollutant concentrations and loads than non-urban highway sites. Also, sites in relatively dry semi-arid regions had higher concentrations of many pollutants than sites in humid regions.

Final Stormwater Regulations

EPA issued final regulations in 1990 establishing a process for stormwater permit application, the required components of municipal stormwater management plans, and a permitting strategy for stormwater discharges associated with industrial activities (55 Fed. Reg. 222, 47992, November 16, 1990). Stormwater discharges associated with industrial activity that discharge to MS4s were required to obtain separate individual or general NPDES permits. Nevertheless, EPA recognized that medium and large MS4s had a significant role to play in source identification and the development of pollution controls for industry, and thus municipalities were obligated to require the implementation of controls under local government authority for stormwater discharges associated with industrial activity in their stormwater management program. The final regulations also established minimum sampling requirements during permit application for medium and large MS4s (serving a population based on the 1990 census of 100,000 to 250,000, and 250,000 or more, respectively). MS4s were required to submit a two-part application over two years with the first part describing the existing program and resources and the second part providing representative stormwater quality discharge data and a description of a proposed stormwater management program, after which individual MS4 NPDES permits would be issued for medium and large MS4s.

In addition, the regulations identified ten industry groups and construction activity disturbing land area five acres or greater as being subject to stormwater NPDES permits. These industries were classified as either heavy industry or light industry where industrial activities are exposed to stormwater, based on the Office of Management and Budget Standard Industrial Classifications (SIC). The main industrial sectors subject to the stormwater program are shown in Table 2-3 and include 11 regulatory categories: (i) facilities with effluent limitations, (ii) manufacturing, (iii) mineral, metal, oil and gas, (iv) hazardous waste treatment, storage, or disposal facilities, (v) landfills, (vi) recycling facilities, (vii) steam electric plants, (viii) transportation facilities, (ix) treatment works, (x) construction activity, and (xi) light industrial activity.

The second phase of final stormwater regulations promulgated on December 8, 1999 (64 Fed. Reg. 68722) required small MS4s to obtain permit coverage for stormwater discharges no later than March 10, 2003. A small MS4 is defined as an MS4 not already covered by an MS4 permit as a medium or large MS4, or is located in "urbanized areas" as defined by the Bureau of the Census (unless waived by the NPDES permitting authority), or is designated by the NPDES permitting authority on a case-by-case basis if situated outside of urbanized areas. Further, the regulations lowered the construction activities regulatory threshold for permit coverage for stormwater discharges from five acres to one acre.

To give an idea of the administrative burden associated with the stormwater program and the different types of permits, Table 2-4 shows the number of regulated entities in the Los Angeles region that fall under either individual or general permit categories. Industrial and construction greatly outweigh municipal permittees, and stormwater permittees are vastly more numerous that traditional wastewater permittees.

TABLE 2-3 Sectors of Industrial Activity Covered by the EPA Stormwater Program

Category (see page 69)	Sector	SIC Major Group	Activity Represented
(i)	A	24	Timber products
(ii)	B	26	Paper and allied products
(ii)	C	28 and 39	Chemical and allied products
(i), (ii)	D	29	Asphalt paving and roofing materials and lubricants
(i) (ii)	E	32	Glass, clay, cement, concrete, and gypsum products
(i) (iii)	F	33	Primary metals
(i), (iii)	G	10	Metal mining (ore mining and dressing)
(i), (iii)	H	12	Coal mines and coal mining-related facilities
(i), (iii)	I	13	Oil and gas refining
(i), (iii)	J	14	Mineral mining and dressing
(iv)	K	HZ	Hazardous waste, treatment, storage, and disposal
(v)	L	LF	Landfills, land application sites, and open dumps
(vi)	M	50	Automobile salvage yards
(vii)	N	50	Scrap recycling facilities
(vii)	O	SE	Steam electric generating facilities
(viii)	P	40, 41, 42, 43, 51	Land transportation and warehousing
(viii)	Q	44	Water transportation
(viii)	R	37	Ship and boat building or repairing yards
(viii)	S	45	Air transportation
(ix)	T	TW	Treatment works
(xi)	U	20, 21	Food and kindred products
(xi)	V	22, 23, 31	Textile mills, apparel, and other fabric product manufacturing, leather and leather products
(xi)	W	24, 25	Furniture and fixtures
(xi)	X	27	Printing and publishing
(xi)	Y	30, 39, 34	Rubber, miscellaneous plastic products, and miscellaneous manufacturing industries
(xi)	AB	35, 37	Transportation equipment, industrial or commercial machinery
(xi)	AC	35, 36, 38	Electronic, electrical, photographic, and optical goods
(x)			Construction activity
	AD		Non-classified facilities designated by Administrator under 40 CFR §122.26(g)(1)(l)

SOURCE: 65 Fed. Reg. 64804, October 30, 2000.

TABLE 2-4 Number of NPDES Wastewater and Stormwater Entities Regulated by the CalEPA, Los Angeles Regional Water Board, as of May 2007

Waste Type	Individual Permittees	General Permittees
Wastewater and Non-stormwater Industry	103	574
Combined Wastewater and Stormwater	23	0
Stormwater (pre-1990)	45	0
Industrial Stormwater (post-1990)	0	2990
Construction Stormwater (post-1990)	0	2551
Municipal Stormwater (post-1990)	100	0
Total	271	6215

Municipal Permits

States with delegated NPDES permit authority (all except Alaska, Arizona, Idaho, Massachusetts, New Hampshire, and New Mexico) issued the first large and medium MS4 permits beginning in 1990, some of which are presently in their fourth permit term. These MS4 permits require large and medium municipalities to implement programmatic control measures (the six minimum measures) in the areas of (1) public education and outreach, (2) public participation and involvement, (3) illicit discharge detection and elimination, (4) construction site runoff control, (5) post-construction runoff control, and (6) pollution prevention and good housekeeping—all to reduce the discharge of pollutants in stormwater to the *maximum extent practicable*. Efforts to meet the six minimum measures are documented in a stormwater management plan. Non-stormwater discharges to the MS4 are prohibited unless separately permitted under the NPDES, except for certain authorized non-stormwater discharges, such as landscape irrigation runoff, which are deemed innocuous nuisance flows and not a source of pollutants. MS4 permits generally require analytic monitoring of pollutants in stormwater discharges for all Phase I medium and large MS4s from a subset of their outfalls that are 36 inches or greater in diameter or drain 50 acres or more. These data, at the discretion of the permitting authority, may be compared with water quality standards and considered (by default) to be effluent limitations, which refer to any restriction, including schedules of compliance, established by a state or the Administrator pursuant to CWA Section 304(b) on quantities, rates, and concentrations of chemical, physical, biological, and other constituents discharged from point sources into navigable waters, the waters of the contiguous zone, or the ocean (40 CFR §401.11). A future exceedance of an effluent limitation constitutes a permit violation. However, permitting authorities have so far not taken this approach to interpreting MS4 stormwater discharge data.

The Phase I stormwater regulations require medium and large MS4s to inspect "high-risk" industrial facilities and construction sites within their jurisdictions. Certain industrial facilities and construction sites of a minimum acreage are also subject to separate EPA/state permitting under the industrial and construction general permits (see below). While EPA envisioned a partnership with municipalities on these inspections in its Phase I Rule Making, it provided no federal funding to build these partnerships. Both industry and municipalities have argued that the dual inspection responsibilities are duplicative and redundant. Municipalities have further contended that the inspection of Phase I industrial facilities and construction sites are solely an EPA/state obligation, although state and federal courts have ruled otherwise. In the committee's experience, many MS4s do not oversee or regulate industries within their boundaries.

As part of the Phase II program, small MS4s are covered under general permits and are required to implement a stormwater management program to meet the six minimum measures mentioned above. Unlike with Phase I, Phase II MS4 stormwater discharge monitoring was made discretionary, and inspection of industrial facilities within the boundary of a Phase II MS4 is not required.

Industrial Permits

EPA issued the first nationwide multi-sector industrial stormwater general permit (MSGP) on September 29, 1995 (60 Fed. Reg. 50804), which was reissued on October 30, 2000 (65 Fed. Reg. 64746). A proposed new MSGP was released for public comment in 2005 (EPA, 2005b). The proposed MSGP requires that industrial facility operators prepare a stormwater pollution prevention plan (similar to an MS4's stormwater management plan) that documents the SCMs that will be implemented to reduce pollutants in stormwater discharges. They must achieve technology-based requirements using BAT or BCT or water quality-based effluent limits, which is the same requirement as for process wastewater permits.

All industrial sectors covered under the MSGP must conduct visual monitoring four times a year. The visual monitoring is performed by collecting a grab sample within the first hour of stormwater discharge and observing its characteristics qualitatively. A subset of MSGP industrial categories is required to perform analytical monitoring for benchmark pollutant parameters four times in Year 2 of permit coverage and again in Year 4 if benchmarks were exceeded in Year 2. The benchmark pollutant parameters, listed in Table 2-5, were selected based on the sampling data included with group permit applications submitted after the EPA issued its stormwater regulations in 1990. To comply with the benchmark monitoring requirements, a grab sample must be collected within the first hour of stormwater discharge after a rainfall event of 0.1 inch or greater and with an interceding dry period of at least 72 hours. A benchmark exceedance is not a permit violation, but rather is meant to trigger the facility operator to investigate SCMs and make necessary improvements.

TABLE 2-5 Industry Sectors and Sub-Sectors Subject to Benchmark Monitoring

MSGP Sector	Industry Sub-sector	Required Parameters for Benchmark Monitoring
C	Industry organic chemicals Plastics, synthetic resins, etc. Soaps, detergents, cosmetics, perfumes Agricultural chemicals	Al, Fe, nitrate and nitrite N Zn Zn, nitrate and nitrite N Pb, Fe, Zn, P, nitrate and nitrite N
D	Asphalt paving and roofing materials	TSS
E	Clay products Concrete products	Al TSS and Fe
F	Steel works, blast furnaces, rolling and finishing mills Iron and steel foundries Non-ferrous rolling and drawing Non-ferrous foundries (casting)	Al, Zn Al, Cu, Fe, Zn, TSS Cu, Zn Cu, Zn
G	Copper ore mining and dressing	COD, TSS, nitrate and nitrite N
H	Coal mines and coal mining related facilities	TSS
J	Dimension stone, crushed stone, and non-metallic minerals (except fuels) Sand and gravel mining	TSS, Al, Fe Nitrate and nitrite N, TSS
K	Hazardous waste treatment, storage, or disposal	NH$_3$, Mg, COD, Ar, Cd, CN, Pb, Hg, Se, Ag
L	Landfills, land application sites, and open dumps	Fe, TSS
M	Automobile salvage yards	TSS, Al, Fe, Pb
N	Scrap recycling	Cu, Al, Fe, Pb, Zn, TSS, COD
O	Steam electric generating facilities	Fe
Q	Water transportation facilities	Al, Fe, Pb, Zn
S	Airports with deicing activities	BOD, COD, NH$_3$, pH
U	Grain mill products Fats and oils	TSS BOD, COD, nitrate and nitrite N, TSS
Y	Rubber products	Zn
AA	Fabricated metal products except coating Fabricated metal coating and engraving	Fe, Al, Zn, nitrate and nitrite N Zn, nitrate and nitrite N

NOTE: BOD, biological oxygen demand; COD, chemical oxygen demand; TSS, total suspended solids.
SOURCE: 65 Fed. Reg. 64817, October 30, 2000.

EPA had already established technology-based effluent limitations for stormwater discharges for eight subcategories of industrial discharges prior to 1987, namely, for cement manufacturing, feedlots, fertilizer manufacturing, petroleum refining, phosphate manufacturing, steam electric, coal mining, and ore mining and dressing (see Table 2-6). Most of these facilities were covered under individual permits prior to 1987 and are generally required to stay covered under individual stormwater permits. Facilities in these sub-categories that had not been issued a stormwater discharge permit prior to 1992 are allowed to be covered under the MSGP, but they still have analytical monitoring requirements that must be compared to effluent limitation guidelines. An exceedance of the effluent limitation constitutes a permit violation.

TABLE 2-6 Select Stormwater Effluent Limitation Guidelines for Illustrative Purposes

Discharges	Design Storm	Pollutant Parameters	Effluent Limitations (max per day)
Phosphate Fertilizer Manufacturing Runoff (40 C.F.R. 418)	Not specified	Total P Fluoride	105 mg/L 75 mg/L
Petroleum Refining (40 C.F.R. 419)	Not specified	O&G TOC BOD5 COD Phenols Cr Hex Cr pH	15 mg/L 110 mg/L 48 kg/1000 m^3 flow 360 mg/1000 m^3 flow 0.35 mg/1000 m^3 flow 0.73 mg/1000 m^3 flow 0.062 mg/1000 m^3 flow 6–9
Asphalt Paving and Roofing Emulsion Products Runoff (40 C.F.R. 443)	Not specified	TSS O&G pH	0.023 kg/m^3 0.015 kg/m^3 6.0–9.0
Cement Manufacturing Material Storage Piles Runoff (40 C.F.R. 411)	10 yr, 24 hour	TSS pH	50 mg/L 6.0–9.0
Coal Mining (40 C.F.R. 434 Subpart B)	1 yr, 24 hour	Fe Mn TSS pH	7.0 mg/L 4 mg/L 70 mg/L 6.0–9.0
Steam Electric Power Generating (40 C.F.R. 423)	10 yr, 24 hour	TSS pH PCBs	50 mg/L 6.0–9.0 No discharge

NOTE: BOD5, biological oxygen demand; COD, chemical oxygen demand; O&G, oil and grease; PCBs, polychlorinated biphenyls; TOC, total organic carbon; TSS, total suspended solids. SOURCE: 40 C.F.R.

At the issuance of the Final Storm Water Rule in 1990, EPA envisioned the use of a mix of general permits and individual permits to better manage the administrative burden associated with permitting thousands of industrial stormwater point sources. In its original permitting strategy for industrial stormwater discharges, EPA articulated a four-tier strategy with the nationwide general permits: Tier 1 was baseline permitting, Tier 2 would incorporate watershed permits, Tier 3 would be industry category-specific permitting, and Tier 4 would encompass facility-specific individual permits. In reality, individual permits, which would allow for the crafting of permit conditions to be better structured to the specific industrial facility based on its higher potential risk to water quality, and could include adequate monitoring for purposes of compliance and enforcement, have been sparsely used. Similarly, neither the watershed permitting strategy nor the industry category-specific permitting strategy has found favor in the absence of better federal guidance and funding.

Industrial stormwater general permits are issued by the State NPDES Permitting Authority in NPDES-delegated states, and may be in the form a single statewide permit covering thousands of industrial permittees or sector-specific stormwater general permits covering less than a hundred facilities. EPA Regions issue the MSGP in states without NPDES-delegated authority and for facilities on Native Indian and Tribal Lands. EPA's nationwide 2000 MSGP presently covers 4,102 facilities.

Construction Permits

EPA issued the first nationwide construction stormwater general permit (CGP) in February 1998 (63 Fed. Reg. 7858). The permits are valid for five-year terms. The most recent CGP was issued in 2005 (68 Fed. Reg. 39087), and the EPA in 2008 administratively continued the CGP until the end of 2009, when it is expected to have developed effluent guidelines for construction activity (73 Fed. Reg. 40338). The EPA is presently under court order to develop effluent limitation guidelines for stormwater discharges from the construction and land development industry. The construction general permit requires the implementation of stormwater pollution prevention plans to prevent erosion, control sediment in stormwater discharges, and manage construction waste materials. Operators of the construction activity are required to perform visual inspections regularly, but no sampling of stormwater discharge during rainfall events is required. As with the industrial and municipal permittees, an exceedance of an effluent limitation incorporated in a permit would be a violation of the CWA and is subject to penalties.

EPA's CGP covers construction activity in areas where EPA is the permitting authority, including Indian lands, Puerto Rico, the District of Columbia, Massachusetts, New Hampshire, New Mexico, Idaho, Arizona, and Alaska. All other states have been delegated the authority to issue NPDES permits, and these states issue CGPs based on the EPA model but with subtle variations. For example the California and Georgia CGPs include monitoring requirements for construction sites discharging to sediment-impaired waterbodies. Wisconsin requires weekly inspections and an inspection within 24 hours of a rain event of 0.5 inches or greater. Georgia imposes discharge limits of an increase of no more than 10 Nephelometric Turbidity Units (NTU) above background in trout streams and no more than 25 NTU above background in other types of streams.

Permit Creation, Administration, and Requirements

For individual permits, the entity seeking coverage submits an application and one permit is issued. The conditions of the permit are based on an analysis of information provided in a rather lengthy permit application by the facility operator about the facility and the discharge. Generally, it takes six to 18 months for the permittee to compile the application information and for the permitting authority to finalize the permit. Individual permits are common for medium and large MS4s (Phase I), small MS4s in a few states (Phase II), and a few industrial activities.

General permits, on the other hand, are issued by the permitting authority, and interested parties then submit an Notice of Intent (NOI) to be covered. This mechanism is used where large numbers of dischargers require permit coverage, such as construction activities, most industrial activities, and most small MS4s (Phase II). The permit must identify the area of coverage, the sources covered, and the process for obtaining coverage. Once the permit is issued, a permittee may submit a NOI and receive coverage either immediately or within a very short time frame (e.g., 30 days).

All permits contain "effluent limitations" or "effluent guidelines," adherence to which is required of the permittee. However, the terms (which are synonymous) are agonizingly broad and encompass (1) meeting numeric pollutant limits in the discharge, (2) using certain SCMs, and (3) meeting certain design or performance standards. Effluent limitations may be expressed as SCMs when numeric limits are infeasible or for stormwater discharges where monitoring data are insufficient to carry out the purposes and intent of the CWA [122.44(k)]. If EPA has promulgated numerical "effluent guidelines" for existing and new stormwater sources under CWA Sections 301, 304, or 306, then the permits must incorporate the "effluent guidelines" as permit limits.

Effluent limitations can be either technology-based or water quality-based requirements. Technology-based requirements establish pollutant limits for discharges on what the best pollution control technology installed for that industry would normally accomplish. Water-quality based requirements, by contrast, look to the receiving waters to determine the level of pollution reduction needed for individual sources. There are national technology-based standards available for many categories of point sources, including many industrial sectors and municipal wastewater treatment plants. In the absence of national standards, technology-based requirements are developed on a case-by-case basis using best professional judgment. In general, BAT is the standard for toxic and non-conventional pollutants, while BCT is the standard for conventional pollutants. Water quality-based effluent limitations are required where technology-based limits are found to be insufficient to achieve applicable water quality standards, including restoring impaired waters, preventing impairments, and protecting high-quality waters. Limitations must control all pollutants or pollutant parameters that are or may be discharged at a level which will *cause, have reasonable potential to cause, or contribute to* an excursion above any applicable water quality standard. To distinguish between technology-based and water quality-based effluent limits, consider that a permittee is required to meet a numeric pollutant limit in their stormwater discharge. A technology-based limit would be based on studies of effluent concentrations coming from that technology, while a water quality-based limit would be based on some assessment of the impact of the discharge on a nearby receiving water (with the applicable water quality standard being the most conservative choice).

EPA is presently writing stormwater "effluent guidelines" for airport de-icing operations and construction/development activity, with an estimated final action date of December 2009.

Permits Prior to 1990

A limited number of individual stormwater permits (perhaps in the low thousands) were first issued prior to 1990, the period before EPA promulgated regulations specific to stormwater discharges, and before EPA first received the authority to issue general NPDES permits. These individual NPDES permits for industrial stormwater discharges, like traditional individual wastewater NPDES permits, incorporate numerical effluent limits and they impose discharge monitoring requirements to demonstrate compliance. These facilities were selected for permitting before 1990, presumably because of the risk they presented to causing or contributing to the exceedance of water quality standards.

Do Permittees Have to Meet Water Quality Standards in their Effluent?

It is unclear as to whether municipal, industrial, and construction stormwater discharges must meet water quality standards. Furthermore, even if such discharges were required to meet water quality standards, the absence of monitoring found within the permits means that enforcement of the requirement would be difficult at best. Nonetheless, some sources suggest that, with the exception of Phase II MS4 discharges, EPA's intent is that stormwater discharges comply with water quality standards, especially where a TMDL is in place.

First, the EPA Office of General Counsel issued a memorandum in 1991 stating that municipal stormwater permits must require that MS4s reduce stormwater pollutant discharges to the maximum extent practicable and must also comply with water quality standards. Recognizing the complexity of stormwater, EPA's 1996 Interim Permitting Approach for Water Quality-Based Effluent Limitations in Storm Water Permits (61 Fed. Reg. 43761) stated that stormwater permits should use SCMs in first-term stormwater permits and expanded or better-tailored SCMs in subsequent term permits to provide for the attainment of water quality standards. However, where adequate information existed to develop more specific conditions or limitations to meet water quality standards, these conditions or limitations are to be incorporated into stormwater permits as necessary and appropriate.

As permitting authorities began to develop TMDL waste load allocations to address impaired receiving waters, and waste load allocations were assigned to stormwater discharges, EPA issued a TMDL Stormwater Policy. It stated that stormwater permits must include permit conditions consistent with the assumptions and requirements of available waste load allocations (EPA, 2002b). Since waste load allocations derive directly from water quality standards, this could be interpreted as saying that stormwater discharges must meet water quality standards. However, EPA expected that most water quality-based effluent limitations for NPDES-regulated stormwater discharges that implement TMDL waste load allocations would be expressed as SCMs, and that numeric limits would be used only in rare instances. This is understandable, given that storm events are dynamic and variable and it would be expensive to monitor all storm events and discharge points, particularly for MS4s, to demonstrate compliance with a waste load allocation expressed as a numeric effluent limitation. Effluent limitations expressed as SCMs appear to be the best interim approach to demonstrate compliance with TMDLs, provided that these SCMs are reasonably expected to satisfy the waste load allocation in the TMDL. As part of the TMDL, the NPDES permit must also specify the monitoring necessary to determine compliance with effluent limitations. Where effluent limits are specified as SCMs, the permit should specify the monitoring necessary to assess if the load reductions expected from SCM implementation are achieved (e.g., SCM performance data).

Implementation of the Stormwater Program by States and Municipalities

NPDES-delegated states and Indian Tribes generally utilize the CGP and the MSGP as model templates for adopting their respective general permits to regulate stormwater discharges associated with industrial activity, including construction, within their jurisdictions. Nevertheless, some variations exist. For example, the California CGP requires sampling of stormwater at construction sites that discharge to surface waters that are listed as being impaired for sediment. Connecticut's MSGP regulates stormwater discharges associated with commercial activity, in addition to industrial activity. With respect to the municipal permits, the variability with which the stormwater program is implemented reflects the flexibility inherent in the MEP standard. In the absence of a definite description of MEP or nationwide effluent guidelines issued by EPA, states and municipalities have not been very rigorous in determining what constitutes an adequate level of compliance. This self-defined compliance threshold has been translated into a wide range of efforts at program implementation.

A number of MS4 programs have been leaders in some areas of program implementation. For example, Prince George's County, Maryland, was a pioneer in implementing low impact development (LID) techniques. Notable efforts have been made by states and municipalities in the Pacific Northwest, such as Oregon and Washington. California and Florida also are in the forefront of implementing comprehensive and progressive stormwater programs.

Greater implementation is evident in states that had state stormwater regulations in place prior to the advent of the national stormwater program (GAO, 2007). Some states issued early MS4 permits (e.g., California, Florida, Washington, and Wisconsin) prior to the promulgation of the national stormwater program, while a number of MS4s (e.g., Austin, Texas,; Santa Monica, California; and Bellevue, Washington) were already implementing comprehensive stormwater management programs. In addition, some MS4s conducted individual stormwater management activities, such as street-sweeping, household hazardous waste collection, construction site plan review, and inspections, prior to the national stormwater program. These areas are more likely than areas without a stormwater program that predated the EPA program to be successfully meeting the requirements of the current program.

One of the obvious differences is the level of interest and effort exercised by coastal communities or communities in close proximity to a water resource that have immediate access to the beneficial uses of those resources but also have an immediate view of the impacts of polluted runoff. That interest may contrast with the less active posture of upstream or further inland communities that may not be as sensitive and willing to implement more stringent stormwater programs. A recent report has found that programs with more specific permit requirements generally result in more comprehensive and progressive stormwater management programs (TetraTech, 2006a). The report concluded that permittees should be required to develop measurable goals based on the desired outcomes of the stormwater program. Furthermore, additional stormwater permit requirements can be expected as more TMDLs are developed and wasteload allocations must be translated into permit conditions.

GAO Report on Current Status of Implementation

In 2007, the GAO issued a report to determine the impact of EPA's Stormwater Program on communities (GAO, 2007). Some of the relevant findings are that urban stormwater runoff continues to be a major contributor to the nation's degraded waters and that stormwater program implementation has been slow for both Phase I and Phase II communities, with almost 11 percent of all communities not yet permitted as of fall 2006. Litigation, among other reasons, delayed the issuance of some permits for years after the application deadlines. As a result, almost all Phase II and some Phase I communities are still in the early stages of program implementation although deadlines for permit applications were years ago—16 years for Phase I and six years for Phase II. EPA has acknowledged that it does not currently have a system in place to measure the success of the Phase I program on a national scale (EPA, 2000b). Therefore, it is reasonable to conclude that the level of implementation of the stormwater program ranges widely, from municipalities having completed a third-term permit (such as Los Angeles County MS4 permit) to municipalities not yet covered by a Phase II MS4 permit.

The GAO report also indicates that communities' inconsistent reporting of activities makes it difficult to evaluate program implementation nationwide. Based on the report's findings it seems that little auditing activity has been performed to gauge the status of implementation and effectiveness in achieving water quality improvements. Most often cited is the effort by EPA's Region 9 and the State of California auditors that recently discovered, among other things, that some MS4s (1) had not developed stormwater management plans, (2) were not properly performing an adequate number of inspections to enforce their stormwater ordinances, and (3) were lax in implementing SCMs at publicly owned construction sites. They also found that some MS4s were not adequately controlling stormwater runoff at municipally owned and operated facilities, such as maintenance yards. In response to these findings, EPA issued in January 2007 an MS4 Program Evaluation Guidance document (EPA, 2007b).

In the absence of a nationwide perspective of the implementation of the stormwater program, it is hard to make a determination about the program's success. There are communities and states that seem to have made great strides in implementing progressive stormwater programs, but it also seems that overall many programs are still in the early stages of implementation, while a number of communities are still waiting to obtain coverage under the MS4 permits. In addition, it appears that there is no national uniform system of tracking success or cost data. All these unknowns make it very difficult to formulate any definite statements about how successful the implementation of the program is on a national perspective.

Committee Survey

In order to get a better understanding of how the stormwater program is implemented by the states, during 2007 the committee conducted two surveys asking states about their monitoring requirements, compliance determination, and other facts for each program (municipal, industrial, and construction). For the larger survey, 18 states representing all ten EPA regions responded to the survey. Both surveys and all responses are found in Appendix C.

As expected, the responding states reported that Phase I MS4s are required to sample their stormwater discharges for pollutants, although the frequency of sampling and the number of pollutants being sampled tended to vary. No state reported requiring Phase II MS4s to sample stormwater discharges. Monitoring requirements for industrial stormwater varied by state from none in Minnesota, Nebraska, and Maine to benchmark monitoring required under the MSGP in Virginia, New York, and Wyoming. California, Connecticut, and Washington require all industrial facilities to monitor for select chemical pollutants. Connecticut, additionally, requires sampling for aquatic toxicity. Most of the responding states do not require construction sites to do much more than visual monitoring periodically and after rain events. Georgia and Washington require construction sites to monitor for parameters such as turbidity and pH. California and Oregon require sampling when the discharge is to a waterbody impaired by sediment.

As mentioned previously, Phase I MS4s (but not Phase II MS4s) are required to address industrial dischargers within their boundaries. There was considerable variability regarding the survey questions of whether MS4s can conduct inspections of industrial facilities and what industries are considered high risk. In all of the responding states except Virginia, the responders think that MS4s have the authority to inspect industries within their boundaries, although the extent to which this is done is not clear and, in the committee's experience, is quite rare. Many of the responding states have not identified "high-risk" facilities and targeted them for compliance scrutiny, although certain categories were felt to be problematic by the state employee responding to the survey, such as metal foundries, auto salvage yards, metal recyclers, cement plants, and saw mills. In California and Washington, however, some of the Phase I MS4 permits have identified high-risk facilities for the municipal permittee to inspect.

Georgia, Maine, Minnesota, Nevada, New York, Vermont, and Washington have State Guidance Manuals for MS4 implementation, while in California a coalition of municipalities and the California Department of Transportation have developed MS4 guidance manuals. The rest of the responding states rely on general guidance provided by the EPA. State guidance manuals for the implementation of the industrial stormwater program were less common than guidance manuals for construction activity, with only California and Washington having such guidance manuals. In contrast, except for Nebraska and Oklahoma, statewide guidance manuals for erosion and sediment control were available. This may have resulted from the fact that many states had laws in place that required erosion and sediment control practices during land development, timber harvesting, and agricultural farming that predated the EPA stormwater regulations.

In an attempt to determine the level of oversight that a state provides for industrial and construction operations, the survey asked whether and to whom stormwater pollution prevention plans (SWPPPs) are submitted. Most of the responding states require the stormwater pollution prevention plans that industrial facilities prepare to be retained at the facility and produced when requested by the state. Only Oregon, Vermont, Washington, and Hawaii required industrial SWPPPs to be submitted to the state when seeking coverage under the MSGP. The practice for the submittal of construction SWPPPs was similar, except that some states required that SWPPPs for large construction projects be submitted to the state.

Compliance with the MS4 permit in the responding States is mainly determined through the evaluation of annual reports and program audits, although no indication was given of the frequency of audits. Regulators in Maine have monthly meetings with municipalities. The responding states evaluate compliance with the MSGP by reviewing annual monitoring reports and conducting inspections of industrial facilities. Connecticut characterized its industrial inspections as "regular," Maine inspects industrial facilities twice per five-year permit cycle, while Vermont performs visual inspections four times a year. No other responding states specified the frequency of inspections. Inspections and reviews of the SWPPPs constitute the main ways for responding states to determine the compliance of sites and facilities covered under the CGP.

With respect to the extent of actual compliance, few states have such information, partly because it has not routinely been collected and analyzed. West Virginia has found that, of the 871 permitted industrial facilities in the state, 576 were delinquent in submitting the results of their benchmark monitoring. Several case studies of compliance rates for municipal, industrial, and construction sites in Southern California are presented in Box 2-4. The data suggest that compliance in all three groups is poor, particularly for industrial sites. This may be partly explained by the preponderance of small businesses covered by the MSGP, whose operators may have financial difficulty in committing funds to SCMs, or lack a recognition and knowledge of the stormwater program and its requirements.

Another aspect of compliance is the extent to which industrial facilities have identified themselves and applied for coverage under the state MSGP. Six states responded to the committee's survey about that topic; only two of the six (California and Vermont) have made efforts to determine the numbers of non-filers of an NOI to be covered by the MSGP. In both cases, the efforts, which involved mailings, telephone calls, and file review, found that the number of non-filing facilities that should be subject to the MSGP was substantial (see Box 2-5 for California's data). Duke and Augustenborg (2006) studied this level of compliance (whether industries are filing an NOI for permit coverage) and found incomplete compliance that is variable among states and urbanized areas. Texas and Oklahoma had higher levels of permit coverage than California or Florida.

BOX 2-4
Compliance with Stormwater Permits in Southern California

Construction General Permits

In order to determine the compliance of construction sites with the general stormwater permit, data were collected and analyzed from three sources: (1) an audit performed in June 2004 of the development construction program of five cities that are permittees in the Los Angeles County MS4 permit (about 44 sites), (2) an audit performed in February 2002 of the development construction program (among others) of five Ventura County MS4 permittees (about 32 sites), and (3) a review and inspection of 24 large construction sites (50 acres or greater of disturbed land). These sites accounted for about 5 percent of all construction sites in the region at the time, and they represent both small and large construction sites. The most common violations on construction sites were paper violations, such as incomplete SWPPPs and a lack of record keeping. Forty (40) percent of the sites had some type of paper deficiency. A close second is the absence of erosion and/or sediment control, observed on 30 percent of the sites. SOURCE: TetraTech (2002, 2006b,c).

Industrial Multi-Sector General Permit

For industrial sites, information was obtained from the following sources: (1) a review of SCM inspections performed in February 2005 which consisted of 38 sites in the transportation sector; (2) a review of inspections and non-filer identification information in the plastics sector performed in 2007, which consisted of about 100 permitted sites among a large number of non-filer sites; and (3) a review of 13 area airport inspections and 55 port tenant inspections at the ports of Los Angeles and Long Beach. The sites are about 6 percent of the total number of permittees covered by California's MSGP and represent some of the major regulated industrial sectors. The most common violations observed at industrial sites were the lack of implementation of SCMs such as overhead cover, secondary containment and/or spill control. Sixty (60) percent of the sites had poor housekeeping problems. This was followed by incomplete stormwater pollution prevention plans (40 percent). (SOURCE: E. Solomon, California EPA, Los Angeles Regional Water Board, personal communication, 2008).

In another study, the California Water Boards with the assistance of an EPA contractor conducted inspections of 1,848 industrial stormwater permittees (21 percent of permitted facilities) between 2001 and 2005 (TetraTech, 2006d). Seventy-one (71) percent of the industrial facilities inspected were not in compliance with the MSGP and 18 percent were identified as a threat to water quality. Fifty-six (56) percent of facilities that collected one or more water quality samples reported an exceedance of a benchmark. Facility follow-up inspections indicated that field presence of the California Water Boards inspectors improved facility compliance with the MSGP.

Municipal Permits

An audit similar to the TetraTech study described above was conducted for 84 Phase I and Phase II MS4s in California during the same period (TetraTech, 2006e). The audits found that municipal maintenance facilities were often deficient in implementing SCMs, MS4 permittees did not obtain adequate legal authority to implement the program, they were not inspecting industrial facilities and construction sites or were inspecting them inadequately, and they were unable to evaluate program effectiveness in improving water quality. Overall, the audits found that programs with more specific permit requirements

continues next page

BOX 2-4 Continued

generally resulted in more comprehensive and progressive stormwater management programs. For example, the Los Angeles or San Diego MS4 permits enumerate in detail the permit tasks such as the frequency of inspection, the types of facilities, and the SCMs to be inspected that permittees must perform in implementing their stormwater program. The auditors concluded that the specificity of the provisions enabled the permitting authorities to enforce the MS4 permits and improve the quality of MS4 discharges.

Compliance with Industrial Permits within MS4s

The EPA and the California EPA Los Angeles Regional Water Board conducted a limited audit of the inspection program requirements of the Los Angeles County MS4 Permit and the City of Long Beach MS4 Permit in conjunction with industrial facilities covered under the MSGP within the Ports of Los Angeles and Long Beach (EPA, 2007c). The Port of Long Beach is covered under a single NOI for its 53 tenant facilities that discharge stormwater associated with industrial activity, while 137 industrial facilities within the Port of Los Angeles file independent NOIs. At the Port of Los Angeles, of the 23 facilities that were inspected, 30 percent were judged to pose a significant threat to water quality, 43 percent were determined to have some violations with regard to implementation of SCMs or paperwork requirements, and 26 percent appeared to be in compliance with the MSGP. At the Port of Long Beach, of the 21 tenant facilities that were inspected, 14 percent were judged to pose a significant threat to water quality, 52 percent were determined to have some deficiencies with regard to implementation of SCMs or paperwork requirements, and 33 percent appeared to be in full compliance with general permit requirements. The Port of Long Beach had a more comprehensive stormwater monitoring program which indicated that several pollutant parameters were above EPA benchmark values. Communication between the MS4 departments and the ports in both programs appeared deficient. The EPA issued 20 compliance orders for violations of the MSGP, but it did not pursue any action against the MS4s overseeing the industries because it was outside the scope of the EPA audit.

LOCAL CODES AND ORDINANCES THAT AFFECT STORMWATER MANAGEMENT

Zoning and building standards, codes, and ordinances have been the basis for city building in the United States for almost a century. They define how to build to protect the health, safety, and welfare of the public, and to establish a predictable, although often lengthy and cumbersome, process for ensuring that built improvements become a well-integrated part of the larger urban environment. Review processes can be as simple as a walk-through in a local building department for a minor house remodeling project. In other cases, extended rezoning processes for larger projects can require several years of planning; multiple public meetings; multiple reviews by city, state, and federal agencies; and specialized studies to determine impacts on the natural environment and water, sewer, and transportation systems.

BOX 2-5
Searching for Non-Filers Under the Industrial MSGP in Southern California

The California Water Boards conducted an industrial non-filer identification study between 1995 and 1998 (CA SWB, 1999). The study had three components: (1) to develop a mechanism to identify facilities subject to the industrial stormwater general permit that had not filed an NOI, which involved a comparison of commercially available and agency databases with that maintained by the California Water Boards; (2) to communicate with operators of these facilities to inform them of their responsibility to comply, which was done using post-mail, telephone calls, and filed verification; and (3) to refer responses to the communication efforts to the Water Boards for any appropriate follow-up.

About 9 percent of the potential non-filers submitted an NOI after the initial mail contact. About 52 percent of facilities indicated that they were exempt. About 37 percent failed to respond and 16 percent of mailed packages were returned unopened. A follow-up on facilities that claimed they were exempt indicated that 16 percent of them indeed needed to comply. Similarly 33 percent of facilities that failed to respond were determined as needing to file NOIs. The study suggested that only half of facilities considered heavy industrial had filed NOIs through the first five years of the program (Duke and Shaver, 1999).

The California EPA Los Angeles Regional Water Board and the City of Los Angeles conducted a study in the City of Los Angeles between January 1998 and June 2000 to identify non-filers and evaluate compliance by door-to-door visits in industrially zoned areas of the city (Swamikannu et al., 2001). The field investigations covered industrial zones totaling about 4.2 square miles, or about 22 percent of the area in the City of Los Angeles zoned for industrial land use. A total of 1,103 of suspected non-filer facilities were subject to detailed on-site facility investigation. Ninety-three (93) were determined to have already have submitted NOIs, and 436 were determined not to be subject to the industrial stormwater general permit. The site visits identified 223 potential non-filers, or industrial facilities where site-visit evidence suggested the facilities probably needed to comply with relevant regulations but that had not filed NOIs or recognized their duty to comply at the time of the visit. Of the facilities identified as potential non-filers, 202 were identified during detailed on-site investigations, or 18 percent of facilities inspected with that methodology; and 21 were identified during the less-detailed non-filer assessment visits, or 6 percent of the 379 facilities inspected with that methodology. In total, 295 of the 1,103 facilities visited under the project (about 27 percent) were known or suspected to be required to file NOIs under the permit, including 93 facilities that had previously filed NOIs and 202 facilities identified as probably required to file NOIs based on visual evidence of industrial activities exposed to stormwater. Thus, prior to the project, only 31 percent of all facilities in the project area needing to comply had submitted an NOI.

There is an overlapping and conflicting maze of codes, regulations, ordinances, and standards that have a profound influence on the ability to implement stormwater control measures, although they can be loosely categorized into three areas. Land-use zoning is the first type of control. Zoning, which was developed in response to unsanitary and unhealthy living conditions in 19th-century cities, prescribes permitted land uses, building heights, setbacks, and the arrangement of different types of land uses on a given site. Zoning often requires improvements that enhance the aesthetic and functional qualities of communities. For example, ordinances prescribing landscaping, minimum parking requirements, paving types, and related requirements have been developed to improve the livability of cities. These ordinances have a significant impact on both how stormwater affects waterbodies and on attempts to mitigate its impacts.

The second category involves the design and construction of buildings. National and international building codes and standards, such as the International Building Code, and Uniform Plumbing, Electrical, and Fire Codes, for example, allow local governments to establish minimum requirements for building construction. Because these controls primarily affect building construction, they have less effect on stormwater discharges than zoning.

The third category includes engineering and infrastructure standards and practices that govern the design and maintenance of the public realm—streets, roads, utilities rights-of-way, and urban waterways. Roadway design standards and emergency access requirements have resulted in contemporary cities that are 30 percent or more pavement, just to accommodate the movement and storage of vehicles in the public right-of-way. The standards for the construction of deep utilities—water and sewer lines that are typically located underneath streets—are often the reason that streets are wider than necessary to safely carry traffic.

Over time, these codes, standards, and practices have become more complex, and they may no longer support the latest innovations in planning practices. The past 10 to 20 years have seen a number of innovations in zoning and related building standards. Mixed-use, mixed-density communities that incorporate traditional patterns of community development (often described as "New Urbanism"), low impact development (LID), and transit-oriented development are examples of building patterns that challenge traditional zoning and city design standards. With the exception of LID, proposed new patterns of development and regulations connected with their implementation rarely incorporate specific guidelines for innovations in stormwater management, other than to have general references to environmental responsibility, ecological restoration, and natural area protection.

The following sections describe in more detail the codes, ordinances, and standards that affect stormwater and our ability to control it, and alternative approaches to developing new standards and practices that support and encourage effective stormwater management.

Zoning

The primary, traditional purpose of zoning has been to segregate land uses thought to be incompatible. In practice, zoning is used as a permitting system to prevent new development from harming existing residents or businesses. Zoning is commonly controlled by local governments such as counties or cities, though the specifics of the zoning regime are determined primarily by state planning laws (see Box 2-6 for a discussion of land use acts in Oregon and Washington).

BOX 2-6
Growth Management in the Pacific Northwest

In Oregon, the 1973 Legislative Assembly enacted the Oregon Land Use Act, which recognized that the uncoordinated use of lands threatens orderly development of the environment, the health, safety, order, convenience, prosperity and welfare of the people of Oregon. The state required all of Oregon's 214 cities and 36 counties to adopt comprehensive plans and land-use regulations. It specified planning concerns that had to be addressed, set statewide standards that local plans and ordinances had to meet, and established a review process to ensure that those standards were met. Aims of the program are to conserve farm land, forest land, coastal resources, and other important natural resources; encourage-efficient development; coordinate the planning activities of local governments and state and federal agencies; enhance the state's economy; and reduce the public costs that result from poorly planned development. Setting urban growth boundaries is a major mechanism for implementing the act.

The Washington State Legislature followed in 1990 with the Growth Management Act (GMA), adopted on grounds similar to Oregon's act. The GMA requires state and local governments to manage Washington's growth by identifying and protecting critical areas and natural resource lands, designating urban growth areas, preparing comprehensive plans, and implementing them through capital investments and development regulations. Similar again to Oregon, rather than centralize planning and decision-making at the state level, the GMA established state goals, set deadlines for compliance, offered direction on how to prepare local comprehensive plans and regulations, and set forth requirements for early and continuous public participation. Urban growth areas (UGAs) are those areas, designated by counties pursuant to the GMA, "within which urban growth shall be encouraged and outside of which growth can occur only if it is not urban in nature." Within these UGAs, growth is encouraged and supported with adequate facilities. Areas outside of the UGAs are reserved for primarily rural and resource uses. Urban growth areas are to be based on population forecasts made by counties, which are required to have a 20-year supply of land for future residential development inside the boundary—a time frame also pertaining in the Oregon system. In both states urban growth boundaries are reconsidered and sometimes adjusted to meet this criterion.

It is important to note that the growth management efforts in the two states have no direct relationship to stormwater management. Rather, the laws control development density, which has implications for how stormwater should be managed (see discussion in Chapter 5). The local jurisdictions in Washington have reacted in different ways to link growth management and stormwater management. For example, the King County, Washington, stormwater code requires drainage review to evaluate and deal with stormwater impacts for development that adds 2,000 square feet or more of impervious surface or clears more than 7,000 square feet. For rural residential lots outside the UGA, the impervious threshold is reduced to 500 square feet.

Sources:
http://bluebook.state.or.us/state/executive/Land_Conservation/land_conservation_history.htm
http://www.oregonmetro.gov/index.cfm/go/by.web/id=277
http://www.gmhb.wa.gov/gma/ and *http://www.mrsc.org/Subjects/Planning/compfaqs.aspx*

Zoning involves regulation of the kinds of activities that will be acceptable on particular lots (such as open space, residential, agricultural, commercial or industrial), the densities at which those activities can be performed (from low-density housing such as single-family homes to high-density housing such as high-rise apartment buildings), the height of buildings, the amount of space structures may occupy, the location of a building on the lot (setbacks), the proportions of the types of space on a lot (for example, how much landscaped space and how much paved space), and how much parking must be provided. Thus, zoning can have a significant impact on the amount of impervious area in a development and on what constitutes allowable stormwater management.

As an example, local parking ordinances are often found within zoning that govern the size, number, and surface material of parking spaces, as well as the overall geometry of the parking lot as a whole. The parking demand requirements are tied to particular land uses and zoning categories, and can create needless impervious cover. Most local parking codes are overly generous and have few, if any, provisions to treat stormwater at the source (Wells, 1995). For example, in a co-housing project under construction in Fresno, California, current city codes require 27-foot-long parking spaces. The developer, in an effort to reduce construction costs, requested that the length of spaces be reduced to 24 feet. The city agreed to the smaller spaces if the developer would sign an indemnity clause guaranteeing that the local government would not be sued in case of an accident (Wenz, 2008).

Similarly, landscaping ordinances apply to certain commercial and institutional zoning categories and specify that a fixed percentage of site area be devoted to landscaping, screening, or similar setbacks. These codes may require as much as 5 to 10 percent of the site area to be landscaped, but seldom reference opportunities to capture and store runoff at the source, despite the fact that the area devoted to landscaping is often large enough to meet some or all of their stormwater treatment needs.

Zoning codes have evolved over the years as urban planning theory has changed, legal constraints have fluctuated, and political priorities have shifted. The various approaches to zoning can be divided into four broad categories: Euclidean, performance, planned unit development, and form-based.

Euclidean Zoning

Named for the type of zoning code adopted in the town of Euclid, Ohio, Euclidean zoning codes are by far the most prevalent in the United States, used extensively in small towns and large cities alike. Euclidean zoning is characterized by the segregation of land uses into specified geographic districts and dimensional standards stipulating limitations on the magnitude of development activity that is allowed to take place on lots within each type of district. Typical land-use districts in Euclidean zoning are residential (single- or multi-family), commercial, and industrial. Uses within each district are usually heavily prescribed to exclude other types of uses (for example, residential districts typically disallow commercial or industrial uses). Some "accessory" or "conditional" uses may be allowed in order to accommodate the needs of the primary uses. Dimensional standards apply to any structures built on lots within each zoning district and typically take the form of setbacks, height limits, minimum lot sizes, lot coverage limits, and other limitations on the building envelope.

Although traditional Euclidean zoning does not include any significant requirements for stormwater drainage, there is no reason that it could not. Modern Euclidean ordinances include a broad list of "development standards" that address topics like signage, lighting, steep slopes, and other topics, and that list could be expanded to included stormwater standards for private development.

Euclidean zoning is used almost universally across the country (with rare exceptions) because of its relative effectiveness, ease of implementation (one set of explicit, prescriptive rules), long-established legal precedent, and familiarity to planners and design professionals. However, Euclidean zoning has received heavy criticism for its unnecessary separation of land uses, its lack of flexibility, and its institutionalization of now-outdated planning theory. . In response, variances and other methods have been used to modify Euclidean zoning so that it is better adapted to localized conditions and existing patterns of development. The sections below briefly describe a range of innovations in local zoning regulations that have potential for incorporating stormwater controls into existing regulations.

Incentive Zoning. Incentive zoning systems are typically an add-on to Euclidean zoning systems. First implemented in Chicago and New York City in 1961, incentive zoning is intended to provide a reward-based system to encourage development that meets established urban development goals. Typically, a base level of prescriptive limitations on development will be established and an extensive list of incentive criteria with an associated reward scale will be established for developers to adopt at their discretion. Common examples include floor-area-ratio bonuses for affordable housing provided on-site and height-limit bonuses for the inclusion of public amenities on-site.

With incentive zoning, developers are awarded additional development capacity in exchange for a public benefit, such as a provision for low- or moderate-income housing, or an amenity, such as additional open space. Incentive zoning is often used in more highly urbanized areas. Consideration for water quality treatment and innovative SCMs fits well within the incentive zoning model. For example, redevelopment sites in urbanized areas are often required to incorporate stormwater control measures into developments to minimize impacts on aging, undersized stormwater systems in that area, and to meet new water quality requirements. An incentive could be to allow greater building height, and therefore higher density, than under existing zoning, freeing up land area for SCMs that could also serve as a passive park area. Another example would be to allow a higher density on the site and to require not an on-site system but a cash payment to the governing entity to provide for consolidated stormwater management and treatment. Off-site consolidated systems, discussed more extensively in Chapter 5, may require creation of a localized maintenance district or an increase in stormwater maintenance fees to offset long-term maintenance costs.

Incentive zoning could be used to preserve natural areas or stream corridors as part of a watershed enhancement strategy. For example, transferrable development rights (TDR) could be used in the context of the urban or semi-urban interface with rural lands. Many of the formal TDR programs in Colorado (such as Fruita/Mesa County and Aspen/Pitkin) involve cities or counties seeking to preserve sensitive areas in the county, or outlying areas of the city, including the floodplain, in exchange for urban-level density on a more appropriate site (David D. Smith, Garfield & Hecht P.C., personal communication, 2008).

Incentive zoning allows for a high degree of flexibility, but it can be complex to administer. The more a proposed development takes advantage of incentive criteria, the more closely it has to be reviewed on a discretionary basis. The initial creation of the incentive structure can also be challenging and often requires extensive ongoing revision to maintain balance between incentive magnitude and value given to developers.

Performance Zoning

Performance zoning uses performance-based or goal-oriented criteria to establish review parameters for proposed development projects in any area of a municipality. At its heart, performance zoning deemphasizes the specific land uses, minimum setbacks, and maximum heights applicable to a development site and instead requires that the development meet certain performance standards (usually related to noise, glare, traffic generation, or visibility). Performance zoning sometimes utilizes a "points-based" system whereby a property developer can apply credits toward meeting established zoning goals through selecting from a menu of compliance options (some examples include mitigation of environmental impacts, providing public amenities, and building affordable housing units). Additional discretionary criteria may also be established as part of the review process.

The appeal of performance zoning lies in its high level of flexibility, rationality, transparency, and accountability. Because performance zoning is grounded in specific and in many cases quantifiable goals, it better accommodates market principles and private property rights with environmental protection. However, performance zoning can be extremely difficult to implement and can require a high level of discretionary activity on the part of the supervising authority. City staff must often be trained to use specialized equipment to measure the performance of the development, and sometimes those impacts cannot be measured until the building is completed and the activity operating, by which time it may be difficult and expensive to modify a building that turns out not to meet the required performance standards. Because stormwater performance is measurable (especially the amounts of water retained/detained and rates and amounts of water discharge), stormwater regulations could be integrated into a performance zoning system. As with other topics, however, it might be time-consuming or require special equipment to measure compliance (particularly before the building is built).

Planned Unit Development (Including Cluster Development and Conservation Design)

A planned unit development (PUD) is generally a large area of land under unified control that is planned and developed as a whole through a single development operation or series of development phases, in accord with a master plan. In California, these are known as Specific Plans. More specialized forms of PUDs include clustered subdivisions where density limitations apply to the development site as a whole but provide flexibility in the lot size, setback, and other standards that apply to individual house lots. These PUDs provide considerable flexibility in locating building sites and associated roads and utilities, allowing them to be concentrated in parts of the site, with the remaining land use for agriculture, recreation, preservation of sensitive areas, or other open-space purposes.

PUDs are typically, although not exclusively, found in new development areas and have significant open space and park areas that are often 25 percent or more of the total land area. This large amount of open space provides considerable opportunity for the use of consolidated, multifunctional stormwater controls.

Form-Based Zoning

Form-based zoning relies on rules applied to development sites according to both prescriptive and potentially discretionary criteria. These criteria are typically dependent on lot size, location, proximity, and other various site- and use-specific characteristics. Form-based codes offer considerably more flexibility in building uses than do Euclidean codes, but, as they are comparatively new, may be more challenging to create. When form-based codes do not contain appropriate illustrations and diagrams, they are criticized as being difficult to interpret.

One example of a recently adopted code with form-based features is the Land Development Code adopted by Louisville, Kentucky, in 2003. This zoning code creates "form districts" for Louisville Metro. Each form district intends to recognize that some areas of the city are more suburban in nature, while others are more urban. Building setbacks, heights, and design features vary according to the form district. As an example, in a "traditional neighborhood" form district, a maximum setback might be 15 feet from the property line, while in a suburban "neighborhood" there may be no maximum setback. Narrower setbacks allow increased density, requiring less land area for the same number of housing units and resulting in a smaller development footprint.

In rural and suburban areas, form-based codes can often reinforce the "open" character of development by preserving open site areas, which could be used for on-site stormwater management. In denser, urban areas, however, some form-based ordinances favor shorter, more pedestrian-scale buildings that cover more of the site than taller buildings of the same square footage, on the basis that keeping activity closer to the ground and enclosing street frontages results in a better pedestrian environment and urban form. One result of this preference is that there may be less of the site left potentially available for on-site stormwater detention or infiltration. Integrating stormwater management considerations into form-based codes may require a cash payment system where the developer contributes to financing of a district or regional stormwater treatment facility because on-site solutions are not available.

Building Codes

Building codes define minimum standards for the construction of virtually all types and scales of structures. With a few exceptions, building codes have limited direct impact on stormwater management. The main example is where structural and geotechnical design standards, which stem from the need to protect buildings and infrastructure from water damage, discourage or prohibit the potential infiltration of water adjacent to building foundations. Such standards can make it difficult to use landscape-based SCMs, such as porous pavement, bioinfiltration, and extended detention. There is a need to examine and redefine structural and geotechnical "standards of care" that ensure the structural integrity of buildings and other infrastructure like buried utilities, in order for landscaped areas adjacent to structures to be utilized more effectively for SCMs. For example, a developer building a mixed-use, medium-density infill development in Denver intended to incorporate innovative approaches to stormwater management by infiltrating stormwater in a number of areas around the site. The standard of care for the geotechnical design of building foundations typically requires that positive drainage be maintained a minimum of 5 feet from the building edge. The geotechnical engineer required, when informed that water might be infiltrated in the area of the building and without further study, that the minimum distance to an infiltration area must be at least to 20 feet from the building, greatly limiting the potential for using the building landscape areas as SCMs. The City of Los Angeles is in the process of updating its Building Code, but it is not clear if it will be sufficiently comprehensive to address the use of some LID practices, such as on-site infiltration. The 2002 Building Code now in effect is written to require the builder to convey water away from the building using concrete or some other "non-erosive device."

Engineering and Infrastructure Standards and Practices

Engineering standards and practices for public rights-of-way complement building and zoning codes which control development on private property. Engineering standards and practices typically describe requirements for public utilities such as stormwater and wastewater, roadways, and related basic services. For example, there are standards for parking and roadway design that typically describe the specific type of roadway and parking surfacing requirements. Regulations and standards often require minimum gradients for surface drainage, site grading, and drainage pipe size, all of which play an important role in how stormwater is transported. There are also often landscape planting requirements, including the requirement to mound landscape areas to screen cars, which can preclude the opportunity to incorporate SCMs into landscape areas.

Unless right-of-way improvements are constructed as part of the subdivision process by private developers, improvements in the right-of-way are typically provided for by city government and public agencies. Because engineering standards are often based on decades of refinement and have evolved regionally and nationally, they are difficult to change. For example, street widths are determined more by the ability to maneuver emergency equipment and to accommodate water and sewer easements than the need for adequate lane widths for vehicles. Street lane-width requirements might be as narrow as 11 feet for each travel lane, resulting in a street width of 22 to 24 feet. This could accommodate emergency vehicle access, which typically can require a minimum of 20 feet of unobstructed street. However, because most streets also include potable water distribution lines and easement requirements for the lines, which are a minimum of 30 feet in width, this results in a minimum roadway width of 30 feet.

Local drainage codes govern the disposal of stormwater and essentially dictate the nature and capacity of the stormwater infrastructure from the roof to the floodplain. Like many codes, they were developed over time to address problems such as basement flooding, nuisance drainage problems, maintenance of floodplain boundaries, and protection of infrastructure such as bridges and sewers from storm damage. Local drainage codes, many of which predate the EPA's stormwater program, often involve peak discharge control requirements for a series of design storm events ranging from the 2-year storm up to the 100-year event. Traditional drainage codes can often conflict with effective approaches to reducing runoff volume or removing pollutants from stormwater. Examples of such codes include requirements for positive drainage, directly connected roof leaders, curbs and gutters, lined channels, storm-drain inlets, and large-diameter storm-drain pipes discharging to a downstream detention or flood control basins.

Often, standards have been tested through legal precedent, and case law has developed around certain standards of care, which can further deter innovation. Changes in design standards could result in unknown legal exposure and liability. Specific types of equipment, maintenance protocols and procedures, and extensive training further discourage changes in established standards and procedures.

Innovations in Codes and Regulations to Promote Better Stormwater Management

A number of innovations have been developed in the previously described zoning, building codes, and infrastructure and engineering standards that make them more amenable to stormwater management. These are described in detail below.

Separate Ordinances for New and Infill Development

Redevelopment of existing urban areas is almost universally more difficult and expensive than Greenfield development because of the deconstruction costs of the former, higher costs of designing around existing infrastructure, upgrading existing infrastructure, and higher costs and risks associated with assuming liability of pre-existing problems (contamination, etc). Redevelopment often occurs in areas of medium to high levels of impervious surface (e.g., downtown areas). Such severely space-limited areas with high land costs drive up stormwater management costs. Consequently, holding developers of such areas to the same stormwater standard as for Greenfield developments creates a financial disincentive for redevelopment. Without careful application, stormwater requirements may discourage needed redevelopment in existing urban areas. This would be unfortunate because redevelopment can take pressure off of the development of lands at the urban fringe, it can accommodate growth without introducing new impervious surfaces, and it can bring improvements in stormwater management to areas that had previously had none.

Stormwater planning can include the development of separate ordinances for infill and new developments. Wisconsin has administrative rules that establish specific requirements for stormwater management based on whether the site is new development, redevelopment, or infill. Requirements for new development include reducing total suspended solids (TSS) by 80 percent, maintaining the pre-development peak discharge for the 2-year, 24-hour storm, infiltrating 90 percent of the pre-development infiltration volume for residential areas, and infiltrating 60 percent of the pre-development infiltration volume for non-residential areas. Redevelopment varies from new development only in that the TSS requirement is less at 40 percent reduction. Requirements for existing developed areas in incorporated cities, villages, and towns do not include peak flow reduction or infiltration performance standards, but the municipalities must achieve a 40 percent reduction in their TSS load by 2013. Other requirements unique to developed areas include public education activities, proper application of nutrients on municipality property, and elimination of illicit discharges (*www.dnr.state.wi.us/org/water/wm/nps/stormwater/post-constr/*). Chapter 5 makes recommendations for the specific types of SCMs that should be used for new, low-density residential development as opposed to redevelopment of existing urban and industrial areas.

Integrated Stormwater Management and Growth Policies

In the city of San Jose, California, an approach was taken to link water quality and development policies that emphasized higher density in-fill development and performance-based approaches to achieving water quality goals. The city's approach encourages stormwater practices such as minimizing impervious surface and incorporating swales as the preferred means of conveyance and treatment. In urbanized areas, the policy then goes on to define criteria to determine the practicability of meeting numeric sizing requirements for stormwater control measures, and identifies Equivalent Alternative Compliance Measures for cases where on-site controls are impractical. Equivalent Measures can include regional stormwater treatment and other specific projects that "count" as SCMs, including certain affordable and senior housing projects, significant redevelopment within the urban core, and Brownfield projects. This is similar to in lieu fee programs that are sometimes implemented by municipalities to provide additional regulated parties with compliance options (see discussion in Chapter 6).

This approach is a breakthrough in terms of measuring environmental performance, which is now focused only on what happens within the boundaries of a site for a project. This myopic view tends to allow many environmentally unfriendly projects that encourage sprawl and expand the city's boundaries to qualify as "low impact," while more intense projects on a small footprint appear to have a much higher impact because they cover so much of the site. San Jose brought several other layers of review, including location in the watershed (close to other uses or not) as a means of estimating performance. A PowerPoint presentation describing their approach in greater detail is linked here (*http://www.cmcgc.com/media/handouts/260126/THR-PDF/040-Ketchum.PDF*, Lisa Nisenson, Nisenson Consulting, LLC, personal communication, May 8, 2007).

Unified Development Codes

A unified development code (UDC) consolidates development-related regulations into a single code that represents a more consistent, logical, integrated, and efficient means of controlling development. UDCs integrate zoning and subdivision regulations, simplifying development controls that are often conflicting, confusing, and that require multiple layers of review and administration. UDC development standards may include circulation standards that address how vehicles and pedestrians move, including provision for adequate emergency access. Utility standards are described for water distribution and sewage collection, and necessary utility easements are prescribed. Because of the integrated nature of the code, efficiencies in requirements for right-of-way can reduce street widths or the reduction in setbacks, for example, resulting in more compact development.

Design Review Incentives to Speed Permitting

A number of incentives have been put in place to promote innovative stormwater control measures in cities such as Portland and Chicago, where environmental concerns have been identified as a key goal for development and redevelopment. Practices such as the waiver or reduction of development fees, preferential treatment and review and approval of innovative plans, reduction in stormwater fees, and related incentives encourage the use of innovative stormwater practices. In Chicago, the Green Permit Program initiated in April 2005 has proven attractive to many developers as it speeds up the permitting process. Under the Green Permit Program, a green building adviser reviews design plans under an aggressive schedule long before a permit application is submitted. There is one point of contact with intimate knowledge about the project to help speed up the permit process. Projects going through the Green Permit Program receive benefits based on their "level of green." Tier I commercial projects are designed to be Leadership in Energy and Environmental Design (LEED) certified (see Box 2-7). Tier II projects must obtain LEED silver rating. At this level, outside consultant review fees, which range from $5,000 to $50,000, are waived. Tier III projects must earn LEED gold. The goal for a Tier III project is to issue a permit in three weeks for a small project such as a 12-unit condo building. Thus, there is both time and money saved. Private developers are interested in the time savings because they can pay less interest on their construction loans by completing the building faster. By the end of 2005, 19 green permits were issued. The program's director estimated that about 50 would be issued in 2006, which exceeds the city's goal of 40.

In Portland, Oregon, the city's Green Building Program is considering instituting a new High-Performance Green Building Policy. Along with goals for reducing global warming pollution, it proposes (1) waiving development fees if goals are exceeded by specified percentages and (2) eligibility for cash rewards and qualification for state and federal financial incentives and tax credits if even higher goals are achieved. Developers can earn credits by incorporating enhanced stormwater management and water conservation features into their projects, including the use of green roofs (Wenz, 2008).

***

BOX 2-7
Innovative Building Codes

An increased interest in energy conservation and more environmentally friendly building practices in general has led to various methods by which buildings can be evaluated for environmentally friendly construction, in addition to conventional code compliance. The most popular system in the United States is the Leadership in Energy and Environmental Design (LEED) system developed in 2000.

The LEED Green Building Rating System is a voluntary, consensus-based national rating system for developing high-performance, sustainable buildings. LEED addresses all building types and emphasizes state-of-the-art strategies in five areas: sustainable site development, water savings, energy efficiency, materials and resources selection, and indoor environmental quality. The U.S. Green Building Council is a 501(c)(3) nonprofit organization that certifies sustainable businesses, homes, hospitals, schools, and neighborhoods.

The LEED system encourages progressive stormwater management practices as part of its rating system. The LEED system has identified specific criteria, with points assigned to each of the criteria, to assess the success of stormwater strategies. Generally, the criteria are based on LID principles and practices and relate directly to the *Better Site Design Handbook* of the Center for Watershed Protection (CWP, 1998). The system identifies eight categories by which building sites and site-planning practices are evaluated. Of the 69 points possible to achieve the highest LEED rating, 16 points are directly related to innovative site design and stormwater management practices. Six of the eight criteria describing sound site-planning practices relate directly to good stormwater practices, including the following:

- Erosion and sediment control;
- Site selection to protect farmland, wetlands, and watercourses;
- Site design to encourage denser infill development to protect Greenfield sites;
- Limitations on site disturbance;
- Specific requirements for the management of stormwater rate and quantity; and
- Specific requirements for the treatment of stormwater for TSS and phosphorous removal.

The LEED rating system has been criticized because it focuses on individual buildings in building sites. A new category, LEED neighborhood development, was developed in response to consider the interrelationship of buildings and building sites and connections to existing urban infrastructure. The category is currently in pilot testing. Evaluation criteria related directly to stormwater include:

- All requirements of the original site design criteria,
- A reduced requirement for parking based on access to transit and reduced auto use, and
- Site planning that emphasizes compact development.

There are parallel challenges in the realm of community development and city building that tend to discourage innovative stormwater management policies and practices. Building codes and zoning have evolved to reflect the complex relationship of legal, political, and social processes and frequently do not promote or allow the most innovative stormwater management. Engineering standards and practices that guide the development of roads and utilities present equal and possibly greater challenges, in that legal and technical precedents and large investments in public equipment and infrastructure present even more intractable reasons to resist change.

The difficulty of implementing stormwater control measures cannot be attributed to an individual code, standard, or regulation. It is important to unravel the complexities of codes, regulations, ordinances, and standards and practices that discourage innovative stormwater management and target the particular element (or multiple elements) that is a barrier to innovation. Elements that are barriers might not have been considered previously. For example, roadway design is controlled more by access for emergency equipment and utilities rights-of-way than by the need for wide travel lanes; it is the fire marshal and the water department that should be the focus of attention, rather than the transportation engineer.

LIMITATIONS OF THE FEDERAL STORMWATER PROGRAM

The regulation of stormwater discharges seems an inevitable next step to the CWA's objective of "restoring the nation's waters," and EPA's stormwater program is still evolving. Yet, in its current configuration EPA's approach seems inadequate to overcome the unique challenges of stormwater and therefore runs the risk of only being partly effective in meeting its goals. A number of regulatory, institutional, and societal obstacles continue to hamper stormwater management in the United States, as described below.

The Poor Fit Between the Clean Water Act's Regulatory Approach and the Realities of Stormwater Management

Controlling stormwater discharges with the CWA introduces a number of obstacles to effective stormwater regulation. Unlike traditional industrial effluent, stormwater introduces not only contaminants but also surges in volume that degrade receiving waterbodies; yet the statute appears focused primarily on the "discharge" of "pollutants." Moreover, unlike traditional effluent streams from manufacturing processes, the pollutant loadings in stormwater vary substantially over time, making effluent monitoring and the development of enforceable control requirements considerably more challenging. Traditional use of end-of-pipe control technologies and automated effluent monitors used for industrial effluent do not work for the episodic and variable loading of pollutants in stormwater unless they account for these eccentricities by adjustments such as flow-weighted measurements. Finally, at the root of the stormwater problem is increasingly intensive land use. Yet the CWA contains little authority for regulators to directly limit land development, even though the discharges that result from these developments increase stormwater loading at a predictably rapid pace. The CWA thus expects regulators to reduce stormwater loadings, but gives them incomplete tools for effectuating this goal.

A more straightforward way to regulate stormwater contributions to water-body impairment would be to use flow or a surrogate, like impervious cover, as a measure of stormwater loading (such as in the Barberry Creek TMDL [Maine DEP, 2003, pp. 16–20] or the Eagle Brook TMDL [Connecticut DEP, 2007, pp. 8–10]). Flow from individual stormwater sources is easier to monitor, model, and even approximate as compared to calculating the loadings of individual con-taminants in stormwater effluent. Efforts to reduce stormwater flow will auto-matically achieve reductions in pollutant loading. Moreover, flow is itself re-sponsible for additional erosion and sedimentation that adversely impacts sur-face water quality. Flow provides an inexpensive, convenient, and realistic means of tracking stormwater contributions to surface waters. Congress itself recently underscored the usefulness of flow as a measure for aquatic impair-ments by requiring that all future developments involving a federal facility with a footprint larger than 5,000 square feet ensure that the development achieves predevelopment hydrology to the maximum extent technically feasible "with regard to the temperature, rate, volume, and duration of flow" (Energy Inde-pendence and Security Act of 2007, § 438). Several EPA regions have also used flow in modeling stormwater inputs for TMDL purposes (EPA, 2007a, Potash Brook TMDL, pp. 12–13).

Permitting and Enforcement

For industrial wastewater discharged directly from industrial operations (rather than indirectly through stormwater), the CWA requirements are rela-tively straightforward. In these traditional cases, EPA essentially identifies an average manufacturer within a category of industry, like iron and steel manufac-turers engaged in coke-making, and then quantifies the pollutant concentrations that would result in the effluent if the industry installed the best available pollu-tion control technology. EPA promulgates these effluent standards as national, mandatory limits (e.g., see Table 2-7).

TABLE 2-7 Effluent Limits for Best Available Technology Requirements for By-product Coke-making in Iron and Steel Manufacturing

Regulated Parameter	Maximum Daily[1]	Maximum Monthly Average[1]
Ammonia-N	0.00293	0.00202
Benzo(a)pyrene	0.0000110	0.00000612
Cyanide	0.00297	0.00208
Naphthalene	0.0000111	0.00000616
Phenols (4AAP)	0.0000381	0.000238

[1]pounds per thousand pound of product.
SOURCE: 40 C.F.R. § 420.13(a).

By contrast, the uncertainties and variability surrounding both the nature of the stormwater discharges and the capabilities of various pollution controls for any given industrial site, construction site, or municipal storm sewer make it much more difficult to set precise numeric limits in advance for stormwater sources. The quantity and quality of stormwater are quite variable over time and vary substantially from one property to another. Natural causes of variation in the pollutant loads in stormwater runoff include the topography of a site, the soil conditions, and of course, the nature of storm flows in intensity, frequency, and volume. In addition, the manner in which the facility stores and uses materials, the amount of impervious cover, and sometimes even what materials the facility uses can vary and affect pollutant loads in runoff from one site to another. Together, these sources of variability, particularly the natural features, make it much more difficult to identify or predict a meaningful "average" pollutant load of stormwater runoff from a facility. As a result, EPA generally leaves it to the regulated facilities, with limited oversight from regulators, to identify the appropriate SCMs for a site. Unfortunately, this deferential approach makes the permit requirements vulnerable to significant ambiguities and difficult to enforce, as discussed below for each permit type.

Municipal Stormwater Permits. MS4 permits are difficult to enforce because the permit requirements have not yet been translated into standardized procedures to establish end-of-pipe numerical effluent limits for MS4 stormwater discharges. CWA Section 402(p) requires that pollutants in stormwater discharges from the MS4 be reduced to the maximum extent practicable and comply with water quality standards (when so required by the permitting authority). However, neither EPA nor NPDES-delegated states have yet expressed these criteria for compliance in numerical form.

The EPA has not yet defined MEP in an objective manner that could lead to convergence of MS4 programs to reduce stormwater pollution. Thus, at present MS4 permittees have no more guidance on the level of effort expected other than what is stated in the CWA:

> [S]hall require controls to reduce the discharge of pollutants to the maximum extent practicable, including management practice, control techniques and system, design and engineering methods, and such other provisions as the Administrator or the State determines appropriate for the control of such pollutants. [CWA Section 402(p)(3)(B)(iii)]

A legal opinion issued by the California Water Board's Office of Chief Counsel in 1993 stated that MEP would be met if MS4 permittees implemented technically feasible SCMs, considering costs, public acceptance, effectiveness, and regulatory compliance (Memorandum from Elizabeth Miller Jennings, Office of Chief Counsel, to Archie Matthews, Division of Water Quality, California Water Board, February 11, 1993). In its promulgation of the Phase II Rule in 1999, the EPA described MEP as a flexible site-specific standard, stating that:

The pollutant reductions that represent MEP may be different for each [MS4 Permittee] given the unique local hydrological and geological concerns that may exist and the differing possible pollutant control strategies. (64 Fed. Reg. 68722, 68754)

As matters stand today, MS4 programs are free to choose from the EPA's menu of SCMs, with MEP being left to the discretionary judgment of the implementing municipality. Similarly, there are no clear criteria to be met for industrial facilities that discharge to MS4s in order for the MS4s to comply with MEP. The lack of federal guidance for MS4s is understandable. A stormwater expert panel convened by the California EPA State Water Board in 2006 (CA SWB, 2006) concluded that it was not yet feasible to establish strictly enforceable end-of-pipe numeric effluent limits for MS4 discharges. The principal reasons cited were (1) the lack of a design storm (because in any year there are few storms sufficiently large in volume and/or intensity to exceed the design volume capacity or flow rates of most treatment SCMs) and (2) the high variability of stormwater quality influenced by factors such as antecedent dry periods, extent of connected impervious area, geographic location, and land use.

Industrial and Construction Stormwater Permits. The industrial and construction stormwater programs suffer from the same kind of deficiencies as the municipal stormwater program. These stormwater discharges are not bound by the MEP criterion, but they are required to comply with either technology-based or, less often, water quality-based effluent limitations. In selecting SCMs to comply with these limitations, the industrial discharger or construction operator similarly selects from a menu of options devised by the EPA or, in some cases, the states or localities for their particular facility (EPA, 2006, p. 15). For example, the regulated party will generally identify structural SCMs, such as fences and impoundments that minimize runoff, and describe how they will be installed. The SWPPP must also include nonstructural SCMs, like good housekeeping practices, that require the discharger to minimize the opportunity for pollutants to be exposed to stormwater. The SWPPP and the accompanying SCMs constitute the compliance requirements for the stormwater discharger and are essentially analogous to the numeric effluent limits listed for industrial effluents in the Code of Federal Regulations.

This set of requirements leaves considerable discretion to regulated parties in several important ways. First, the regulations require the discharger to evaluate the site for problematic pollutants; but where the regulated party does not have specific knowledge or data, they need only offer "estimates" and "predictions" of the types of pollutants that might be present at the site (EPA, 1996a, pp. IV-3, V-3). With the exception of visible features, the deferential site investigation requirements allow regulated parties to describe site conditions in ways that may effectively escape accountability unless there is a vigorous regulatory presence.

Second, dischargers enjoy considerable discretion in drafting the SWPPP (EPA, 1996a, p. IV-3). Despite EPA's instructions to consider a laundry list of considerations that will help the facility settle on the most effective plan (EPA, 2006, p. 20), rational operators may take advantage of the wiggle room and develop ambiguous requirements that leave them with considerable discretion in determining whether they are in compliance (EPA, 2006, pp. 15, 20, 132). Indeed, the federal regulations do little to prevent regulated parties from devising requirements that maximize their discretion. Instead, EPA describes many of the permit requirements in general terms. For example, in its industrial stormwater permit program the EPA commands the regulated party to "implement any additional SCMs that are economically reasonable and appropriate in light of current industry practice, and are necessary to eliminate or reduce pollutants in . . . stormwater discharges" (EPA, 2006, p. 23).

EPA's program provides few rewards or incentives for dischargers to go beyond the federal minimum and embrace rigorous or innovative SCMs. In fact, if the regulated party invests resources to measure pollutant loads on their property, they are creating a paper trail that puts them at risk of greater regulation. Under the EPA's regulations, a regulated party "must provide a summary of existing stormwater discharge sampling data previously taken at [its] facility," but if there are no data or sampling efforts, then the facility is off the hook (EPA, 2006, p. 20). Quantitative measures can thus be incriminating, particularly in a regulatory setting where the regulator is willing to settle for estimates.

Dilemma of Self-Monitoring

Unlike the wastewater program where there are relatively rigid self-monitoring requirements for the end-of-pipe effluent, self-monitoring is much more difficult to prescribe for stormwater discharges, which are variable over time and space. [For example, *compare* 33 U.S.C. § 1342(a)(2)-(b)(2) (2000) (outlining requirements for compliance under NPDES) *with* EPA, 2006, p. 26 (outlining requirements for self-compliance under EPA regulations.)] EPA's middle ground, in response to these challenges, requires self-monitoring of select chemicals in stormwater for only a subset of regulated parties—Phase I MS4 permittees and a limited number of industrial facilities (see Table 2-8, EPA, 2006, pp. 93-94). Yet even for these more rigid monitoring requirements, the discharger enjoys some discretion in sampling. The EPA's sampling guidelines do prescribe regular intervals for sampling but ultimately must defer to the discharger insofar as requiring only that the samples should be taken within 30 minutes after the storm begins, and only if it is the first storm in three days (EPA, 2006, p. 33).

TABLE 2-8 Effluent Monitoring Requirements for Various Dischargers of Stormwater

Source Category	Type of Effluent Monitoring Required by EPA
Phase I MS4	Municipality must develop a monitoring plan that provides for representative data collection. This requires the municipality, at the very least, to select at least 5 to 10 of its most representative outfalls for regular sampling and sample for selected conventional pollutants and heavy metals in its effluent.
Phase II MS4	None
Small subset of highest risk industries, like hazardous waste landfills	Must conduct compliance monitoring as specified in effluent guidelines and ensure compliance with these effluent limits. Must also conduct visual monitoring and benchmark monitoring.
Larger subset of higher risk industrial dischargers	Benchmark monitoring: Must conduct analytic monitoring to determine whether effluent exceeds numeric benchmark values; compliance with the numeric values is not required, however. Must also conduct visual monitoring.
Remaining set of industry except construction	Visual monitoring: Must take four grab samples of stormwater effluent each year during first 30 minutes of a storm event and inspect the sample visually for contamination.
Construction (larger than 5 acres)	Visual monitoring: Must take four grab samples of stormwater effluent each year during first 30 minutes of a storm event and inspect the sample visually for contamination.
Construction (between 1 and 5 acres)	Visual monitoring: Must take four grab samples of stormwater effluent each year during first 30 minutes of a storm event and inspect the sample visually for contamination.

Note: State regulators can and sometimes do require more—see Appendix C.

Moreover, while the monitoring itself is mandatory, the legal consequences of an exceedance of a numerical limit vary and may be quite limited. For a small number of identified industries, exceedances of effluent limits established by EPA are considered permit violations (65 Fed. Reg. 64766). For the other high-risk industries subject to benchmark monitoring requirements (see Table 2-5), the analytical limits do not lead to violations per se, but only serve to "flag" the discharger that it should consider amending its SWPPP to address the problematic pollutant (EPA, 2006, pp. 10, 30, 34). Although municipalities are required to do more extensive sampling of stormwater runoff and enjoy less sampling discretion, even municipalities are allowed to select what they believe are their most representative outfalls for purposes of monitoring pollutant loads (EPA, 1996a. p. VIII-1).

A large subset of dischargers—the remaining industrial dischargers and construction sites—are subject to much more limited monitoring requirements. They are not required to sample contaminant levels, but instead are required only to conduct a visual inspection of a grab sample of their stormwater runoff on a quarterly basis and describe the visual appearance of the sample in a document that is kept on file at the site (EPA, 2006, p. 28). Certainly a visual sample is better than nothing, but the requirement allows the discharger not only some discretion in determining how and when to take the sample (explained below), but also discretion in how to describe the sample.

A final set of regulated parties, the Phase II MS4s, are not required to per-form any quantitative monitoring of runoff to test the effectiveness of SCMs (EPA, 1996a, p. 3).

Making matters worse, in some states there appear to be limited regulatory resources to verify compliance with many of these permit requirements. Thus, even though monitoring plans are subject to review and approval by permitting agencies, there may be insufficient resources to support this level of oversight. As shown in Appendix C, the total number of staff associated with state storm-water programs is usually just a handful, except in cases of larger states (Cali-fornia and Georgia) or those where there is a longer history of stormwater man-agement (Washington and Minnesota). In its survey of state stormwater pro-grams, the committee asked states how they tracked sources' compliance with the stormwater permits. For the 18 states responding to the questionnaire, re-view of (1) monitoring data, (2) annual reports, and (3) SWPPP as well as on-site inspections were the primary mechanisms. However, several states indi-cated that they conduct an inspection only after receiving complaints. West Virginia tracked whether industrial facilities submitted their required samples and followed up with a letter if they failed to comply, but in 2006 it found that over 65 percent of the dischargers were delinquent in their sampling. Although the states were not asked in the survey to estimate the overall compliance rate, Ohio admitted that at least for construction, "the general sense is that no site is 100 percent in compliance with the Construction General Permit" (see Appendix C).

Even where considerable regulatory resources are dedicated to ensuring that dischargers are in compliance, it is not clear how well regulators can independ-ently assess compliance with the permit requirements. For example, some of the permits will require "good housekeeping" practices that should take place daily at the facility. Whether or how well these practices are followed cannot be as-sessed during a single inspection. While a particularly non-compliant facility might be apparent from a brief visual inspection, a facility that is mildly sloppy, or at least has periods during which it is not careful, can escape detection on one of these pre-announced audits. Facilities also know best the pollutants they gen-erate and how or whether those pollutants might make contact with stormwater. Inspectors might be able to notice some of these problems, but because they do not have the same level of information about the operations of the facility, they can be expected to miss some problems.

Identifying Potentially Regulatable Parties

Evidence suggests that a sizable percentage of industrial and construction stormwater dischargers are also failing to self-identify themselves to regulators, and hence these unreported dischargers remain both unpermitted and unregu-lated (GAO, 2005; Duke and Augustenborg, 2006). In contrast to industrial pipes that carry wastes from factories out to receiving waters, the physical presence of stormwater dischargers may be less visible or obvious. Thus, particularly for some industries and construction, if a stormwater discharger does not apply for a permit, the probability of detecting it is quite low.

In Maine, less than 20 percent of the stormwater dischargers that fall within the regulatory jurisdiction of the federal stormwater program actually applied for permits before 2005—more than a decade after the federal regulations were promulgated (Richardson, 2005). Yet there is no record of enforcement action taken by Maine against the unpermitted dischargers during that interim period. Indeed, in the one enforcement action brought by citizens in Maine for an unpermitted discharge, the discharger claimed ignorance of the stormwater program. In Washington, the State Department of Ecology speculates that between 10 and 25 percent of all businesses that should be covered by the federal stormwater permit program are actually permitted (McClure, 2004). In a four-state study, Duke and Augustenborg (2006) found a higher percentage of stormwater dischargers—between 50 and 80 percent—had applied for permits by 2004, but they concluded that this was still "highly incomplete" compliance for an established permit program.

In 2007, the committee sent a short survey to each state stormwater program inquiring as to whether and how they tracked non-filing stormwater dischargers, but only six states replied to the questions and only two of the six states had any methods for tracking non-filers or conducting outreach to encourage all covered parties to apply for permits (see Appendix C). While the low response rate cannot be read to mean that the states do not take the stormwater program seriously, the responses that were received lend some support to the possibility that there is substantial noncompliance at the filing stage.

In response to this problem of unpermitted discharges, the EPA appears to be targeting enforcement against stormwater dischargers that do not have permits. In several cases, the EPA pursued regulated industries that failed to apply for stormwater permits (EPA Region 9, 2005; Kaufman et al., 2005). The EPA has also brought enforcement actions against at least three construction companies for failing to apply for a stormwater permit for their construction runoff (EPA Region 1, 2004). Such enforcement actions help to make the stormwater program more visible and give the appearance of a higher probability of enforcement associated with non-compliance. Nevertheless, the non-intuitive features of needing a permit to discharge stormwater, coupled with a rational perception of a low probability of being caught, likely encourage some dischargers to fail to enter the regulatory system.

Absence of Regulatory Prioritization

Many states have been overwhelmed with the sheer numbers of permittees, particularly industry and construction sites, and lack a prioritization strategy to identify high-risk sources in particular need of rigorous and enforceable permit conditions. For example, in California major facilities like the Los Angeles International Airport and the Los Angeles and Long Beach ports are covered under California's MSGP along with a half-acre metal plating facility in El Segundo—all subject to the same level of compliance scrutiny even after nearly two decades of implementation! Similarly, a multiphase, 20-year, thousand-acre residential development such as Newhall Land Development in North Los Angeles County is covered by the same California CGP as a one-acre residential home construction project in West Los Angeles, and subject to the same level of compliance scrutiny. The lack of an EPA strategy to identify and address high-risk industrial facilities and construction sites (i.e., those that pose the greatest risk of discharging polluted stormwater) remains an enormous deficiency. Phase I MS4s, for example, are left to their own devices to determine how to identify the most significant contributors to their stormwater systems (Duke, 2007).

Limited Public Participation

Public participation is more limited in the stormwater program in comparison to the wastewater permit program, providing less citizen-based oversight over stormwater discharges. Typically, during the issuance of an individual NPDES permit (for either wastewater or stormwater) the public has a chance to comment and review the draft permit requirements that are specifically prescribed for a certain site and discharge. While the same is true about the public participation during the adoption of a general stormwater permit, those general permits contain only the framework of the requirements and the menu of conditions, but do not prescribe specific requirements. Instead, it is up to the permittee to tailor the compliance to the specific conditions of the site in the form of a SWPPP. However, at this phase neither the public nor the regulators have access to the site-specific plan developed by the permittee to comply with the obligations of the permit. In the case of general permits, then, the discharger has enormous flexibility in designing its compliance activities.

Citizens also encounter difficulties in enforcing stormwater permit requirements. Citizens have managed to sue facilities for unpermitted stormwater discharges: this is a straightforward process because citizens need only verify that the facility should be covered and lacks a permit (Richardson, 2005). Overseeing facility compliance with stormwater permit requirements is a different story, however, and citizens are stymied at this stage of ensuring facility compliance. Citizens can access a facility's SWPPP, but only if they request the plan from the facility in writing (EPA, 2006, p. 25). Moreover, the facility is given the authority to make a determination—apparently without regulator oversight—of whether the plan contains confidential business information and thus cannot be disclosed to citizens (EPA, 2006, p. 26). But, even if the facility sends the plan to the citizens, it will be nearly impossible for them to independently assess whether the facility is in compliance unless the citizens station telescopes, conduct air surveillance of the site, or are allowed to access the facility's records of its own self-inspections. Moreover, to the extent that the stormwater outfalls are on the facility's property, citizens might not be able to conduct their own sampling without trespassing.

Not surprisingly, significant progress has nevertheless been made in reducing stormwater pollution when stormwater becomes a visible public issue. This increased visibility is often accomplished with the help of local environmental advocacy groups who call attention to the endangered species, tourism, or drinking water supplies that are jeopardized by stormwater contamination. Box 2-8 describes two cases of active public participation in the management of stormwater.

BOX 2-8
Citizen Involvement/Education in Stormwater Regulations

The federal Clean Water Act, under Section 505, authorizes citizen groups to bring an action in U.S. or state courts if the EPA or a state fails to enforce water quality regulations. Unsurprisingly, the few areas nationally where stormwater quality has become a visible public issue and significant progress has been made in reducing stormwater pollution have prominent local environmental advocacy groups actively involved.

Heal the Bay, Santa Monica, California. In Southern California, Santa Monica-based Heal the Bay has utilized research, education, community action, public advocacy, and political activism to improve the quality of stormwater discharges from MS4s in Southern California. Heal the Bay operates an aquarium to educate the public, conducts stream teams to survey local streams, posts a beach report card on the web to inform swimmers on beach quality, appears before the California Water Boards to comment on NPDES stormwater permits, and works with lawmakers to sponsor legislative bills that protect water quality.

In 1998, the organization helped co-author legislation to notify the public when shoreline water samples show that water may be unsafe for swimming. California regulations (AB411) require local health agencies (county or city) to monitor water quality at beaches that are adjacent to a flowing storm drain and have 50,000 visitors annually (from April 1 to October 31). At a minimum, these beaches are tested on a weekly basis for three specific bacteria indicators: total coliform, fecal coliform, and *enterococcus*. Local health officials are required to post or close the beach, with warning signs, if state standards for bacterial indicators are exceeded. The monitoring data collected are available to the public.

In order to better inform and engage the public, Heal the Bay has followed up with a web-based Weekly Beach Report Card (*http://healthebay.org/brc/statemap.asp*) and the release of an Annual California Beach Report Card assigning an "A" to "F" letter grade to more than 500 beaches throughout the state based on their levels of bacterial pollution. Heal the Bay's Annual Beach Report Card is a comprehensive evaluation of California coastal water quality based on daily and weekly samples gathered at beaches from Humboldt County to the Mexican border. A poor grade means beachgoers face a higher risk of contracting illnesses such as stomach flu, ear infections, upper respiratory infections, and skin rashes than swimmers at cleaner beaches.

Heal the Bay was instrumental in passing Proposition O in the City of Los Angeles which sets aside half a billion dollars to improve the quality of stormwater discharges. In the 2007 term of the California Legislature, the organization has sponsored five legislative bills to address marine debris, including plastic litter transported in stormwater runoff, that

continues next page

Box 2-8 Continued

foul global surface waters (*Currents*, Vol. 21, No. 2, p.8, 2007). Heal the Bay also coordinates its actions and partners with other regional and national environmental organizations, such as the WaterKeepers and the NRDC, in advancing water quality protection nationally.

Save Our Springs, Austin, Texas. Citizen groups have played a very influential role in the development of a rigorous stormwater control program in the City of Austin, Texas. Catalyzed in 1990 by a proposal for extensive development that threatened the fragile Barton Springs area, a citizens group named Save Our Springs Legal Defense Fund (later renamed Save our Springs Alliance) formed to oppose the development. It orchestrated an infamous all-night council meeting, with 800 citizens registering in opposition to the proposed development and ultimately led to the City Council's rejection of the 4,000-acre proposal and the formulation of a "no degradation" policy for the Barton Creek watershed. The nonprofit later sponsored the Save Our Springs Ordinance, a citizen initiative supported by 30,000 signatures, which passed by a 2 to 1 margin in 1992 to further strengthen protection of the area. The Save Our Springs Ordinance limits impervious cover in the Barton Springs watershed to a maximum of between 15 and 25 percent, depending on the location of the development in relation to the recharge and contributing zones. The ordinance also mandates that stormwater runoff be as clean after development as before. The ordinance was subject to a number of legal challenges, all of which were successfully defended by the nonprofit in a string of court battles.

Since its initial formation in 1990, the Save Our Springs Alliance has continued to serve a vital role in educating the community about watershed protection and organizing citizens to oppose development that threatens Barton Springs. The organization has also been instrumental in working with a variety of government and nonprofit organizations to set aside large areas of parkland and open spaces within the watershed. Other citizen groups, like the Save Barton Creek Association, also play a very active, complementary role to the Save Our Springs Alliance in protecting the watershed. These other nonprofits are sometimes allied and sometimes diverge to take more moderate stances to development proposals. The resulting constellation of citizen groups, citizen outreach, and community participation is very high in the Austin area and has unquestionably led to a much more informed citizenry and a more rigorous watershed protection program than would exist without such grassroots leadership.

Accounting for Future Land Use

One of the challenges of managing stormwater from urban watersheds thus involves anticipating and channeling future urban growth. Currently, the CWA does little to anticipate and control for future sources of stormwater pollution in urban watersheds. Permits are issued individually on a technology-based basis, allowing for uncontrolled cumulative increases in pollutant and volume loads over time as individual sources grow in number. The TMDL process in theory requires states to account for future growth by requiring a "margin of safety" in loading projections. However, it is not clear how frequently future growth is included in individual TMDLs or how vigorous the growth calculations are (for example, see EPA [2007a, pp. 12, 37], mentioning considerations of future land use as a consideration in stormwater related TMDLs for only a few—Potash Brook and the lower Cuyahoga River—of the 17 TMDLs described in the report). In any event, as already noted a TMDL is generally triggered only after waters have been impaired, which does nothing to anticipate and channel land development before waters become degraded.

The fact that stormwater regulation and land-use regulation are largely decoupled in the federal regulatory system is understandable given the CWA's industrial and municipal wastewater focus and concerns about federalism, but this limited approach is not a credible approach to stormwater management in the future. Federal incentives must be developed to encourage states and municipalities to channel growth in a way that acknowledges, estimates, and minimizes stormwater problems.

Picking up the Slack at the Municipal and State Level

Because it involves land use, any stormwater discharge program strikes at a target that is traditionally within the province of state and even more likely local government regulation. Indeed, it is possible that part of the reason for the EPA's loosely structured permit program is its concern about intruding on the province of state and local governments, particularly given their superior expertise in regulating land-use practices through zoning, codes, and ordinances.

In theory, it is perfectly plausible that some state and local governments will step into the void and overcome some of the problems that afflict the federal stormwater discharge program. If local or state governments required mandatory monitoring or more rigorous and less ambiguous SCMs, they would make considerable progress in developing a more successful stormwater control program. In fact, some states and localities have instituted programs that take these steps. For example, Oregon has established its own benchmarks based on industrial stormwater monitoring data, and it uses the benchmark exceedances to deny industries coverage under Oregon's MSGP. In such cases, the facility operator must file for an individual stormwater discharge NPDES permit. Some municipalities are also engaging in these problems, such as the City of Austin and its ban on coal tar sealants.

Despite these bursts of activity, most state and local governments have not taken the initiative to fill the gaps in the EPA's federal program (see Tucker [2005] for some exceptions). Because they involve some expense, stormwater discharge requirements can increase resident taxes, anger businesses, and strain already busy regulatory staff. Moreover, if the benefits of stormwater controls are not going to materialize in waters close to or of value to the community instituting the controls, then the costs of the program from the locality's standpoint are likely to outweigh its benefits. Federal financial support for state and local stormwater programs is very limited (see section below). Until serious resources are allocated to match the seriousness and complexity of the problem and the magnitude of the caseload, it seems unlikely that states and local communities will step in to fill the gaps in EPA's program. These impediments help explain why there appear to be so many stormwater sources out of compliance with the stormwater discharge permit program as discussed above, at least in the few states that have gone on record.

Funding Constraints

Without a doubt, the biggest challenge for states, regions, and municipalities is having adequate fiscal resources dedicated to implement the stormwater program. Box 2-9 highlights the costs of the program for the State of Wisconsin, which has been traditionally strong in stormwater management. Phase I regulations require that a brief description of the annual proposed budget for the following year be included in each annual report, but this requirement has been dispensed with entirely for Phase II.

Ever since the promulgation of the stormwater amendments to the CWA and the issuance of the stormwater regulations, the discharger community pointed out that this statutory requirement had the flavor of an unfunded mandate. Unlike the initial CWA that provided significant funding for research, design, and construction of wastewater treatment plants, the stormwater amendments did not provide any funding to support the implementation of the requirements by the municipal operators. The lack of a meaningful level of investment in addressing the more complex and technologically challenging problem of cleaning up stormwater has left states and municipalities in the difficult position of scrambling for financial support in an era of multiple infrastructure funding challenges.

While a number of communities have passed stormwater fees linked to water quality as described below, a significant number of communities still do not have that financial resource. Municipalities that have not formed utility districts or imposed user fees have had to rely on general funds, where stormwater permit compliance must compete with public safety, fire protection, and public libraries. This circumstance explains why elected local government officials have been reluctant to embrace the stormwater program. Stormwater quality management is often not regarded as a municipal service, unlike flood control or wastewater conveyance and treatment. A concerted effort will need to be made by all stakeholders to make the practical and legal case that stormwater quality management is truly another municipal service like trash collection, wastewater treatment, flood control, etc. Even in states that do collect fees to finance stormwater permit programs, the programs appear underfunded relative to other types of water pollution initiatives. Table 2-10 shows the water quality budget of the California EPA, Los Angeles Regional Water Board. The amount of money per regulated entity (see Table 2-4) dedicated to the stormwater program pales in comparison to the wastewater portion of the NPDES program, and it has

BOX 2-9
Preliminary Cost Estimates for Complying with
Stormwater Discharge Permits in Wisconsin

The Wisconsin Department of Natural Resources (WDNR) was delegated authority under the CWA to administer the stormwater permit program under Chapter NR 216. There are 75 municipalities regulated under individual MS4 permits and 141 MS4s regulated under a general permit for a total of 216 municipalities with stormwater discharge permits.

As part of the "pollution prevention" minimum measure the municipalities are required to achieve compliance with the developed urban area performance standards in Chapter NR 151.13. By March 10, 2008, municipalities subject to a municipal stormwater permit under NR 216 must reduce their annual TSS loads by 20 percent. These same permitted municipalities are required to achieve an annual TSS load reduction of 40 percent by March 10, 2013. The reduction in TSS is compared to no controls, and any existing SCMs will be given credit toward achieving the 20 or 40 percent. As part of their compliance with NR151.13 developed area performance standards, the municipalities are preparing stormwater plans describing how they will achieve the 20 and 40 percent TSS reduction. They are required to use an urban runoff model, such as WinSLAMM or P8, to do the pollutant load analysis.

As the permitted municipalities comply with the six minimum control measures and submit the stormwater plans for their developed area urban areas, the WDNR is learning how much it is going to cost to achieve the requirements in the stormwater discharge permits. Some cities have already been submitting annual reports that include the cost of the six minimum measures. Nine of the permitted municipalities in the southeast part of Wisconsin have been submitting their annual reports for at least four years. The average population of these nine communities is 17,700 with a range of about 6,000 to 65,000. The average cost of the six minimum measures in 2007 for the nine municipalities is $162,900 with a range of $11,600 to $479,000. These costs have not changed significantly from year to year. The average per capita cost is $9 with a range of $1 to $16 per person. Street cleaning and catch basin cleaning (Figures 2-3 and 2-4) cost are included in the cost for the pollution prevention measure, and most of the cities were probably incurring costs for these two activities before the issuing of the permit. On average the street cleaning and catch basin cleaning represent about 40 percent of the annual cost for the six minimum measures. These two activities will help the cities achieve the 20 and 40 percent TSS performance standards for developed urban areas.

Information is available on the preliminary cost of achieving the 40 percent TSS performance standard for selected cities in Wisconsin. The costs were prepared for 15 municipalities by Earth Tech Inc. in Madison, Wisconsin. Areas of the municipality developed after October 2004 are not included in the TSS load analysis. At this point in the preparation of the stormwater plans the costs are just capital cost estimates done at the planning level (Table 2-9). Because the municipalities receive credit for their existing practices, these capital costs represent the additional practices needed to achieve the annual 40 percent TSS reduction. The costs per capita appear to decline for cities with a population over 50,000. All of the costs in Table 2-9 will increase when other costs, such as maintenance and land cost, are included.

For most of the 15 municipalities, the capital costs are for retrofitting dry ponds with permanent pools, installing new wet detention ponds, and improved street cleaning capabilities. Because of their lower cost, the regional type practices have received more attention in the stormwater plans than the source area practices, such as proprietary devices and biofilters. Municipalities with a higher percentage of newer areas will usually have lower cost because the newer developments tend to have stormwater control measures designed to achieve a high level of TSS control, such as wet detention ponds. Older parts of a municipality are usually limited to practices with a lower TSS reduction, such as street cleaning and catch basin cleaning. Of course, retrofitting older areas with higher efficiency practices is expensive,

continues next page

BOX 2-9 Continued

and the cost can go higher than expected when unexpected site limitations occur, such as the presence of underground utilities.

Over the next five years all of the 15 municipalities must budget the costs in Table 2-9. It is not clear yet how much of a burden these costs represent to the taxpayers in each municipality. All the permits will be reviewed for compliance with the performance standards in 2013.

TABLE 2-9 Planning-Level Capital Cost Estimate to Meet 40 Percent TSS Reduction

Population	Number of Cities	Average Cost ($)	Minimum Cost ($)	Maximum Cost ($)	Avg. Cost per Capita per Year over 5 Years ($)
5,000 to 10,000	5	1,380,000	425,000	2,800,000	34
10,000 to 50,000	6	4,600,00	2,700,00	9,200,000	35
50,000 to 100,000	4	9,200,000	7,000,000	12,500,000	26

SOURCE: Reprinted, with permission, from James Bachhuber, Earth Tech Inc., personnel communication (2008). Copyright 2008 by James Bachhuber, Earth Tech Inc.

FIGURE 2-3 Catch basin cleaning. SOURCE: Robert Pitt, University of Alabama.

FIGURE 2-4 Street cleaning. SOURCE: Courtesy of the U.S. Geological Survey.

TABLE 2-10 Comparison of Fiscal Year (FY) 02–03 Budget with FY 06–07 Budget for Water Quality Programs at the California EPA, Los Angeles Regional Water Board

Program	Funding Source	2002–2003	2006–2007
NPDES[1]	Federal	$2.8 million	$2.6 million
Stormwater	State	$2.3 million	$2.1 million
TMDLs	Federal	$1.47 million	$1.38 million
Spills, Leaks, Investigation Cleanup	State	$1.32 million	$2.87 million
Underground Storage Tanks	State	$2.78 million	$2.74 million
Non-Chapter 15 (Septics)	State	$0.93 million	$0.93 million
Water Quality Planning	Federal	$0.2 million	$0.21 million
Well Investigation	State	$1.36 million	$0.36 million
Water Quality Certification	Federal	$0.2 million	$0.23 million
Total		$17.1 million	$15.82 million

[1]The NPDES row is entirely wastewater funding, as there is no federal money for implementing the stormwater program. Note that the stormwater program in the table is entirely state funded.

declined over time. Furthermore, of the more than $5 billion dollars in low-interest loans provided in 2006 for investments in water quality improvements, 96 percent of that total funding went to wastewater treatment (EPA, 2007d).

There are a number of potential methods that agencies can use to collect stormwater quality management fees, as described more extensively in Chapter 5. A number of states now levy permit fees, with some permits costing in excess of $10,000, to help defray the costs of implementation and enforcement of their stormwater programs. The State of Colorado, for example, has developed an elaborate fee structure for separate types of general permits for industry and construction, as well as MS4s (see *http://www.cdphe.state.co.us/wq/permitsunit/stormwater/StormwaterFees.pdf*). The ability of a state agency to collect fees generally must first be authorized by the state legislatures (see, e.g., Revised Code of Washington 90.48.465, providing the state agency with the authority to "collect expenses for issuing and administering each class of permits"). The lack of state legislative authorization may limit some state agencies from creating such programs on their own. In fact, in those states where fees cannot be levied against permittees, the stormwater programs appear to be both underfinanced and understaffed. Some municipalities have even experienced political backlash because of the absence of a strong state or federal program requiring them to engage in rigorous stormwater management (see Box 2-10).

Stormwater Management Expertise

Historically, engineering curriculum dealt with stormwater management by focusing on the flood control aspects, with little attention given to the water quality aspects. Thus, there has been a significant gap in knowledge and a lack

BOX 2-10
A City's Ability to Pay for Stormwater, Water, and Sewage Utility Fees

With the implementation of the stormwater permit program of the CWA, stormwater utilities are becoming more common as a way to jointly address regional stormwater quality and drainage issues. One such program is the Jefferson County, Alabama, Storm Water Management Authority (SWMA), formed in 1997 under state legislation that enables local governments to pool their resources in a regional stormwater authority to meet regulations required by the CWA. Jefferson County, the City of Birmingham, and 22 other regional municipalities in Jefferson, part of Shelby and part of St. Clair counties, Alabama, were required to comply with CWA regulations. The act gave the stormwater program the ability to develop a funding mechanism for the program and to form a Public Corporation.

Over the years, SWMA has been responsible for many activities. One of their first goals was to develop a comprehensive GIS database to map outfalls, land uses, stormwater practices, and many other features that were required as part of the permit program. Another major activity conducted by SWMA was the collection of water samples from about 150 sites in the authority's jurisdiction, both during wet and dry weather. SWMA also inspects approximately 4,000 outfalls during dry weather to check for inappropriate connections to the storm drainage system. SWMA coordinates public volunteer efforts with local environmental groups, including the Alabama Water Watch, the Alabama River Alliance, the Black Warrior Riverkeeper, and the Cahaba River Society. SWMA also inspects businesses and industries (including construction sites) within their jurisdictions that are not permitted by the Alabama Department of Environmental Management (ADEM). SWMA does not enforce rules or issue fines, although it can report violators to the state. In its most famous case, it reported McWane Inc. for pollution that led to investigations by the state and the federal government, and ultimately a trial and criminal convictions.

The Birmingham News (Bouma, 2007) reported that from 1997 to 2005, SWMA's responsibilities under the CWA increased substantially, although their fees did not rise. In late 2005, SWMA proposed that member cities increase their stormwater charges from $5 a year to $12 a year per household for residences and from $15 to $36 per year for businesses. At that point, the Business Alliance for Responsible Development (BARD), a group of large businesses, utilities, mining interests, developers and landowners, began to argue that the group was financially irresponsible, and its attorneys convinced member cities that they could save money by withdrawing from SWMA. Even though SWMA withdrew its fee increase request, many local municipalities have pulled out of SWMA, significantly reducing the agency's budget and ability to conduct comprehensive monitoring and reporting. BARD claims the pollution control programs of the ADEM are sufficient. In their countersuit, several environmental groups maintain that ADEM has failed to adequately protect the state's waters because the agency is underfunded, understaffed, and ineffective at enforcement. Much of the Cahaba and Black Warrior River systems within Jefferson County have such poor water quality that they frequently violate water quality standards (*http://www.southernenvironment.org*). SWMA has been significantly impaired in its ability to monitor and report water quality violations with the withdrawal of many of its original member municipalities and the associated reduced budget.

At the same time, the sewer bill for a family of four in the region is expected to be about $63 per month in 2008. Domestic water rates have also increased, up to about $32 per month (*The Birmingham News*, Barnett Wright, December 30, 2007). Domestic water rates have increased in recent years in attempts to upgrade infrastructure in response to widespread and long-lasting droughts and to cover rising fuel costs. It is ironic that stormwater management agency fees are very small compared to these other urban water agency fees per household by orders of magnitude. The $12 per year stormwater fee was used to justify the dismantling of an agency that was doing its job and identifying CWA violators. In order to bring some reasonableness to the stormwater management situation and expected fees, it may be possible for the EPA to re-examine its guidelines of 2 percent of the household income for sewer fees to reflect other components of the urban water system, and to ensure adequate enforcement of existing regulations, especially by underfunded state environmental agencies.

of qualified personnel. In areas where SCMs are just beginning to be introduced, many municipalities, industrial operators, and construction site operators are not prepared to address water quality issues; the problem is especially difficult for smaller municipalities and operators. The profession and academia are moving to correct this shortfall. Professional associations such as the Water Environment Federation (WEF) and the American Society for Civil Engineers (ASCE) are co-authoring an update of the WEF/ASCE Manual of Practice "Design of Urban Runoff Controls" that integrates quality and quantity, after years of issuing separate manuals of design and operation for the water quality and water quantity elements of stormwater management.

The split between water quantity and quality is evident in municipal efforts that have focused primarily on flood control issues and design of appropriate appurtenances tailored for this purpose. As discussed earlier, most municipal codes specify practices to collect and move water away as fast as possible from urbanized areas. Very little focus has been put on practices to mitigate the quality of the stormwater runoff. This is especially true in urbanized areas with separate municipal storm sewer systems. Even the designation "sewer" is borrowed from the sanitary sewer conveyance system terminology. In arid or semiarid areas, these flood control systems have been maximally engineered such that river beds have become concrete channels. A typical example is the Los Angeles River, which most of the year resembles an empty freeway. This analysis does not intend to minimize the engineering feat of designing a robust and reliable flood control system. For example, during the unusually wet 2005 season in Southern California, the Los Angeles area did not have any major flooding incidents. However, based on recent studies (Stein and Ackerman, 2007) up to 80 percent of the annual metals loading from six watersheds in the Los Angeles area was transported by stormwater events.

Because of the historical lack of focus on stormwater quality, municipal departments in general are not designed to address the issue of pollution in urban runoff. Just recently and due to the stormwater regulations, cities have been adding personnel and creating new sections to deal with the issue. However, because of the complexities of the task, many duties are spread among various municipal departments, and more often than not coordination is still lacking. Perhaps most problematic is the fact that the local governmental entities in charge of stormwater management are often different from those that oversee land-use planning and regulation. This disconnect between land-use planning and stormwater management is especially true for large cities. It is not unusual for program responsibilities to be compartmentalized, with industrial aspects of the program handled by one group, construction by another, and planning and public education by other distinct units. Smaller cities may have one person handling all aspects of the program assisted by a consulting firm. While coordination may be ensured, the task can be overwhelming for a single staff person.

Beyond water quality issues, training to better understand the importance of volume control and the role of LID has not yet reached many practitioners.

Many established practices and industry standards in the fields of civil, geotechnical, and structural engineering were developed prior to the introduction of the current group of SCMs and can unnecessarily limit their use. Indeed, certain SCMs such as porous landscape detention, extended detention, and vegetated swales require special knowledge about soils and appropriate plant communities to ensure their longevity and ease of maintenance.

Supplementing the Clean Water Act with Other Federal Authorities that Can Control Stormwater Pollutants at the Source

EPA does have other supplemental authorities that are capable of making significant progress in reducing or even eliminating some of the problematic stormwater pollutants at the national level. Under both the Federal Insecticide, Fungicide, and Rodenticide Act (FIFRA) and the TSCA, for example, EPA could restrict some of the most problematic pollutants at their source by requiring labels that alert consumers to the deleterious water quality impacts caused by widely marketed chemical products, restricting their use, or even banning them. This source-based regulation bypasses the need of individual dischargers or governments to be concerned with reducing the individual contaminants in stormwater.

The City of Austin's encounter with coal tar-based asphalt sealants provides an illustration of the types of products contributing toxins to stormwater discharges that could be far better controlled at the production or marketing stage. Through detective work, the City of Austin learned that coal tar-based asphalt sealants leach high levels of polycyclic aromatic hydrocarbons (PAHs) into surface waters (Mahler et al., 2005; Van Metre et al., 2006). The city discovered this because the PAHs were found in sediments in Barton Springs, which were in turn leading to the decline of the endangered Barton Creek salamander (Richardson, 2006). By tracing upstream, the city was able to find the culprit—a parking lot at the top of the hill that was recently sealed with coal tar sealant and produced very high PAH readings. Further tests revealed that coal tar sealants typically leach very high levels of PAHs, but other types of asphalt sealants that are not created from coal tar are much less toxic to the environment and are no more expensive than the coal tar-based sealants (City of Austin, 2004). As a result of its findings, the City of Austin banned the use of coal tar-based asphalt sealants. Several retailers, including Lowes and Home Depot followed the city's lead and refused to carry coal tar sealants. Dane County in the State of Wisconsin has now also banned coal tar sealants[1].

---

[1] See, e.g., Coal Tar-based pavement sealants studied, Science Daily, February 12, 2007, available at *http://www.sciencedaily.com/upi/index.php?feed=Science&article=UPI-1-2007 0212-10255500-bc-us-sealants.xml*; Matthew DeFour, Dane County bans Sealants with Coal Tar, Wisconsin State Journal, April 6, 2007, available at *http://www.madison.com/ wsj/home/local/index.php?ntid=128156&ntpid=5.*

For reasons that appear to inure to the perceived impotency of TSCA and the enormous burdens of restricting chemicals under that statute, EPA declined to take regulatory action under TSCA against coal tar sealants (Letter from Brent Fewell, Acting Assisting Administrator, U.S. EPA, to Senator Jeffords, October 16, 2006, p. 3). Yet, it had authority to consider whether this particular chemical mixture presents an "unreasonable risk" to health and the environment, particularly in comparison to a substitute product that is available at the same or even lower price [15 U.S.C. § 2605(a); *Corrosion Proof Fittings vs. EPA*, 947 F.2d 1201 (5th Cir. 1991)]. Indeed, if EPA had undertaken such an assessment, it might have even discovered that the coal tar sealants are not as inferior as Austin and others have concluded; alternatively it could reveal that these sealants do present an "unreasonable risk" since there are substantial risks from the sealant without corresponding benefits, given the availability of a less risky substitute.

A similar situation holds for other ubiquitous stormwater pollutants, such as the zinc in tires, roof shingles, and downspouts; the copper in brake pads; heavy metals in fertilizers; creosote- and chromated copper arsenate (CCA)-treated wood; and de-icers, including road salt. Each of these sources may be contributing toxins to stormwater in environmentally damaging amounts, and each of these products might have less deleterious and equally cost-effective substitutes available, yet EPA and other federal agencies seem not to be undertaking any analysis of these possibilities. The EPA's phase-out of lead in gasoline in the 1970s, which led to measurable declines in the concentrations of lead in stormwater by the mid-1980s (see Figure 2-5), may provide a model of the type of gradual regulatory ban EPA could use to reduce contaminants in products that are non-essential.

Some states are taking more aggressive forms of product regulation. For example, in the mid-1990s, numerous scientific studies conducted in California by stormwater programs, wastewater treatment plants, the University of California, California Water Boards, the U.S. Geological Survey, and EPA showed widespread toxicity in local creeks, stormwater runoff, and wastewater treatment plant effluent from pesticide residues, particularly diazinon and chlopyrifos (which are commonly used organophosphate pesticides available in hundreds of consumer products) (Kuivila and Foe, 1995; MacCoy et al., 1995). As a result, the California Water Boards and EPA listed many waters in urban areas of California as being impaired in accordance with CWA Section 303(d). Many cities and counties were required to implement expensive programs to control the pollution under the MS4 NPDES permits to restore the designated beneficial uses of pesticide-impaired waters. Figure 2-6 shows the results of one such action—a ban on diazinon.

In sum, even though there are a number of sources of pollutants—from roof tiles to asphalt sealants to de-icers to brake linings—that could be regulated more restrictively at the product and market stage, EPA currently provides little meaningful regulatory oversight of these sources with regard to their contri-

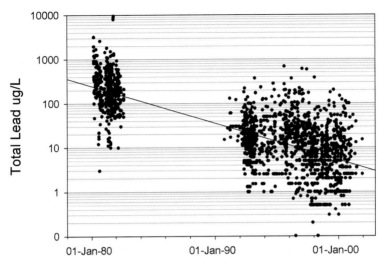

FIGURE 2-5 Trend of lead concentrations in stormwater in EPA rain zone 2 from 1980 to 2001. Although the range of lead concentrations for any narrow range of years is quite large, there is a significant and obvious trend in concentration for these 20 years. SOURCE: National Stormwater Quality Database (version 3).

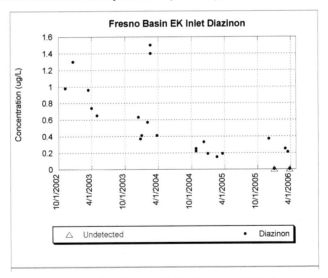

FIGURE 2-6 Trend of the organophosphate pesticide diazinon in MS4 discharges that flow into a stormwater basin in Fresno County, California, following a ban on the pesticide. The figure shows the significant drop in the diazinon concentration in just four years to levels where it is no longer toxic to freshwater aquatic life. EPA prohibited the retail sale of diazinon for crack and crevice and virtually all indoor uses after December 31, 2002, and non-agriculture outdoor use was phased out by December 31, 2004. Restricted use for agricultural purposes is still allowed. SOURCE: Reprinted, with permission, from Brosseau (2007). Copyright 2006 by Fresno Metropolitan Flood Control District.

bution to stormwater pollution. The EPA's authority to prioritize and target products that increase pollutants in runoff, both for added testing and regulation, seems clear from the broad language of TSCA [15 U.S.C. § 2605(a)]. The underutilization of this national authority to regulate environmentally deleterious stormwater pollutants thus seems to be a remediable shortcoming of EPA's current stormwater regulatory program.

CONCLUSIONS AND RECOMMENDATIONS

In an ideal world, stormwater discharges would be regulated through direct controls on land use, strict limits on both the quantity and quality of stormwater runoff into surface waters, and rigorous monitoring of adjacent waterbodies to ensure that they are not degraded by stormwater discharges. Future land-use development would be controlled to prevent increases in stormwater discharges from predevelopment conditions, and impervious cover and volumetric restrictions would serve as a reliable proxy for stormwater loading from many of these developments. Large construction and industrial areas with significant amounts of impervious cover would face strict regulatory standards and monitoring requirements for their stormwater discharges. Products and other sources that contribute significant pollutants through stormwater—like de-icing materials, urban fertilizers and pesticides, and vehicular exhaust—would be regulated at a national level to ensure that the most environmentally benign materials are used when they are likely to end up in surface waters.

In the United States, the regulation of stormwater looks quite different from this idealized vision. Since the primary federal statute—the CWA—is concerned with limiting pollutants into surface waters, the volume of discharges are secondary and are generally not regulated at all. Moreover, given the CWA's focus on regulating pollutants, there are few if any incentives to anticipate or limit intensive future land uses that generate large quantities of stormwater. Most stormwater discharges are regulated instead on an individualized basis with the demand that existing point sources of stormwater pollutants implement SCMs, without accounting for the cumulative contributions of multiple sources in the same watershed. Moreover, since individual stormwater discharges vary with terrain, rainfall, and use of the land, the restrictions governing regulated parties are generally site-specific, leaving a great deal of discretion to the dischargers themselves in developing SWPPPs and self-monitoring to ensure compliance. While states and local governments are free to pick up the large slack left by the federal program, there are effectively no resources and very limited infrastructure with which to address the technical and costly challenges faced by the control of stormwater. These problems are exacerbated by the fact that land use and stormwater management responsibilities within local governments are frequently decoupled. The following conclusions and recommendations are made.

EPA's current approach to regulating stormwater is unlikely to produce an accurate or complete picture of the extent of the problem, nor is it likely to adequately control stormwater's contribution to waterbody impairment. The lack of rigorous end-of-pipe monitoring, coupled with EPA's failure to use flow or alternative measures for regulating stormwater, make it difficult for EPA to develop enforceable requirements for stormwater dischargers. Instead, under EPA's program, the stormwater permits leave a great deal of discretion to the regulated community to set their own standards and self-monitor.

Implementation of the federal program has also been incomplete. Current statistics on the states' implementation of the stormwater program, discharger compliance with stormwater requirements, and the ability of states and EPA to incorporate stormwater permits with TMDLs are uniformly discouraging. Radical changes to the current regulatory program (see Chapter 6) appear necessary to provide meaningful regulation of stormwater dischargers in the future.

Future land development and its potential increases in stormwater must be considered and addressed in a stormwater regulatory program. The NPDES permit program governing stormwater discharges does not provide for explicit consideration of future land use. Although the TMDL program expects states to account for future growth in calculating loadings, even these more limited requirements for degraded waters may not always be implemented in a rigorous way. In the future, EPA stormwater programs should include more direct and explicit consideration of future land developments. For example, stormwater permit programs could be predicated on rigorous projections of future growth and changes in impervious cover within an MS4. Regulators could also be encouraged to use incentives to lessen the impact of land development (e.g., by reducing needless impervious cover within future developments).

Flow and related parameters like impervious cover should be considered for use as proxies for stormwater pollutant loading. These analogs for the traditional focus on the "discharge" of "pollutants" have great potential as a federal stormwater management tool because they provide specific and measurable targets, while at the same time they focus regulators on water degradation resulting from the increased volume as well as increased pollutant loadings in stormwater runoff. Without these more easily measured parameters for evaluating the contribution of various stormwater sources, regulators will continue to struggle with enormously expensive and potentially technically impossible attempts to determine the pollutant loading from individual dischargers or will rely too heavily on unaudited and largely ineffective self-reporting, self-policing, and paperwork enforcement.

Local building and zoning codes, and engineering standards and practices that guide the development of roads and utilities, frequently do not promote or allow the most innovative stormwater management. Fortunately, a variety of regulatory innovations—from more flexible and thoughtful zoning to using design review incentives to guide building codes to having separate ordinances for new versus infill development can be used to encourage more effective stormwater management. These are particularly important to promoting redevelopment in existing urban areas, which reduces the creation of new impervious areas and takes pressure off of the development of lands at the urban fringe (i.e., reduces sprawl).

EPA should provide more robust regulatory guidelines for state and local government efforts to regulate stormwater discharges. There are a number of ambiguities in the current federal stormwater program that complicate the ability of state and local governments to rigorously implement the program. EPA should issue clarifying guidance on several key areas. Among the areas most in need of additional federal direction are the identification of industrial dischargers that constitute the highest risk with regard to stormwater pollution and the types of permit requirements that should apply to these high-risk sources. EPA should also issue more detailed guidance on how state and local governments might prioritize monitoring and enforcement of the numerous and diverse stormwater sources within their purview. Finally, EPA should issue guidance on how stormwater permits could be drafted to produce more easily enforced requirements that enable oversight and enforcement not only by government officials, but also by citizens. Further detail is found in Chapter 6.

EPA should engage in much more vigilant regulatory oversight in the national licensing of products that contribute significantly to stormwater pollution. De-icing chemicals, materials used in brake linings, motor fuels, asphalt sealants, fertilizers, and a variety of other products should be examined for their potential contamination of stormwater. Currently, EPA does not apparently utilize its existing licensing authority to regulate these products in a way that minimizes their contribution to stormwater contamination. States can also enact restrictions on or tax the application of pesticides or even ban particular pesticides or other particularly toxic products. Austin, for example, has banned the use of coal-tar sealants within city boundaries. States and localities have also experimented with alternatives to road salt that are less environmentally toxic. These local efforts are important and could ultimately help motivate broader scale, federal restrictions on particular products.

The federal government should provide more financial support to state and local efforts to regulate stormwater. State and local governments do not have adequate financial support to implement the stormwater program in a rigorous way. At the very least, Congress should provide states with financial support for engaging in more meaningful regulation of stormwater discharges. EPA should also reassess its allocation of funds within the NPDES program. The agency has traditionally directed funds to focus on the reissuance of NPDES wastewater permits, while the present need is to advance the NPDES stormwater program because NPDES stormwater permittees outnumber wastewater permittees more than five fold, and the contribution of diffuse sources of pollution to degradation of the nation's waterbodies continues to increase.

REFERENCES

Athayde, D. N., P. E. Shelly, E. D. Driscoll, D. Gaboury, and G. Boyd. 1983. Results of the Nationwide Urban Runoff Program—Volume 1, Final report. EPA WH-554. Washington, DC: EPA.

Barbour, M. T., J. Diamond, B. Fowler, C. Gerardi, J. Gerritsen, B. Snyder, and G. Webster. 1999a. The status and use of biocriteria in water quality monitoring. Project 97-IRM-1. Alexandria, VA: Water Environment Research Foundation (WERF).

Barbour, M. T., J. Gerritsen, B. D. Snyder, and J. B. Stribling. 1999b. Rapid Bioassessment Protocols for Use in Streams and Wadeable Rivers: Periphyton, Benthic Macroinvertebrates, and Fish. Second Edition. EPA/841-B-99-002. Washington, DC: EPA Office of Water.

Barbour, M. T., M. J. Paul, D. W. Bressler, A. H. Purcell, V. H. Resh, and E. Rankin. 2006. Bioassessment: a tool for managing aquatic life uses for urban streams. Research Digest #01-WSM-3 for the WERF.

Beckman, D. 2007. Presentation to the NRC Committee on Reducing Stormwater Discharge Contributions to Water Pollution, December 17, 2007.

Bouma, K. 2007. Agency may lose 40% of budget. The Birmingham News. January 18, 2007.

Brosseau, G. 2007. Presentation to the NRC Committee on Reducing Stormwater Pollution, December 17, 2007, Irvine, CA. Summarizing the Report: Fresno-Clovis Stormwater Quality Monitoring Program—Evaluation of Basin EK Effectiveness—Fresno Metropolitan Flood Control District, November 2006.

CA SWB (California State Water Board). 1999. Storm Water General Industrial Permit Non-Filer Identification and Communication Project: Final Report.

CA SWB. 2006. Storm Water Panel Recommendations—The Feasibility of Numeric Effluent Limits Applicable to Discharges of Storm Water Associated with Municipal, Industrial, and Construction Activities.

City of Austin. 2004. The Coal Tar Facts, available at *http://www.ci.austin.tx.us/watershed/downloads/coaltarfacts.pdf*. Last accessed August 20, 2008.

Connecticut DEP. 2007. A Total Maximum Daily Load Analysis for Eagleville Brook, Mansfield, CT, Final. 27 pp.

CWP (Center for Watershed Protection). 1998. Better Site Design: A Handbook for Changing Development Rules in Your Community. Ellicott City, MD: CWP.

Davies, S. P., L. Tsomides, J. L. DiFranco, and D. L. Courtemanch. 1999. Biomonitoring Retrospective: Fifteen Year Summary for Maine Rivers and Streams. Maine DEP (DEP LW1999-26).

Davies, S. P., and S. K. Jackson. 2006. The biological condition gradient: a descriptive model for interpreting change in aquatic ecosystems. Ecological Applications 16(4):1251–1266.

Davies, S. P., and L. Tsomides. 1997. Methods for Biological Sampling and Analysis of Maine's Inland Waters. MDEP, revised June 1997.

Davis, W. S. (ed.). 1990. Proceedings of the 1990 Midwest Pollution Control Biologists Meeting. U. S. Environmental Protection Agency, Chicago, Illinois. EPA 905/9-89-007.

Davis, W. S. 1995. Biological assessment and criteria: building on the past. Pp. 15–*30 In:* Biological Assessment and Criteria: Tools for Water Resource Planning and Decision Making for Rivers and Streams. W. Davis and T. Simon (eds.). Boca Raton, FL: Lewis Publishers.

DeShon, J. D. 1995. Development and application of the invertebrate community index (ICI). Pp. 217–*243 In:* Biological Assessment and Criteria: Tools for Risk-Based Planning and Decision Making. W. S. Davis and T. Simon (eds.). Boca Raton, FL: Lewis Publishers.

Driscoll, E. D., P. E. Shelley, and E. W. Strecker. 1990. Pollutant loadings and impacts from highway stormwater runoff volume III: analytical investigation and research report. Federal Highway Administration Final Report FHWA-RD-88-008, 160 pp.

Duke, D. 2007. Industrial Stormwater Runoff Pollution Prevention: Regulations and Implementation. Presentation to the NRC Committee on Reducing Stormwater Discharge Contribution to Water Pollution, Seattle, WA, August 22, 2007.

Duke, L. D., and C. A. Augustenborg. 2006. Effectiveness of self-identified and self-reported environmental regulations for industry: the case of stormwater runoff in the U.S. Journal of Environmental Planning and Management 49:385.

Duke, L. D., and K. A. Shaver. 1999. Industrial storm water discharger identification and compliance evaluation in the City of Los Angeles. Environmental Engineering Science 16:249–263.

EPA (U.S. Environmental Protection Agency). 1996a. Overview of the Stormwater Program. EPA 833-R-96-008. Available at *http://www.epa.gov/npdes/pubs/owm0195.pdf.* Last accessed August 20, 2008.

EPA. 1996b. Biological Criteria: Technical Guidance for Streams and Rivers. EPA 822-B-94-001. Washington, DC: EPA Office of Science and Technology.

EPA. 2000a. Stressor Identification Guidance Document. EPA 822-B-00-025. Washington, DC: EPA Offices of Water and Research and Development.

EPA. 2000b. Report to Congress on the Phase I Storm Water Regulations. EPA-833-R-00-001. Washington, DC: EPA Office of Water.

EPA. 2002a. Summary of Biological Assessment Programs and Biocriteria Development for States, Tribes, Territories, and Interstate Commissions: Streams and Wadeable Rivers. EPA-822-R-02-048. Washington, DC: EPA.

EPA. 2002b. Establishing Total Maximum Daily Load (TMDL) Wasteload Allocations (WLAs) for Storm Water Sources and NPDES Permit Requirements Based on Those WLAs. Memorandum dated November 22, 2002, from R. H. Wayland, Director, Office of Wetlands, Oceans, and Watersheds, and J. A. Hanlon, Director, Office of Wastewater Management, to Water Division Directors.

EPA. 2005a. Use of Biological Information to Better Define Designated Aquatic Life Uses in State and Tribal Water Quality Standards: Tiered Aquatic Life Uses. EPA-822-R-05-001. Washington, DC: Health and Ecological Criteria Division Office of Science and Technology, Office of Water.

EPA. 2005b. Proposed National Pollutant Discharge Elimination System (NPDES) General Permit for Stormwater Discharges from Industrial Activities, 70 Fed. Reg. 72116.

EPA. 2006. National Pollution Discharge Elimination System (NPDES), Proposed 2006 MSGP. Available at *http://www.epa.gov/npdes/pubs/msgp 2006_all-proposed.pdf.* Last accessed August 20, 2008.

EPA. 2007a. TMDLs with Stormwater Sources: A Summary of 17 TMDLs. Washington, DC: EPA.

EPA. 2007b. MS4 Program Evaluation Guidance Document. EPA-833-R-07-003. Washington, DC: EPA Office of Wastewater Management.

EPA. 2007c. Port of Los Angeles, Port of Long Beach Municipal Separate Storm Sewer System and California Industrial General Storm Water Permit Audit Report.

EPA. 2007d. Clean Water State Revolving Fund Programs: 2006 Annual Report. EPA-832-R-07-001. Washington, DC: EPA Office of Water.

EPA. 2008. Memorandum from Linda Boornazian, Water Permits Division, EPA Headquarters, to Water Division Directors, Regions 1–10. Clarification on which stormwater infiltration practices/technologies have the potential to be regulated as "Class V" wells by the Underground Injection Control Program. June 13, 2008. Washington, DC: EPA Office of Water.

EPA Region 1. 2004. Three NH Companies Agree to Pay Fine to Settle EPA Complaint; Case is Part of EPA Push to Improve Compliance with Stormwater Regulations. Press Release 04-08-05.

EPA Region 3. 2003. Nutrient and Siltation TMDL Development for Wissahickon Creek, Pennsylvania, Final Report. Available at *http://www.epa. gov/reg3wapd/tmdl/pa_tmdl/wissahickon/index.htm.* Last accessed August 20, 2008.

EPA Region 9. 2005. Press Release: EPA Orders Oakland Facility to Comply with its Stormwater Permit.

Gallant, A. L., T. R. Whittier, D. P. Larson, J. M. Omernik, and R. M. Hughes. 1989. Regionalization as a Tool for Managing Environmental Resources. EPA/600/3089/060. Corvallis, OR: EPA Environmental Research Laboratory.

General Accounting Office (GAO). 1989. EPA Action Needed to Improve the Quality of Heavily Polluted Waters. GAO/RCED 89-38. Washington, DC: GAO.

GAO. 2000. Water quality: Key EPA and State Decisions Limited by Inconsistent and Incomplete Data. GAO/RCED-00-54. Washington, DC: GAO.

GAO. 2005. Storm Water Pollution: Information Needed on the Implications of Permitting Oil and Gas Construction Activities. GAO-05-240. Washington, DC: GAO. Available at *http://www.gao.gov/new.items/d05240.pdf.* Last accessed August 20, 2008.

GAO. 2007. Report to Congressional Requesters, Further Implementation and Better Cost Data Needed to Determine Impact of EPA's Storm Water Program on Communities. GAO-07-479. Washington, DC: GAO.

Hawkins, C. P., R. H. Norris, J. N. Hogue, and J. W. Feminella. 2000. Development and evaluation of predictive models for measuring the biological integrity of streams. Ecological Applications 10:1456–1477.

Hill, B. H., F. H. McCormick, A. T. Herlihy, P. R. Kaufmann, R. J. Stevenson, and C. Burch Johnson. 2000. Use of periphyton assemblage data as an index of biotic integrity. Journal of the North American Benthological Society 19(1):50–67.

Hilsenhoff, W. L. 1987. An improved biotic index of organic stream pollution. Great Lakes Entomology 20:31.

Hilsenhoff, W. L. 1988. Rapid field assessment of organic pollution with a family-level biotic index. Journal of the North American Benthological Society 7:65.

Houck, O. 1999. TMDLs IV: the final frontier. Environmental Law Reporter 29:10469.

Hughes, R. M., D. P. Larsen, and J. M. Omernik. 1986. Regional reference sites: a method for assessing stream potentials. Environmental Management 10:629–635.

Karr, J. R. 1981. Assessment of biotic integrity using fish communities. Fisheries 6(6):21–27.

Karr, J. R., and E. W. Chu. 1999. Restoring Life In Running Waters: Better Biological Monitoring. Washington, DC: Island Press.

Karr, J. R., and D. R. Dudley. 1981. Ecological perspective on water quality goals. Environmental Management 5:55–68.

Karr, J. R., K. D. Fausch, P. L. Angermeier, P. R. Yant, and I. J. Schlosser. 1986. Assessing biological integrity in running waters: a method and its rationale. Illinois Natural History Survey Special Publication 5.

Kaufman, B., L. R. Liebesman, and R. Peterseen. 2005. Regulation of Stormwater Pollution: An Area of Increasing Importance to the Construction Industry. Mondaq Business Briefing, October 14, 2005. Available at *http://www.mondaq.com/article.asp?articleid=35276&lastestnews=1*. Last accessed October 31, 2008.

Kerans, B. L., and J. R. Karr. 1994. A benthic index of biotic integrity (B-IBI) for rivers of the Tennessee Valley. Ecological Applications 4(4):768–785.

Klemm, D. J., K. A. Blocksom, F. A. Fulk, A. T. Herlihy, R. M. Hughes, P. R. Kaufmann, D. V. Peck, J. L. Stoddard, W. T. Thoeny, M. B. Griffith, and W. S. Davis. 2003. Development and evaluation of macroinvertebrate biotic integrity index (MBII) for regionally assessing Mid-Atlantic highlands streams. Environ. Mgt. 31(5):656-669.

Kuivila, K., and C. Foe. 1995. Concentrations, transport and biological effects of dormant spray pesticides in the San Francisco estuary, California. Environmental Toxicology and Chemistry 14(7):1141–1150.

MacCoy, D., K. L. Crepeau, and K. M. Kuivila. 1995. Dissolved pesticide data for the San Joaquin River at Vemalis and the Sacramento River at Sacramento, California, 1991-94. U.S. Geological Survey Report 95-1 10.

Mack, J. J. 2007. Developing a wetland IBI with statewide application after multiple testing iterations. Ecological Indicators 7(4):864–881.

Mahler, B. J., P.C. Van Metre, T.J.Bashara, J.T. Wilson, and D.A. Johns. 2005. Parking lot sealcoat: an unrecognized source of urban polycyclic aromatic hydrocarbons. Environmental Science and Technology 39:5560.

Maine DEP. 2006. Barberry Creek TMDL. 41 pp.

McClure, R. 2004. Stormwater Bill Raises Concern, Seattle Post-Intelligencer, February 25, p. B1.

National Wildlife Federation. 2000. Pollution Paralysis II: Red Code for Watersheds 1-2.

NRC (National Research Council). 2001. Assessing the TMDL Approach to Water Quality Management. Washington, DC: National Academies Press.

Ohio EPA (Ohio Environmental Protection Agency). 1987. Biological Criteria for the Protection of Aquatic Life: Volume II. Users Manual for Biological Field Assessment of Ohio Surface Waters. Division of Water Quality Monitoring and Assessment, Surface Water Section, Columbus, OH.

Ohio EPA. 1989a. Biological criteria for the protection of aquatic life: Volume III. Standardized biological field sampling and laboratory methods for assessing fish and macroinvertebrate communities. Division of Water Quality Monitoring and Assessment, Columbus, Ohio.

Ohio EPA. 1989b. Addendum to biological criteria for the protection of aquatic life: Volume II. Users manual for biological field assessment of Ohio surface waters. Division of Water Quality Monitoring and Assessment, Surface Water Section, Columbus, Ohio.

Ohio EPA. 2007. Compensatory mitigation requirements for stream impacts in the State of Ohio and 3745-1-55 Compensatory mitigation requirements for wetlands. Division of Surface Water, Ohio EPA, Columbus, OH.

Omernik, J. M. 1987. Ecoregions of the conterminous United States. Map (scale 1:7,500,000). Annals of the Association of American Geographers 77(1):118–125.

Omernik, J. M. 1995. Ecoregions: a framework for environmental management. *In*: Biological Assessment and Criteria: Tools for Water Resource Planning and Decision Making. W. Davis and T. Simon (eds.). Chelsea, MI: Lewis Publishers.

Richardson, D. C. 2006. Parking lot sealants. Stormwater May/June:40.

Richardson, J. 2005. Maine makes it clear: watch your stormwater; Businesses are being warned about meeting the rules on polluted runoff. Portland Press Herald, November 28, p. A1.

Shannon, C. E., and W. Weaver. 1949. The Mathematical Theory of Communication. Urbana, IL: University of Illinois Press.

Simpson, J., and R. H. Norris. 2000. Biological assessment of water quality: development of AUSRIVAS models and outputs. Pp. 125–*142 In:* Assessing the Biological Quality of Fresh Waters: RIVPACS and Other Techniques. J. F. Wright, D. W. Sutcliffe, and M. T. Furse (eds.). Ambleside, UK: Freshwater Biological Association.

Stein, E. D., and D. Ackerman. 2007. Dry weather water quality loadings in arid, urban watersheds of the Los Angeles Basin, California, USA. Journal of the American Water Resources Association 43(2):398–413.

Stoddard, J. L., D. P. Larsen, C. P. Hawkins, R. K. Johnson, and R. H. Norris. 2006. Setting expectations for the ecological condition of streams: the concept of reference condition. Ecological Applications 16(4):1267–1276.

Swamikannu, X., M. Mullin, and L. D. Duke. 2001. Final Report: Industrial Storm Water Discharger Identification and Compliance Evaluation in the City of Los Angeles. California State Water Resources Control Board Contract No. 445951-LD-57453. Santa Monica Bay Restoration Project, Los Angeles, CA. 44 pp.

TetraTech. 2002. Ventura County Program Evaluation. Prepared for the California Water Board, Los Angeles Region, and EPA.

TetraTech. 2006a. Assessment Report on Tetra Tech's Support of California's MS4 Stormwater Program. Produced for U.S. EPA Region IX California State and Regional Water Quality Control Boards.

TetraTech. 2006b. Los Angeles Construction Program Review. Tetra Tech, California Water Board, Los Angeles Region, and EPA, January.

TetraTech. 2006c. Review of Best Management Practices at Large Construction Sites. TetraTech, California Water Board Los Angeles Region, and EPA, August.

TetraTech. 2006d. Assessment Report of Tetra Tech's Support of California's Industrial Stormwater Program.

TetraTech. 2006e. Assessment Report of Tetra Tech's Support of California's Municipal Stormwater Program.

Trautman, M. 1957. The Fishes of Ohio. Columbus, OH: Ohio State University Press.

Tucker, L. 2005. Oregon and Washington to release tougher standards for stormwater permits. Daily Journal of Commerce (Portland, OR), December 12, p. 2.

Van Metre, P. C., B.J. Mahler, T.J. Bashara, J.T. Wilson, and D.A. Johns. 2005. Parking lot sealcoat: a major source of polycyclic aromatic hydrocarbons (PAHs) in urban and suburban environments. Environmental Science & Technology 39: 5560-5566.

Washington, H. G. 1984. Diversity, biotic and similarity indices: a review with special relevance to aquatic ecosystems. Water Research 18:653–694.

Weisberg, S. B., J. A. Ranasinghe, D. M. Dauer, L. C. Schaffner, R. J. Diaz, and J. B. Frithsen. 1997. An estuarine benthic index of biotic integrity (B-IBI) for Chesapeake Bay estuaries. Estuaries and Coasts 20(1):149–158.

Wells, C. 1995. Impervious cover reduction study: final report. City of Olympia Public Works Department. Water Resources Program. Olympia, WA. 206 pp.

Wenz, P. S. 2008. Greening Codes. Planning Magazine 74(6):12–16.

Wright, J. F., M. T. Furse, and P. D. Armitage. 1993. RIVPACS—a technique for evaluating the biological quality of rivers in the UK. European Water Pollution Control 3(4):15–25.

Wright, T., J. Tomlinson, T. Schueler, K. Cappiella, A. Kitchell, and D. Hirschman. 2006. Direct and Indirect Impacts of Urbanization on Wetland Quality. Wetlands & Watersheds Article #1. Ellicott City, MD: Center for Watershed Protection.

Yoder, C. O. 1978. A proposal for the evaluation of water quality conditions in Ohio's rivers and streams. Division of Industrial Wastewater, Columbus, OH. 43 pp. NR 216, 2004, Storm Water Discharge Permits, Administrative Rules No. 583.

Yoder, C. O., and E. T. Rankin. 1995. Biological criteria program development and implementation in Ohio. Pp. 109–*152 In:* Biological Assessment and Criteria: Tools for Water Resource Planning and Decision Making. W. S. Davis and T. P. Simon (eds.). Boca Raton, FL: CRC Press.

Yoder, C. O., and E. T. Rankin. 1998. The role of biological indicators in a state water quality management process. Environmental Monitoring and Assessment 51(1–2):61–88.

3
Hydrologic, Geomorphic, and Biological Effects of Urbanization on Watersheds

A watershed is defined as the contributing drainage area connected to an outlet or waterbody of interest, for example a stream or river reach, lake, reservoir, or estuary. Watershed structure and composition include both naturally formed and constructed drainage networks, and both undisturbed areas and human dominated landscape elements. Therefore, the watershed is a natural geographic unit to address the cumulative impacts of urban stormwater. Urbanization has affected change to natural systems that tends to occur in the following sequence. First, land use and land cover are altered as vegetation and topsoil are removed to make way for agriculture or subsequently buildings, roads, and other urban infrastructure. These changes, and the introduction of a built drainage network, alter the hydrology of the local area, such that receiving waters in the affected watershed can experience radically different flow regimes than they did prior to urbanization. This altered hydrology, when combined with the introduction of pollutant sources that accompany urbanization (such as people, domesticated animals, industries, etc.), has led to water quality degradation of many urban streams.

This chapter first discusses the typical land-use and land-cover composition of urbanized watersheds. This is followed by a description of changes to the hydrologic and geomorphic framework of the watershed that result from urbanization, including altered runoff, streamflow mass transport, and stream-channel stability. The chapter then discusses the characteristics of stormwater runoff, including its quantity and quality from different land covers, as well as the characteristics of dry weather runoff. Finally, the effects of urbanization on aquatic ecosystems and human health are explored.

LAND-USE CHANGES

Land use has been described as the human modification of the natural environment into the built environment, such as fields, pastures, and settlements. Important characteristics of different land uses are the modified surface characteristics of the land and the activities that take place within that land use. From a stormwater viewpoint, land uses are usually differentiated by building density and comprised of residential, commercial, industrial, institutional, recreational, and open-space land uses, among others. Each of these land uses usually has distinct activities taking place within it that affect runoff quality. In addition, each land use is comprised of various amounts of surface land cover, such as roofs, roads, parking areas, and landscaped areas. The amount and type of each cover also affect the quality and quantity of runoff from urban areas. Changes in land use and in the land covers within the land uses associated with development and redevelopment are therefore important considerations when studying local receiving water problems, the sources of these problems within the watershed, and the stormwater control opportunities.

Land-Use Definitions

Although there can be many classifications of residential land use, a crude and common categorization is to differentiate by density. High-density residential land use refers to urban single-family housing at a density of greater than 6 units per acre, including the house, driveway, yards, sidewalks, and streets. Medium density is between 2 and 6 units per acre, while low density refers to areas where the density is 0.7 to 2 units per acre. Another significant residential land use is multiple-family housing for three or more families and from one to three stories in height. These units may be adjoined up-and-down, side-by-side, or front-and-rear.

There are a variety of commercial land uses common in the United States. The strip commercial area includes those buildings for which the primary function is the sale of goods or services. This category includes some institutional lands found in commercial strips, such as post offices, court houses, and fire and police stations. This category does not include warehouses or buildings used for the manufacture of goods. Shopping centers are another common commercial area and have the unique distinction that the related parking lot that surrounds the buildings is at least 2.5 times the area of the building roof area. Office parks are a land use on which non-retail business takes place. The buildings are usually multi-storied and surrounded by larger areas of lawn and other landscaping. Finally, downtown central business districts are highly impervious areas of commercial and institutional land use.

Industrial areas can be differentiated by the intensity of the industry. For example, "manufacturing industrial" is a land use that encompasses those buildings and premises that are devoted to the manufacture of products, with many of the operations conducted outside, such as power plants, steel mills, and cement plants. Institutional areas include a variety of buildings, for example schools, churches, and hospitals and other medical facilities that provide patient overnight care.

Roads constitute a very important land use in terms of pollutant contributions. The "freeway" land use includes limited-access highways and the interchange areas, including any vegetated rights-of-ways. Finally, there are a variety of open-space categories, such as cemeteries, parks, and undeveloped land. Parks include outdoor recreational areas such as municipal playgrounds, botanical gardens, arboretums, golf courses, and natural areas. Undeveloped lands are private or publicly owned with no structures and have a complete vegetative cover. This includes vacant lots, transformer stations, radio and TV transmission areas, water towers, and railroad rights-of-way.

The preceding land-use descriptions are the traditional categories that make up the vast majority of the land in U.S. cities. However, there are emerging categories of land use, such as those espoused under the term New Urbanism, which combine several area types (such as commercial and high-density residential areas). Although land use can be broadly and generally categorized, local variations can be extremely important such that locally available land-use data and definitions should always be used. For example, local planning agencies typically do not separate the medium-density residential areas into subcategories. However, this may be necessary to represent different development trends that have occurred with time, and to represent newly emerging types of land uses for an area. Box 3-1 discusses the subtle influence that tree canopy could have on the residential land-use classification.

Trends in Urbanization

Researchers at Columbia University (de Sherbinin, 2002) state that 83 percent of the Earth's land surface has been affected by human settlements and activities, with the urbanized areas comprising about 4 percent of the total land use of the world. Urban areas are expanding world-wide, especially in developing countries. The United Nations Population Division estimates suggest that the

---

BOX 3-1
The Role of Tree Cover in Residential Land Use

Figure 3-1 shows two medium-density residential neighborhoods, one older and one newer. Tree canopy is obviously different in each case, and it may have an effect on seasonal organic debris in an area and possibly on nutrient loads (although nutrient discharges appear to be more related to homeowner fertilizer applications). Increased tree canopy cover also has a theoretical benefit in reducing runoff quantities due to increased interception losses. In both cases, however, monitoring data to quantify these benefits are sparse. Xiao (1998) examined the effect urban tree cover had on the rainfall volume striking the ground in Sacramento, California. The results indicated that the type of tree or type of canopy cover affected the amount of rainfall reduction measured during a rain event, such that large broad-leafed evergreens and conifers reduced the rainfall that reached the ground by 36 percent, while medium-sized conifers and deciduous trees reduced the rainfall by 18 percent. Cochran (2008) compared the volume and intensity of rain that reached the ground in an open area (no canopy cover) versus two areas with intact canopy covers in Shelby County, Alabama, over a year. The sites were sufficiently close to each other to assume that the rainfall characteristics were the same in terms of the intensity and the variation of intensity and volume during the storm. Rainfall "throughfall" was reduced by about 13.5 percent during the spring and summer months when heavily wooded cover existed. The rainfall characteristics at the leafless tree sites (winter deciduous trees) were not significantly different from the parking lot control sites. In many locations around the county, very high winds are associated with severe storms, significantly decreasing the interception losses. Of course, mature trees are known to provide other benefits in urban areas, including shading to counteract stormwater temperature increases and massive root systems that help restore beneficial soil structure conditions. Additional research is needed to quantify the benefits of urban trees through a comprehensive monitoring program.

continues next page

BOX 3-1 Continued

FIGURE 3-1 Two medium-density residential areas (no alleys); the area below is older. SOURCE: Robert Pitt, University of Alabama.

world's population will become mostly urbanized by 2010, whereas only 37 percent of the world's population was urbanized in 1970. De Sherbinin (2002) concludes that although the extent of urban areas is not large when compared with other land uses (such as agriculture or forestry) their environmental impact is significant. Population densities in the cities are large, and their political, cultural, and economic influence is great. Most industrial activity is also located near cities. The influence of urban areas extends beyond their boundaries due to the need for large amounts of land for food and energy production, to generate raw materials for industry, for building water supplies, for obtaining other resources such as construction materials, and for recreational areas. One study estimated that the cities of Baltic Europe require from 500 to more than 1,000 times the urbanized land area (in the form of forests, agricultural, marine, and wetland areas) to supply their resources and to provide for waste disposal (de Sherbinin, 2002).

Currently, considerable effort is being spent investigating land-use changes world-wide and in the United States in support of global climate change research. The U.S. Geological Survey (USGS, 1999) has prepared many research reports describing these changes; Figure 3-2 shows the results for one study in the Chicago and Milwaukee areas, and Figure 3-3 shows the results for a study in the Chesapeake Bay area. These maps graphically show the dramatic rate of change in land use in these areas. The very large growth in urban areas during the 20 years between 1975 and 1995 is especially astonishing. By 1995, Milwaukee and Chicago's urbanized areas more than doubled in size from prior years. Even more rapid growth has occurred in the Washington, D.C.–Baltimore area.

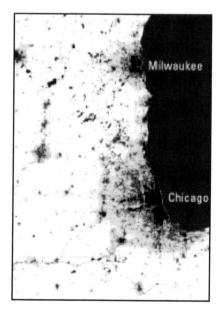

FIGURE 3-2 The extent of urban land in Chicago and Milwaukee in 1955 (black), 1975 (medium gray), and 1995 (light gray). SOURCE: USGS (1999).

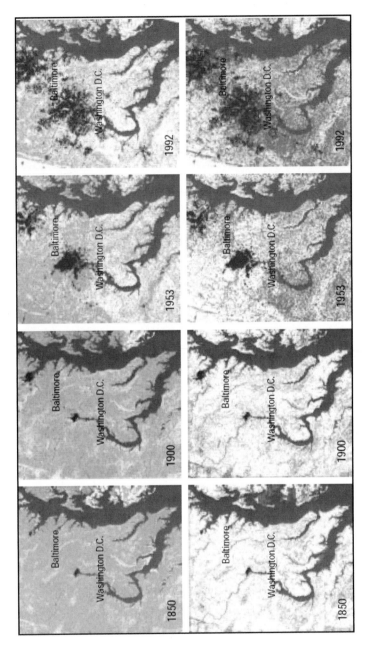

FIGURE 3-3 This series of maps compares changes in urban, agricultural, and forested lands in the Patuxent River watershed over the past 140 years. The top series shows the extent of urban areas (black) along with agriculture (gray), which was at its peak in the mid- to late 1800s. The bottom series show that urban (black) and forested land (gray) have increased since 1900. SOURCE: USGS (1999).

Many different metrics can be used to measure the rate of urbanization in the United States, including the number of housing starts and permits and the level of new U.S. development. The latter is tracked by the U.S. Department of Agriculture's (USDA) National Resources Inventory (USDA, 2000). The inventory, conducted every five years, covers all non-federal lands in the United States, which is 75 percent of the U.S. total land area. The inventory uses land-use information from about 800,000 statistically selected locations. From 1992 to 1997, about 2.2 million acres per year were converted from non-developed to developed status. According to the USDA (2000), the per capita developed land use (acres per person, a classical measure of urban sprawl) has increased in the United States between the years of 1982 and 1997 from about 0.43 to about 0.49 acres per person. The smallest amount of developed land used per person was for New York and Hawaii (0.15 acres), while the largest land consumption rate was for North Dakota, at about 10 times greater. Surprisingly, Los Angeles is the densest urban area in the country at 0.11 acres per person. The amount of urban sprawl is also directly proportionate to the population growth. According to Beck et al. (2003):

> In the 16 cities that grew in population by 10 percent or less between 1970 and 1990 (but whose population did not decline), developed area expanded 38 percent—more than in cities that declined in population but considerably less than in the cities where population increased more dramatically. Cities that grew in population by between 10 and 30 percent sprawled 54 percent on average. Cities that grew between 31 and 50 percent sprawled 72 percent on average. Cities that grew in population by more than 50 percent sprawled on average 112 percent. These findings confirm the common sense, but often unacknowledged proposition, that there is a strong positive relationship between sprawl and population growth.

In most areas, the per capita use of developed land has increased, along with the population growth. However, even some cities that had no population growth or had negative growth, such as Detroit, still had large amounts of sprawl (increased amounts of developed land used per person), but usually much less than cities that had large population growth. Los Angeles actually had an 8 percent decreased rate of land consumption per resident during this period, but the city still experienced tremendous growth in land area due to its very large population growth. The additional 3.1 million residents in the Los Angeles area during this time resulted in the development of almost an additional 400 square miles.

Land-Cover Characteristics in Urban Areas

As an area urbanizes, the land cover changes from pre-existing rural surfaces, such as agricultural fields or forests, to a combination of different surface types. In municipal areas, land cover can be separated into various common categories—pictured and described in Box 3-2—that include roofs, roads, parking areas, storage areas, other paved areas, and landscaped or undeveloped areas.

Most attention is given to impervious cover, which can be easily quantified for different types of land development. Given the many types of land cover described in Box 3-2, impervious cover is composed of two principal components: building rooftops and the transportation system (roads, driveways, and parking lots). Compacted soils and unpaved parking areas and driveways also have "impervious" characteristics in that they severely hinder the infiltration of water, although they are not composed of pavement or roofing material. In terms of total impervious area, the transportation component often exceeds the rooftop component (Schueler, 1994). For example, in Olympia, Washington, where 11 residential multifamily and commercial areas were analyzed in detail, the areas associated with transportation-related uses comprised 63 to 70 percent of the total impervious cover (Wells, 1995). A significant portion of these impervious areas—mainly parking lots, driveways, and road shoulders—experience only minimal traffic activity. Most retail parking lots are sized to accommodate peak parking usage, which occurs only occasionally during the peak holiday shopping season, leaving most of the area unused for a majority of the time. On the other hand, many business and school parking areas are used to their full capacity nearly every work day and during the school year. Other differences at parking areas relate to the turnover of parking during the day. Parked vehicles in business and school lots are mostly stationary throughout the work and school hours. The lighter traffic in these areas results in less vehicle-associated pollutant deposition and less surface wear in comparison to the greater parking turnover and larger traffic volumes in retail areas (Brattebo and Booth, 2003).

As described in Box 1-1, impervious cover is broken down into two main categories: directly connected impervious areas (or effective impervious area) and non-directly connected (disconnected) impervious areas (Sutherland, 2000; Gregory et al., 2005) (although it is recognized that these two states are end-members of a range of conditions). Directly connected impervious area includes impervious surfaces which drain directly to the sealed drainage system without flowing appreciable distances over pervious surfaces (usually a flow length of less than 5 to 20 feet over pervious surfaces, depending on soil and slope characteristics and the amount of runoff). Those areas are the most important component of stormwater runoff quantity and quality problems. Approximately 80 percent of directly connected impervious areas are associated with vehicle use such as streets, driveways, and parking (Heaney, 2000).

Values of imperviousness can vary significantly according to the method used to estimate the impervious cover. In a detailed analysis of urban imperviousness in Boulder, Colorado, Lee and Heaney (2003) found that hydrologic modeling of the study area resulted in large variations (265 percent difference)

BOX 3-2
Land Cover in Urban Areas

For any given land use, there is a range of land covers that are typical. Common land covers are described below, along with some indication of their contribution to stormwater runoff and their pollutant-generating ability.

Roofs. These are usually either flat or pitched, as both have significantly different runoff responses. Flat roofs can have about 5 to 10 mm of detention storage while pitched roofs have very little detention storage. Roofing materials are also usually quite different for these types of roofs, further affecting runoff quality. In addition, roof flashing and roof gutters may be major sources of heavy metals if made of galvanized metal or copper. Directly connected roofs have their roof drains efficiently connected to the drainage system, such as direct connections to the storm drainage itself or draining to driveways that lead to the drainage system. These directly connected roofs have much more of their runoff waters reaching the receiving waters than do partially connected roofs, which drain to pervious areas.

A directly connected roof drain

A disconnected roof drain (drains to pervious area)

Parking Areas. These can be asphalt or concrete paved (impervious surface) or unpaved (traditionally considered a pervious surface) and are either directly connected or drain to adjacent pervious areas. Areas that have rapid turnover of parked cars throughout the day likely have greater levels of contamination due to the frequent starting of the vehicles, an expected major source of pavement pollutants. Unpaved parking areas actually should be considered impervious surfaces, as the compacted surface does not allow any infiltration of runoff. Besides automobile activity in the parking areas, other associated activities contribute to contamination. For example, parked cars in disrepair awaiting service can contribute to parking area runoff contamination. In addition, maintenance of the pavement surface, such as coal-tar seal coating, can be significant sources of polycyclic aromatic hydrocarbons (PAHs) to the runoff.

continues next page

BOX 3-2 Continued

Paved parking area with frequent automobile movement

Contamination of paved parking areas due to commercial activities

Storage Areas. These can also be paved, unpaved, directly connected, or drained to pervious areas. As with parking areas, unpaved storage areas should not be considered pervious surfaces because the compacted material effectively hinders infiltration. Detention storage runoff losses from unpaved storage areas can be significant. In storage areas (especially in commercial and industrial land uses), activities in the area can have significant effects on runoff quality.

Contaminated paved storage area at vehicle junk yard

Heavy equipment storage area on concrete surface

Streets. Streets in municipal areas are usually paved and directly connected to the storm drainage system. In municipal areas, streets constitute a significant percentage of all impervious surfaces and runoff flows. Features that affect the quality of runoff from streets include the varying amounts of traffic on different roads and the amount and type of roadside vegetation. Large seasonal phosphorus loads can occur from residential roads in heavily wooded areas, for example.

Wide arterial street with little roadside vegetation (left) and narrow residential street with substantial vegetation (top, right)

Other Paved Areas. Other paved areas in municipal regions include driveways, playgrounds, and sidewalks. Depending on their slopes and local grading, these areas may drain directly to the drainage system or to adjacent pervious areas. In most cases, the runoff from these areas contributes little to the overall runoff for an area, and the runoff quality is of relatively better quality than from the other "hard" surfaces.

Landscaped and Turf Areas. Although these are some of the only true pervious surfaces in municipal areas, disturbed urban soils can be severely compacted, with much more reduced infiltration rates than are assumed for undisturbed regional soils. Besides the usually greater than expected quantities of runoff of pervious surfaces in urban areas, they can also contribute high concentrations of various pollutants. In areas with high rain intensities, erosion of sediment can be high from pervious areas, resulting in much higher concentrations of total suspended solids (TSS) than from paved areas. Also, landscaping chemicals, including fertilizers and pesticides, can be transported from landscaped urban areas. Undeveloped woods in urban areas can have close to natural runoff conditions, but many parks and other open-space areas usually have degraded runoff compared to natural conditions. Turf grass has unique characteristics compared to other landscaped areas in that the soil structure is usually more severely degraded compared to natural conditions. The normally shallower root systems are not as effective in restoring compacted soils and they can remain compacted due to some activities (pathways, parked cars, playing fields, etc.) that do not occur on areas planted with shrubs and trees.

continues on next page

BOX 3-2 Continued

Soil erosion from turf areas with fine-grained soils during periods of high rain intensities

Undeveloped Areas. Undeveloped areas in otherwise urban locations differ from natural areas. In many situations, they can be previously disturbed (cleared and graded) areas that have not been sold or developed. They may be overgrown with various local vegetation types that thrive in disturbed locations. In other situations, undeveloped areas may be small segments of natural areas that have not been disturbed or revegetated. In this case, their stormwater characteristics may approach natural conditions but still be degraded due to adjacent activities and atmospheric deposition.

SOURCE: Pitt and Voorhees (1995, 2002). Photographs courtesy of Robert Pitt, University of Alabama.

in the calculations of peak discharge when impervious surface areas were determined using different methods. They concluded that the main focus should be on effective impervious area (EIA) when examining the effects of urbanization on stormwater quantity and quality.

Runoff from disconnected impervious areas can be spread over pervious surfaces as sheet flow and given the opportunity to infiltrate before reaching the drainage system. Therefore, there can be a substantial reduction in the runoff volume and a delay in the remaining runoff entering the storm drainage collection system, depending on the soil infiltration rate, the depth of the flow, and the available flow length. Examples of disconnected impervious surfaces are roof-tops that discharge into lawns, streets with swales, and parking lots with runoff directed to adjacent open space or swales. From a hydrologic point of view, road-related imperviousness usually exerts a larger impact than rooftop-related imperviousness, because roadways are usually directly connected whereas roofs can be disconnected (Schueler, 1994).

Methods for Determining Land Use and Land Cover

Historically, land-use and land-cover information was acquired by a combi-nation of field measurements and aerial photographic analyses—methods that required intensive interpretation and cross validation to guarantee that the ana-lyst's interpretations were reliable (Goetz et al., 2003). Figure 3-4 is an example of a high-resolution panchromatic aerial photograph that was taken from an air-plane in Toronto and used for measurements of urban surfaces (Pitt and McLean, 1986). Most recently, satellite images have become available at high spatial resolution for many areas (<1 to 5 m resolution) and have the advantage of digital multi-spectral information more complete than even that provided by digital orthophotographs. Minnesota has one of the longest records (over 20 years) of continuously recorded statistics on land cover and impervious surfaces derived from satellite images—information which has been incorporated into the

FIGURE 3-4 Example of a high-resolution panchromatic aerial photograph of an industrial area used for measurements of ur-ban surfaces. SOURCE: Pitt and McLean (1986).

Minnesota Statewide Conservation and Preservation Plan. Some of the remaining problems to be overcome with satellite imagery include difficulties in obtaining consistent sequential acquisition dates, intensive computer processing time requirements, and large computer storage space requirements to store massive amounts of image information.

The recommended approach for conducting a survey of land uses and development characteristics (land cover and activities) for an area is to use both aerial photography and site surveys. Aerial photography has improved greatly in recent years, but it is still not suitable for obtaining all the information needed for developing a comprehensive stormwater management plan. Initially, aerial photos should be used to identify the locations and extents of the various land uses in the study area. Neighborhoods representing homogenous land uses should then be identified for site surveys. Usually, about 10 to 15 neighborhoods for each land use are sufficient for a community being studied (Burton and Pitt, 2002). After the field surveys are conducted, the aerials are again used to measure the actual areas associated with land surface cover. This information can be used with field survey data to separate the surfaces into the appropriate categories for analyses and modeling.

Box 3-3 presents a detailed study of land cover for several land uses in the southern United States using satellite imagery and ground surveys (Bochis, 2007; Bochis et al., 2008). The results presented here have been found to be broadly similar to other areas studied in the United States, although few studies have been as detailed, and there are likely to be regional differences.

The general conclusion of many land-use and land-cover studies is that in urban areas, the amount of impervious surfaces has increased since the early years of the 20th century because of the tendency toward increased automobile use and bigger houses, which is associated with an increase in the facilities necessary to accommodate them (wider streets, more parking lots, and garages). As shown in later sections of this report, the construction of impervious surfaces leads to multiple impacts on stream systems. Therefore, future development plans and water resource protection programs should consider reducing impervious cover in the potential expansion of communities. Wells (1995), Booth (2000), Stone (2004), and Gregory et al. (2005) show that reducing the size and dimensions of residential parcels, promoting cluster developments (clustered medium-density residential areas in conjunction with open space, instead of large tracts of low-density areas), building taller buildings, reducing the residential street width (local access streets), narrowing the width and/or building one-side sidewalks, reducing the size of paved parking areas to reflect the average parking needs instead of peak needs, and using permeable pavement for intermittent/overflow parking can reduce the traditional impervious cover in communities by 10 to 50 percent. Many of these benefits can also be met by paying better attention to how the pavement and roof areas are connected to the drainage system. Impervious surfaces that are "disconnected" by allowing their drainage water to flow to adjacent landscaped areas can result in reduced runoff quantities.

BOX 3-3
Land Use and Land Cover for the Little Shades Creek Watershed

Data collected by Bochis-Micu and Pitt (2005) and Bochis (2007) for the Little Shades Creek watershed near Birmingham, Alabama, were acquired using IKONOS satellite imagery (provided by the Jefferson County Storm Water Management Authority) as an alternative to classical aerial photography to map the characteristics of the land uses in the monitored watershed areas, supplemented with verified ground truth surveys. IKONOS is the first commercially owned satellite that provides 1-m-resolution panchromatic image data and 4-m multi-spectral imagery (Goetz et al., 2003).

This project was conducted to evaluate the effects of variable site conditions associated with each land-use category. About 12 homogeneous neighborhoods were investigated in each of the 16 major land uses in this 2,500-hectare watershed. Detailed land-cover measurements were made using a variety of techniques, as listed above, including field surveys for small details that were not visible with remote sensing tools (such as roof drain connectiveness, pavement texture, and landscaping maintenance practices). Each of these individual neighborhoods was individually modeled to investigate the resultant variability in runoff volume and pollutant discharges. These were statistically evaluated to determine if the land-use categories properly stratified these data by explaining significant fractions of the variability. Bochis-Micu and Pitt (2005) and Bochis (2007) concluded that land-use categories were an appropriate surrogate that can be used to describe the observed combinations of land surfaces. However, proper stormwater modeling should examine the specific land surfaces in each land-use category in order to better understand the likely sources of the pollutants and the effectiveness of candidate stormwater control measures (SCMs).

This watershed has an overall impervious cover of about 35 percent, of which about 25 percent is directly connected to the drainage system. Table 3-1 shows the average land covers for each of the surveyed land uses, along with the major source areas in each of the directly connected and disconnected impervious and pervious surface categories. The impervious covers include streets, driveways, parking, playgrounds, roofs, walkways, and storage areas. The directly connected areas are indicated as "connected" or "draining to impervious" and do not include the pervious area or the impervious areas that drain to pervious areas. As expected, the land uses with the least impervious cover are open space (vacant land, cemeteries, golf courses) and low-density residential, and the land uses with the largest impervious covers are commercial areas, followed by industrial areas. For a typical high-density residential land use in this region (having 15 or more units per hectare), the major land cover was found to be landscaped areas, subdivided into front- and backyard categories, while 25 percent of this land-use area is covered by impervious surfaces broken down into three major subcategories: roofs, streets, and driveways. The subareas making up each land use show expected trends, with roofs and streets being the predominant directly connected impervious covers in residential areas, and parking and storage areas also being important in commercial and industrial areas.

continues next page

BOX 3-3 Continued

TABLE 3-1 Little Shades Creek Watershed Land Cover Information (percent and the predominant land cover)

Land Use	Directly Connected Impervious Cover (%)	Disconnected Impervious Cover (%)	Pervious Cover (%)
High-Density Residential	14 (streets and roof)	10 (roofs)	76 (front and rear landscaping)
Medium-Density Residential (<1960 to 1980)	11 (streets and roofs)	8 (roofs)	81 (front and rear landscaping)
Medium-Density Residential (>1980)	14 (streets and roofs)	5 (roofs)	80 (front and rear landscaping)
Low-Density Residential	6 (streets)	4 (roofs)	89 (front and rear landscaping)
Apartments	21 (streets and parking)	22 (roofs)	58 (front and rear landscaping)
Multiple Families	28 (roofs, parking, and streets)	7 (roofs)	65 (front and rear landscaping)
Offices	59 (parking, streets, and roofs)	3 (parking)	39 (front and rear landscaping)
Shopping Centers	64 (parking, roofs, and streets)	4 (roofs)	31 (front landscaping)
Schools	16 (roofs and parking)	20 (playground)	64 (front and rear landscaping, large turf)
Churches	53 (parking and streets)	7 (parking)	40 (front landscaping)
Industrial	39 (storage, parking, and streets)	18 (storage and roofs)	44 (front and rear landscaping)
Parks	32 (streets and parking)	33 (playground)	34 (large turf and undeveloped)
Cemeteries	7 (streets)	15 (parking)	78 (large turf)
Golf Courses	2 (streets)	4 (roofs)	95 (large turf)
Vacant	5 (streets)	1 (driveways)	94 (undeveloped and large turf)

SOURCE: Bochis-Micu and Pitt (2005) and Bochis (2007). Reprinted, with permission, from Bochis (2007). Copyright 2007 by Celina Bochis.

HYDROLOGIC AND GEOMORPHIC CHANGES

The watershed provides an organizing framework for the management of stormwater because it determines the natural patterns of water flow as well as the constituent sediment, nutrient, and pollutant loads. In undeveloped watersheds, hillslope hydrologic flow-path systems co-evolve with microclimate, soils, and vegetation to form topographic patterns within which ecosystems are spatially arranged and adjusted to the long-term patterns of water, energy, and nutrient availability. The landforms that comprise the watershed include the network patterns of streams, rivers, and their associated riparian zones and floodplains, as well as component freshwater lakes, reservoirs, wetlands, and estuaries.

This section starts with a discussion of precipitation measurement and characteristics before turning to the typical changes in hydrology and geomorphology of the watershed brought on by urbanization. In both the terrestrial and aquatic phases, retention and residence time of sediment and solutes decreases with increasing flow volume and velocity. This results in relatively high retention and low export of water and nutrients in undeveloped watersheds compared to decreasing retention and greater pollutant export in disturbed or developed systems.

The Storm in Stormwater

The magnitude and frequency of stormwater discharges are not just determined by rainfall. Instead, they are the combined product of storm and interstorm characteristics, land use, the natural and built drainage system, and any stormwater control measures (SCMs) that have been implemented. The total volume and peak discharge of runoff, as well as the mobilization and transport of pollutants, are dependent on all aspects of the storm magnitude, catchment antecedent moisture conditions, and the interstorm period. Therefore, information on the frequency distribution of storm events and properties is an important aspect of understanding the distribution of pollutant concentrations and loads in stormwater discharges. In northern climates, runoff production from precipitation can be significantly delayed by the accumulation, ripening, and melt of snowpacks, such that much of the annual load of certain pollutants may be mobilized in peak flow from snowmelt events. Therefore, measurement of precipitation and potential accumulation in both liquid and solid form is critical for stormwater assessment.

Precipitation Measurements

Any given storm is characterized by the storm's total rainfall (depth), its duration, and the average and peak intensity. A *storm hyetograph* depicts measured precipitation depth (or intensity) at a precipitation gauge as a function of time; an example is shown in Figure 3-5. This figure illustrates the typical high degree of variability of precipitation over the total duration of a storm. In this example, the total storm depth is 50.9 mm, the duration is 19 hours, and the peak intensity is 0.56 mm/minute (peak depth of 2.79 mm divided by the measurement increment of 5 minutes). The average intensity is 0.045 mm/minute, quite a bit lower than the peak intensity, since the storm duration is punctuated by periods of low and no measurable precipitation.

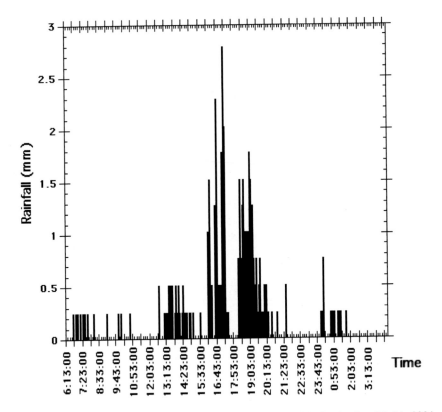

FIGURE 3-5 Example of a storm hyetograph at location RG2, September 20–21, 2001, Valley Creek watershed, Chester County, Pennsylvania. The time increment of measurement is 5 minutes, while the entire duration of this storm is about 16 hours.

In addition to measurements of individual storm events, precipitation data are routinely collected for longer time periods and compiled and analyzed annually when trying to understand local rainfall patterns and their impact on baseflow, water quality, and infrastructure design. Figure 3-6 shows the rainfall during 2007 at both humid (Baltimore) and arid (Phoenix) locations. Especially apparent in the Baltimore data is the fact that the majority of storm events are less than 20 mm in depth.

Several networks of precipitation gauges are available in the United States; gauge data are available online from the National Climatic Data Center (NCDC) (*http://ncdc.nws.noaa.gov*). High-resolution precipitation data (i.e., with measurement intervals of an hour or less) are typically not recorded except at primary weather service meteorological stations, while daily precipitation records are more extensively collected and available through the Cooperative Weather Observer Program (*http://www.nws.noaa.gov/om/coop/*). This distinction is important to stormwater managers because most stormwater applications require short-duration measurements or model results (minutes to hours). Fortunately, a combination of precipitation gauges and precipitation radar estimates are available to estimate precipitation depth and duration, as well as additional methods to estimate snowfall and snowpack water equivalent depth and conditions. (A thorough description of precipitation measurement by radar is given by Krajewski and Smith [2001]). While most of the conterminous United States is covered by NEXRAD radar for estimation of high-temporal-resolution precipitation at current resolutions of ~4 km, the radar backscatter information requires calibration and correction with precipitation gauge data, and satellite estimates

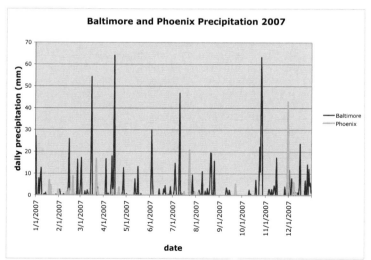

FIGURE 3-6 Daily precipitation totals for the Baltimore-Washington and Phoenix airports for 2007. SOURCE: Data from the National Weather Service.

of precipitation are generally not sufficiently reliable for stormwater applications. It goes without saying that the measurement, quality assurance, and maintenance of long-term precipitation records are both vital and nontrivial to stormwater management.

Precipitation Statistics

The basic characterization of precipitation is by depth-duration-frequency curves, which describe the return period, recurrence interval, and exceedance probability (terms all denoting frequency) of different precipitation intensities (depths) over different durations. The methodology for determining the curves is described in Box 3-4. Precipitation durations of interest in stormwater management range from a few minutes (important for determining peak discharge from small urban drainage areas) to a year (where the interest is in the total annual volume of runoff production). As an example, one might be interested in the return period of the 1-inch, 1-hour event, or the 1-inch, 24-hour event; the latter would have a much shorter return period, because accumulating an inch of rain over a day is much more common than accumulating the same amount over just an hour.

BOX 3-4
Determining Depth-Duration-Frequency Curves

Depth-duration-frequency curves are developed from precipitation records using either annual maximum data series or annual exceedance data series. Annual maximum data series are calculated by extracting the annual maximum precipitation depths of a chosen duration from a record. In cases where there are only a few years of data available (less than 20 to 25 years), then an annual exceedance series (a type of "partial duration series") for each storm duration can be calculated, where N largest values from N years are chosen. An annual maximum series excludes other extreme values of record that may occur in the same year. For example, the second highest value on record at an observing station may occur in the same year as the highest value on record but will not be included in the annual maximum series. The design precipitation depths determined from the annual exceedance series can be adjusted to match those derived from an annual maximum series using empirical factors (Chow et al., 1988; NOAA Atlas data series, see *http://www.weather.gov/oh/hdsc/currentpf.htm*, e.g., Bonnin et al., 2006). Hydrologic frequency analysis is then applied the data series to determine desired return periods by fitting a probability distribution to the data to determine the return periods[1] of interest. The process is repeated for other chosen storm durations.

---

[1]Analysis of annual maximum series produces estimates of the average period between years when a particular value is exceeded ("average recurrence interval"). Analysis of partial duration (annual exceedance) series gives the average period between cases of a particular magnitude ("annual exceedance probability"). The two results are numerically similar at rarer average recurrence intervals but differ at shorter average recurrence intervals (below about 20 years). NOAA (e.g., Bonnin et al., 2006) notes that the use of the terminology "average recurrence interval" and "annual exceedance probability" typically reflects the analysis of the two different series, but that sometimes the term "average recurrence interval" is used as a general term for ease of reference.

The National Weather Service has developed an online utility to estimate the return period for a range of depth–duration events for any place in the conterminous United States (*http://hdsc.nws.noaa.gov/hdsc/pfds/*). Figures 3-7 and 3-8 show examples of precipitation depth-duration-frequency curves for a humid location (Baltimore, Maryland) and an arid site (Phoenix, Arizona). As an illustration of the climatic influence on the depth-duration-frequency curves, the 2-year, 1-hour storm is associated with a depth of 1.2 inches of precipitation in Baltimore, whereas this same recurrence interval and duration are associated with a depth of only 0.6 inch of precipitation in Phoenix. Durations from 5 minutes to one day are shown because this is the range typically used in the design of stormwater management facilities. The shorter durations provide expected magnitude and frequency for brief but significant precipitation intensity peaks that can mobilize and transport large amounts of pollutants and erode soil, and they are used in high-resolution stormwater models. More commonly, however, stormwater regulations are written for 24-hour durations at 2-, 10-, 25-, 50-, or 100-year recurrence intervals.

Because storm magnitudes and frequencies vary by climatic region, it is reasonable to expect them to change during recurring climate events (e.g., El Niño) or over the long term by climate change. Alteration in convective precipitation by major urban centers has been documented for some time (Huff and Changnon, 1973). Some evidence exists that precipitation regimes are shifting systematically toward an increase in more intense rainfall events, which is consistent with modeled projections of global climate change increases in hydrologic extremes. Kunkel et al. (1999) analyzed precipitation data from 1,295 weather stations from 1931 to 1996 across the contiguous United States and found that storms with extreme levels of precipitation have increased in frequency. The analysis considered short-duration events (1, 3, and 7 days) of 1-year and 5-year return intervals. A linear trend analysis using Kendall's slope estimator statistic indicated that the overall trend in 7-day, 1-yr events for the conterminous United States is upward at a rate of about 3 percent per decade for 1931 to 1996; the upward trend in 7-day, 5-year events is about 4 percent per decade. These two time series are shown in Figure 3-9. An increased frequency of intense precipitation events will shift depth-frequency-duration curves for a given location, with a given return period being associated with a more intense event. Alternatively, the return period for a given intensity (or depth) of an event will be reduced if the event is occurring more frequently. In light of climate change, depth-duration-frequency curves will need to be updated regularly in order to ensure that stormwater management facilities are not underdesigned for an increasing intensity of precipitation. Additional implications of climate change for stormwater management are discussed in Box 3-5.

Precipitation Depth-Duration-Frequency - BWI

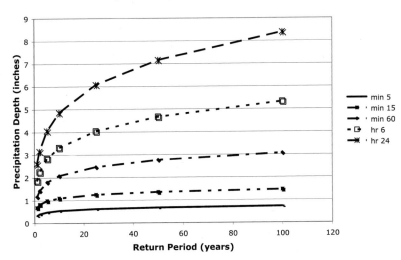

FIGURE 3-7 Depth-duration-frequency curves for Baltimore, Maryland. SOURCE: Data from the National Weather Service.

Precipitation Depth-Duration-Frequency - Phoenix Airport

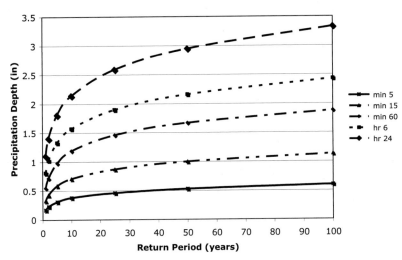

FIGURE 3-8 Depth-duration-frequency curves for Phoenix, Arizona. SOURCE: Data from the National Weather Service.

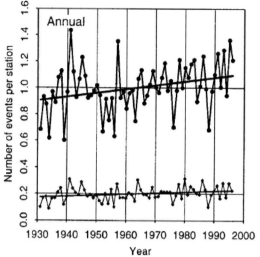

FIGURE 3-9 Nationally averaged annual U.S. time series of the number of precipitation events of 7-day duration exceeding 1-year (dots) and 5-year (diamonds) recurrence intervals. SOURCE: Reprinted, with permission, from Kunkel et al. (1999). Copyright 1999 by American Meteorological Society.

BOX 3-5
Climate Change and Stormwater Management

An ongoing report series issued by the U.S. Climate Change Science Program and the Subcommittee on Global Change Research summarizes the evidence for climate change to date and expected impacts of climate change, including impacts on the water resources sector (*http://www.climatescience.gov/*). According to the Intergovernmental Panel on Climate Change (IPCC 2007), annual precipitation will likely increase in the northeastern United States and will likely decrease in the southwestern United States over the next 100 years. In the western United States, precipitation increases are projected during the winter, whereas decreases are projected for the summer. As temperatures warm, precipitation will increasingly fall as rain rather than snow, and snow season length and snow depth are very likely to decrease in most of the country. More extreme precipitation events are also projected, which, when coupled with an anticipated increase in rain-on-snow events, would contribute to more severe flooding due to increases in extreme stormwater runoff.

The predictions for increases in the intensity and frequency of extreme events have significant implications for future stormwater management. First, many of the design standards currently in use will need to be revised, since they are based on historical data. For example, depth-duration-frequency curves used for design storm data will need to be updated, because the magnitude of the design storms will change. Even with revised design standards, in light of future uncertainty, new SCMs will need to be designed conservatively to allow for additional storage that will be required for regions with predicted trends in increased precipitation. In addition, existing SCM designs based on old standards may prove to be undersized in the future. Implementation of a monitoring program to check existing SCM inflows against original design inflows may be prudent to aid in judging whether retrofit of existing facilities or additional stormwater infrastructure is needed.

Design Storms

Given that only daily precipitation records are widely available, but short-duration data are required for stormwater analysis and prediction, *design storms* have been developed for the different regions of the United States by different state and federal resource agencies. A *design storm* is a specified temporal pattern of rainfall at a location, created using an overall storm duration and frequency relevant to the design problem at hand. Examples of design storms include the 24-hour, 100-year event for flood control and the 24-hour, 2-year event for channel protection. The magnitude of the design storm can be derived from data at a single gauge, or from synthesized regional data published by state or federal agencies. The simplest form of a design storm is a triangular hyetograph where the base is the duration and the height is adjusted so that the area under the curve equals the total precipitation. In instances where the hyetograph is to be used to estimate sequences of shorter duration intensities (i.e., minutes to a few hours) within larger duration events, depth-duration-frequency curve data can be used to synthesize a design storm hyetograph (see Chow et al., 1988). An example design storm for the 100-year storm event for St. Louis based on NOAA Atlas 14 depth-duration-frequency data is shown in Figure 3-10.

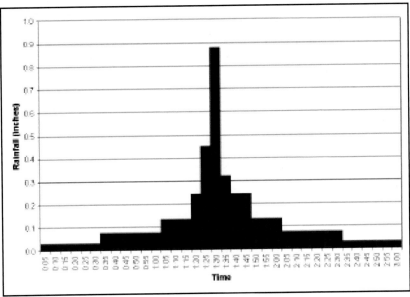

FIGURE 3-10 Hundred-year design storm for St. Louis based on NOAA Atlas 14 data.
SOURCE: Hoblit et al. (2004) based on data from Bonnin et al. (2003).

Conversion of Precipitation to Runoff

Dynamics of Watershed Flowpaths

Precipitation falling on the land surface is subject to evaporative loss to the atmosphere by vegetation canopy and leaf litter interception, evaporation directly from standing water on the surface and upper soil layers or impervious surfaces, and later transpiration through root uptake by vascular plants. Snowpack is also subject to sublimation (conversion of snow or ice directly to vapor), which results in the loss of a portion of the snow prior to melt. The rate of evaporative loss depends on local weather conditions (temperature, humidity, wind speed, solar radiation) and the rate and duration of precipitation. Precipitation (or snowmelt) in excess of interception and potential evaporative loss rates is then partitioned into infiltration and direct runoff.[1]

There is a gradation of flowpaths transporting water, sediment, and solutes through a watershed, ranging from rapid surface flowpaths through generally slower subsurface flowpaths. Residence times generally increase from surface to subsurface flowpaths, with rapid surface flow providing the major contribution to flood flow while subsurface flowpaths contribute to longer-term patterns of surface wetness. Watershed characteristics that influence the relative dominance of surface versus subsurface flowpaths include infiltration capacity as affected by land cover, soil properties, and macropores; subsurface structure or soil horizons with varying conductivity; antecedent soil moisture and groundwater levels; and the precipitation duration and intensity for a particular storm.

The distribution and activity of flowpaths result in changing patterns of soil moisture and groundwater depth, which result in patterns of soil properties, vegetation, and microbial communities. These ecosystem patterns, in turn, can have strong influences on the hydraulics of flow and biogeochemical transformations within the flowpaths, with important implications for sources, sinks, and transport of solutes and sediment in the watershed. Riparian areas, wetlands, and the benthos of streams and waterbodies are nodes of interaction between surface and groundwater flowpaths, yielding reactive environments in which "hot spots" of biogeochemical transformation develop (McClain et al., 2003). Thus, any alteration of surface and subsurface hydrologic flowpaths, for

---

[1] The term *runoff* is often used in two senses. For a given precipitation event, direct *storm runoff* refers to the rainfall (minus losses) that is shed by the landscape to a receiving waterbody. In an area of 100 percent imperviousness, the runoff nearly equals the rainfall (especially for larger storms). Over greater time and space scales, *surface water runoff* refers to streamflow passing through the outlet of a catchment, including base flow from groundwater that has entered the stream channel. The raw units of runoff in either case are volume per time, but the volumetric flowrate (discharge) is often divided by contributing area to express runoff in units of depth per time. In this way, unit runoff rates from various-sized watersheds can be compared to account for differences other than the contributing area.

example due to urbanization, not only alters the properties of soil and vegetation canopy but also reforms the ecosystem distribution of biogeochemical transformations.

Runoff Measurements

Surface water runoff for a given area is measured by dividing the discharge at a given point in the stream channel by the contributing watershed area. The basic variables describing channel hydraulics include width, mean depth, slope, roughness, and velocity. Channel discharge is the product of width, depth, and velocity and is typically estimated by either directly measuring each of these three components, or by development of a rating curve of measured discharge as a function of water depth, or stage relative to a datum, of the channel that is more easily estimated by a staff gauge or pressure transducer. The establishment of a gauging station to measure discharge typically requires a stable cross section so that stage can be uniquely related to discharge. Maintenance of reliable, long-term gauge sites is expensive and requires periodic remeasurement to update rating curves, as well as to remove temporary obstructions that may raise stage relative to unobstructed conditions.

Most stream gauging in the United States is carried out by the USGS, and can be found on-line at *http://waterdata.usgs.gov/nwis*. Recent reviews of standard methods of stream gauging and the status of the USGS stream gauging network are given by the USGS (1998) and the National Research Council (NRC, 2004). A major concern is the overall decline in the number of active gauges, particularly long-term gauges, as well as the representativeness of the stream gauge network relative to the needs of stormwater permitting. For example, restored streams typically lack any gauged streamflow or water quality information prior to or following restoration. This makes it very difficult to assess both the potential for successful restoration and whether project goals are met.

Support of existing and development of new gauges is often in collaboration through a co-funding mechanism with other agencies. Municipal co-funding for stations in support of National Pollutant Discharge Elimination System (NPDES) permitting is common and has tended to shift the concentration of active gauges toward more urban areas. Note that the USGS river monitoring system was originally designed for resource inventory, and therefore did not originally sample many headwater streams, particularly intermittent and ephemeral channels that are typically most proximal to stormwater discharges. While this is beginning to change with municipal co-funding, headwater streams are still underrepresented in the National Water Information System relative to their ecological significance.

Reliable records for stream discharge are vital because the frequency distribution and temporal trends of flows must be known to evaluate long-term loading to waterbodies. Magnitude and frequency analysis of sediment and other stream constituent loads consists of a transport equation as a function of discharge, integrated over the discharge frequency distribution (e.g., Wolman and Miller, 1960). Different constituent loads have different forms of dependency on discharge, but are often nonlinear such that long-term or expected loads cannot be simply evaluated from mean flow conditions. Similar to precipitation, discharge levels often follow an Extreme Value distribution, dependent on climate, land use, and hydrogeology, but which is typically dampened compared to precipitation due to the memory effects of subsurface storage and flows (e.g., Winter, 2007).

Impacts of Urbanization on Runoff

Shift from Infiltration and Evapotranspiration to Surface Runoff

Replacement of vegetation with impervious or hardened surfaces affects the hydrologic budget—the quantity of water moving through each component of the hydrologic cycle—in a number of predictable ways. As the percent of the landscape that is paved over or compacted is increased, the land area available for infiltration of precipitation is reduced, and the amount of stormwater available for direct surface runoff becomes greater, leading to increased frequency and severity of flooding. Reduced infiltration of precipitation leads to reduced recharge of the groundwater reservoir; absent new sources of recharge, this can lead to reduction in base flow of streams (e.g., Simmons and Reynolds, 1982; Rose and Peters, 2001). Vegetation removal also results in a lower amount of evapotranspiration compared to undeveloped land. This can have particularly profound hydrologic effects in those regions of the country where a significant percent of precipitation is evapotranspired, such as the arid Southwest (Ng and Miller, 1980). Figure 3-11 illustrates the changes to these components of the hydrologic budget as the percent of impervious area is increased.

It should be noted that the conversion in hydrology from infiltrated water to surface runoff following urbanization is not entirely straightforward in all cases. Leaking pressurized water supply pipes and sanitary sewers, subsurface discharge of septic system effluent (Burns et al., 2005), infiltration of stormwater from unlined detention ponds, and lawn irrigation can offset reduced infiltration of precipitation, such that stream baseflow levels may actually be increased, especially during low base flow months, when such effects would be most pronounced (Konrad and Booth, 2005; Meyer, 2005). Cracks in sealed surfaces can also provide concentrated points of infiltration (Sharp et al., 2006).

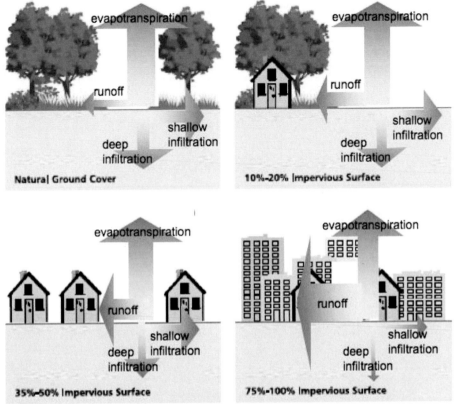

FIGURE 3-11 As land cover changes from vegetated and undeveloped (upper left) to developed with increased connected impervious surfaces (lower right), the partitioning of precipitation into other components of the hydrologic cycle is shifted. Evapotranspiration and shallow and deep infiltration are reduced, and surface runoff is increased. SOURCE: Adapted from the Federal Interagency Stream Restoration Working Group (FISRWG, 2000).

Relationship Between Imperviousness, Drainage Density, and Runoff

Excess runoff due to urbanization is a direct reflection of the land uses onto which the precipitation falls, as well as the presence of drainage systems that receive stormwater from many separate source areas before it enters receiving waters. Thus, a functional way of partitioning urban areas is by the nature of the impervious cover and by its connection to the drainage system, underlying the differentiation of total impervious area and effective impervious area discussed in Box 1-2.

As examples of how runoff changes with urbanization, Figure 3-12 shows daily stream flow values for a low-density suburban catchment and a high-density urban catchment in the Baltimore, Maryland area. The low-density site (Figure 3-12A) shows a strong seasonal signal and a marked decline in flow during an extreme drought in 2002. In contrast, the more densely urbanized catchment (Figure 3-12B) shows a much greater variability in flow that is dominated by impervious surface runoff, and a dampened response to the drought because natural groundwater flow is a much smaller component of the total discharge.

The percentage of time a discharge level is equaled or exceeded is displayed by flow duration curves, which show the cumulative frequency distributions of flows for a given duration. Examples for three catchments in the Baltimore area are given in Figure 3-13, showing the tendency for urban areas to produce high flows with much longer aggregate durations.

As another example of how runoff changes with imperviousness, a locally calibrated version of WinSLAMM was used to investigate the relationships between watershed and runoff characteristics for 125 individual neighborhoods in

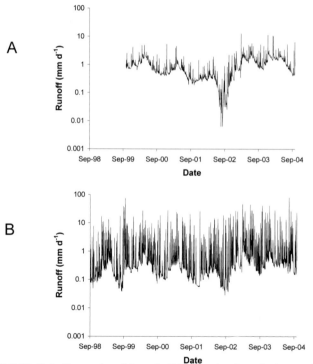

FIGURE 3-12 Daily time series of flows in (A) a low-density suburban and forested catchment (Baisman Run, *http://waterdata.usgs.gov/md/nwis/uv/?site_no=01583580*) and (B) a catchment dominated by medium- to high-density residential and commercial land uses (Dead Run, *http://waterdata.usgs.gov/md/nwis/uv/?site_no=01589330*). Both lie within the Piedmont physiographic province.

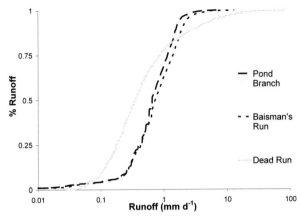

FIGURE 3-13 Flow duration curves for three watersheds with distinct land use in the Baltimore, Maryland area. Pond branch is a forested reference site, Baisman's Run is ex-urban, and Dead Run is urban. Urban areas have flashier runoff with greater frequency of low and high extreme flows.

Jefferson County, Alabama (Bochis-Micu and Pitt, 2005). Figure 3-14 shows the relationships between the directly connected impervious area values and the calculated volumetric runoff coefficient (R_v, which is the volumetric fraction of the rainfall that occurs as runoff), based on 43 years of local rain data. As expected, there is a strong relationship between these parameters for both sandy and clayey soil conditions. It is interesting to note that the R_v values are relatively constant until values of directly connected impervious cover of 10 to 15 percent are reached (at R_v values of about 0.07 for sandy soil areas and 0.16 for clayey soil areas)—the point where receiving water degradation typically has been observed to start (as discussed later in the chapter). The 25 to 30 percent directly connected impervious levels (where significant degradation is usually observed) is associated with R_v values of about 0.14 for sandy soil areas and 0.25 for clayey soil areas; this is where the curves start to greatly increase in slope.

Relationship Between Runoff and Rainfall Conditions

The runoff that results from various land uses also varies depending on rainfall conditions. For small rain depths, almost all the runoff originates solely from directly connected impervious areas, as disconnected areas have most of their flows infiltrated (Pitt, 1987). For larger storms, both directly connected

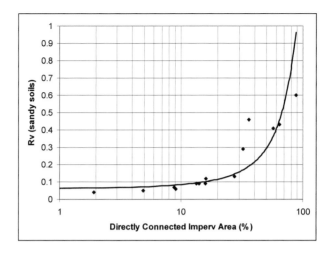

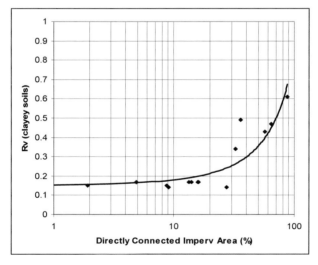

FIGURE 3-14 Relationships between the directly connected impervious area (%) and the calculated volumetric runoff coefficients (R_v) for sandy soil (top) and clayey soil (bottom). SOURCE: Reprinted, with permission, from Bochis-Micu and Pitt (2005). Copyright 2005 by Water Environment Federation, Alexandria, Virginia.

and disconnected impervious areas contribute runoff to the stormwater management system. For example, Figure 3-15 (created using WinSLAMM; Pitt and Voorhees, 1995) shows the relative runoff contributions for a large commercial/mall area in Hoover, Alabama, for different rains (Bochis, 2007). In this example, about 80 percent of the runoff originates from the parking areas for the smallest runoff-producing rains. This contribution decreases to about 55 percent at rain depths of about 0.5 inch (13 mm). This decrease in the importance of parking areas as a source of runoff volume is associated with an increase in runoff contributions from streets and directly connected roofs. In many areas, pervious areas are not hydrologically active until the rain depths are relatively large and are not significant runoff contributors until the rainfall exceeds about 25 mm for many land uses and soil conditions. However, compacted urban soils can greatly increase the flow contributions from pervious areas during smaller rains. Burges and others (1998), for example, found that more than 60 percent of the storm runoff in a suburban development in western Washington State originated from nominally "green" parts of the landscape, primarily lawns.

A further example illustrating the relationship between rainfall and runoff is given for Milwaukee, summarized in Box 3-6. The two curves of Figure 3-16 show a relationship between rainfall and runoff that is typical of urban areas. Very small storms (< 0.05 inch) produce no measurable runoff, owing to removal by interception storage and evaporation. Storms that deposit up to one inch of rainfall constitute about 90 percent of the storm events in this region, but these events produced only about 50 percent of the runoff. Very large events (greater than 3 inches of precipitation) are rare and destructive, accounting for only a few percent of the annual rainfall events.

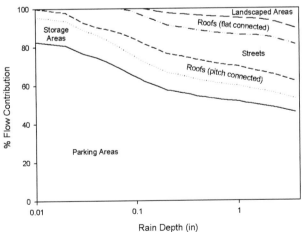

FIGURE 3-15 Surfaces contributing to runoff for a commercial/mall area. SOURCE: Reprinted, with permission, from Bochis (2007). Copyright 2007 by Celina Bochis.

BOX 3-6
Example Rainfall and Runoff Distributions

Figure 3-16 is an example of rainfall and runoff observed at Milwaukee, Wisconsin (Bannerman et al., 1983), as monitored during the Nationwide Urban Runoff Program (NURP) (EPA, 1983). This observed distribution is interesting because of the unusually large rains that occurred twice during the monitoring program. These two major rains would be in the category of design storms for conventional drainage systems. These plots indicate that these very large events, in the year they occurred, caused a measureable fraction of the annual pollutant loads and runoff volume discharges, but smaller events were responsible for the vast majority of the discharges. In typical years, when these rare design events do not occur, their pro-rated contributions would be even smaller.

More than half of the runoff from this typical medium-density residential area was associated with rain events that were smaller than 0.75 inch. Two large storms (about 3 and 5 inches in depth), which are included in the figure, distort this figure because, on average, the Milwaukee area only expects one 3.5-inch storm about every five years, and 5-inch storms even less frequently. If these large rains did not occur, such as for most years, then the significance of the smaller rains would be even greater. The figure also shows the accumulated mass discharges of different pollutants (suspended solids, chemical oxygen demand [COD], phosphates, and lead) monitored during the Milwaukee NURP project. When these figures are compared, it is seen that the runoff and pollutant mass discharge distributions are very similar and that variations in the runoff volume are much more important than variations in pollutant concentrations (the mass divided by the runoff volume) for determining pollutant mass discharges.

These rainfall and runoff distributions for Milwaukee can thus be divided into four regions:

• **Less than 0.5 inch.** These rains account for most of the events, but little of the runoff volume, and they are therefore easiest to control. They produce much less pollutant mass discharge and probably have less receiving water effects than other rains. However, the runoff pollutant concentrations likely exceed regulatory standards for several categories of critical pollutants (bacteria and some total recoverable heavy metals). They also cause large numbers of overflow events in uncontrolled combined sewers. These rains are very common, occurring once or twice a week (accounting for about 60 percent of the total rainfall events and about 45 percent of the total runoff-generating events), but they only account for about 20 percent of the

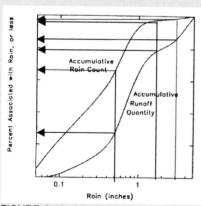
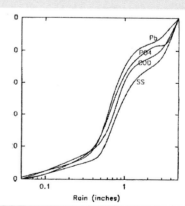

FIGURE 3-16 Milwaukee rainfall and runoff probability distributions, and pollutant mass discharge probability distributions (1981 to 1983). Rain count refers to the number of rain events. SOURCE: Data from Bannerman et al. (1983).

continues next page

BOX 3-6 Continued

annual runoff and pollutant discharges. Rains less than about 0.05 inch did not produce notice-able runoff.

• **0.5 to 1.5 inches**. These rains account for the majority of the runoff volume (about 50 percent of the annual volume for this Milwaukee example) and produce moderate to high flows. They account for about 35 percent of the annual rain events, and about 20 percent of the annual runoff events, by number. These rains occur on average about every two weeks from spring to fall and subject the receiving waters to frequent high pollutant loads and moderate to high flows.

• **1.5 to 3 inches**. These rains produce the most damaging flows from a habitat destruction standpoint and occur every several months (at least once or twice a year). These recurring high flows, which were historically associated with much less frequent rains, establish the energy gradient of the stream and cause unstable streambanks. Only about 2 percent of the rains are in this category, but they are responsible for about 10 percent of the annual runoff and pollutant discharges.

• Greater than **3 inches**. The rains in this category are included in design storms used for traditional drainage systems in Milwaukee, depending on the times of concentration and rain intensities. These rains occur only rarely (once every several years to once every several decades, or less frequently) and produce extremely large flows that greatly exceed the capacities of the storm drainage systems, causing extensive flooding. The monitoring period during the Milwaukee NURP was unusual in that two of these events occurred. Less than 2 percent of the rains were in this category (typically <<1 percent would be in this category), and they produced about 15 percent of the annual runoff quantity and pollutant discharges. However, when they do occur, substantial property and receiving water damage results (mostly associated with habitat destruction, sediment scouring, and the flushing of organisms great distances downstream and out of the system). The receiving water can conceivably recover naturally to pre-storm conditions within a few years. These storms, while very destructive, are sufficiently rare that the resulting environmental problems do not justify the massive controls that would be necessary to decrease their environmental effects.

Alteration of the Drainage Network

As shown in Figure 3-17, urbanization disrupts natural systems in ways that further complicate the hydrologic budget, beyond the imperviousness effects on runoff discussed earlier. As an area is urbanized, lower-order stream channels are typically re-routed or encased in pipes and paved over, resulting in a highly altered drainage pattern. The buried stream system is augmented by an extensive system of storm drains and pipes, providing enhanced drainage density (total lengths of pipes and channels divided by drainage area) compared to the natural system. Figure 3-18 shows how the drainage density of Baltimore today compares to the natural watershed before the modern stormwater system was fully developed. The artificial drainage system occupies a greater percentage of the landscape compared to natural conditions, permanently altering the terrestrial component of the hydrologic cycle.

The Urban Water Cycle

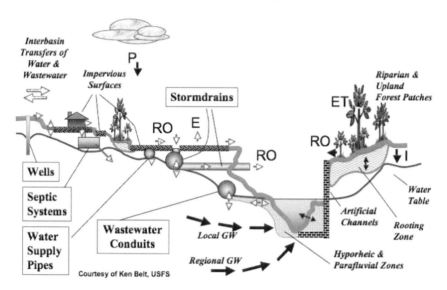

Courtesy of Ken Belt, USFS

FIGURE 3-17 Alteration of the natural hydrologic cycle by the presence of piped systems. Black arrows represent the natural system; outlined arrows indicate short-circuiting due to piped systems. Note that several elements of the water cycle shown in this diagram are not considered in this report, such as septic systems, interbasin transfers of water and wastewater, and the influence of groundwater withdrawals. SOURCE: Courtesy of Kenneth Belt, USDA Forest Service, Baltimore, Maryland.

Flowpaths are altered in other ways by urban infrastructure. Buried stormwater and sewer pipes can act as infiltration galleries for groundwater, causing shortened groundwater flowpaths between groundwater reservoirs and stream systems. Natural surface water pathways are often interrupted or reversed, as shown by the blue lines in Figure 3-19 for a drainage system in Baltimore. Understanding how the system operates as a whole can often require knowledge of the history of construction conditions and field verification of the actual flow paths.

Large-scale infrastructure such as dams, ponds, and bridges can also have a major impact on stormwater flows. Figure 3-20 illustrates the interruption of the drainage network by bridges and culverts, even in places where there have been attempts to keep excessive development out of the riparian corridor. Simulations and post-flood mapping in areas around Baltimore have shown that bridge abutments such as those shown in Figure 3-20 can slow down channel floodwaters during storms. This is because water backs up behind bridges constructed

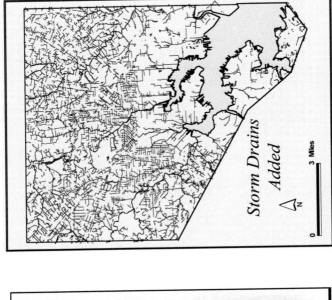

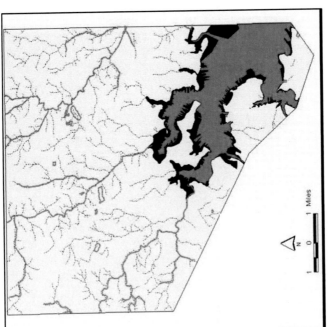

FIGURE 3-18 Baltimore City before and after development of its stormwater system. The left-hand panel shows first- and second-order streams lost to development. The right-hand panel shows the increase in drainage density resulting from construction of the modern storm-drain network. SOURCE: Courtesy of William Stack, Baltimore Department of Public Works.

FIGURE 3-19 Dead Run drainage system, Baltimore, Maryland. Black lines indicate surface (daylighted) drainage; dark grey indicates the subsurface storm-drain system. The surface drainage system is highly disconnected. From the coverage it is difficult to impossible to discern the flow direction of some of the surface drainage components. SOURCE: Reprinted, with permission, from Meierdierks et al. (2004). Copyright 2004 by the American Geophysical Union.

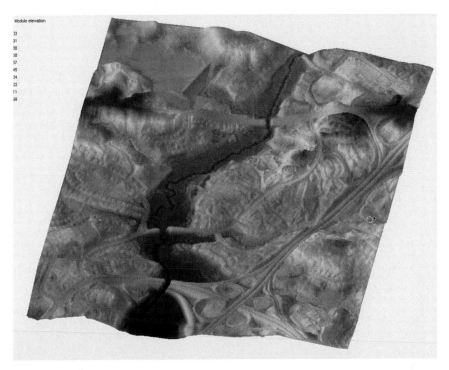

Module elevation
23
31
90
38
57
45
34
22
11
39

FIGURE 3-20 Shaded-relief lidar image of a portion of the Middle Patuxent River valley in Howard County, Maryland, showing the pervasive interruption of the drainage network by bridges and culverts, even in places where there is an attempt to keep excessive development out of the riparian corridor. SOURCE: Reprinted, with permission, from Miller, University of Maryland, Baltimore County. Copyright 2006 by Andrew J. Miller.

across the floodplain and spreads out over land surfaces and then flows back into channels as floodwaters subside. Although reducing the severity of downstream flooding, this phenomenon also interrupts the transport of sediment, leading to local zones of both enhanced deposition and downstream scour.

Alteration of Travel Times

The combination of impervious surface and altered drainage density provides significantly more rapid hydraulic pathways for stormwater to enter the nearest receiving waterbody compared to a natural landscape. This is illustrated quantitatively by Figure 3-21, which shows that the lag time—the difference in time between the center of mass of precipitation and the center of mass of the storm response hydrograph—is reduced for an urbanized landscape compared to a natural one.

The increase in surface runoff volumes and reduction in lag times between

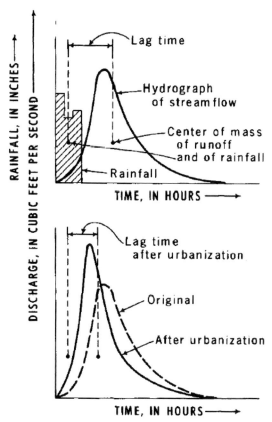

FIGURE 3-21 Illustration of the effect of urbanization on storm hydrograph lag time, the difference in time between the center of mass of rainfall and runoff response before and after urbanization. SOURCE: Leopold (1968).

precipitation and a waterbody's response give rise to greater velocities and volumetric discharges in receiving waters. Storm hydrographs in a developed setting peak earlier and higher than they do in undeveloped landscapes. This altered flow regime is of concern to property owners because upstream development can increase the probability of a flood-prone property being inundated. Properties in the floodplain and near stream channels are particularly susceptible to flooding from upstream development. Such increased flood risk is accompanied by associated potential property damages and costs of replacement or repair.

Various descriptors can be used to quantify the effects of urbanization on streamflow including flood frequency, flow duration, mean annual flood, discharge at bankfull stage, and frequency of bankfull stage. The "classic" view of urban-induced changes to runoff was presented by Leopold (1968), who provided several quantitative descriptors of the effects of urbanization on the mean annual flood. For example, Figure 3-22 shows the ratio of discharge before and after urbanization for the mean annual flood for a 1-square-mile area as a function of percentage of impervious area and percentage area served by a storm-drain system. This shows that for unsewered areas, increases from 0 to 100 percent impervious area will increase the peak discharge by a factor of 2.5. However, for 100 percent sewered areas, the ratio of peak discharges ranges from 1.7 to 8 for 0 to 100 percent impervious area. Clearly both impervious surfaces and the presence of a storm-drain system combine to increase discharge rates in receiving waters. Combining this information with regional flood frequency data, a discharge–frequency relationship can be developed that shows the expected discharge and recurrence interval for varying degrees of storm-drain coverage and impervious area coverage. An example is shown in Figure 3-23, using data from the Brandywine Creek watershed in Pennsylvania (Leopold, 1968). Bankfull flow for undeveloped conditions in general has a recurrence interval of about 1.5 years (which, in the particular case of the Brandywine, was 67 cubic feet per second); with 40 percent of the watershed area paved, this discharge would occur about three times as often.

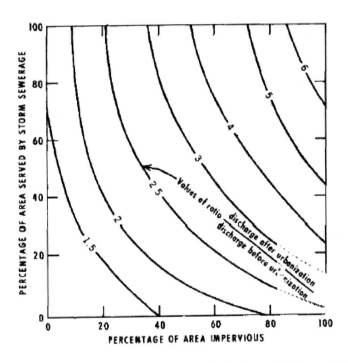

FIGURE 3-22 Ratio of peak discharge after urbanization to peak discharge before urbanization for the mean annual flood for a 1-square-mile drainage area, as a function of percent impervious surface and percent area drained by storm sewers. SOURCE: Leopold (1968).

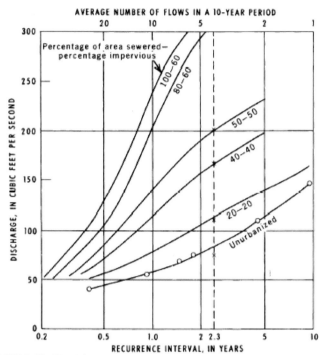

FIGURE 3-23 Flood frequency curves as a function of percent impervious area and percent of area serviced by storm sewers. The unurbanized data are from Brandywine Creek, Pennsylvania. SOURCE: Leopold (1968).

Over the past four decades since this first quantitative characterization of urban hydrology, a much greater variety of hydrologic changes resulting from urbanization has been recognized. Increases in peak discharge are certainly among those changes, and they will always gather attention because of their direct impact on human infrastructure and potential for more frequent and more severe flooding. The extended duration of flood flows, however, also affects natural channels because of the potential increase in erosion. Ecological effects of urban-altered flow regimes are even more diverse, because changes in the sequence and frequency of high flows, the rate of rise and fall of the hydrograph, and even the season of the year in which high flows can occur all have significant ecological effects and can be dramatically altered by watershed urbanization (e.g., Rose and Peters, 2001; Konrad et al., 2005; Roy et al., 2005; Poff et al., 2006).

***

The overarching conclusion of many studies is that the impact of urbanization on the hydrologic cycle is dramatic. Increased impervious area and drainage connectedness decreases stormwater travel times, increases flow rates and volumes, and increases the erosive potential of streams. The flooding caused by increased flows can be life-threatening and damaging to property. As described below, changes to the hydrologic flow regime also can have deleterious effects on the geomorphic form of stream channels and the stability of aquatic ecosystems. Although these impacts are commonly ignored in efforts to improve "water quality," they are inextricably linked to measured changes in water chemistry and must be part of any attempt to recover beneficial uses that have been lost to upstream urbanization.

Geomorphology

Watershed geomorphology is determined by the arrangement, interactions, and characteristics of component landforms, which include the stream-channel network, the interlocking network of ridges and drainage divides, and the set of hillslopes between the channel (or floodplain) and ridge. The stream and ridge systems define complementary networks, with the ridge (or drainage divide) network separating the drainage areas contributing to each reach in the stream network. At the hillslope scale, the ridges provide upper boundaries of all surface flowpaths which converge into the complementary stream reaches. A rich literature describes the topology and geometry of stream and ridge networks (e.g., Horton, 1945; Strahler, 1957, 1964; Shreve, 1966, 1967, 1969; Smart, 1968; Abrahams, 1984; Rodriguez-Iturbe et al., 1992).

Besides stream channels, a variety of other water features and landforms make up a watershed. Fresh waterbodies (ponds, lakes, and reservoirs) are typically embedded within the stream network, while wetlands may be either embedded within the stream network or separated and upslope from the channels. Estuaries represent the interface of the stream network with the open ocean. Additional fluvial and colluvial landforms include alluvial fans, landslide features, and a set of smaller features within or near the channels and floodplains including bar deposits, levees, and terraces. Each of these landforms are developed and maintained by the fluvial and gravitational transport and deposition of sediment, and are therefore potentially sensitive to disruption or alteration of flowpaths, hydrologic flow regimes, and sediment supply.

Stream Network Form and Ordering Methods

Most watersheds are fully convergent, with tributary streams combining to form progressively larger channels downstream. The manner is which streams from different source areas join to produce mainstreams strongly influences the propagation of stormwater discharge and pollutant concentrations, and the consequent level of ecological impairment in the aquatic ecosystem.

Methods for indexing the topologic position of individual reaches within the drainage network have been introduced by Horton (1945), Strahler (1957), Shreve (1966, 1967) and others. All stream topologic systems are dependent on the identification of first-order streams—the most upstream element of the network—and their lengths and drainage areas. Unfortunately, no universal standards exist to define where the stream head is located, or whether perennial, intermittent, and ephemeral channels should be considered in this determination. While this may seem like a trivial process, the identification and delineation of these sources effectively determines what lengths and sections of channels are defined to be *waterbodies* and, thus, the classification of all downstream waterbodies.

Nadeau and Rains (2007) have recently reviewed stream-channel delineation in the United States using standardized maps and hydrographic datasets to better relate climate to the extent of perennial, intermittent, and ephemeral channel types. Because this may influence the set of stream channels that are regulated by the Clean Water Act (CWA), it is the subject of current legal arguments in courts up to and including the Supreme Court (e.g., *Solid Waste Agency of Northern Cook County v. U.S. Army Corps of Engineers,* 531 U.S. 159 [2001], *John A. Rapanos et al. vs. United States* [U.S., No. 04-1034, 2005]). In addition to the stream-channel network, additional features (discussed below) that are embedded in or isolated from the delineated stream network (lakes, ponds, and wetlands) are subject to regulation under the CWA based on their proximity or interaction with the defined stream and river network. Therefore, definition of the extent and degree of connectivity of the nation's stream network, with an emphasis on the headwater region, is a critical determinant of the set of waterbodies that are regulated for stormwater permitting (Nadeau and Rains, 2007).

Stream Reach Geomorphology

Within the channel network, stream reaches typically follow a regular pattern of changes in downstream channel form. Hydraulic geometry equations, first introduced by Leopold and Maddock (1953), describe the gross geomorphic adjustment of the channel (in terms of average channel depth and width) to the flow regime and sometimes the sediment supply. Within this general pattern of larger flows producing larger channels, variations in channel form are evident, particularly the continuum among straight, meandering, or braided patterns. These forms are dependent on the spatial and temporal patterns of discharge, sediment supply, transport capacity, and roughness elements.

Most natural channels have high width-to-depth ratios and complexity of channel form compared with engineered channels. Meanders are ubiquitous self-forming features in channels, created as accelerated flow around the outside of the meander entrains and transports more sediment, producing greater flow depths and eroding the bank, while decelerated flow on the inside of the meander results in deposition and the formation of lower water depth and bank gradients. These channels typically show small-scale alternation between larger cross sections with lower velocities and defining pools, and smaller cross sections with higher velocity flow in riffles. Braided streams form repeated subdivision and reconvergence of the channel in multiple threads, with reduced specific discharge compared to a single channel. Natural obstructions including woody debris, boulders, and other large (relative to channel dimensions) features all contribute to hydraulic and habitat heterogeneity. The complexity of these channel patterns contributes to hydraulic roughness, further dissipating stream energy by increasing the effective wetted perimeter of the channel through a valley and deflecting flow between banks.

Embedded Standing Waterbodies

Standing waterbodies include natural, constructed, or modified ponds and lakes and are characterized by low or near-zero lateral velocity. They can be thought of as extensions of pools within the drainage network, although there is no clear threshold at which a pool can be defined as a pond or lake. When they are embedded within the channel network, they are characterized with much greater cross-sectional area (width x depth), lower surface water slopes (approaching flat), and lower velocities than a stream reach of similar length. Therefore, standing waterbodies function as depositional zones, have higher residence times, and provide significant storage of water, sediment, nutrients, and other pollutants within the stream network.

Riparian Zone

The riparian area is a transitional zone between the active channel and the uplands, and between surface water and groundwater. The area typically has shallower groundwater levels and higher soil moisture than the surrounding uplands, and it may support wetlands or other vegetation communities that require higher soil moisture. Riparian zones provide important ecosystem functions and services, such as reducing peak flood flows, transforming bioavailable nutrients into organic matter, and providing critical habitat.

In humid landscapes, a functioning riparian area commonly is an area where shallow groundwater forms discharge seeps, either directly to the surface and then to the stream channel or through subsurface flowpaths to the stream channel. The potential for high moisture and organic material content provides an environment conducive to anaerobic microbial activity, which can provide effective sinks for inorganic nitrogen by denitrification, reducing nitrate loading to the stream channel. However, the width of the effective riparian zone depends on local topographic gradients, hydrogeology, and the channel geomorphology (Lowrance et al., 1997). In steeply incised channels and valleys, or areas with deeper flowpaths, the riparian zone may be narrow and relatively well drained.

Under more arid conditions with lower groundwater levels, riparian areas may be the only areas within the watershed with sufficient moisture levels to support significant vegetation canopy cover, even though saturation conditions may occur only infrequently. Subsurface flowpaths may be oriented most commonly from the channel to the bed and banks, forming the major source of recharge to this zone from periodic flooding. In monsoonal climates in the U.S. southwest, runoff generated in mountainous areas or from storm activity may recharge riparian aquifers well downstream from the storm or snowmelt activity. Channelization that reduces this channel-to-riparian recharge may significantly impair riparian and floodplain ecosystems that provide critical habitat and other ecosystem services (NRC, 2002).

Floodplains

The presence and distribution of alluvial depositional zones, including floodplains, is dependent on the distribution and balance of upstream sediment sources and sediment transport capacity, the temporal and spatial variability of discharge, and any geological structural controls on valley gradient. Lateral migration of streams contributes to the development of floodplains as the outer bank of the migrating channel erodes sediment and deposition occurs on the opposite bank. This leads to channels that are closely coupled to their floodplains, with frequent overbank flow and deposition, backwater deposits, wetlands, abandoned channels, and other floodplain features. During major events, overbank flooding and deposition adds sediment, nutrients, and contaminants to the floodplain surface, and may significantly rework preexisting deposits and drainage patterns. Constructional landforms typical of urbanized watersheds, such as levees, tend to disconnect streams from their floodplains.

Changes in Geomorphology from Urbanization

Changes to channel morphology are among the most common and readily visible effects of urban development on natural stream systems (Booth and Henshaw, 2001). The actions of deforestation, channelization, and paving of the uplands can produce tremendous changes in the delivery of water and sediment into the channel network. In channel reaches that are alluvial, the responses are commonly rapid and often dramatic. Channels widen and deepen, and in some cases may incise many meters below the original level of their beds. Alternatively, channels may fill with sediment derived from farther upstream to produce a braided form where a single-thread channel previously existed.

The clearest single determinant of urban channel change is the alteration of the hydrologic response of an urban watershed, notably the increase in streamflow discharges. Increases in runoff mobilize sediment both on the land surface and within the stream channel. Because transport capacity increases nonlinearly with flow velocity (Vogel et al., 2003), much greater transport will occur in higher flow events. However, the low frequency of these events may result in decreasing cumulative sediment transport during the highest flows, as described by standard magnitude and frequency analysis (Wolman and Miller, 1960), such that the maximum time-integrated sediment transport occurs at moderate flows (e.g., bankfull stage in streams in the eastern United States).

If the increase in sediment transport caused by the shift in the runoff regime is not matched by the sediment supply, channel bed entrenchment and bank erosion and collapse lead to a deeper, wider channel form. Increases in channel dimensions caused by increased discharges have been observed in numerous studies, including Hammer (1972), Hollis and Luckett (1976), Morisawa and LaFlure (1982), Neller (1988), Whitlow and Gregory (1989), Moscrip and Montgomery (1997), and Booth and Jackson (1997). MacRae (1997), reporting on other studies, found that channel cross-sectional areas began to enlarge after about 20 to 25 percent of the watershed was developed, commonly corresponding to about 5 percent impervious cover. When the watersheds were completely developed, the channel enlargements were about 5 to 7 times the original cross-sectional areas. Channel widening can occur for several decades before a new equilibrium is established between the new cross-section and the new discharges.

Construction results in a large—but normally temporary—increase in sediment load to aquatic systems (e.g., Wolman and Schick, 1967). Indeed, erosion and sediment transport rates can reach up to more than 200 Mg/ha/yr on construction sites, which is well in excess of typical rates from agricultural land (e.g., Wolman and Schick, 1967; Dunne and Leopold, 1978); rates from undisturbed and well-vegetated catchments are negligible (e.g., <<1 Mg/ha/yr). The increased sediment loads from construction exert an opposing tendency to channel erosion and probably explain much of the channel narrowing or shallowing that is sometimes reported (e.g., Leopold, 1973; Nanson and Young, 1981; Ebisemiju, 1989; Odemerho, 1992).

Additional sediment is commonly introduced into the channel network by the erosion of the streambank and bed itself. Indeed, this source can become the largest single fraction of the sediment load in an urbanizing watershed (Trimble, 1997). For example, Nelson and Booth (2002) reported on sediment sources in the Issaquah Creek watershed, an urbanizing, mixed-use watershed in the Pacific Northwest. Human activity in the watershed, particularly urban development, has caused an increase of nearly 50 percent in the annual sediment yield, now estimated to be 44 tons/km^2/yr[1]. The main sources of sediment in the watershed are landslides (50 percent), channel-bank erosion (20 percent), and stormwater discharges (15 percent).

The higher flow volumes and peak discharge caused by urbanization also tend to preferentially remove fine-grained sediment, leaving a lag of coarser bed material (armoring) or removing alluvial material entirely and eroding into the geologic substrate (Figure 3-24). The geomorphic outcome of these changes is a

FIGURE 3-24 Example of an urban stream that has eroded entirely through its alluvium to expose the underlying consolidated geologic stratum below (Thornton Creek, Seattle, Washington). SOURCE: Derek Booth, Stillwater Sciences, Inc.

mix of erosional enlargement of some stream reaches, significant sedimentation in others, and potential head-ward downcutting of tributaries as discharge levels from small catchments increase. The collective effects of these processes have been described by Walsh et al. (2005) as "Urban Stream Syndrome," which includes not only the visible alteration of the physical form of the channel but also the consequent deterioration of stream biogeochemical function and aquatic trophic structures.

Other changes also accompany these geomorphic changes. Episodic inundation of the floodplain during floods may be reduced in magnitude and frequency, depending on the increases in peak flow relative to the deepening and resultant increase in flow capacity of the channel. Where deeply entrenched, this channel morphology will lower the groundwater level adjacent to the channel. The effectiveness of riparian areas in filtering or removing solutes is thus reduced because subsurface water may reach the channel only by flowpaths now well below the organic-rich upper soil horizons. Removal of fine-grained stream-bottom sediment, or erosion down to bedrock, may substantially lower the exchange of stream water with the surrounding groundwater of the hyporheic zone.

In addition to these indirect effects on the physical form of the stream channel, urbanization also commonly modifies streams directly to improve drainage, applying channel straightening and lining to reduce friction, increase flow capacity, and stabilize channel position (Figure 3-25). The enlarged and often

FIGURE 3-25 Example of a channelized urban stream for maximized flood conveyance and geomorphic stability (Los Angeles River, California). SOURCE: Robert Pitt, University of Alabama.

lined and straightened stream-channel cross section reduces the complexity of the bed and the contact between the stream and floodplain, and increases transport efficiency of sediment and solutes to receiving waterbodies. Enhanced sedimentation of receiving waterbodies, in turn, reduces water clarity, decreases depth, and buries the benthic environment.

POLLUTANT LOADING IN STORMWATER

Hydrologic flowpaths influence the production of particulate and dissolved substances on the land surface during storms, as well as their delivery to the stream-channel network. Natural watersheds typically develop a sequence of ecosystem types along hydrologic flowpaths that utilize available limiting resources, thereby reducing their export farther downslope or downstream, such that in-stream concentrations of these nutrients are low. As a watershed shifts from having mostly natural pervious surfaces to having heavily disturbed soils, new impervious surfaces, and activities characteristic of urbanization, the runoff quality shifts from relatively lower to higher concentrations of pollutants. Anthropogenic activities that can increase runoff pollutant concentrations in urban watersheds include application of chemicals for fertilization and pest control; leaching and corrosion of pollutants from exposed materials; exhaust emissions, leaks from, and wear of vehicles; atmospheric deposition of pollutants; and inappropriate discharges of wastes.

Most lands in the United States that have been developed were originally grasslands, prairies, or forest. About 40 percent of today's developed land went through an agricultural phase (cropland or pastureland) before becoming urbanized, while more than half of today's developed land area has been a direct conversion of natural covers (USDA, 2000). Agricultural land can produce stormwater runoff with high pollutant concentrations via soil erosion, the introduction of chemicals (fertilizers, pesticides, and herbicides), animal operations that are major sources of bacteria in runoff, and forestry operations. Indeed, urban stormwater may actually have slightly lower pollutant concentrations than other nonpoint sources of pollution, especially for sediment and nutrients. The key difference is that urban watersheds produce a much larger annual volume of runoff waters, such that the mass of pollutants discharged is often greater following urbanization. Some of the complex land-use–pollutant loading relationships are evident in Box 3-7, which shows the measured annual mass loads of nitrogen and phosphorus in four small watersheds of different land use monitored as part of the Baltimore Long-Term Ecological Research program. Depending on the nutrient and the year, the agricultural and urban watersheds had a higher nutrient export rate than the forested subwatershed.

Table 3-3 summarizes the comparative importance of urban land-use types in generating pollutants of concerns that can impact receiving waters (Burton and Pitt, 2002). This summary is highly qualitative and may vary depending on the site-specific conditions, regional climate, activities being conducted in each land use, and development characteristics. It should be noted that the rankings in Table 3-3 are relative to one another and classified on a per-unit-area basis. Furthermore, this table shows the parameters for each land-use category, such that the effects for a community at large would be dependent on the areas of each land use shown. Thus, although residential land use is shown to be a relatively smaller source of many pollutants, it is the largest fraction of land use in most communities, typically making it the largest stormwater source on a mass pollutant discharge basis. Similarly, freeway, industrial, and commercial areas can be very significant sources of many stormwater problems, and their discharge significance is usually much greater than their land area indicates. Construction sites are usually the overwhelming source of sediment in urban areas, even though they make up very small areas of most communities. A later table (Table 3-4) presents observed stormwater discharge concentrations for selected constituents for different land uses.

The following section describes stormwater characteristics associated with urbanized conditions. At any given time, parts of an urban area will be under construction, which is the source of large sediment losses, flow path disruptions, increased runoff quantities, and some chemical contamination. Depending on the time frame of development, increased stormwater pollutant discharges associated with construction activities may last for several years until land covers are stabilized. After construction has been completed, the characteristics of urban

BOX 3-7
Comparison of Nitrogen and Phosphorus Export from
Watersheds with Different Land Uses

Land use is a significant influence on nutrient export as controlled by impervious area, sanitary infrastructure, fertilizer application, and other determinants of input, retention, and stormwater transport. Tables 3-2A and 3-2B compare dissolved nitrate, total nitrogen, phosphate, and total phosphorus loads exported from forest catchments with catchments in different developed land uses studied by the Baltimore Ecosystem Study (Groffman et al., 2004). Loads were computed with the Fluxmaster system (Schwarz et al., 2006) from weekly samples taken at outlet gauges. In these sites in Baltimore County, the forested catchment, Pond Branch, has nitrogen loads one to two orders of magnitude lower than the developed catchments. Baisman Run, with one-third of the catchment in low-density, septic-served suburban land use, has nitrogen export exceeding Dead Run, an older, dense urban catchment. In this case, nutrient load does not follow the direct variation of impervious area because of the switch to septic systems and greater fertilizer use in lower density areas. However, Figure 3-26 shows that as impervious area increases, a much greater proportion of the total nitrogen load is discharged in less frequent, higher runoff events (Shields et al., 2008), reducing the potential to decrease loads by on-site SCMs. Total phosphorus loads were similarly as low (0.05–0.6 kg P/ha/yr) as nitrogen in the Pond Branch catchment (forest) over the 2000–2004 time period, and one to two orders of magnitude lower compared to agricultural and residential catchments.

It should be noted that specific areal loading rates, even in undeveloped catchments, can vary significantly depending on rates of atmospheric deposition, disturbance, and climate conditions. The hydrologic connectivity of nonpoint pollutant source areas to receiving waterbodies is also a critical control on loading in developed catchments (Nadeau and Rains, 2007) and is dependent on both properties of the pollutant as well as the catchment hydrology. For example, total nitrogen was high in both the agricultural and low-density suburban sites. Total phosphorus, on the other hand, was high in the Baltimore Ecosystem Study agricultural catchment, but close to the concentration of the forest site in the low-density suburban site serviced by septic systems. This is because septic systems tend to retain phosphorus, while septic wastewater nitrogen is typically nitrified in the unsaturated zone below a spreading field and efficiently transported in the groundwater to nearby streams.

TABLE 3-2A   Dissolved Nitrate and Total Nitrogen Export Rates from Forest and Developed Land-Use Catchments in the Baltimore Ecosystem Study

Catchment	Land Use	Nitrate (kg N/ha/yr)			Total N (kg N/ha/yr)		
		2000	2001	2002	2000	2001	2002
Pond Branch	Forest	0.11	0.08	0.04	0.47	0.37	0.17
McDonogh	Agriculture	17.6	12.9	4.3	20.5	14.5	4.5
Baisman Run	Mixed Forest and Suburban	7.2	3.8	1.5	8.2	4.2	1.7
Dead Run	Urban	3.0	2.9	2.9	5.6	5.3	4.2

TABLE 3-2B Dissolved Phosphate and Total Phosphorus Export Rates from Forest and Developed Land-Use Catchments in the Baltimore Ecosystem Study

Catchment	Land Use	Phosphate (kg P/ha/yr)			Total P (kg P/ha/yr)		
		2000	2001	2002	2000	2001	2002
Pond Branch	Forest	0.009	0.007	0.003	0.02	0.014	0.006
McDonogh	Agriculture	0.12	0.080	0.022	0.22	0.14	0.043
Baisman Run	Mixed Forest and Suburban	0.009	0.005	0.002	0.02	0.011	0.004
Dead Run	Urban	0.039	0.037	0.03	0.10	0.10	0.08

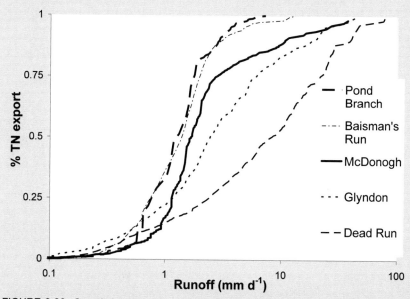

FIGURE 3-26 Cumulative transport of total nitrogen at increasing flow levels from catchments in Baltimore City and County including dominantly forest (Pond Branch), low-density development on septic systems and forest (Baisman Run), agricultural (McDonogh), medium-density suburban development on separate sewers (Glyndon), and higher-density residential, commercial, and highway land cover (Dead Run). SOURCE: Reprinted, with permission, from Shields et al. (2008). Copyright 2008 by the American Geophysical Union.

TABLE 3-3 Relative Sources of Parameters of Concern for Different Land Uses in Urban Areas

Problem Parameter	Residential	Commercial	Industrial	Freeway	Construction
High flow rates (energy)	Low	High	Moderate	High	Moderate
Large runoff volumes	Low	High	Moderate	High	Moderate
Debris (floatables and gross solids)	High	High	Low	Moderate	High
Sediment	Low	Moderate	Low	Low	Very high
Inappropriate discharges (mostly sewage and cleaning wastes)	Moderate	High	Moderate	Low	Low
Microorganisms	High	Moderate	Moderate	Low	Low
Toxicants (heavy metals/organics)	Low	Moderate	High	High	Moderate
Nutrients (eutrophication)	Moderate	Moderate	Low	Low	Moderate
Organic debris (SOD and DO)	High	Low	Low	Low	Moderate
Heat (elevated water temperature)	Moderate	High	Moderate	High	Low

NOTE: SOD, sediment oxygen demand; DO, dissolved oxygen.
SOURCE: Summarized from Burton and Pitt (2002), Pitt et al. (2008), and CWP and Pitt (2008).

runoff are controlled largely by the increase in volume and the washoff of pollutants from impervious surfaces. Stormwater in this phase is associated with increases in discharges of most pollutants, but with less sediment washoff than from construction and likely less sediment and nutrient discharges compared to any pre-urbanization agricultural operations (although increased channel erosion may increase the mass of sediment delivered in this phase; Pitt et al., 2007). A third significant urban land use is industrial activity. As described later, industrial site stormwater discharges are highly variable, but often greater than other land uses.

Construction Site Erosion Characteristics

Problems associated with construction site runoff have been known for many years. More than 25 years ago, Willett (1980) estimated that approximately 5 billion tons of sediment reached U.S. surface waters annually, of which 30 percent was generated by natural processes and 70 percent by human activities. Half of this 70 percent was attributed to eroding croplands. Although construction occurred on only about 0.007 percent of U.S. land in the 1970s, it accounted for approximately 10 percent of the sediment load to all U.S. surface waters and equaled the combined sediment contributions of forestry, mining, industrial, and commercial land uses (Willett, 1980).

Construction accounts for a much greater proportion of the sediment load in urban areas than it does in the nation as a whole. This is because construction sites have extremely high erosion rates and because urban construction sites are efficiently drained by stormwater drainage systems installed early during the construction activities. Construction site erosion losses vary greatly throughout the nation, depending on local rain, soil, topographic, and management conditions. As an example, the Birmingham, Alabama, area may have some of the highest erosion rates in the United States because of its combination of very high-energy rains, moderately to severely erosive soils, and steep slopes (Pitt et al., 2007). The typically high erosion rates mean that even a small construction project may have a significant detrimental effect on local waterbodies.

Extensive evaluations of urban construction site runoff problems have been conducted in Wisconsin for many years. Data from the highly urbanized Menomonee River watershed in southeastern Wisconsin indicate that construction sites have much greater potentials for generating sediment and phosphorus than do other land uses (Chesters et al., 1979). For example, construction sites can generate approximately 8 times more sediment and 18 times more phosphorus than industrial sites (the land use that contributes the second highest amount of these pollutants) and 25 times more sediment and phosphorus than row crops. In fact, construction sites contributed more sediment and phosphorus to the Menomonee River than any other land use, although in 1979, construction comprised only 3.3 percent of the watershed's total land area. During this early study, construction sites were found to contribute about 50 percent of the suspended sediment and total phosphorus loading at the river mouth (Novotny and Chesters, 1981).

Similar conclusions were reported by the Southeastern Wisconsin Regional Planning Commission (SEWRPC) in a 1978 modeling study of the relative pollutant contributions of 17 categories of point and nonpoint pollution sources to 14 watersheds in the southeast Wisconsin regional planning area (SEWRPC, 1978). This study revealed construction as the first or second largest contributor of sediment and phosphorus in 12 of the 14 watersheds. Although construction occupied only 2 percent of the region's total land area in 1978, it contributed approximately 36 percent of the sediment and 28 percent of the total phosphorus load to inland waters, making construction the region's second largest source of these two pollutants. The largest source of sediment was estimated to be cropland; livestock operations were estimated to be the largest source of phosphorus. By comparison, cropland comprised 72 percent of the region's land area and contributed about 45 percent of the sediment and only 11 percent of the phosphorus to regional watersheds. When looking at the Milwaukee River watershed as a whole, construction is a major sediment contributor, even though the amount of land under active construction is very low. Construction areas were estimated to contribute about 53 percent of the total sediment discharged by the

Milwaukee River in 1985 (total sediment load of 12,500 lb/yr), while croplands contributed 25 percent, streambank erosion contributed 13 percent, and urban runoff contributed 8 percent.

Line and White (2007) recently investigated runoff characteristics from two similar drainage areas in the Piedmont region of North Carolina. One of the drainage areas was being developed as part of a large residential subdivision during the course of the study, while the other remained forested or in agricultural fields. Runoff volume was 68 percent greater for the developing compared with the undeveloped area, and baseflow as a percentage of overall discharge was approximately zero compared with 25 percent for the undeveloped area. Overall annual export of sediment was 95 percent greater for the developing area, while export of nitrogen and phosphorus forms was 66 to 88 percent greater for the developing area.

The biological stream impact of construction site runoff can be severe. For example, Hunt and Grow (2001) describe a field study conducted to determine the impact to a stream from a poorly controlled construction site, with impact being measured via fish electroshocking and using the Qualitative Habitat Evaluation Index. The 33-acre construction site consisted of severely eroded silt and clay loam subsoil and was located within the Turkey Creek drainage, Scioto County, Ohio. The number of fish species declined (from 26 to 19) and the number of fish found decreased (from 525 to 230) when comparing upstream unimpacted reaches to areas below the heavily eroding site. The Index of Biotic Integrity and the Modified Index of Well-Being, common fisheries indexes for stream quality, were reduced from 46 to 32 and 8.3 to 6.3, respectively. Upstream of the area of impact, Turkey Creek had the highest water quality designation available, but fell to the lowest water quality designation in the area of the construction activity. Water quality sampling conducted at upstream and downstream sites verified that the decline in fish diversity was not due to chemical affects alone.

Municipal Stormwater Characteristics

The suite of stormwater pollutants generated by municipal areas is expected to be much more diverse than construction sites because of the greater variety of land uses and pollutant source areas found within a typical city. Many studies have investigated stormwater quality, with the U.S. Environmental Protection Agency's (EPA's) NURP (EPA, 1983) being the best known and earliest effort to collect and summarize these data. Unfortunately, NURP was limited in that it did not represent all areas of the United States or all important land uses. More recently, the National Stormwater Quality Database (NSQD) (CWP and Pitt, 2008; Pitt et al., 2008 for version 3) has been compiling data from the EPA's NPDES stormwater permit program for larger Phase I municipal separate storm sewer system (MS4) communities. As a condition of their Phase I permits, municipalities were required to establish a monitoring program to characterize their local stormwater quality for their most important land uses discharging to the

MS4. Although only a few samples from a few locations were required to be monitored each year in each community, the many years of sampling and large number of communities has produced a database containing runoff quality information for nearly 8,000 individual storm events over a wide range of urban land uses. The NSQD makes it possible to statistically compare runoff from different land uses for different areas of the country.

A number of land uses are represented in MS4 permits and also the database, including industrial stormwater discharges to an MS4. However, there is no separate compilation of quantitative mass emissions from specific industrial stormwater sources that may have been collected under industrial permit monitoring efforts. The observations in the NSQD were all obtained at outfall locations and do not include snowmelt or construction erosion sources. The most recent version of the NSQD contains stormwater data from about one-fourth of the total number of communities that participated in the Phase I NPDES stormwater permit monitoring activities. The database is located at *http://unix.eng.ua.edu/~rpitt/Research/ms4/mainms4.shtml.*

Table 3-4 is a summary of *some* of the stormwater data included in NSQD version 3, while Figure 3-27 shows selected plots of these data. The table describes the total number of observations, the percentage of observations above the detection limits, the median, and coefficients of variation for a few of the major constituents for residential, commercial, industrial, institutional, freeway, and open-space land-use categories, although relatively few data are available for institutional and open-space areas. It should be noted that even if there are significant differences in the median concentrations by the land uses, the range of the concentrations within single land uses can still be quite large. Furthermore, plots like Figure 3-27 do not capture the large variability in data points observed at an individual site.

There are many factors that can be considered when examining the quality of stormwater, including land use, geographical region, and season. The following is a narrative summary of the entire database and may not reflect information in Table 3-4 and Figure 3-29, which show only subsets of the data. First, statistical analyses of variance on the NSQD found significant differences among land-use categories for all of the conventional constituents, except for dissolved oxygen. (Turbidity, total solids, total coliforms, and total *E. coli* did not have enough samples in each group to evaluate land-use differences.) Freeway sites were found to be significant sources of several pollutants. For example, the highest TSS, COD, and oil and grease concentrations (but not necessarily the highest *median* concentrations) were reported for freeways. The median ammonia concentration in freeway stormwater is almost three times the median concentration observed in residential and open-space land uses, while freeways have the lowest orthophosphate and nitrite–nitrate concentrations—half of the concentration levels that were observed in industrial land uses.

In almost all cases the median metal concentrations at the industrial areas were about three times the median concentrations observed in open-space and residential areas. The highest lead and zinc concentrations (but not necessarily

TABLE 3-4 Summary of Selected Stormwater Quality Data Included in NSQD, Version 3.0

	TSS (mg/L)	COD (mg/L)	Fecal Colif. (mpn/100 mL)	Nitrogen, Total Kjeldahl (mg/L)	Phosphorus, Total (mg/L)	Cu, Total (µg/L)	Pb, Total (µg/L)	Zn, Total (µg/L)
All Areas Combined (8,139)								
Coefficient of variation (COV)	2.2	1.1	5.0	1.2	2.8	2.1	2.0	3.3
Median	62.0	53.0	4300	1.3	0.2	15.0	14.0	90.0
Number of samples	6780	5070	2154	6156	7425	5165	4694	6184
% samples above detection	99	99	91	97	97	88	78	98
All Residential Areas Combined (2,586)								
COV	2.0	1.0	5.7	1.2	1.6	1.9	2.1	3.3
Median	59.0	50.0	4200	1.2	0.3	12.0	6.0	70.0
Number of samples	2167	1473	505	2026	2286	1640	1279	1912
% samples above detection	99	99	89	98	98	88	77	97
All Commercial Areas Combined (916)								
COV	1.7	1.0	3.0	0.9	1.2	1.4	1.7	1.4
Median	55.0	63.0	3000	1.3	0.2	17.9	15.0	110.0
Number of samples	843	640	270	726	920	753	605	839
% samples above detection	97	98	89	98	95	85	79	99
All Industrial Areas Combined (719)								
COV	1.7	1.3	6.1	1.1	1.4	2.1	2.0	1.7
Median	73.0	59.0	2850	1.4	0.2	19.0	20.0	156.2
Number of samples	594	474	317	560	605	536	550	596
% samples above detection	98	98	94	97	95	86	76	99

All Freeway Areas Combined (680)

COV	2.6	1.0	2.7	1.2	5.2	2.2	1.1	1.4
Median	53.0	64.0	2000	1.7	0.3	17.8	49.0	100.0
Number of samples	360	439	67	430	585	340	355	587
% samples above detection	100	100	100	99	99	99	99	99

All Institutional Areas Combined (24)

COV	1.1	1.0	0.4	0.6	0.9	0.6	1.0	0.9
Median	18.0	37.5	3400	1.1	0.2	21.5	8.6	198.0
Number of samples	23	22	3	22	23	21	21	22
% samples above detection	96	91	100	91	96	57	86	100

All Open-Space Areas Combined (79)

COV	1.8	0.6	1.2	1.2	1.5	0.4	0.9	0.8
Median	10.5	21.3	2300	0.4	0.0	9.0	48.0	57.0
Number of samples	72	12	7	50	77	15	10	16
% samples above detection	97	83	100	96	97	47	20	50

NOTE: The complete database is located at: http://unix.eng.ua.edu/~rpitt/Research/ms4/mainms4.shtml. SOURCE: National Stormwater Quality Database.

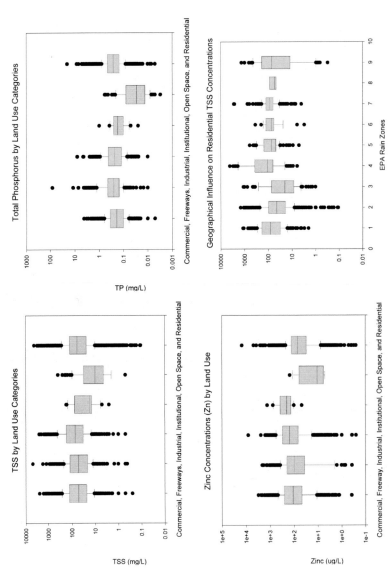

FIGURE 3-27 Grouped box and whisker plots of data from the NSQD. The median values are indicated with the horizontal line in the center of the box, while the ends of the box represent the 25th and 75th percentile values. The whickers extend to the 5th and 95th percentile values, and values outside of these extremes are indicated with separate dots. These groups were statistically analyzed and were found to have at least one group that is significantly different from the other groups. The ranges of the values in each group are large, but a very large number of data points is available for each group. The grouping of the data into these categories helps explain much of the total variability observed, and the large number of samples in each category allows suitable statistical tests to be made. Many detailed analyses are presented at the NSQD website (Maestre and Pitt, 2005).

the highest *median* concentrations) were found in industrial land uses. Lower concentrations of TDS, five-day biological oxygen demand (BOD_5), and fecal coliforms were observed in industrial land-use areas. By contrast, the highest concentrations of dissolved and total phosphorus were associated with residential land uses. Fecal coliform concentrations are also relatively high for residential and mixed residential land uses. Open-space land-use areas show consistently low concentrations for the constituents examined. There was no significant difference noted for total nitrogen among any of the land uses monitored.

In terms of regional differences, significantly higher concentrations of TSS, BOD_5, COD, total phosphorus, total copper, and total zinc were observed in arid and semi-arid regions compared to more humid regions. In contrast, fecal coliforms and total dissolved solids were found to be higher in the upper Midwest. More detailed discussions of land use and regional differences in stormwater quality can be found in Maestre et al. (2004) and Maestre and Pitt (2005, 2006). In addition to the information presented above, numerous researchers have conducted source area monitoring to characterize sheet flows originating from urban surfaces (such as roofs, parking lots, streets, landscaped areas, storage areas, and loading docks). The reader is referred to Pitt et al. (2005a,b,c) for much of this information.

Industrial Stormwater Characteristics

The NSQD, described earlier, has shown that industrial-area stormwater has higher concentrations of most pollutants compared to other land uses, although the variability is high. MS4 monitoring activities are usually conducted at outfalls of drainage systems containing many individual industrial activities, so discharge characteristics for specific industrial types are rarely available. This discussion provides some additional information concerning industrial stormwater beyond that included in the previous discussion of municipal stormwater. In general, there is a profound lack of data on industrial stormwater compared to municipal stormwater, and a correspondingly greater uncertainty about industrial stormwater characteristics.

The first comprehensive monitoring of an industrial area that included stormwater, dry weather base flows, and snowmelt runoff was conducted in selected Humber River catchments in Ontario (Pitt and McLean, 1986). Table 3-5 shows the annual mass discharges from the monitored industrial area in North York, along with ratios of these annual discharges compared to discharges from a mixed commercial and residential area in Etobicoke. The mass discharges of heavy metals, total phosphorus, and COD from industrial stormwater are three to six times that of the mixed residential and commercial areas.

Hotspots of contamination on industrial sites are a specific concern. Stormwater runoff from "hotspots" may contain loadings of hydrocarbons, trace metals, nutrients, pathogens and/or other toxicants that are greater than the loadings of "normal" runoff. Examples of these hotspots include airport de-icing

TABLE 3-5 Annual Storm Drainage Mass Discharges from Toronto-Area Industrial Land Use

Measured Parameter	Units	Annual Mass Discharges from Industrial Drainage Area	Stormwater Annual Discharge Ratio (Industrial Compared to Residential and Commercial Mixed Area)
Runoff volume	m³/hr/yr	6,580	1.6
total solids	kg/ha/yr	6,190	2.8
total phosphorus	kg/ha/yr	4,320	4.5
TKN	g/ha/yr	16,500	1.2
COD	kg/ha/yr	662	3.3
Cu	g/ha/yr	416	4.0
Pb	g/ha/yr	595	4.2
Zn	g/ha/yr	1,700	5.8

SOURCE: Pitt and McLean (1986).

facilities, auto recyclers/junkyards, commercial garden nurseries, parking lots, vehicle fueling and maintenance stations, bus or truck (fleet) storage areas, industrial rooftops, marinas, outdoor transfer facilities, public works storage areas, and vehicle and equipment washing/steam cleaning facilities (Bannerman et al., 1993; Pitt et al., 1995; Claytor and Schueler, 1996).

The elevated concentrations and mass discharges found in stormwater at industrial sites are associated with both the activities that occur and the materials used in industrial areas, as discussed in the sections that follow.

Effects of Roofing Materials on Stormwater Quality

The extensive rooftops of industrial areas can be a significant pollutant source area. A summary of the literature on roof-top runoff quality, including both roof surfaces and underlying materials used as subbases (such as treated wood), is presented in Table 3-6. Good (1993) found that dissolved metals' concentrations and toxicity remained high in roof runoff samples, especially from rusty galvanized metal roofs during both first flush and several hours after a rain has started, indicating that metal leaching continued throughout the events and for many years. During pilot-scale tests of roof panels exposed to rains over a two-year period, Clark et al. (2008) found that copper roof runoff concentrations for newly treated wood panels exceeded 5 mg/L (a very high value compared to median NSQD stormwater concentrations of about 10 to 40 µg/L for different land uses) for the first nine months of exposure. These results indicated that copper continued to be released from these wood products at levels high enough to exceed aquatic life criteria for long periods after installation, and were not simply due to excess surface coating washing off in the first few storms after installation.

TABLE 3-6 Roof Runoff Analysis—A Literature Summary

Roof Type	Location	Water Quality Parameter								Reference
		Cu (µg/L)	Zn (µg/L)	Pb (µg/L)	Cd (µg/L)	As (µg/L)	pH	NH$_4^+$ (mg/L)	NO$_3^-$ (mg/L)	
Polyester	Duebendorf, Switzerland	6817	2076	510	3.1					Boller (1997)
Tile		1905	360	172	2.1					
Flat gravel		140	36	22	0.2					
Plywood w/ roof paper/tar	Washington	166^T/128^D	877^T/909^D	11^T/<5^D			4.3			Good (1993)
Rusty galvanized metal		20^T/2^D	12200^T/1190^D	302^T/35^D			5.9			
Old metal w/Al paint		11^T/7^D	0^D	10^T/<5^D			4.8			
Flat tar surface w/fibrous reflective Al paint		25^T/14^D	1980^T/1610^D	10^T/5^D			4.1			
New anodized Al		16^T/7^D	297^T/257^D 101^T/82^D	15^T/<5^D			5.9			
Zinc-galvanized Fe	Dunedin City, New Zealand	560 µg/g	5901 µg/g	670 µg/g						Brown & Peake (2006)
Fe-Zn sheets	Ile-Ife, Nigeria						6.77	0.06	1.52	Adeniyi and Olabanji (2005)
Concrete slate tiles							7.45	0.05	3.34	
Asbestos cement sheets							7.09	0.06	2.26	
Aluminum sheets							6.68	0.05	6.18	
Cu panels	Munich, Germany	200–11100					6.7–7.0			Athanasiadis et al. (2006)
Galvanized metals (primarily Galvalume®)	Seattle, WA	10–1400	420–14700	ND						Tobiason (2004)
CCA wood Untreated wood	Florida					1200–1800 2–3				Khan et al. (2006)

Note: D, dissolved; T, total; ND, not detected.
SOURCE: Reprinted, with permission, from Clark et al. (2008). Copyright 2008 by American Society of Civil Engineers.

Traditional unpainted or uncoated hot-dip galvanized steel roof surfaces can also produce very high zinc concentrations. For example, pilot-scale tests by Clark et al. (2008) indicated that zinc roof runoff concentrations were 5 to 30 mg/L throughout the first two years of monitoring of a traditional galvanized metal panel. These are very high values compared to median stormwater values reported in the NSQD of 60 to 300 μg/L for different land uses. Factory-painted aluminum–zinc alloy panels had runoff zinc levels less than 250 μg/L, which were closer to the reported NSQD median values. The authors concluded that traditional galvanized metal roofing contributed the greatest concentrations of many metals and nutrients. In addition, they found that pressure-treated and waterproofed wood contributed substantial copper loads. The potential for nutrient release exists in many of the materials tested (possibly as a result of phosphate washes and binders used in the material's preparation or due to natural degradation).

Other researchers have investigated the effects of industrial rooftop runoff on receiving waters and biota. Bailey et al. (1999) investigated the toxicity to juvenile rainbow trout of runoff from British Columbia sawmills and found that much of the toxicity may have been a result of divalent cations on the industrial site, especially zinc from galvanized roofs.

Effects of Pavement and Pavement Maintenance on Stormwater Quality

Pavement surfaces can also have a strong influence on stormwater runoff quality. For example, concrete is often mixed with industrial waste sludges as a way of disposing of the wastes. However, this can lead to stormwater discharges high in toxic compounds, either due to the additives themselves or due to the mobilization of compounds via the additives. Salaita and Tate (1998) showed that high levels of aluminum, iron, calcium, magnesium, silicon, and sodium were seen in the cement-waste samples. A variety of sands, including waste sands, have been suggested as potential additives to cement and for use as fill in roadway construction. Wiebusch et al. (1998) tested brick sands and found that the higher the concentration of alkaline and alkaline earth metals in the samples, the more easily the heavy metals were released. Pitt et al. (1995) also found that concrete yard runoff had the highest toxicity (using Microtox screening methods) observed from many source areas, likely due to the elevated pH (about 11) from the lime dust washing off from the site.

The components of asphalt have been investigated by Rogge et al. (1997), who found that the majority of the elutable organic mass that could be identified consisted of *n*-alkanes (73 percent), carboxylic acids such as *n*-alkanoic acids (17 percent), and benzoic acids. PAHs and thiaarenes were 7.9 percent of the identifiable mass. In addition, heterocyclic aromatic hydrocarbons containing sulfur (S-PAH), such as dibenzothiophene, were identified at concentration levels similar to that of phenanthrene. S-PAHs are potentially mutagenic (similar to other PAHs), but due to their slightly increased polarity, they are more soluble in water and more prone to aquatic bioaccumulation.

In addition to the bitumens and asphalts, other compounds are added to paving (and asphaltic roofing) materials. Chemical modifiers are used both to increase the temperature range at which asphalts can be used and to prevent stripping of the asphalt from the binder. A variety of fillers may also be used in asphalt pavement mixtures. The long-term environmental effects of these chemicals in asphalts are unknown. Reclaimed asphalt pavements have also been proposed for use as fill materials for roadways. Brantley and Townsend (1999) performed a series of leaching tests and analyzed the leachate for a variety of organics and heavy metals. Only lead from asphalt pavements reclaimed from older roadways was found to be elevated in the leachate.

Stormwater quality from asphalt-paved surfaces seems to vary with time. Fish kills have been reported when rains occur shortly after asphalt has been installed in parking areas near ponds or streams (Anonymous, 2000; Perez-Rivas, 2000; Kline, 2002). It is expected that these effects are associated with losses of the more volatile and toxic hydrocarbons that are present on new surfaces. It is likely that the concentrations of these materials in runoff decrease as the pavement ages. Toxicity tests conducted on pavements several years old have not indicated any significant detrimental effects, except for those associated with activities conducted on the surface (such as maintenance and storage of heavy equipment; Pitt et al., 1995, 1999). However, pavement maintenance used to "renew" the asphalt surfaces has been shown to cause significant problems, which are summarized below.

A significant source of PAHs in the Austin, Texas, area (and likely elsewhere) has been identified as coal-tar sealants commonly used to "restore" asphalt parking lots and storage areas. Mahler et al. (2005) found that small particles of sealcoat that flake off due to abrasion by vehicle tires have PAH concentrations about 65 times higher than for particles washed off parking lots that are not seal coated. Unsealed parking lots receive PAHs from the same urban sources as do sealed parking lots (e.g., tire particles, leaking motor oil, vehicle exhaust, and atmospheric fallout), and yet the average yield of PAHs from the sealed parking lots was found to be 50 times greater than that from the control lots. The authors concluded that sealed parking lots could be the dominant source of PAHs in watersheds that have seal-coated surfaces, such as many industrial, commercial, and residential areas. Consequently, the City of Austin has restricted the use of parking lot coal-tar sealants, as have several Wisconsin communities.

Stored Materials Exposed to Rain

Although roofing and pavement materials make up a large fraction of the total surface covers and can have significant effects on stormwater quality, leaching of rain through stored materials may also be a significant pollutant source at industrial sites. Exposed metals in scrap yards can result in very high concentrations of heavy metals. For example, Table 3-7 summarizes data from three metals recycling facilities/scrap yards in Wisconsin and shows the large fraction of metals that are either dissolved in the runoff or associated with very fine particulate matter. For most of these metals, their greatest abundance is associated with the small particles (<20 μm in diameter), and relatively little is associated with the filterable fraction. These metals concentrations (especially zinc, copper, and lead) are also very high compared to that of most outfall industrial stormwater.

OTHER SOURCES OF URBAN RUNOFF DISCHARGES

Wet weather stormwater discharges from separate storm sewer outfalls are not the only discharges entering receiving waters from these systems. Dry weather flows, snowmelt, and atmospheric deposition all contribute to the pollutant loading of urban areas to receiving waters, and for some compounds may be the largest contributor. Many structural SCMs, especially those that rely on sedimentation or filtration, have been designed to function primarily with stormwater and are not nearly as effective for dry weather discharges, snowmelt, or atmospheric deposition because these nontraditional sources vary considerably in key characteristics, such as the flow rate and volume to be treated, sediment concentrsations and particle size distribution, major competing ions, association of pollutants with particulates of different sizes, and temperature. Information on the treatability of stormwater vs. snowmelt and other nontraditional sources of urban runoff can be found in Pitt and McLean (1986), Pitt et al. (1995), Johnson et al. (2003), and Morquecho (2005).

TABLE 3-7 Metal Concentration Ranges Observed in Scrapyard Runoff

Particle Size	Iron (mg/L)	Aluminum (mg/L)	Zinc (mg/L)
Total	20 – 810	15 – 70	1.6 – 8
< 63 μm diameter	22 – 767	15 – 58	1.5 – 7.6
< 38 μm diameter	21 – 705	15 – 58	1.4 – 7.4
< 20 μm diameter	15 – 534	12 – 50	1.1 – 7.2
< 0.45 μm diameter (filterable fraction)	0.1 – 38	0.1 – 5	0.1 – 6.7
	Copper (mg/L)	Lead (mg/L)	Chromium (mg/L)
Total	1.1 – 3.8	0.6 – 1.7	0.1 – 1.9
< 63 μm diameter	1.1 – 3.6	0.1 – 1.6	0.1 – 1.6
< 38 μm diameter	1.1 – 3.3	0.1 – 1.6	0.1 – 1.4
< 20 μm diameter	1.0 – 2.8	0.1 – 1.6	0.1 – 1.2
< 0.45 μm diameter (filterable fraction)	0.1 – 0.3	0.1 – 0.3	0.1 – 0.3

SOURCE: Reprinted, with permission, from Clark et al. (2000). Copyright 2000 by Shirley Clark.

Dry Weather Flows

At many stormwater outfalls, discharges occur during dry weather. These may be associated with discharges from leaking sanitary sewer and drinking water distribution systems, industrial wastewaters, irrigation return flows, or natural spring water entering the system (Figures 3-28 to 3-33). Possibly 25 percent of all separate stormwater outfalls have water flowing in them during dry weather, and as much as 10 percent are grossly contaminated with raw sewage, industrial wastewaters, and so forth (Pitt et al., 1993). These flow contributions can be significant on an annual mass basis, even though the flow rates are relatively small, because they have long duration. This is particularly true in arid areas, where dry weather discharges can occur daily. For example, despite the fact that rain is scarce from May to September in Southern California, an estimated 40 to 90 million liters of discharge flow per day into Santa Monica Bay through approximately 70 stormwater outlets that empty onto or across beaches (LAC DPW, 1985; SMBRP, 1994), such that the contribution of dry weather flow to the total volume of runoff into the bay is about 30 percent (NRC, 1984). Furthermore, in the nearby Ballona Creek watershed, dry weather discharges of trace metals were found to comprise from 8 to 42 percent of the total annual loading (McPherson et al., 2002). Stein and Tiefenthaler (2003) further found that the highest loadings of metals and bacteria in this watershed discharging during dry weather can be attributed to a few specific stormwater drains.

In many cases, stormwater managers tend to overlook the contribution of dry weather discharges, although the EPA's NPDES Stormwater Permit program requires municipalities to conduct stormwater outfall surveys to identify, and then correct, inappropriate discharges into separate storm sewer systems. The role of inappropriate discharges in the NPDES Stormwater Permit program, the developed and tested program to identify and quantify their discharges, and an extensive review of these programs throughout the United States can be found in the recently updated report prepared for the EPA (CWP and Pitt, 2004).

FIGURE 3-28 Washing of vehicle engine and allowing runoff to enter storm drainage system. SOURCE: Robert Pitt, University of Alabama.

FIGURE 3-29   Contamination of storm drainage with inappropriate disposal of oil.
SOURCE: Courtesy of the Center for Watershed Protection.

FIGURE 3-30 Dry weather flows from Toronto industrial area outfall. SOURCE: Pitt and
McLean (1986).

FIGURE 3-31 Sewage from clogged system overflowing into storm drainage system. SOURCE: Robert Pitt, University of Alabama.

FIGURE 3-32 Failing sanitary sewer, causing upwelling of sewage through soil, and draining to gutter and then to storm drainage system. SOURCE: Robert Pitt, University of Alabama.

FIGURE 3-33 Dye tests to confirm improper sanitary sewage connection to storm drainage system. SOURCE: Robert Pitt, University of Alabama.

Snowmelt

In northern areas, snowmelt runoff can be a significant contributor to the annual discharges from urban areas through the storm drainage system (see Figure 3-34). In locations having long and harsh winters, with little snowmelt until the spring, pollutants can accumulate and be trapped in the snowpack all winter until the major thaw when the contaminants are transported in short-duration events to the outfalls (Jokela, 1990). The sources of the contaminants accumulating in snowpack depend on the location, but they usually include emissions from nearby motor vehicles and heating equipment and industrial activity in the neighborhood. Dry deposition of sulfur dioxide from industrial and power plant smokestacks affects snow packs over a wider area and has frequently been studied because of its role in the acid deposition process (Cadle, 1991). Pollutants are also directly deposited on the snowpack. The sources of directly deposited pollutants include debris from deteriorated roadways, vehicles depositing petroleum products and metals, and roadway maintenance crews applying salt and anti-skid grit (Oberts, 1994). Urban snowmelt, like rain runoff, washes some material off streets, roofs, parking and industrial storage lots, and drainage gutters. However, snowmelt runoff usually has much less energy than striking rain and heavy flowing stormwater. Novotny et al. (1986) found that urban soil erosion is reduced or eliminated during winter snow-cover conditions. However, erosion of bare ground at construction sites in the spring due to snowmelt can still be very high (Figure 3-35).

FIGURE 3-34 Snowmelt photos. SOURCE: Roger Bannerman, Wisconsin Department of Natural Resources.

FIGURE 3-35 Construction site in early spring after snowmelt showing extensive sediment transport. SOURCE: Roger Bannerman, Wisconsin Department of Natural Resources.

Sources of Contaminants in Snowmelt

Several mechanisms can bring about contamination of snow and snowmelt waters. Initially, air pollutants can be incorporated into snowflakes as they form and fall to the ground. After it falls to the ground and accumulates, the snow can become further contaminated by dry atmospheric deposition, deposition of nearby lost fugitive dust materials (usually blown onto snow packs near roads by passing vehicles), and wash off of particulates from the exposed ground surfaces as it melts and flows to the drainage system.

Snowflakes can remove particulates and gases from the air by in-cloud or below-cloud capture. In-cloud capture of pollutants can occur during snowflake formation as super-cooled cloud water condenses on particles and aerosols that act as cloud condensation nuclei. This is known as nucleation scavenging and is a major pathway for air pollution to be incorporated into snow. Particles and gases may also be scavenged as snowflakes fall to the ground. Gases can also be absorbed as snow falls. Snowflakes are more effective below-cloud scavengers than raindrops because they are bigger and fall slower. Barrie (1991) reports that large snowflakes capture particles in the 0.2- to 0.4-µm-diameter range, not by impaction but by filtering the air that moves through the snow flakes as they fall to the ground.

Most of the contamination of snow in urban areas likely occurs after it lands on the ground. Table 3-8 shows the flow-weighted mean concentrations of pollutants found in undisturbed falling snow compared to snow found in urban snow cover (Bennett et al., 1981). Pitt and McLean (1986) also measured snowpack contamination as a function of distance from a heavily traveled road passing through a park. The contaminants in the snow were at much greater concentrations near the road (the major source of blown contamination on the snow) than farther away. (The pollutant levels in the fresh fallen snow are generally a small fraction of the levels in the snow collected from urban study areas.) Pierstorff and Bishop (1980) also analyzed freshly fallen snow and compared the quality to snow stored at a snow dump site. They concluded that "pollutant levels at the dump site are the result of environmental input occurring after the snow falls." Some pollutants in snowmelt have almost no atmospheric

TABLE 3-8 Comparison of Flow-Weighted Pollutant Concentration Means of Snow Samples from Boulder, Colorado

	Fresh Fallen	High Density Land Use	Low Density Land Use
COD	10	402	54
TS	86	2000	165
SS	16	545	4.5
TKN	0.19	2.69	2
NO$_3$	0.15	0	0
P	—	0.66	0.017
Pb	—	0.95	—

Note: The units are mg/L. SOURCE: Reprinted, with permission, from Bennett et al. (1981). Copyright 1981 by Water Pollution Control Federation.

sources. For example, Oliver et al. (1974) found negligible amounts of chlorides in samples of snow from rooftops, indicating that the high chloride level found in the snowmelt runoff water comes almost entirely from surface sources (i.e., road salting). Similar roadside snowpack observations along city park roads by Pitt and McLean (1986) also indicated the strong association of road salt with snowpack chloride levels.

Runoff and Pollutant Loading from Snowmelt

Snowmelt events can exhibit a first flush, in which there are higher concentrations of contaminants at the beginning compared to the total event averaged concentration. The enrichment of the first portion of a snowmelt event by soluble pollutants may be due to snowpack density changes, where water percolation and melt/freeze events that occur in the snowpack cause soluble pollutants to be flushed from throughout the snowpack to concentrate at the bottom of the pack (Colbeck, 1981). This concentrated layer leaves the snowpack as a highly concentrated pulse, as snow melts from the bottom due to warmth from the ground (Oberts, 1994).

When it rains on snow, heavy pollutant loads can be produced because both soluble and particulate pollutants are melted from the snowpack simultaneously. Also, the large volume of melt plus rain can wash off pollutants that have accumulated on various surfaces such as roads, parking lots, roofs, and saturated soil surfaces. The intensity of runoff from a rain-on-snow event can be greater than a summer thunderstorm because the ground is saturated or frozen and the rapidly melting snowpack provides added runoff volume (Oberts, 1994).

Figure 3-36 compares the runoff volumes associated with snowmelts alone to those associated with snowmelts mixed with rain from monitoring at an industrial area in Toronto (Pitt and McLean, 1986). Rain with snowmelt contributes over 80 percent of the total cold-weather event runoff volume.

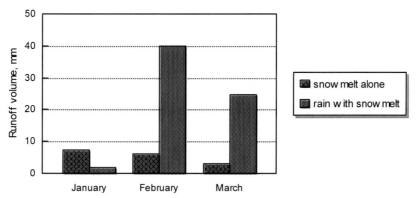

FIGURE 3-36 Runoff volumes for snowmelt events alone and when rain falls on melting snow packs (Toronto industrial area). SOURCE: Pitt and McLean (1986).

Whether pollutant loadings are higher or lower for snowmelt than for rainfall depends on the particular pollutant and its seasonal prevalence in the environment. For example, the high concentrations of dissolved solids found in snowmelt are usually caused by high chloride concentrations that stem from the amount of de-icing salt used. Figure 3-37 is a plot of the chloride concentrations in the influent to the Monroe Street detention pond in Madison, Wisconsin. Chloride levels are negligible in the non-winter months but increase dramatically when road salting begins in the fall, and remain high through the snow melting period, even extending another month or so after the snowpack in the area has melted. Bennett et al. (1981) found that suspended solids and COD loadings for snowmelt runoff were about one-half of those for rainfall. Nutrients were much lower for snowmelt, while the loadings for lead were about the same for both forms of precipitation. Oberts (1994) reports that much of the annual pollutant yields from event flows in Minneapolis is accounted for by end-of-winter major melts. End-of-winter melts yielded 8 to 20 percent of the total phosphorous and total lead annual load in Minnesota. Small midwinter melts accounted for less than 5 percent of the total loads. Box 3-8 shows mass pollutant discharges for a study site in Toronto and emphasizes the significance of snowmelt discharges on the total annual storm drainage discharges.

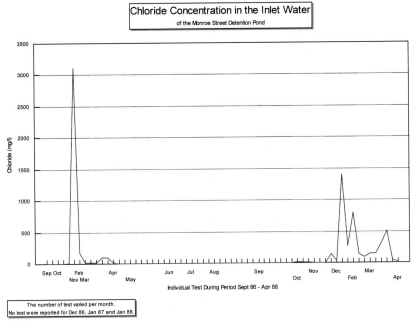

FIGURE 3-37 Monroe Street detention pond chloride concentration of influent (1986–1988). SOURCE: House et al. (1993).

BOX 3-8
The Contribution of Dry Weather Discharges and
Snowmelt to Overall Runoff in Toronto, Ontario

An extensive analysis of all types of stormwater flow—for both dry and wet weather—was conducted in Toronto in the mid-1980s (Pitt and McLean, 1986). The Toronto Area Watershed Management Strategy study included comprehensive monitoring in a residential/commercial area and an industrial area for summer stormwater, warm season dry weather flows, snowmelt, and cold season dry weather flows. In addition to the outfall monitoring, detailed source area sheet flow monitoring was also conducted during rain and snowmelt events to determine the relative magnitude of pollutant sources. Particulate accumulation and wash-off tests were also conducted for a variety of streets in order to better determine their role in contaminant contributions.

Tables 3-9 and 3-10 summarize Toronto residential/commercial and industrial urban runoff median concentrations during both warm and cold weather, respectively. These tables show the relative volumes and concentrations of wet weather and dry weather flows coming from the different land uses. The bacteria densities during cold weather are substantially less than during warm weather, but are still relatively high; similar findings were noted during the NURP studies (EPA, 1983). However, chloride concentrations and dissolved solids are much higher during cold weather. Early spring stormwater events also contain high dissolved solids concentrations. Cold weather runoff accounted for more than half of the heavy metal discharges in the residential/commercial area, while warm weather discharges of zinc were much greater than the cold weather discharges for the industrial area. Warm weather flows were also the predominant sources of phosphorus for the industrial area.

One of the interesting observations is that, at these monitoring locations, warm weather stormwater runoff only contributed about 20 to 30 percent of the total annual flows being discharged from the separate stormwater outfalls. The magnitudes of the base flows were especially surprising, as these monitoring locations were research sites to investigate stormwater processes and were carefully investigated to ensure that they did not have significant inappropriate discharges before they were selected for the monitoring programs.

In comparing runoff from the industrial and residential catchments, Pitt and McLean (1986) observed that concentrations of most constituents in runoff from the industrial watershed were typically greater than the concentrations of the same constituents in the residential runoff. The only constituents with a unit-area yield that were lower in the industrial area were chlorides and total dissolved solids, which was attributed to the use of road de-icing salts in residential areas. Annual yields of several constituents (total solids, total dissolved solids, chlorides, ammonia nitrogen, and phenolics) were dominated by cold weather flows, irrespective of the land use.

A comparison of the Toronto sheet flow data from the different land-use areas indicated that the highest concentrations of lead and zinc were found in samples collected from paved areas and roads during both rain runoff and snowmelt (Pitt and McLean, 1986). Fecal coliform values were significantly higher on sidewalks and on, or near, roads during snowmelt sampling, likely because these areas are where dogs would be walked in winter conditions. In warm weather, dog walking would be less concentrated into these areas. The concentrations for total solids from grass or bare open areas were reduced dramatically during snowmelt compared to rain runoff, an indication of the reduced erosion and the

continues next page

BOX 3-8 Continued

poor delivery of particulate pollutants during snowmelt periods. Cold weather sheet flow median concentrations of particulate solids for the grass and open areas (80 mg/L) were much less than the TSS concentrations observed during warm weather runoff (250 mg/L) for these same areas. Snowmelt total solids concentrations also increased in areas located near roads due to the influence of road salting on dissolved solids concentrations. In the residential areas, streets were the most significant source of snowmelt solids, while yards and open areas were the major sources of nutrients. Parking and storage areas contrib-

TABLE 3-9 Median Pollutant Concentrations Observed at Toronto Outfalls during Warm Weather[1]

	Baseflow		Stormwater	
Measured Parameter	Residential	Industrial	Residential	Industrial
Stormwater volume (m³/ha/season)	—	—	950	1500
Baseflow volume (m³/ha/season)	1700	2100	—	—
Total residue	979	554	256	371
Total dissolved solids	973	454	230	208
Suspended solids	<5	43	22	117
Chlorides	281	78	34	17
Total phosphorus	0.09	0.73	0.28	0.75
Phosphates	<0.06	0.12	0.02	0.16
Total Kjeldahl nitrogen (organic N plus NH_3)	0.9	2.4	2.5	2.0
Ammonia nitrogen	<0.1	<0.1	<0.1	<0.1
Chemical oxygen demand	22	108	55	106
Fecal coliform bacteria (#/100 mL)	33,000	7,000	40,000	49,000
Fecal strep. bacteria (#/100 mL)	2,300	8,800	20,000	39,000
Pseudomonas aeruginosa bacteria (#/100 mL)	2,900	2,380	2,700	11,000
Cadmium	<0.01	<0.01	<0.01	<0.01
Chromium	<0.06	0.42	<0.06	0.32
Copper	0.02	0.05	0.03	0.06
Lead	<0.04	<0.04	<0.06	0.08
Zinc	0.04	0.18	0.06	0.19
Phenolics (μg/L)	<1.5	2.0	1.2	5.1
α-BHC (ng/L)	17	<1	1	3.5
γ-BHC (lindane) (ng/L)	5	<2	<1	<1
Chlordane (ng/L)	4	<2	<2	<2
Dieldrin (ng/L)	4	<5	<2	<2
Pentachlorophenol (ng/L)	280	50	70	705

[1]Values are in mg/L unless otherwise indicated. Warm weather samples were obtained during the late spring, summer, and early fall months when the air temperatures were above freezing and no snow was present.
SOURCE: Pitt and McLean (1986).

uted the most snowmelt pollutants in the industrial area. An analysis of snow samples taken along a transect of a snowpack adjacent to an industrial road showed that the pollutant levels decreased as a function of distance from the roadway. At distances greater than 3 to 5 meters from the edge of the snowpack, the concentrations were relatively constant. Novotny et al. (1986) sampled along a transect of a snowpack by a freeway in Milwaukee. They also found that the concentration of constituents decreased as the distance from the road increased. Most of the measured constituents, including total solids and lead, were at or near background levels at 30 meters or more from the road.

TABLE 3-10 Median Pollutant Concentrations Observed at Toronto Outfalls during Cold Weather[1]

Measured Parameter	Baseflow		Snowmelt	
	Residential	Industrial	Residential	Industrial
Stormwater volume (m³/ha/season)	—	—	1800	830
Base flow volume (m³/ha/season)	1100	660	—	—
Total residue	2230	1080	1580	1340
Total dissolved solids	2210	1020	1530	1240
Suspended solids	21	50	30	95
Chlorides	1080	470	660	620
Total phosphorus	0.18	0.34	0.23	0.50
Phosphates	<0.05	<0.02	<0.06	0.14
Total Kjeldahl nitrogen (organic N plus NH_3)	1.4	2.0	1.7	2.5
Ammonia nitrogen	<0.1	<0.1	0.2	0.4
Chemical oxygen demand	48	68	40	94
Fecal coliform bacteria (#/100 mL)	9800	400	2320	300
Fecal strep bacteria (#/100 mL)	1400	2400	1900	2500
Pseudomonas aeruginosa bacteria (#/100 mL)	85	55	20	30
Cadmium	<0.01	<0.01	<0.01	0.01
Chromium	<0.01	0.24	<0.01	0.35
Copper	0.02	0.04	0.04	0.07
Lead	<0.06	<0.04	0.09	0.08
Zinc	0.07	0.15	0.12	0.31
Phenolics (mg/L)	2.0	7.3	2.5	15
α-BHC (ng/L)	NA	3	4	5
γ-BHC (lindane) (ng/L)	NA	NA	2	1
Chlordane (ng/L)	NA	NA	11	2
Dieldrin (ng/L)	NA	NA	2	NA
Pentachlorophenol (ng/L)	NA	NA	NA	40

[1]Values are in mg/L unless otherwise indicated. Cold weather samples were obtained during the winter months when the air temperatures were commonly below freezing. Snowmelt samples were obtained during snowmelt episodes and when rain fell on snow.
NA, not analyzed
SOURCE: Pitt and McLean (1986).

Atmospheric Deposition

The atmosphere contains a diverse array of contaminants, including metals (e.g., copper, chromium, lead, mercury, zinc), nutrients (nitrogen, phosphorus), and organic compounds (e.g., PAHs, polychlorinated biphenyls, pesticides). These contaminants are introduced to the atmosphere by a variety of sources, including local point sources (e.g., power plant stacks) and mobile sources (e.g., motor vehicles), local fugitive emissions (e.g., street dust and wind-eroded materials), and transport from non-local areas. These emissions, composed of gases, small particles (aerosols), and larger particles, become entrained in the atmosphere and subject to a complex series of physical and chemical reactions (Schueler, 1983).

Atmospheric contaminants are deposited on land and water in two ways—termed wet deposition and dry deposition. Wet deposition (or wetfall) involves the sorption and condensation of pollutants to water drops and snowflakes followed by deposition with precipitation. This mechanism dominates the deposition of gases and aerosol particles. Dry deposition (or dryfall) is the direct transfer of contaminants to land or water by gravity (particles) or by diffusion (vapor and particles). Dry deposition occurs when atmospheric turbulence is not sufficient to counteract the tendency of particles to fall out at a rate governed, but not exclusively determined, by gravity (Schueler, 1983).

As atmospheric contaminants deposit, they can exert an influence on stormwater in several ways. Contaminants deposited by wetfall are directly conveyed to stormwater while those in dryfall can be washed off the land surface. For both processes, the atmospheric load of contaminants is strongly influenced by characteristics such as the amount of impervious surface, the magnitude and proximity of emission sources, wind speed and direction, and precipitation magnitude and frequency (Schueler, 1983). Deposition rates can depend on the type of contaminant and can be site-specific. The relationships between atmospheric deposition and stormwater quality are, however, not well understood and difficult to determine. Following are a few illustrative examples.

Southern California

Several studies have addressed atmospheric deposition in Southern California (e.g., Lu et al., 2003; Harris and Davidson, 2005; Stolzenbach et al., 2007). Stolzenbach et al. and Lu et al. conclude the following *for this region:*

- the major source of contaminants to the atmosphere in this region is associated with resuspended dust, primarily from roads,
- contaminants in resuspended dust may reflect historical as well as current sources and distant as well as local sources,
- atmospheric loadings to the receiving water are primarily the result of chronic daily dry deposition of large particles greater than 10 μm in size on the

watershed rather than directly on a waterbody,

•   significant spatial variability occurs in trace metal mass loadings and deposition fluxes, particularly along transportation corridors along the coast and the mountain slopes of the airshed,

•   significant diurnal and seasonal variations occur in the deposition of trace metals, and

•   atmospheric deposition of metals is a significant component of contaminant loading to waterbodies in the region relative to other point and nonpoint sources.

Harris and Davidson (2005) have reported that traditional sources of lead to the south coast air basin of California accounted for less than 15 percent of the lead exiting the basin each year. They resolve this difference by considering that lead particles deposited during the years of leaded gasoline use are resuspended as airborne lead at this time, some decades after their original deposition. This result indicates that lead levels in the soil will remain elevated for decades and that resuspension of this lead will remain a major source of atmospheric lead well into the future.

Sabin et al. (2005) assessed the contribution of trace metals (chromium, copper, lead, nickel, and zinc) from atmospheric deposition to stormwater runoff in a small impervious urban catchment in the Los Angeles area. Dry deposition contributed 90 percent or more of the total deposition inside the catchment, indicating the dominance of dry deposition in semi-arid regions such as Los Angeles. Deposition potentially accounted for from 57 to 90 percent of the total trace metals in stormwater in the study area, demonstrating that atmospheric deposition can be an important source of trace metals in stormwater near urban centers.

San Francisco

Dissolved copper is toxic to phytoplankton, the base of the aquatic food chain. Copper and other metals are released in small quantities when drivers depress their brakes.   The Brake Pad Partnership (*http://www.suscon.org/brakepad/index/asp*) has conducted studies to determine how much copper is released as wear debris, and how it travels through the air and streets to surface waters. A comprehensive and complex model of copper loads to and of transport and reactions in San Francisco Bay was developed (Yee and Franz, 2005). Objectives were to provide daily loadings of flow, TSS, and copper to the bay and to estimate the relative contribution of brake pad wear debris to copper in the bay. The modeling results (Rosselot, 2006a) indicated that an estimated 47,000 kg of copper was released to the atmosphere in the Bay Area in 2003. Of this amount, 17,000 kg Cu/yr was dry-deposited in subwatersheds; 3,200 kg Cu/yr was wet-deposited in subwatersheds; 1,200 kg Cu/yr was dry-deposited directly to bay waters; and 1,300 kg Cu/yr was wet-deposited directly to bay waters. The remaining 24,000 kg Cu/yr remained airborne until it left the Bay

Area. The contribution of copper from brake pads to the bay is estimated to range from 10 to 35 percent of the total copper input, with the best estimate being 23 percent (Rosselot, 2006a,b).

Washington, D.C., Metropolitan Area

Schueler (1983) investigated the atmospheric deposition of several contaminants in Washington, D.C., and its surrounding areas in the early 1980s. The contaminants assessed included trace metals (cadmium, copper, iron, lead, nickel, and zinc), nutrients (nitrogen and phosphorus), solids, and organics as measured collectively by BOD and COD. Dryfall solids loading increased progressively from rural to urban sites. A similar trend was observed for total phosphorus, total nitrogen, and trace metal dry deposition rates. Wet deposition rates exhibited few consistent regional patterns.

The relative importance of wet and dry deposition varied considerably with each contaminant and each site. For example, most of the nitrogen was supplied by wet deposition while most of the phosphorus was delivered via dry deposition. If a contaminant is deposited primarily by wet deposition, it is likely that a major fraction of it will be rapidly entrained in urban runoff.

Atmospheric sources were estimated to contribute from 70 to 95 percent of the total nitrogen load to urban runoff and 20 to 35 percent of the total phosphorus load. Overall, atmospheric deposition appeared to be a moderate source of pollutants in urban runoff. However, with the exception of nitrogen, atmospheric deposition was not the major source.

Average annual atmospheric deposition rates suggested a general trend toward greater deposition rates from rural to suburban to urban sites. This pattern was most pronounced for dry deposition. Wet deposition was the most important deposition mechanism for total nitrogen, nitrate, organic nitrogen, COD, copper, and zinc. Dry deposition was most important for most soil-related constituents, such as total solids, iron, lead, total phosphorus, and orthophosphate.

Measurements of rainfall pH showed median values between 4.0 and 4.1 at all stations and during all seasons. Increased mobilization of trace metals from urban surfaces caused by acid rain was noted at several monitoring sites.

***

Relationships between atmospheric deposition rates and the quality of urban stormwater are complex and cannot be generalized regionally or temporally. Site-specific measurements or reliable estimates of (1) contaminant sources, (2) atmospheric particle size and contaminant concentrations, (3) deposition rates and mechanisms, (4) land surface characteristics, (5) local and regional hydrology and meteorology, and (6) contaminant concentrations in stormwater are needed to assess management decisions to improve stormwater quality. Transportation is a major source of metals (lead in gasoline, zinc in tires, copper in brake pads). The results of the modeling of copper in San Francisco and its watershed demonstrate the feasibility of modeling the impact of a source, in this case copper input by atmospheric deposition, on water quality in a receiving waterbody.

BIOLOGICAL RESPONSES TO URBANIZATION

As discussed in Chapter 1, the biological integrity of aquatic ecosystems is influenced by five major categories of environmental stressors: (1) chemical, (2) hydrologic, (3) physical (e.g., habitat), (4) biological (e.g., disease, alien species), and (5) energy-related factors (e.g., nutrient dynamics). Recent studies on biological assemblages in urban or urbanizing waters have begun to examine how stormwater stressors limit biological potential along various urban gradients (Horner et al., 2003; Carter and Fend, 2005; Meador et al., 2005; Barbour et al., 2008; Purcell et al., 2009). Advances in biological monitoring and assessment over the past two decades have enabled much of this research. Today, many states and tribes use biological data to directly measure their aquatic life beneficial uses and have developed numeric biocriteria that are institutionalized in their water quality standards. Most of these approaches compare biology and stressors to suites of reference sites (Hughes, 1995; Stoddard et al., 2006), which can vary from near-pristine areas to agricultural landscapes. While this section focuses on streams because of the wealth of data, similar work is being performed on other waterbody types such as wetlands (Mack and Micacchion, 2007) and estuaries, both of which are susceptible to stormwater pollutants such as metals because of their depositional nature (Morrisey et al., 2000).

Aquatic life beneficial uses are based on achieving aquatic *potential* given feasible restorative actions. Because such potential may vary substantially across a region depending on land use and other factors, some states have adopted tiered aquatic life uses (see Box 2-1). The potential of many urban streams is likely to be something less than "biological integrity" (the ultimate goal of the CWA) or even "fishable–swimmable" goals, which are the interim goals of the CWA. Indeed, there is a near-universal, negative association between biological assemblages in streams and increasing urbanization, to the extent that it has been termed the "Urban Stream Syndrome" (Walsh et al., 2005). Recent investigations that have quantified the responses of macroinvertebrates and other biological assemblages along multiple measures of urban/stormwater stressors have discussed how best to set aquatic life goals for urban streams (Booth and Jackson, 1997; Bernhardt and Palmer, 2007). One of the most important contributions to this debate has been the development of the Biological Condition Gradient (BCG) concept by EPA. The BCG is an attempt to anchor and standardize interpretations of biological conditions and to unify biological monitoring results across the United States in order to advance the use of tiered aquatic life beneficial uses. This section summarizes the characteristic biological responses to urban gradients, within the framework of the BCG, and it reviews evidence of biological responses within the aforementioned five major categories of environmental stressors.

Biological Condition Gradient

The BCG framework is an ecological model of how structural and functional components of biological assemblages change along gradients of increasing stressors of many kinds (Davies and Jackson, 2006). Ecological systems have some common general attributes related to their structure and function that form the basis for how biological organisms respond to stressors in the environment. Over the past 20 years, development of biological indicators nationwide has taken advantage of these repeatable biological responses to stress; however, state benchmarks often have varied substantially, even between adjacent states. To gain consistency, the EPA convened a national workgroup of EPA Regions, States, and Tribes to develop the BCG—a standardized, nationally applicable model that defines important attributes of biological assemblages and describes how these attributes change along a gradient of increasing stress from pristine environments to severely impaired conditions (Figure 3-38; Davies and Jackson, 2006). The goals of this work were to improve national consistency in the rating and application of biological assessment tools for all types of waterbodies and to provide a baseline for the development of tiered aquatic life uses.

To date, the BCG has been applied to assemblages including aquatic macroinvertebrates, fish, Unionid mussels, and algae in streams, but it could be applied to any organism group in any type of waterbody. The BCG is derived by applying a suite of ten ecological attributes that allows biological condition to be interpreted independently of assessment method (Table 3-11; Davies and Jackson, 2006). The first five attributes focus on taxa sensitivity, an important component of tools such as multimetric indices (e.g., the Index of Biotic Integrity [IBI], the Invertebrate Community Index [ICI]; see Box 2-3) used in the United States and Europe. Many indicator taxa have been widely studied, and, for groups such as fish, historical data often exist. Most states have established lists of tolerant and intolerant species as part of their use of biological indices (Simon and Lyons, 1995). The relatively large literature on species population and distribution changes in response to stressors and landscape condition offers insight into the mechanisms for population shifts, some of which are summarized in this section.

The first two attributes of the BCG relate to those streams that are closest to natural or pristine, with most taxa "as naturally occur." Attribute 1 and 2 taxa are the most sensitive species that typically disappear with even minor stress. Table 3-12 lists some example attribute 1 taxa for four different regions of the United States. Attribute 3 reflects more ubiquitous, but still sensitive, species that can provide information as human influence on the landscape becomes more obvious, but is not yet severe. Attributes 5 and 6 are taxa that increase in abundance and distribution with increasing stress. The organism condition at-

The Biological Condition Gradient: Biological Response to Increasing Levels of Stress

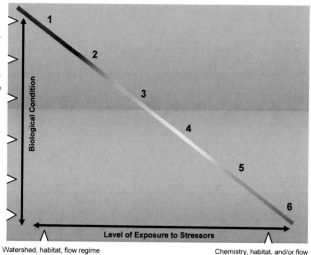

Levels of Biological Condition

Natural structural, functional, and taxonomic integrity is preserved.

Structure & function similar to natural community with some additional taxa & biomass; ecosystem level functions are fully maintained.

Evident changes in structure due to loss of some rare native taxa; shifts in relative abundance; ecosystem level functions fully maintained.

Moderate changes in structure due to replacement of sensitive ubiquitous taxa by more tolerant taxa; ecosystem functions largely maintained.

Sensitive taxa markedly diminished; conspicuously unbalanced distribution of major taxonomic groups; ecosystem function shows reduced complexity & redundancy.

Extreme changes in structure and ecosystem function; wholesale changes in taxonomic composition; extreme alterations from normal densities.

Watershed, habitat, flow regime and water chemistry as naturally occurs.

Chemistry, habitat, and/or flow regime severely altered from natural conditions.

FIGURE 3-38 The Biological Condition Gradient (BCG) and summaries of biological condition along tiers of this gradient. SOURCE: Modified from Davies and Jackson (2006) by EPA.

TABLE 3-11 Ecological attributes that comprise the basis for the BCG

1.	Historically documented, sensitive, long-lived or regionally endemic taxa
2.	Sensitive-rare taxa
3.	Sensitive-ubiquitous taxa
4.	Taxa of intermediate tolerance
5.	Tolerant taxa
6.	Non-native or introduced taxa
7.	Organism condition
8.	Ecosystem functions
9.	Spatial and temporal extent of detrimental effects
10.	Ecosystem connectance

SOURCE: EPA (2005).

TABLE 3-12 Example of Taxa that *Might Serve* as Attribute 1: "Historically Documented, Sensitive, Long-Lived, Regionally Endemic Taxa for Streams in Four Regions of the United States"

State and Taxon	Taxa Representative of Attribute I
Maine	
Mollusks	brook floater (*Alasmodonta varicosa*), triangle floater (*Alasmodonta undulata*), yellow lampmussel (*Lampsilis cariosa*)
Fishes	brook stickleback (*Culaea inconstans*), swamp darter (*Etheostoma fusiforme*)
Washington	
Fishes	steelhead (*Oncorhynchus mykiss*)
Amphibians	spotted frog (*Rana pretiosa*)
Arizona	
Mollusks	spring snails (*Pyrgulopsis* spp.)
Fishes	Gila trout (*Oncorhynchus gilae*), Apache trout (*Oncorhynchus apache*), cutthroat trout (endemic strains) (*Oncorhynchus clarki*)
Amphibians	Chihuahua leopard frog (*Rana chiricahuensis*)
Kansas	
Mollusks†	hickorynut (*Obovaria olivaria*), black sandshell (*Ligumia recta*), ponderous campeloma (*Campeloma crassulum*)
Fishes	Arkansas River shiner (*Notropis girardi*), Topeka shiner (*Notropis topeka*), Arkansas darter (*Etheostoma cragini*), Neosho madtom (*Noturus placidus*), flathead chub (*Platygobio gracilisa*)
Other invertebrates	ringed crayfish (*Orconectes neglectus neglectus*), Plains sand-burrowing mayfly (*Homoeoneuria ammophila*)
Amphibians	Plains spadefood toad (*Spea bombifrans*), Great Plains toad (*Bugo cognatus*), Great Plains narrowmouth toad (*Gastrophryne olivaceae*), Plains leopard frog (*Rana blairi*)

†Although not truly endemic to the central plains, these regionally extirpated mollusks were widely distributed in eastern Kansas prior to the onset of intensive agriculture.
SOURCE: Table 7 from Davies and Jackson (2006). Reprinted, with permission, from Davies and Jackson (2006). Copyright 2006 by Ecological Society of America.

tribute (7) includes the presence of anomalies (e.g., tumors, lesions, eroded fins, etc.) or the presence of large or long-lived individuals in a population. Most natural streams typically have few or incidental rates of "anomalies" associated with disease and stress. Natural waterbodies typically also have the entire range of life stages present, as would be expected. However, as stress is increased, larger individuals may disappear or emigrate, or reproductive failure may occur. Ecosystem function (attribute 8) is very difficult to measure directly (Davies and Jackson, 2006). However, certain functions can be inferred from structural measures common to various multimetric indices, examples of which are listed in Table 3-13. The last two attributes (9 and 10) may be of particular importance with regard to stormwater and urban impacts. Cumulative impacts are a characteristic of urbanization, and biological organisms typically integrate the effects of many small insults to the landscape. Additionally, most natural systems often have strong "connectance," such that aquatic life often has stages that rely on migrating across multiple types or sizes of waterbodies. Urbanized streams can decrease connectance by creating migration blocks, including vertical barriers at road crossings and small dams (Warren and Pardew, 1998).

TABLE 3-13 Function Ecological Attributes or Process Rates and Their Structural Indicators

Biotic Level and Function or Process	Structural Indicator
Individual level	
Fecundity	Maximum individual size, number of eggs
Growth and metabolism	Length/mass (condition)
Morbidity	Percentage anomalies
Population Level	
Growth and fecundity	Density
Mortality	Size- or age-class distribution
Production	Biomass, standing crop, catch per unit effort
Sustainability	Size- or age-class distribution
Migration, reproduction	Presence or absence, density
Community or assemblage level	
Production/respiration ratio, autotrophy vs heterotrophy	Trophic guilds, indicator species
Primary production	Biomass, ash-free dry mass
Ecosystem level	
Connectivity	Degree of aquatic and riparian fragmentation longitudinally, vertically, and horizontally; presence or absence of diadromous and potadromous species

SOURCE: Table 4 from Davies and Jackson (2006). Reprinted, with permission, from Davies and Jackson (2006). Copyright 2006 by Ecological Society of America.

Construction of a BCG creates a conceptual framework for developing stressor–response gradients for particular urban areas. The initial work done to develop the BCG derived a series of six tiers to describe a gradient of biological condition that is anchored in pristine conditions ("as naturally occurs") and that extends to severely degraded conditions (see Figure 3-38). Exercises done by the national work group to derive such a gradient for macroinvertebrates in wadeable streams showed strong consistency in assigning tiers to datasets using the descriptions of taxa for each attribute along these gradients (Davies and Jackson, 2006). Substantial data already exist to populate many of the attributes of the BCG and to provide mechanistic underpinning for the expected directions of change.

The BCG is not a replacement for assessment tools such as the IBI or multivariate predictive models (e.g., RIVPACS approach), but rather a conceptual overlay for characterizing the anchor point-of-reference conditions and a consistent way to communicate biological condition along gradients of stress. As such, it has strong application to understanding stormwater impacts and to communicating where a goal is located along the gradient of biological condition. While most urban goals may be distant from "pristine" or "natural," the BCG process can dispel misconceptions that alternate urban goals are "dead streams" or unsafe in some manner.

Factors Limiting Aquatic Assemblages in Urban Waters

A slew of recent investigations have quantified the responses of macroinvertebrates and other biological assemblages to multiple measures of urbanization and to stormwater in particular. One important conclusion of some of this work is that declines in the highest biological condition start with low levels of anthropogenic change (e.g., 5 to 25 percent impervious surface); higher levels of urbanization severely alter aquatic conditions (Horner et al., 2003). This has important consequences for protecting sites with the highest biological integrity, as they may be among the most vulnerable. The non-threshold nature of this aquatic response and the typical wedge-shaped response to multiple stressors by aquatic assemblages are discussed in Box 3-9.

BOX 3-9
Non-threshold Nature of the Decline of Biological
Assemblages Along Urban Stressor Gradients

Several recent surveys have demonstrated that biological assemblages begin to decline in condition with even low levels of urban disturbance as measured by various gradients of urbanization (e.g., May, 1996; Horner et al., 1997; May et al., 1997; Horner et al., 2003; Moore and Palmer, 2005; Barbour et al., 2008). This box summarizes the work of Horner et al. (2003) in small streams in three regions: Montgomery County, Maryland; Austin, Texas; and the Puget Sound area of Washington. Geographic Information System (GIS) analyses using information such as land use, total impervious area, and riparian land use were used to develop multi-metric Watershed Condition Indices (WCIs) for each region. These in turn were related to fish and macroinvertebrate indices, e.g., benthic IBIs, (B-IBI, all three regions), a fish IBI (F-IBI for Maryland) and an index that was the ratio of the sensitive coho salmon to the more tolerant cutthroat trout in collections for the Puget Sound lowland area.

In each of these areas, no or extremely low urban development, substantial forest cover, and minimal disturbance of riparian zones characterized sites with the highest biological scores, but these conditions did not guarantee high scores because other impacts could limit biology even with these "natural" characteristics. In all three regions, high urbanization and loss of natural cover always led to biological degradation (Figures 3-39 and 3-40). The results of this study were similar to other recent studies such as Barbour et al. (2008) that identify a "wedge-shaped" relationship or a "polygonal" relationship (Carter and Fend, 2005) between urban gradients and biological condition. These types of relationships have also been termed "factor-ceiling" relationships (Thomson et al., 1996). The outer surface of these wedges or polygons reflects where the urban gradients limit biological assemblages, such that points below this surface typically represent sites affected by other stressors (e.g., combined sewer overflows, discharges, etc.). In all of these studies it is easier to predict loss of biological conditions as the urban gradients (e.g., WCI) worsen than it is to ensure high biological integrity at low proportions of urban stress (because some other stressor may still limit aquatic condition).

continues next page

BOX 3-9 Continued

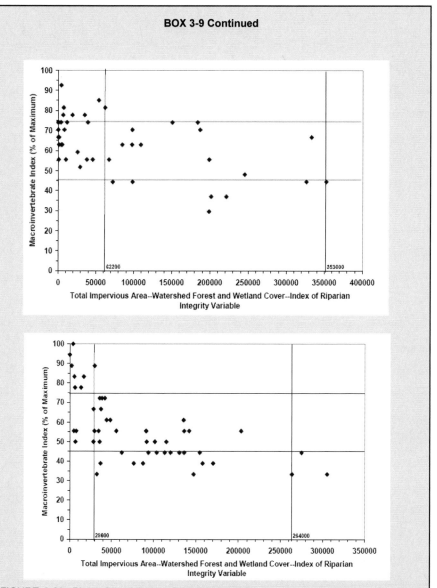

FIGURE 3-39 Plots of a measure of urbanization (TIA + Wetland & Forest Cover + IRI) versus B-IBIs for Austin, Texas (top), and Montgomery County, Maryland (bottom). SOURCE: Horner et al. (2003).

continues next page

BOX 3-9 Continued

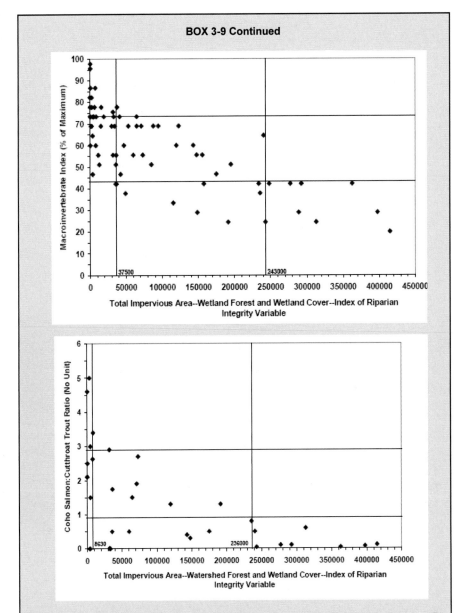

FIGURE 3-40 Plots of a measure of urbanization (TIA + Wetland & Forest Cover + IRI) versus B-IBIs for Puget Sound (top) and versus the ratio of coho salmon to cutthroat trout for Puget Sound (bottom). SOURCE: Horner et al. (2003).

continues next page

BOX 3-9 Continued

Horner et al. (2003) also focused on whether structural SCMs could moderate the effects of urbanization on biological assemblages. They made detailed observations of two subbasins in the Puget Sound lowland area, one with a greater degree of stormwater management than the other (although neither had what would be considered comprehensive stormwater management with a focus on water quality issues). As shown in Figure 3-41, at the highest levels of urbanization (triangles), the subbasin with the more extensive use of structural SCMs did have better biological conditions. There was less evidence of biological benefit in the watershed that used SCMs but it had only moderate urbanization and more natural land cover (squares and diamonds). There were no circumstances where high biological condition was observed along with the use of SCMs because high biological condition only occurred where little human alteration was present, and thus SCMs were not used.

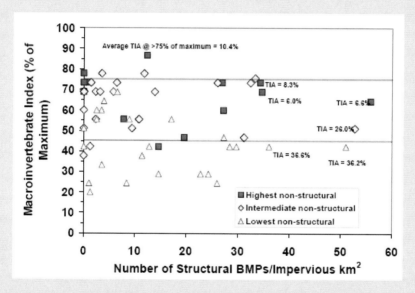

FIGURE 3-41 Macroinvertebrate community index versus structural SCM density with the highest, intermediate, and lowest one-third of natural watershed and riparian cover. The upper and lower horizontal lines represent indices considered to define relatively high and low levels of biological integrity, respectively. SOURCE: Horner et al. (2003).

The sections that follow review the evidence underlying biological responses to each of the major categories of stressors: chemical, hydrologic, physical habitat, biological, and energy-related factors. As will be evident in some of the examples, the stressors themselves can interact (e.g., flow can influence habitat, habitat can influence energy processing, etc.), which increases the complexity of understanding how stormwater affects aquatic ecosystems.

Biological Responses to Toxic Pollutants

The chemical constituents of natural streams vary widely with climatic region, stream size, soil types, and geological setting. Most small natural streams, outside of unique areas wth naturally occurring toxicants, have very low levels of chemicals considered to be toxicants and have relatively low levels of dissolved and particulate materials in general. This applies to chemicals in the water column and in sediments. Increasing amounts of impervious surface in the watershed typically increase the concentrations of many chemical parameters in runoff derived from urban surfaces (e.g., Porcella and Sorenson, 1980; Sprague et al., 2007).

Stormwater concentrations of these pollutants can be variable and sometimes extreme or "toxic" depending on the timing of flows (e.g., first flush), although concentrations at base flows may not routinely exceed water quality benchmarks (Sprague et al., 2007). Historical deposition of toxics in sediments can also be responsible for extremely high pollutant concentrations within waterbodies, even though the stormwater discharges may no longer be active. These situations have been termed "legacy pollution" and are most commonly associated with urban centers that have a history of industrial production.

Natural constituents such as dissolved materials (e.g., chlorides), particulate material (e.g., fine sediments), nutrients (e.g., phosphorus and nitrogen compounds), as well as a myriad of man-made parameters such as heavy metals and organic chemicals (e.g., hydrocarbons, pesticides and herbicides) have been documented to be increased and at times pervasive in stormwater (Heany and Huber, 1984; Paul and Meyer, 2001; Roy et al., 2003; Gilliom et al., 2006) although specific patterns of concentrations can vary with region and ecological setting (Sprague et al., 2007). Water chemistry impacts can also arise from a complex array of permitted discharges, storm sewer discharges, and combined sewer overflows that are treated to certain limits but at times fail to remove all constituents from flows, especially when associated with storm events (Paul and Meyer, 2001).

Streams in urban settings can have increases in toxicant levels compared to background concentrations. In many instances these cases have been associated with loss of aquatic species and impairment of aquatic life goals (EPA, 2002), which are usually explained in terms of typical lethal responses. The complexity of urban systems with regard to pathways, magnitude, duration, and timing of toxicity as well as possible synergistic or antagonistic effects of mixtures of pollutants argues for a broad approach to characterizing effects including not only toxicity testing, but also novel approaches and direct monitoring of biological assemblages (Burton et al., 1999). What is problematic from a traditional management perspective is that aquatic communities may decline before exceedances of water quality criteria are evident (May et al., 1997; Horner et al., 2003).

The first three BCG attributes focus on populations of species of high to very high sensitivity, most of which are uncommon or absent in waters with any substantial level of urbanization. Multi-metric indices such as IBI, which reflect loss of these species, decline at least linearly with increasing urbanization (e.g., Miltner et al., 2004; Meador et al., 2005; Walters et al., 2005). Although toxicity to compounds varies with species, many species of federal and state endangered and threatened aquatic species are more sensitive than "commonly" used test species (Dwyer et al., 2005), such that the loss of aquatic species when toxicant levels exceed criteria are readily explained.

The mechanisms of species population declines in response to chemical contaminants are likely complex and not just limited to direct lethality of the pollutant. Indeed, initial chemical changes may have no "toxic" effects, but rather could change competitive and trophic dynamics by changing primary production and energy dynamics in streams. For example, exposures to aromatic and chlorinated organic compounds from sediments derived from urban areas have been found to increase the susceptibility of salmonids to the bacterial pathogen *Vibrio anguillarum* (Arkoosh et al., 2001). Recent work has found that salmonids show substantial behavioral changes from olfactory degradation related to copper at concentrations as low as 2 μg/L, well below copper water quality criteria and above levels measured in most stormwater-affected streams (Hecht et al., 2007; Sandahl et al., 2007). Salmonid and other fish depend extensively on olfactory cues for feeding, emigration, responding to prey and predators, social and spawning interactions, and other behaviors, such that loss or diminution of such cues may have population-level effects on these species (Sandahl et al., 2007). Copper has been shown to cause olfactory effects on other species (Beyers et al., 2001) and to impair the sensory ability of the fish lateral line (Hernandez et al., 2006), which is nearly ubiquitous in fishes and important for most freshwater species in feeding, schooling, spawning, and other behaviors.

Whole effluent toxicity testing or sediment toxicity testing may misclassify the effects of runoff and effluents in urban settings (Burton et al., 1999). Short-term toxicity tests of stormwater often result in no identified toxicity. However, longer studies (e.g., 30 days) have shown increasing toxicity with time (Masterson and Bannerman, 1994; Ramcheck and Crunkilton, 1995). This suggests that the mechanism of toxicity could be through an ingestion pathway, for example, rather than gill uptake. Metals are often in high concentrations where fine sediments accumulate, and their legacy can extend past the time period of active discharge. Metal concentrations in urban stream sediments have been associated with high rates of fish and invertebrate anomalies such as tumors, lesions, and deformities (Burton, 1992; Ingersoll et al., 1997; Smith et al., 2003).

Biological Responses to Non-Toxicant Chemicals

Non-toxic chemical compounds that occur in stormwater such as nutrients, dissolved oxygen (DO), pH, and dissolved solids as well as physical factors such as temperature can have impacts on aquatic life. The effects of some of these compounds (e.g., DO, pH) have been well documented from other impacts (e.g., wastewater, mining), such that nearly all states have developed water quality criteria for these parameters. For example, nutrient enrichment in stormwater runoff has been associated with declines of biological condition in streams (Miltner and Rankin, 1998). Chloride, sulfate, and other dissolved ions that are often elevated in urban areas can have effects on osmoregulation of aquatic organisms and have been associated with loss of species sensitive to dissolved materials such as mayflies (Kennedy et al., 2004). The concentrations of these compounds can vary regionally (Sprague et al., 2007) and with the degree of urbanization.

Water quality criteria for temperature were spurred by the need for thermal permits for industrial and power plant cooling water discharges. There is a very large literature on the importance of water temperature to aquatic organisms; preference, avoidance, and lethal temperature ranges have been derived for many aquatic species (e.g., Brungs and Jones, 1977; Coutant, 1977; Eaton et al., 1995). In addition, temperature is one of the key classification strata for aquatic life, in that streams are routinely classified as cold water, cool water, or warm water based on the geographic and natural settings of waters. The removal of catchment and riparian vegetation and the general increase in surface runoff from impervious, man-made, and heat-capturing surfaces has been associated with increasing water temperatures in urban waterbodies (Wang and Kanehl, 2003; Nelson and Palmer, 2007). A number of researchers have created models to predict in-stream temperatures based on urban characteristics (Krause et al., 2004; Herb et al., 2008).

Hydrologic Influences on Aquatic Life

The importance of "natural" flow regimes on aquatic life has been well documented (Poff et al., 1997; Richter et al., 1997a, 2003). As watersheds urbanize, flow regimes change from little runoff to over 40 to 90 percent of the rainfall becoming surface runoff (Roesner and Bledsoe, 2003). Flow regimes in urban streams typically are very "flashy," with higher and more frequent peak events, compared to undisturbed systems (Poff et al., 1997; Baker et al., 2004) and well as reduced base flows and more frequent desiccation (Bernhardt and Palmer, 2007). Richter et al. (1996) proposed a series of indicators that could be used to measure hydrologic disturbance, many of which have been used in the recent studies identifying the hydrologic effects of stormwater on aquatic biota (Barbour et al., 2008). Pomeroy et al. (2008) did an extensive review of which flow characteristics appear to have the greatest influence on biological metrics and biological integrity. No single measure of flow was found to be significant in all studies; however, important attributes included flow variability and flashiness, flood frequency, flow volume, flow variability, flow timing, and flow duration.

There are a number of mechanisms that may be responsible for the influence of flow characteristics on aquatic assemblages. Aquatic species vary dramatically in their swimming performance and behaviors, and species are generally adapted to undisturbed flow regimes in an area. Many low- to moderate-gradient small streams in the United States, for example, have strong connections with their flood-prone areas and often possess habitat features that insulate poor swimming species from episodic natural high flows. Undercut banks, rootwads, oxbows, and backwater habitats all can act as refugia from high flows. Some aquatic species are more or less mobile within the sediments, like certain macroinvertebrates (meiofauna or hyporheos) and fish species such as sculpins and madtoms. Secondary impacts from hydrologic changes such as bank erosion and aggradation of fines can render substrates embedded and prohibit organisms, particularly the meiofauna, from moving vertically within the bottom substrates (Schmid-Araya, 2000). Substrate fining has been documented to occur with increasing urbanization, especially in the early stages of development, which can embed spawning habitats and eliminate or reduce spawning success of fish such as salmonids and minnows (Waters, 1995).

Flood flows can cause mortality in the absence of urbanization. For example, flood flows in streams under natural conditions have been documented as a cause of substantial mortality in young or larval fish such as smallmouth bass (Funk and Fleener, 1974; Lorantas and Kristine, 2004). Increased flashiness from urbanization is likely to exacerbate this effect. Thus, increases in the frequency of peak flows during spring will increase the probability of spawning failure, such that sensitive species may eventually be locally extirpated. In urban areas, culverts and other flow obstructions can create conditions that may preclude re-colonization of upstream reaches because weak-swimming fishes cannot move past flow constrictions or leap past vertical drops caused by artificial structures.

Hydrologic simplification and stream straightening that occur in urban streams, often as a result of increased peak flows or as a local management response, typically remove habitat used as temporary refuges from high flows, such as backwater areas, undercut banks, and rootwads. There is a large literature relating populations of fish and macroinvertebrates to various habitat features of streams, rivers, and wetlands. The first two attributes of the BCG identify taxa that are historically documented, sensitive, long-lived, or regionally endemic taxa or sensitive-rare taxa. Many of these taxa are endangered because of large-scale changes in flow-influenced habitats; that is, threats of extinction often center on habitat degradation that influence spawning, feeding, or other aspects of a species life history (Rieman et al., 1993). In contrast, many of the fish and macroinvertebrate taxa that compose regional lists of tolerant taxa are tolerant to habitat changes related to flow disturbance as well as chemical parameters. Understanding the life history attributes of certain species and how they may change with multiple stressors (Power, 1997) is an important tool for understanding complex responses of aquatic ecosystems to urban stressors.

Geomorphic and Habitat Influences on Aquatic Life

In natural waters, geomorphic factors and climate, modified by vegetation and land use, constrain the types of physical habitat features likely to occur in streams (Webster and D'Angelo, 1997). For example, very-low-gradient streams may have few riffles and be dominated by woody debris and bank cover, whereas higher gradient waters may have more habitat types formed by rapidly flowing waters (riffles, runs). Aquatic life in streams is influenced directly by the habitat features that are present, such as substrate types, in-stream structures, bank structure, and flow types (e.g., deep-fast vs. shallow-slow).

As discussed previously, human alteration of landscapes, encroachment on riparian areas, and direct channel modifications (e.g., channelization) that acompany urbanization have often resulted in unstable channels, with negative consequences for aquatic habitat. As urbanization has increased, channel density has declined because streams have been piped, dewatered, and straightened (Meyer and Wallace, 2001; Paul and Meyer, 2001). Changes in the magnitude, relative proportions, and timing of sediment and water delivery have resulted in loss of aquatic life and habitat via a wide range of mechanisms, including changes in channel bed materials, increased suspended sediment loads, loss of riparian habitat due to bank erosion, and changes in the variability of flow and sediment transport characteristics relative to aquatic life cycles (Roesner and Bledsoe, 2003). There are still significant gaps in knowledge about how stormwater stressors can affect stream habitat, especially as one moves from the reach scale to the watershed scale. Understanding the stage and trajectory of channel evolution is critical to understanding channel recovery and expected habitat conditions or in choosing effective restoration options (Simon et al., 2007).

Across much of the United States, stream habitats have been altered to the imperilment of aquatic species (Williams et al., 1989; Richter et al., 1997b; Strayer et al., 2004). A study of rapidly urbanizing streams in central Ohio identified the loss of highly and moderately sensitive species as a key factor the decline in the IBI in these streams (Miltner et al., 2004). These streams had historical fish collections when they were primarily influenced by agricultural land use; sampling after the onset of suburban development documented the loss of many of these species attributable to land-use changes and habitat degradation along these urban streams. Along the BCGs that have been developed for streams, most of the species in attributes 1–3 are specialists requiring very specific habitats for spawning, feeding, and refuge. Habitat alteration, either direct or indirect, creates harsh environments that tend to favor tolerant taxa, which would otherwise be in low abundance. Often these tolerant species are characterized by high reproductive potential, generalist feeding behaviors, tolerance to chemical stressors such as low DO, and pioneering strategies that allow rapid recolonization following acute stressful events.

Altered Energy Pathways in Urban Streams

The pathways of energy flow in streams are an important determinant of aquatic species distributions. In most natural temperate streams, headwaters transform and export energy from stream side vegetation and adjacent land uses into aquatic biomass. The types, amount, and timing of delivery of water, organic material, and debris have important consequences for conditions downstream (Dolloff and Webster, 2000). The energy-transforming aspect of stream ecosystems is difficult to capture directly, so most measures are surrogates, such as the trophic characteristics of assemblages and chemical and physical characteristics consistent with natural energy processes.

An increasingly urban landscape can have a complex array of effects on energy dynamics in streams (Allan, 2004). Loss of riparian areas and changes in riparian vegetation can reduce the supply and quality of coarse organic matter that forms the base of aquatic food webs in most small streams. The reduction in the amount of organic matter with riparian loss is obvious; however, changing species of vegetation (e.g., invasion or planting of exotic species) can affect the quality of organic matter and influence higher trophic levels because, for example, exotic species may have different nutrient values (e.g., C/N ratios, trace chemicals) or process nutrients at a different rate (Royer et al., 1999). Furthermore, native invertebrate taxa may not be adapted to utilize the exotic material (Miller and Boulton, 2005). For example, changes in leaf species in a stream may alter the macroinvertebrate community by favoring species that feed on fast-decaying versus slow-decaying leaves (Smock and MacGregor, 1988; Cummins et al., 1989; Gregory et al., 1991).

Other recent work is examining ways that changes in geomorphology with increasing urbanization can influence trophic structure in streams (Doyle, 2006). Groffman et al. (2005) examined nitrogen processing in stream geomorphic structures such as bars, riffles, and debris dams in suburban and forested areas. Although suburban areas had high rates of production in organic-rich debris dams and gravel bars, higher storm flow effects in urban streams may make these features less stable and able to be maintained (Groffman et al., 2005). Changes in habitat and riparian vegetation may greatly alter trophic patterns of energy transport. For example, local nutrient enrichments combined with reduced riparian vegetation can result in nuisance algal growths in waterbodies that are evidence of simpler energy pathways. Corresponding effects are further water chemistry changes from algal decomposition (e.g.., low DO) or very high algal activity (e.g., high pH) (Ehlinger et al., 2004).

The complexity of energy flow through simple ecosystems is illustrated in Figure 3-42, a "simplified" food web of a headwater stream published by Meyer (1994). The forms in which nutrients are delivered to streams may be more important than actual concentrations as well as the availability of carbon sources essential for nutrient transformation. The nutrient components that form the base of the food web in Figure 3-42 are the FPOM and CPOM boxes. In many natural streams, woody and leafy debris are the most common form of nutrient

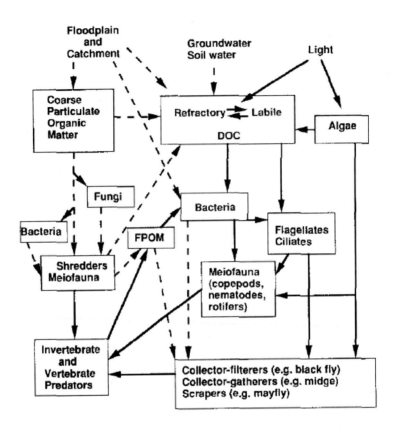

FIGURE 3-42 Simplified diagram of a lotic food web showing sources and major pathways of organic carbon. Dotted lines indicate flows that are a part of the microbial loop in flowing water but not in planktonic systems. SOURCE: Reprinted, with permission, from Meyer (1994). Copyright 1994 by Springer.

input, and changes to urban landscapes often change this to dissolved and finer forms. Urbanization can also reduce the retention of organic debris of streams (Groffman et al., 2005) and the timing of nutrient delivery. Timing can be of crucial importance since species spawning and growth periods may be specifically timed to take advantage of available nutrients.

As important as energy and nutrient dynamics are to stream function, many of the stream characteristics that determine effective energy flow are not typically considered when characterizing stormwater impacts. The best chance for considering these variables and maximizing ecosystem function is through integrated, biologically based monitoring programs that include urban areas (Barbour et al., 2008) and stressor identification procedures (EPA, 2000) to isolate likely causes of impact and to inform the choices of SCMs.

Biological Interactions in Urban Streams

Streams in urbanized environments often are characterized by fewer native and more alien species than natural streams (DeVivo, 1996; Meador et al., 2005). The influence of exotic species is not always predictable and may be most severe in lentic environments (e.g., wetlands, estuaries) and in riparian zones where various exotic aquatic plants can greatly alter natural systems in both structure and function (Hood and Naiman, 2000). Riley et al. (2005) foundthat the presence of alien aquatic amphibians was positively related to degree of urbanization, as was the absence of certain native amphibian species. In a review of possible reasons for this observation, he suggested that altered flow regimes were responsible. In the arid California streams they studied, flow became more constant with urbanization (i.e., natural streams were generally ephemeral), which allowed invasion by exotic species that can prey on, compete with, or hybridize with native species (Riley et al., 2005). The alteration of stream habitat that accompanies urbanization can also lead to predation by domestic cats and dogs or collection by humans, especially where species (e.g., California newts) are large and conspicuous (Riley et al., 2005).

The effects of specific exotic species on aquatic systems has been observed to vary geographically, although recent work has found correlations between total invasion rate and the number of high-impact exotic species (Ricciardi and Kipp, 2008). This suggests that overall efforts to reduce the importation or spread of all alien species should be helpful.

The Role of Biological Monitoring

The preceding sections illustrate the importance of biological data to understanding the complexities associated with urban and stormwater impacts to waterbodies. Although categories of urban stressors have been discussed individually, these stressors routinely, if not universally, co-occur in urban waterbodies. Their cumulative impacts are best measured with biological tools because the biota integrate the influence of all of these stressors.

Many programmatic aspects of the CWA arose as a response to rather obvious impacts of chemical pollutants that were occurring in surface waters during this time. The initial focus of water quality standards was on developing chemical criteria that could serve as engineering endpoints for waste treatment systems (e.g., NPDES permits). Rather general aquatic life goals for streams and rivers that were suitable for the initial focus of the CWA are now considered insufficient to deal with the complex suite of stressors limiting aquatic systems.

To that end, refined aquatic life goals and improved biological monitoring are essential for effective water quality management, including stormwater issues (NRC, 2001). Practical biological and physical monitoring tools have even been developed for very small headwater streams (Ohio EPA, 2002; Fritz et al., 2006), which are particularly affected by stormwater because of their prevalence (greater than 95 percent of channels), their relatively high surface-to-volume ratio, their role in nutrient and material processing, and their vulnerability to direct modification such as channelization and piping (Meyer and Wallace, 2001).

Surrogate indicators of stormwater impacts to aquatic life (such as TSS concentrations) have been widely used because direct biological measures were poorly developed and these surrogates were assumed to be important to pollutant delivery to urban streams. However, biological assessment has rapidly advanced in many states and can be readily applied or if needed modified to be sensitive to stormwater stressors (Barbour et al., 2008). As Karr and Chu (1999) warned, the management of complex systems requires measures that integrate multiple factors. Stormwater permitting is no different, and care must be taken to ensure that permitting and regulatory actions retain ecological relevance. Surrogate measures have an essential role in the assessment of individual SCMs; however, this needs to be kept in context with the entire suite of stressors likely to be important to the aquatic life goals in streams.

Stormwater management programs should not necessarily bear the burden of biological monitoring; rather, well-conceived biological monitoring should be the prevue of state and local government agencies (as discussed more extensively in Chapter 6). Refined aquatic life goals developed for all waters, including urban waters, measured with appropriate biological measures, should be the final endpoint for management. The collection of biological data needs to be closely integrated across multiple disciplines in order to be effective. Pomeroy et al. (2008) describe a multidisciplinary approach to study the effects of stormwater in urban settings, and Scholz and Booth (2001) also propose a monitoring approach for urban watersheds. Such efforts are not necessarily easy, and many institutions find pitfalls when trying to integrate scientific information across disciplines (Benda et al., 2002).

EPA water programs, such as the Total Maximum Daily Load (TMDL) program, have been criticized for having too narrow a focus on a limited number of traditional pollutants to the exclusion of important stressors such as hydrology, habitat alteration, and invasive taxa (Karr and Yoder, 2004)—all serious problems associated with stormwater and urbanization. The science has advanced significantly over the past decade so that biological assessment should be an essential tool for identifying stormwater impacts and informing the choice of SCMs in a region or watershed. Although biological responses to stressors in the ambient environment are by their nature correlative exercises, ecological epidemiology principles or "stressor identification" methods can identify likely causative agents of impairment with relatively high certainty in many instances (Suter, 1993, 2006; EPA, 2000). Coupled with other ambient and source monitoring information, biological information can form the basis for an effective stormwater program. As an example, Box 3-10 introduces the Impervious Cover Model (ICM), which was developed using correlative information on the association between impervious cover and biological metrics. The crux of the ICM is that stormwater management is tailored along a readily measureable gradient (impervious cover) that integrates multiple individual stressor categories that would otherwise be overlooked in the traditional pollutant-based approach to stormwater management. Even the form of the ICM (as conceptualized in Figure 3-43) matches that outlined for the BCG (Figure 3-38). Use of the ICM to improve the MS4 stormwater program is discussed in Chapter 6.

Human Health Impacts

Despite the unequivocal evidence of ecosystem consequences resulting from urban stormwater, a formal risk analysis of the human health effects associated with stormwater runoff is not yet possible. This is because (1) many of the most important waterborne pathogens have not been quantified in stormwater, (2) enumeration methods reported in the current literature are disparate and do not account for particle-bound pathogens, and (3) sampling times during storms have not been standardized nor are known to have occurred during periods of human exposure. Individual studies have investigated the runoff impacts on public health in freshwater (Calderon et al., 1991) and marine waters (Haile et al., 1999; Dwight et al,. 2004; Colford et al., 2007). Although these studies provide ample evidence that stormwater runoff can serve as a vector of pathogens with potential health implications (for example, Ahn et al., 2005, found that fecal indicator bacteria concentrations could exceed California ocean bathing water standards by up to 500 percent in surf zones receiving stormwater runoff), it is difficult to draw conclusive inferences about the specific human health impacts from microbial contamination of stormwater. Calderon et al. (1991) concluded that the currently recommended bacterial indicators are ineffective for predicting potential health effects associated with water contaminated by nonpoint sources of fecal pollution. Furthermore, in a study conducted in Mission Bay, California, which analyzed bacterial indicators using traditional and nontraditional methods (chromogenic substrate and quantitative polymerase chain reaction), as well as a novel bacterial indicator and viruses, traditional fecal indicators were not associated with identified human health risks such as diarrhea and skin rash (Colford et al., 2007).

The Santa Monica Bay study (Haile et al., 1999) indicated that the risks of several health outcomes were higher for people who swam at storm-drain locations compared to those who swam farther from the drain. However, the list of health outcomes that were more statistically significant (fever, chills, ear discharge, cough and phlegm, and significant respiratory) did not include highly

BOX 3-10
The Impervious Cover Model: An Emerging Framework
for Urban Stormwater Management

The Impervious Cover Model (ICM) is a management tool that is useful for diagnosing the severity of future stream problems in a subwatershed. The ICM defines four categories of urban streams based on how much impervious cover exists in their subwatershed: *high-quality streams, impacted streams, non-supporting streams,* and *urban drainage.* The ICM is then used to develop specific quantitative or narrative predictions for stream indicators within each stream category (see Figure 3-43). These predictions define the severity of current stream impacts and the prospects for their future restoration. Predictions are made for five kinds of urban stream impacts: changes in stream hydrology, alteration of the stream corridor, stream habitat degradation, declining water quality, and loss of aquatic diversity.

The general predictions of the ICM are as follows. Stream segments with less than 10 percent impervious cover (IC) in their contributing drainage area continue to function as **Sensitive Streams**, and are generally able to retain their hydrologic function and support good-to-excellent aquatic diversity. Stream segments that have 10 to 25 percent IC in their contributing drainage area behave as **Impacted Streams** and show clear signs of declining stream health. Most indicators of stream health will fall in the fair range, although some segments may range from fair to good as riparian cover improves. The decline in stream quality is greatest toward the higher end of the IC range. Stream segments that range between 25 and 60 percent subwatershed impervious cover are classified as **Non-Supporting Streams** (i.e., no longer supporting their designated uses in terms of hydrology, channel stability habitat, water quality, or biological diversity). These stream segments become so degraded that any future stream restoration or riparian cover improvements are insufficient to fully recover stream function and diversity (i.e., the streams are so dominated by subwatershed IC that they cannot attain predevelopment conditions). Stream segments whose subwatersheds exceed 60 percent IC are physically altered so that they merely function as a conduit for flood waters. These streams are classified as **Urban Drainage** and consistently have poor water quality, highly unstable channels, and very poor habitat and biodiversity scores. In many cases, these urban stream segments are eliminated altogether by earthworks and/or storm-drain enclosure. Table 3-14 shows in greater detail how stream corridor indicators respond to greater subwatershed impervious cover.

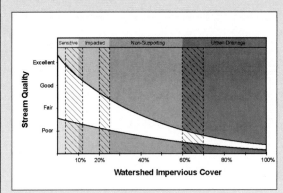

FIGURE 3-43 Changes in Stream Quality with Percent Impervious Cover in the Contributing Watershed. SOURCE: Chesapeake Stormwater Network (2008). Reprinted, with permission, from Chesapeake Stormwater Network (2008). Copyright 2008 by Chesapeake Stormwater Network.

TABLE 3-14 General ICM Predictions Based on Urban Subwatershed Classification (CWP, 2004):

Prediction	Impacted (IC 11 to 25%) [8]	Non-supporting (IC 26 to 60%)	Urban Drainage (IC > 60%)
Runoff as a Fraction of Annual Rainfall [1]	10 to 20%	25 to 60%	60 to 90%
Frequency of Bankfull Flow per Year [2]	1.5 to 3 per year	3 to 7 per year	7 to 10 per year
Fraction of Original Stream Network Remaining	60 to 90%	25 to 60%	10 to 30%
Fraction of Riparian Forest Buffer Intact	50 to 70%	30 to 60%	Less than 30%
Crossings per Stream Mile	1 to 2	2 to 10	None left
Ultimate Channel Enlargement Ration [3]	1.5 to 2.5 larger	2.5 to 6 times larger	6 to 12 times larger
Typical Stream Habitat Score	Fair, but variable	Consistently poor	Poor, often absent
Increased Stream Warming [4]	2 to 4 °F	4 to 8 °F	8+ °F
Annual Nutrient Load [5]	1 to 2 times higher	2 to 4 times higher	4 to 6 times higher
Wet Weather Violations of Bacteria Standards	Frequent	Continuous	Ubiquitous
Fish Advisories	Rare	Potential risk of accumulation	Should be presumed
Aquatic Insect Diversity [6]	Fair to good	Fair	Very poor
Fish Diversity [7]	Fair to good	Poor	Very poor

[1] Based on annual storm runoff coefficient; ranges from 2 to 5% for undeveloped streams.
[2] Predevelopment bankfull flood frequency is about 0.5 per year, or about one bankfull flood every two years.
[3] Ultimate stream-channel cross-section compared to typical predevelopment channel cross section.
[4] Typical increase in mean summer stream temperature in degrees Fahrenheit, compared with shaded rural stream.
[5] Annual unit-area stormwater phosphorus and/or nitrogen load produced from a rural subwatershed.
[6] As measured by benthic index of biotic integrity. Scores for rural streams range from good to very good.
[7] As measured by fish index of biotic integrity. Scores for rural streams range from good to very good.
[8] IC is not the strongest indicator of stream health below 10% IC, so the sensitive streams category is omitted from this table.
SOURCE: Adapted from Schueler (2004).

continues next page

BOX 3-10 Continued

Scientific Support for the ICM

The ICM predicts that hydrological, habitat, water quality, and biotic indicators of stream health first begin to decline sharply at around 10 percent total IC in smaller catchments (Schueler, 1994). The ICM has since been extensively tested in ecoregions around the United States and elsewhere, with more than 200 different studies confirming the basic model for single stream indicators or groups of stream indicators (CWP, 2003; Schueler, 2004). Several recent research studies have reinforced the ICM as it is applied to first- to third-order streams (Coles et al., 2004; Horner et al., 2004; Deacon et al., 2005; Fitzpatrick et al., 2005; King et al., 2005; McBride and Booth, 2005; Cianfrina et al., 2006; Urban et al., 2006; Schueler et al., 2008).

Researchers have focused their efforts to define the specific thresholds where urban stream degradation first begins. There is robust debate as to whether there is a sharp initial threshold or merely a continuum of degradation as IC increases, although the latter is more favored. There is much less debate, however, about the dominant role of IC in defining the hydrologic, habitat, water quality, and biodiversity expectations for streams with higher levels of IC (15 to 60 percent).

Caveats to the ICM

The ICM is a powerful predictor of urban stream quality when used appropriately. The first caveat is that subwatershed IC is defined as total impervious area (TIA) and not effective impervious area (EIA). Second, the ICM should be restricted to first- to third-order alluvial streams with moderate gradient and no major point sources of pollutant discharge. The ICM is most useful in projecting the behavior of numerous stream health indicators, and it is not intended to be accurate for every individual stream indicator. In addition, management practices in the contributing catchment or subwatershed must not be poor (e.g., no deforestation, acid mine drainage, intensive row crops, etc.); just because a subwatershed has less than 10 percent IC does not automatically mean that it will have good or excellent stream quality if past catchment management practices were poor.

ICM predictions are general and may not apply to every stream within the proposed classifications. Urban streams are notoriously variable, and factors such as gradient, stream order, stream type, age of subwatershed development, and past land use can and will make some streams depart from these predictions. Indeed, these "outlier" streams are extremely interesting from the standpoint of restoration. In general, subwatershed IC causes a continuous but variable decline in most stream corridor indicators. Consequently, the severity of individual indicator impacts tends to be greater at the upper end of the IC range for each stream category.

Effects of Catchment Treatment on the ICM

Most studies that investigated the ICM were done in communities with some degree of catchment treatment (e.g., stormwater management or stream buffers). Detecting the effect of catchment treatment on the ICM involves a very complex and difficult paired watershed design. Very few catchments meet the criteria for either full treatment or the lack of it, no two catchments are ever really identical, and individual catchments exhibit great variability from year to year. Not surprisingly, the first generation of research studies has produced ambiguous results. For example, seven research studies showed that ponds and wetlands are unable to prevent the degradation of aquatic life in downstream channels associated with higher levels of IC (Galli, 1990; Jones et al., 1996; Horner and May, 1999; Maxted, 1999; MNCPPC, 2000; Horner et al., 2001; Stribling et al., 2001). The primary reasons cited are stream warming (amplified by ponds), changes in organic matter processing, the increased runoff volumes delivered to downstream channels, and habitat degradation caused by channel enlargement.

Riparian forest cover is defined as canopy cover within 100 meters of the stream, and is measured as the percentage of the upstream network in this condition. Numerous researchers have evaluated the relative impact of riparian forest cover and IC on stream geomorphology, aquatic insects, fish assemblages, and various indices of biotic integrity. As a group, the studies suggest that indicator values for urban streams improve when riparian forest cover is retained over at least 50 to 75 percent of the length of the upstream network (Booth et al., 2002; Morley and Karr, 2002; Wang et al., 2003; Allan, 2004; Sweeney et al., 2004; Moore and Palmer, 2005; Cianfrina et al., 2006; Urban et al., 2006).

Application of the ICM to other Receiving Waters

Recent research has focused on the potential value of the ICM in predicting the future quality of receiving waters such as tidal coves, lakes, wetlands and small estuaries. The primary work on small estuaries by Holland et al. (2004) [references cited in CWP (2003), Lerberg et al. (2000)] indicates that adverse changes in physical, sediment, and water quality variables can be detected at 10 to 20 percent subwatershed IC, with a clear biological response observed in the range of 20 to 30 percent IC. The primary physical changes involve greater salinity fluctuations, greater sedimentation, and greater pollutant contamination of sediments. The biological response includes declines in diversity of benthic macroinvertebrates, shrimp, and finfish.

More recent work by King et al. (2005) reported a biological response for coastal plain streams at around 21 to 32 percent urban development (which is usually about twice as high as IC). The thresholds for important water quality indicators such as bacterial exceedances in shellfish beds and beaches appears to begin at about 10 percent subwatershed IC, with chronic violations observed at 20 percent IC (Mallin et al., 2001). Algal blooms and anoxia resulting from nutrient enrichment by stormwater runoff also are routinely noted at 10 to 20 percent subwatershed IC (Mallin et al., 2004).

The primary conclusion to be drawn from the existing science is that the ICM does apply to tidal coves and streams, but that the impervious levels associated with particular biological responses appear to be higher (20 to 30 percent IC for significant declines) than for freshwater streams, presumably due to their greater tidal mixing and inputs from nearshore ecosystems. The ICM may also apply to lakes (CWP, 2003) and freshwater wetlands (Wright et al., 2007) under carefully defined conditions. The initial conclusion is that the application of the ICM shows promise under special conditions, but more controlled research is needed to determine if IC (or other watershed metrics) is useful in forecasting receiving water quality conditions.

continues next page

BOX 3-10 Continued

Utility of the ICM in Urban Stream Classification and Watershed Management

The ICM is best used as an urban stream classification tool to set reasonable expectations for the range of likely stream quality indicators (e.g., physical, hydrologic, water quality, habitat, and biological diversity) over broad ranges of subwatershed IC. In particular, it helps define general thresholds where water quality standards or biological narrative conditions cannot be consistently met during wet weather conditions (see Table 6-2). These predictions help stormwater managers and regulators to devise appropriate and geographically explicit stormwater management and subwatershed restoration strategies for their catchments as part of MS4 permit compliance. More specifically, assuming that local monitoring data are available to confirm the general predictions of the ICM, it enables managers to manage stormwater within the context of current and future watershed conditions.

---

credible gastrointestinal illness, which is curious because the vast majority of epidemiological studies worldwide suggests a causal dose-related relationship between gastrointestinal symptoms and recreational water quality measured by bacterial indicator counts (Pruss, 1998). Dwight et al. (2004) found that surfers in an urban environment reported more symptoms than their rural counterparts; however, water quality was not specifically evaluated in that study.

To better assess the relationship between swimming in waters contaminated by stormwater, which have not been influenced by human sewage, and the risk of related illness, the California Water Boards and the City of Dana Point have initiated an epidemiological study. This study will be conducted at Doheny Beach, Orange County, California, which is a beach known to have high fecal indicator bacteria concentrations with no known human source. The project will examine new techniques for measuring traditional fecal indicator bacteria, new species of bacteria, and viruses to determine whether they yield a better relationship to human health outcomes than the indicators presently used in California. The study is expected to be completed in 2010. In addition, the State of California is researching new methods for rapid detection of beach bacterial indicators and ways to bring these methods into regular use by the environmental monitoring and public health communities to better protect human health.

CONCLUSIONS AND RECOMMENDATIONS

The present state of the science of stormwater reflects both the strengths and weaknesses of historic, monodisciplinary investigations. Each of the component disciplines—hydrology, geomorphology, aquatic chemistry, ecology, land use, and population dynamics—have well-tested theoretical foundations and useful predictive models. In particular, there are many correlative studies showing how parameters co-vary in important but complex and poorly understood ways (e.g., changes in fish community associated with watershed road density or the percentage of IC). Nonetheless, efforts to create mechanistic links between population growth, land-use change, hydrologic alteration, geomorphic adjustments, chemical contamination in stormwater, disrupted energy flows, and biotic interactions, to changes in ecological communities are still in development. Despite this assessment, there are a number of overarching truths that remain poorly integrated into stormwater management decision making, although they have been robustly characterized and have a strong scientific basis. These are expanded upon below.

There is a direct relationship between land cover and the biological condition of downstream receiving waters. The possibility for the highest levels of aquatic biological condition exists only with very light urban transformation of the landscape. Even then, alterations to biological communities have been documented at such low levels of imperviousness, typically associated with roads and the clearing of native vegetation, that there has been no real "urban development" at all. Conversely, the lowest levels of biological condition are inevitable with extensive urban transformation of the landscape, commonly seen after conversion of about one-third to one-half of a contributing watershed into impervious area. Although not every degraded waterbody is a product of intense urban development, all highly urban watersheds produce severely degraded receiving waters. Because of the close and, to date, inexorable linkage between land cover and the health of downstream waters, stormwater management is an unavoidable offshoot of watershed-based land-use planning (or, more commonly, its absence).

The protection of aquatic life in urban streams requires an approach that incorporates all stressors. Urban Stream Syndrome reflects a multitude of effects caused by altered hydrology in urban streams, altered habitat, and polluted runoff. Focusing on only one of these factors is not an effective management strategy. For example, even without noticeably elevated pollutant concentrations in receiving waters, alterations in their hydrologic regimes are associated with impaired biological condition. Achieving the articulated goals for stormwater management under the CWA will require a balanced approach that incorporates hydrology, water quality, and habitat considerations.

The full distribution and sequence of flows (i.e., the flow regime) should be taken into consideration when assessing the impacts of stormwater on streams. Permanently increased stormwater volume is only one aspect of an urban-altered storm hydrograph. It contributes to high in-stream velocities, which in turn increase streambank erosion and accompanying sediment pollution of surface water. Other hydrologic changes, however, include changes in the sequence and frequency of high flows, the rate of rise and fall of the hydrograph, and the season of the year in which high flows can occur. These all can affect both the physical and biological conditions of streams, lakes, and wetlands. Thus, effective hydrologic mitigation for urban development cannot just aim to reduce post-development peak flows to predevelopment peak flows.

A single design storm cannot adequately capture the variability of rain and how that translates into runoff or pollutant loadings, and thus is not suitable for addressing the multiple objectives of stormwater management. Of particular importance to the types of problems associated with urbanization is the size of rain events. The largest and most infrequent rains cause near-bankfull conditions and may be most responsible for habitat destruction; these are the traditional "design storms" used to design safe drainage systems. However, moderate-sized rains are more likely to be associated with most of the annual mass discharges of stormwater pollutants, and these can be very important to the eutrophication of lakes and nearshore waters. Water quality standards for bacterial indicators and total recoverable heavy metals are exceeded for almost *every* rain in urban areas. Therefore, the whole distribution of storm size needs to be evaluated for most urban receiving waters because many of these problems coexist.

Roads and parking lots can be the most significant type of land cover with respect to stormwater. They constitute as much as 70 percent of total impervious cover in ultra-urban landscapes, and as much as 80 percent of the directly connected impervious cover. Roads tend to capture and export more stormwater pollutants than other land covers in these highly impervious areas because of their close proximity to the variety of pollutants associated with automobiles. This is especially true in areas of the country having mostly small rainfall events (as in the Pacific Northwest). As rainfall amounts become larger, pervious areas in most residential land uses become more significant sources of runoff, sediment, nutrients, and landscaping chemicals. In all cases, directly connected impervious surfaces (roads, parking lots, and roofs that are directly connected to the drainage system) produce the first runoff observed at a storm-drain inlet and outfall because their travel times are the quickest.

Generally, the quality of stormwater from urbanized areas is well characterized, with the common pollutants being sediment, metals, bacteria, nutrients, pesticides, trash, and polycyclic aromatic hydrocarbons. These results come from many thousands of storm events from across the nation, systematically compiled and widely accessible; they form a robust data set of utility to theoreticians and practitioners alike. These data make it possible to accurately estimate pollutant concentrations, which have been shown to vary by land cover and by region across the country. However, characterization data are relatively sparse for individual industrial operations, which makes these sources less amenable to generalized approaches based on reliable assumptions of pollutant types and loads. In addition, industrial operations vary greatly from site to site, such that it may be necessary to separate them into different categories in order to better understand industrial stormwater quality.

Nontraditional sources of stormwater pollution must be taken into consideration when assessing the overall impact of urbanization on receiving waterbodies. These nontraditional sources include atmospheric deposition, snowmelt, and dry weather discharges, which can constitute a significant portion of annual pollutant loadings from storm systems in urban areas (such as metals in Los Angeles). For example, atmospheric deposition of metals is a very significant component of contaminant loading to waterbodies in the Los Angeles region relative to other point and nonpoint sources. Similarly, much of the sediment found in receiving waters following watershed urbanization can come from streambank erosion as opposed to being contributed by polluted stormwater.

Biological monitoring of waterbodies is critical to better understanding the cumulative impacts of urbanization on stream condition. Over 25 years ago, individual states developed the concept of regional reference sites and developed multi-metric indices to identify and characterize degraded aquatic assemblages in urban streams. Biological assessments respond to the range of non-chemical stressors identified as being important in urban waterways including habitat degradation, hydrological alterations, and sediment and siltation impacts, as well as to the influence of nutrients and other chemical stressors where chemical criteria do not exist or where their effects are difficult to measure directly (e.g., episodic stressors). The increase in biological monitoring has also helped to frame issues related to exotic species, which are locally of critical importance but completely unrecognized by traditional physical monitoring programs.

Epidemiological studies on the human health risks of swimming in freshwater and marine waters contaminated by urban stormwater discharges in temperate and warm climates are needed. Unlike with aquatic organisms, there is little information on the health risks of urban stormwater to humans. Standardized watershed assessment methods to identify the sources of human pathogens and indicator organisms in receiving waters need to be developed, especially for those waters with a contact-recreation use designation that have had multiple exceedances of pathogen or indicator criteria in a relatively short period of time. Given their difficulty and expense, epidemiological studies should be undertaken only after careful characterization of water quality and stormwater flows in the study area.

REFERENCES

Abrahams, A. 1984. Channel networks: a geomorphological perspective. Water Resources Research 20:161–168.

Adeniyi, I. F., and I. O. Olabanji. 2005. The physico-chemical and bacteriological quality of rainwater collected over different roofing materials in Ile-Ife, southwestern Nigeria. Chemistry and Ecology 21(3):149–166.

Ahn, J. H., S. B. Grant, C. Q. Surbeck, P. M. DiGiacomo, N. P. Nezlin, and S. Jiang. 2005. Coastal water quality impact of stormwater runoff from an urban watershed in Southern California. Environ. Sci. Technol. 39(16):5940–5953.

Allan, J. D. 2004. Landscapes and riverscapes: the influence of land use on stream ecosystems. Annual Review of Ecology, Evolution, and Systematics 35:257–284.

Anonymous. 2000. Fish kill at Pinch Gut Creek. The Birmingham News (AL). April 18.

Arkoosh, M. R., E. Clemons, P. Huffman, and A. N. Kagley. 2001. Increased susceptibility of juvenile chinook salmon to *Vibriosis* after exposure to chlorinated and aromatic compounds found in contaminated urban estuaries. Journal of Aquatic and Animal Health 13:257–268.

Athansiadis, K., B. Helmreich, and P. A. Wilderer. 2006. Infiltration of a copper roof runoff through artificial barriers. Water Science and Technology 54(6–7):281–289.

Bailey, H. C., J. R. Elphick, A. Potter, and B. Zak. 1999. Zinc toxicity in stormwater runoff from sawmills in British Columbia. Water Research 33(11):2721–2725.

Baker, D. B., R. P. Richards, T. T. Loftus, and J. W. Kramer. 2004. A new flashiness index: Characteristics and applications to midwestern rivers and streams. Journal of the American Water Resources Association 40(2):503–522.

Bannerman, R., K. Baun, M. Bohn, P. E. Hughes, and D. A. Graczyk. 1983. Evaluation of Urban Nonpoint Source Pollution Management in Milwaukee County, Wisconsin, Vol. I. PB 84-114164. EPA, Water Planning Division.

Bannerman, R. T., D. W. Owens, R. B. Dodds, and N. J. Hornewer. 1993. Sources of pollutants in wisconsin stormwater. Water Science and Technology 28(3–5):241–259.

Barbour, M. T., M. J. Paul, D. W. Bressler, A. H. Purcell, V. H. Resh, and E. T. Rankin. 2008. Bioassessment: a tool for managing aquatic life uses for urban streams. Water Environment Research Foundation Research Digest 01-WSM-3.

Barrie, L. A. 1991. Snow formation and processes in the atmosphere that influence its chemical composition. Pp. 1–20 *In:* Seasonal Snowpacks, NATO ASI Series G, Ecological Sciences, Vol. 28. T. D. Davies, M. Tranter, and H. G. Jones (eds.). Springer: Berlin.

Beck, R., L. Kolankiewicz, and S. A. Camarota. 2003. Outsmarting Smart Growth: Population Growth, Immigration, and the Problem of Sprawl. Washington, DC: Center for Immigration Studies. Available at *http://www.cis.org/articles/2003/SprawlPaper.pdf*. Last accessed October 31, 2007.

Benda, L. E., N. L. Poff, C. Tague, M. A. Palmer, J. Pizzuto, S. Cooper, E. Stanley, and G. Moglen. 2002. How to avoid train wrecks when using science in environmental problem solving. BioScience 52(12):1127–1136.

Bennett, E. R., K. D. Linstedt, V. Nilsgard, G. M. Battaglia, and F. W. Pontius. 1981. Urban snowmelt—characteristics and treatment. Journal Water Pollution Control Federation 53(1):119–125.

Bernhardt, E. S., and M. A. Palmer. 2007. Restoring streams in an urbanizing world. Freshwater Biology 52:738–751.

Beyers, D. W., and M. S. Farmer. 2001. Effects of copper on olfaction of Colorado pikeminnow. Environmental Toxicology and Chemistry 20:907–912.

Bochis, C. 2007. The Magnitude of Impervious Surfaces in Urban Areas. Master Thesis. The University of Alabama at Tuscaloosa.

Bochis, C., R. Pitt, and P. Johnson. 2008. Land development characteristics in Jefferson County, Alabama. Pp. 249–281 *In:* Reliable Modeling of Urban Water Systems, Monograph 16. W. James, K. N. Irvine, E. A. McBean, R. E. Pitt, and S.J. Wright (eds.). Guelph, Ontario, Canada: CHI.

Bochis-Micu, C., and R. Pitt. 2005. Impervious Surfaces in Urban Watersheds. Proceedings of the 78thAnnual Water Environment Federation Technical Exposition and Conference in Washington, D.C., October 29–November 2, 2005.

Boller, M. 1997. Tracking heavy metals reveals sustainability deficits of urban drainage systems. Water Science and Technology 35(9):77–87.

Bonnin, G. M., D. Todd, B. Lin, T. Parzybok, M.Yekta, and D. Riley. 2003. Precipitation-Frequency Atlas of the United States. NOAA Atlas 14, Volume 2, Version 2. Silver Spring, MD: NOAA, National Weather Service. Available at *http://hdsc.nws.noaa.gov/hdsc/pfds/*. Last accessed October 19, 2008.

Bonnin, G. M., Martin, D., Lin, B., Parzybok, T., Yekta,M., Riley, D. 2006. NOAA Atlas 14. Precipitation-Frequency Atlas of the United States Volume 3 Version 4.0: Puerto Rico and the U.S. Virgin Islands U.S. Department of Commerce National Oceanic and Atmospheric Administration National Weather Service Silver Spring, Maryland.

Booth, D. 2000. Forest cover, impervious-surface area, and the mitigation of urbanization impacts in King County, Washington. Center for Urban Water Resources Management, Seattle, WA. Available at *http://www.metrokc. gov/ddes/cao/PDFs04ExecProp/BAS-AppendixB-04.pdf.* Last accessed February 14, 2006.

Booth, D. B., and P. C. Henshaw. 2001. Rates of channel erosion in small urban streams. Pp. 17–38 *In*: Land Use and Watersheds: Human Influence on Hydrology and Geomorphology in Urban and Forest Areas. AGU Monograph Series, Water Science and Application, Vol. 2.

Booth, D. B., and C. R. Jackson. 1997. Urbanization of aquatic systems— degradation thresholds, stormwater detention, and limits of mitigation. Journal of the American Water Resources Association 33:1077–1090.

Booth, D. B., D. Hartley, and C. R. Jackson. 2002. Forest cover, impervious-surface area, and the mitigation of stormwater impacts. Journal of the American Water Resources Association 38:835–845.

Brantley, A. S., and T. G. Townsend. 1999. Leaching of pollutants from reclaimed asphalt pavement. Environmental Engineering Science 16(2):105–116.

Brattebo, B. O., and D. B. Booth. 2003. Long-term stormwater quantity and quality performance of permeable pavement systems. Water Research 37:4369–4376.

Brown, J. N., and B. M. Peake. 2006. Sources of heavy metals and polycyclic aromatic hydrocarbons in urban stormwater runoff. Science of the Total Environment 359(1–3):145–155.

Brungs, W. A., and B. R. Jones. 1977. Temperature Criteria for Freshwater Fish: Protocol and Procedures. EPA-600-3-77-061. Duluth, MN: EPA Office of Research and Development.

Burges, S. J., M. S. Wigmosta, and J. M. Meena. 1998. Hydrological effects of land-use change in a zero-order catchment. Journal of Hydraulic Engineering 3(2):86-97.

Burns, D., A. T. Vitvar, J. McDonnell, J. Hassett, J. Duncan, and C. Kendall. 2005. Effects of suburban development on runoff generation in the Croton River Basin, New York USA. Journal of Hydrology 311:266–281.

Burton, G. A., Jr., ed. 1992. Sediment Toxicity Assessment. Chelsea, MI: Lewis Publishers.

Burton, G. A., Jr., and R. Pitt. 2002. Stormwater Effects Handbook: A Tool Box for Watershed Managers, Scientists, and Engineers. Boca Raton, FL: CRC Press, 911 pp.

Burton, G. A., Jr., R. Pitt, and S. Clark. 1999. The role of traditional and novel toxicity test methods in assessing stormwater and sediment contamination. Critical Reviews in Environmental Science and Technology 30(4):413–447.

Cadle, S. H. 1991. Dry deposition to snowpacks. Pp. 21–66 In: Seasaonal Snowpacks, NATO ASI Series G, Ecological Sciences, Vol. 28. T. D. Davies, M. Tranter, and H. G. Jones (eds.). Berlin: Springer.

Calderon, R., E. Mood, and A. Dufour. 1991. Health effects of swimmers and nonpoint sources of contaminated water. International Journal of Environmental Health Research 1:21–31.

Carter, J. L., and S. V. Fend. 2005. Setting limits: the development and use of factor-ceiling distributions for an urban assessment using macroinvertebrates. Pp. 179–192 In: Effects of Urbanization on Stream Ecosystems. L. R. Brown, R. H. Gray, R. M. Hughes, and M. R. Meador (eds.). Bethesda, MD: American Fisheries Society.

Chesapeake Stormwater Network. 2008. CSN Technical Bulletin No. 3. Implications of the Impervious Cover Model: Stream Classification, Urban Subwatershed Management and Permitting. Version 1.0. Baltimore, MD: Chesapeake Stormwater Network.

Chesters, G., J. Konrad, and G. Simsiman. 1979. Menomonee River Pilot Watershed Study—Summary and Recommendations. EPA-905/4-79-029. Chicago: EPA.

Chow, V. T., D. R. Maidment, and L. W. Mays. 1988. Applied Hydrology. McGraw-Hill.

Cianfrani, C. M., W. C. Hession, and D. M. Rizzo. 2006. Watershed imperviousness impacts on stream channel condition in southeastern Pennsylvania. Journal of the American Water Resources Association 42(4):941–956.

Clark, S. 2000. Urban Stormwater Filtration: Optimization of Design Parameters and a Pilot-Scale Evaluation. Ph.D. Dissertation. University of Alabama at Birmingham, 450 pp.

Clark, S. E., K. A. Steeke, J. Spicher, C. Y. S. Siu, M. M. Lalor, R. Pitt, and J. Kirby. 2008. Roofing materials' contributions to stormwater runoff pollution. Journal of Irrigation and Drainage Engineering 134(5):638 - 645.

Claytor, R., and T. Schueler. 1996. Design of stormwater filtering systems. Prepared for Chesapeake Research Consortium, Center for Watershed Protection, Ellicott City, MD.

Cochran, J. W. 2008. Evaluation of the Effect of Canopy Cover on the Volume and Intensity of Rain Throughfall. MSCE Thesis. University of Alabama at Birmingham. 56 pp.

Colbeck, S. C. 1981. A simulation of the enrichment of atmospheric pollutants in snowcover runoff. Water Resources Research 17:1383–1388.

Coles, J. F., T. F. Cuffney, G. McMahon, and K. M. Beaulieu. 2004. The Effects of Urbanization on the Biological, Physical, and Chemical Characteristics of Coastal New England Streams: U.S. Geological Survey Professional Paper 1695, 47 pp.

Colford, J. M., Jr, T. J. Wade, K. C. Schiff, C. C. Wright, J. F. Griffith, S. K. Sandhu, S. Burns, J. Hayes, M. Sobsey, G. Lovelace, and S. Weisberg. 2007. Water quality indicators and the risk of illness at non-point source beaches in Mission Bay, California. Epidemiology 18(1):27–35.

Coutant, C. C. 1977. Compilation of temperature preference data. Journal of Fisheries Research Board of Canada 34:739–745.

Cummins, K. W., M. A.Wilzbach, D. M. Gates, J. B. Perry, and W.B. Taliaferro. 1989. Shredders and riparian vegetation. Bioscience 39:24–30.

CWP (Center for Watershed Protection). 2003. Impacts of impervious cover on aquatic ecosystems. Watershed Protection Techniques Monograph No. 1. Ellicott City, MD: Center for Watershed Protection.

CWP and R. Pitt. 2004. Illicit Discharge Detection and Elimination: A Guidance Manual for Program Development and Technical Assessments. EPA Cooperative Agreement X-82907801-0. Washington, DC: EPA Office of Water.

CWP and R. Pitt. 2008. Monitoring to Demonstrate Environmental Results: Guidance to Develop Local Stormwater Monitoring Studies Using Six Example Study Designs. EPA Cooperative Agreement CP-83282201-0. Washington, DC: EPA Office of Water.

Davies, S. P., and S. K. Jackson. 2006. The biological condition gradient: a descriptive model for interpreting change in aquatic ecosystems. Ecological Applications 16(4):1251–1266.

Deacon, J., S. Soule, and T. Smith. 2005. Effects of urbanization on stream quality at selected sites in the seacoast region in New Hampshire, 2001-

2003. USGS Scientific Investigations Report 2005-5103.

de Sherbinin, A. 2002. A Guide to Land-Use and Land-Cover Change (LUCC). A collaborative effort of SEDAC and the IGBP/IHDP LUCC Project. Center for International Earth Science Information Network, Columbia University, New York. Available at *http://sedac.ciesin.org/tg/guide_frame.jsp?rd=LU&ds=1.* Last accessed October 31, 2007.

DeVivo, J. C. 1996. Fish assemblages as indicators of water quality within the Apalachicola-Chattahoochee-Flint (ACF) River basin. Master's Thesis. University of Georgia, Athens.

Dolloff, C. A., and J. R. Webster. 2000. Particulate organic contributions from forests to streams: debris isn't so bad. Pp. 125–138 *In:* Riparian Management in Forests of the Continental Eastern United States. E. S. Verry, J. W. Hornbeck, and C. A. Dolloff (eds.). Boca Raton, FL: Lewis Publishers.

Doyle, M. W. 2006. A heuristic model for potential geomorphic influences on trophic interactions in streams. Geomorphology 77:235–248.

Dunne, T., and L. B. Leopold. 1978. Water in Environmental Planning. New York: W.H. Freeman.

Dwight, R. H., D. B. Baker, J. C. Semenza, and B. H. Olson. 2004. Health effects associated with recreational coastal water use: urban vs. rural California. American Journal of Public Health 94(4):565–567.

Dwyer, F. J., F. L. Mayer, L. C. Sappington, D. R. Buckler, C. M. Bridges, I. E. Greer, D. K. Hardesty, C. E. Henke, C. G. Ingersoll, J. L. Kunz, D. W. Whites, T. Augspurger, D. R. Mount, K. Hattala, and G. N. Neuderfer. 2005. Assessing contaminant sensitivity of endangered and threatened aquatic species: Part I. Acute toxicity of five chemicals. Archives of Environmental Contamination and Toxicology 48:143–154.

Eaton, J. G., J. H. McCormick, B. E. Goodno, D. G. O'Brien, H. G. Stefan, M. Hondzo, and R. M. Scheller. 1995. A field information-based system for estimating fish temperature tolerances. Fisheries 20(4):10–18.

Ebisemiju, F. S. 1989. The response of headwater stream channels to urbanization in the humid tropics. Hydrological Processes 3:237–253.

Ehlinger, T. J., C. D. Sandgren, and L. Schacht-DeThorne. 2004. Degradation and recovery in urban watersheds: Executive Summary Narrative and Recommendations, February 2004. Submitted to Great Lakes Protection Fund, Grant WR539.

EPA (U.S. Environmental Protection Agency). 1983. Results of the Nationwide Urban Runoff Program. PB 84-185552. Washington, DC: Water Planning Division.

EPA. 2000. Stressor Identification Guidance Document. EPA 822-B-00-025. Washington, DC: EPA, Offices of Water and Research and Development.

EPA. 2002. Summary of Biological Assessment Programs and Biocriteria Development for States, Tribes, Territories, and Interstate Commissions: Streams and Wadeable Rivers. EPA-822-R-02-048. Washington, DC: EPA.

EPA. 2005. Draft: Use of Biological Information to Better Define Designated Aquatic Life Uses in State and Tribal Water Quality Standards: Tiered Aquatic Life Uses—August 10, 2005. EPA-822-E-05-001. Washington, DC: EPA Office of Water.

FISRWG (Federal Interagency Stream Restoration Working Group). 2000. Stream Corridor Restoration: Principles, Processes and Practices. Washington, DC: USDA Natural Resource Conservation Service.

Fitzpatrick, F. A., M. W. Diebel, M. A. Harris, T. L. Arnold, M. A. Lutz, and K. D. Richards. 2005. Effects of urbanization on the geomorphology, habitat, hydrology, and fish index of biotic integrity of streams in the Chicago area, Illinois and Wisconsin. American Fisheries Society Symposium 47:87–115.

Fritz, K. M., B. R. Johnson, and D. M. Walters. 2006. Field Operations Manual for Assessing the Hydrologic Permanence and Ecological Condition of Headwater Streams. EPA/600/R-06/126. Washington, DC: EPA Office of Research and Development.

Funk, J. L., and G. G. Fleener. 1974. The fishery of a Missouri Ozark stream, Big Piney River, and the effects of stocking fingerling smallmouth bass. Transactions of the American Fisheries Society 103(4):757–771.

Galli, J. 1990. Thermal Impacts Associated with Urbanization and Stormwater Best Management Practices. Washington, DC: Metropolitan Washington Council of Governments.

GAO (General Accounting Office). 2003. Water Quality: Improved EPA Guidance and Support Can Help States Develop Standards that Better Target Cleanup Efforts. GAO-03-308. Washington, DC: GAO.

Gilliom, R. J., J. E. Barbash, C. G. Crawford, P. A. Hamilton, J. D. Martin, N. Nakagaki, L. H. Nowell, J. C. Scott, P. E. Stackelberg, G. P. Thelin, and D. M. Wolock. 2006. The Quality of our Nation's Water—Pesticides in the Nation's Streams and Ground Water, 1992–2001. U.S. Geological Survey Circular 1291, 169 pp.

Goetz, S., R. Wright, A. Smith, E. Zinecker, and E. Schaub. 2003. IKONOS imagery for resource management: Tree cover, impervious surfaces, and riparian buffer analyses in the mid-Atlantic region. Remote Sensing in the Environment 88:195–208.

Good, J. C. 1993. Roof runoff as a diffuse source of metals and aquatic toxicity in stormwater. Water Science and Technology 28(3–5):317–321.

Gregory, M., J. Aldrich, A. Holtshouse, and K. Dreyfuss-Wells. 2005. Evaluation of imperviousness impacts in large, developing watersheds. Pp. 115–150 In: Efficient Modeling for Urban Water Systems, Monograph 14. W. James, E. A. McBean, R. E. Pitt, and S. J. Wright (eds.). Guelph, Ontario, Canada: CHI.

Gregory, S. V., F. J. Swanson, W. A. McKee, and K. W. Cummins. 1991. An ecosystem perspective of riparian zones. Bioscience 41:540–551.

Groffman, P. M., N. L. Law, K. T. Belt, L. E. Band, and G. T. Fisher. 2004. Nitrogen fluxes and retention in urban watershed ecosystems. Ecosystems

7:393–403.

Groffman, P. M., A. M. Dorsey, and P. M. Mayer. 2005. N processing within geomorphic structures in urban streams. Journal of the North American Benthological Society 24(3):613–625.

Haile, R.W., J. S. Witte, M. Gold, R. Cressey, C. McGee, R. C. Millikan, A. Glasser, N. Harawa, C. Ervin, P. Harmon, J. Harper, J. Dermand, J. Alamillo, K. Barrett, M. Nides, and G. Wang. 1999. The health effects of swimming in ocean water contaminated by storm drain runoff. Epidemiology 10(4):355–363.

Hammer, T. R. 1972. Stream and channel enlargement due to urbanization. Water Resources Research 8:1530–1540.

Harris, A. R., and C. I. Davidson. 2005. The role of resuspended soil in lead flows in the California South Coast Air Basin. Environmental Science and Technology 39:7410–7415.

Heany, J. P., and W. C. Huber. 1984. Nationwide assessment of urban runoff impact on receiving water quality. Water Resources Bulletin 20:35–42.

Heaney, J. 2000. Principles of integrated urban water management. Pp. 36–73 *In*: Innovative Urban Wet-Weather Flow Management Systems. Lancaster, PA: Technomic Publishing.

Hecht, S. A., D. H. Baldwin, C. A. Mebane, T. Hawkes, S. J. Gross, and N. L. Scholz. 2007. An Overview of Sensory Effects on Juvenile Salmonids Exposed to Dissolved Copper: Applying a Benchmark Concentration Approach to Evaluate Sublethal Neurobehavioral Toxicity. NOAA Tech. Memo. NMFS-NWFSC-83. Washington, DC: U.S. Department of Commerce.

Herb, W. R., B. Janke, O. Mohseni, and H. G. Stefan. 2008. Thermal pollution of streams by runoff from paved surfaces. Hydrological Processes 22(7):987–999.

Hernandez, P. P, V. Moreno, F. A. Olivari, and M. L. Allende. 2006. Sublethal concentrations of waterborne copper are toxic to lateral line neuromasts in zebrafish (Danio rerio). Hearing Research 213:1–10.

Hoblit, B., S. Zelinka, C. Castello, and D. Curtis. 2004. Spatial Analysis of Storms Using GIS. Paper 1891. 24th Annual ESRI International User Conference, San Diego, August 9–13, 2004. Available at *http://gis.esri.com/library/userconf/proc04/docs/pap1891.pdf*. Last accessed October 19, 2008.

Holland, F., D. Sanger, C. Gawle, S. Lerberg, M. Santiago, G. Riekerk, L. Zimmerman, and G. Scott. 2004. Linkages between tidal creek ecosystems and the landscape and demographic attributes of their watersheds. Journal of Experimental Marine Biology and Ecology 298:151–178.

Hollis, G. E., and J. K. Luckett. 1976. The response of natural river channels to urbanization: two case studies from southeast England. Journal of Hydrology 30:351–363.

Hood, W. G., and R. J. Naiman. 2000. Vulnerability of riparian zones to invasion by exotic vascular plants. Plant Ecology 148:105–114.

Horner, R., and C. May. 1999. Regional study supports natural land cover protection as leading management practice for maintaining stream ecological integrity. Pp. 233–247 *In:* Comprehensive Stormwater and Aquatic Ecosystem Management. First South Pacific Conference, February 22–26, Auckland, New Zealand.

Horner, R. R., D. B. Booth, A. Azous, and C. W. May. 1997. Watershed determinants of ecosystem functioning. Pp. 251–274 *In:* Effects of Watershed Development and Management on Aquatic Ecosystems. L.A. Roesner (ed.). New York: American Society of Civil Engineers.

Horner, R., C. May, E. Livingston, D. Blaha, M. Scoggins, J. Tims, and J. Maxted. 2001. Structural and non-structural BMPs for protecting streams. L inking Stormwater BMP Designs and Performance to Receiving Water Impact Mitigation. Pp. 60–77 *In:* Proceedings Engineering Research Foundation Conference, American Society of Civil Engineers.

Horner, R. R., C. W. May, and E. H. Livingston. 2003. Ecological Effects of Stormwater and Stormwater Controls on Small Streams. Prepared for EPA, Office of Water, Office of Science and Technology by Watershed Management Institute, Inc., Cooperative Agreement CX824446.

Horner, R. R., H. Lim, and S. J. Burges. 2004. Hydrologic Monitoring of the Seattle Ultra-Urban Stormwater Management Projects: Summary of the 2000-2003 Water Years. Water Resources Series Technical Report Number 181. Department of Civil and Environmental Engineering, University of Washington, Seattle, WA.

Horton, R. E. 1945. Erosional development of streams and their drainage basins; hydrophysical approach to quantitative morphology. Geological Society of America Bulletin 56:275–370.

House, L. B., R. J. Waschbusch, and P. E. Hughes. 1993. Water Quality of an Urban Wet Detention Pond in Madison, Wisconsin, 1987-1988. USGS, in cooperation with the Wisconsin Department of Natural Resources. USGS Open File Report 93-172.

Huff, F. A., and S. A. Changnon. 1973. Precipitation modification by major urban areas. Bulletin of the American Meteorological Society 54:1220–1232.

Hughes, R. M. 1995. Defining acceptable biological status by comparing with reference conditions. Pp. 31–47 *In:* Biological Assessment and Criteria: Tools for Water Resource Planning and Decision Making for Rivers and Streams. W. Davis and T. Simon (eds.). Boca Raton, FL: Lewis Publishers.

Hughes, R. M., D. P. Larsen, and J. M. Omernik. 1986. Regional reference sites: A method for assessing stream potentials. Environmental Management 10:629–635.

Hunt, C. L., and J. K. Grow. 2001. Impacts of Sedimentation to a Stream by a Non-compliant Construction Site. Proc. ASCE EWRI Conf. - Bridging the Gap: Meeting the World's Water and Environmental Resources Challenges.

Ingersoll, C. G., T. Dillon, and G. R. Biddinger, eds. 1997. Ecological Risk

Assessment of Contaminated Sediments. SETAC Pellston Workshop on Sediment Ecological Risk Assessment. Pensacola, FL: SETAC Press.

Intergovernmental Panel on Climate Change (IPCC). 2007. Summary for Policymakers. Pp. 7-22 *In:* Climate Change 2007: Impacts, Adaptation and Vulnerability. Contribution of Working Group II to the Fourth Assessment Report of the Intergovernmental Panel on Climate Change. Parry, M.L., O.F. Canziani, J.P. Palutikof, P.J. van der Linden, and C.E. Hanson (eds.). Cambridge, UK: Cambridge University Press.

Johnson, P. D., R. Pitt, S. R. Durrans, M. Urrutia, and S. Clark. 2003. Metals Removal Technologies for Urban Stormwater. WERF 97-IRM-2. ISBN: 1-94339-682-3. Alexandria, VA: Water Environment Research Foundation.

Jokela, B. 1990. Water quality considerations. Pp. 349–369 *In:* Cold Regions Hydrology and Hydraulics. W. L. Ryan and R. D. Crissman (eds.). New York: ASCE.

Jones, R. C., A. Via-Norton, and D. R. Morgan. 1996. Bioassessment of BMP effectiveness in mitigating stormwater impacts on aquatic biota. Pp. 402–417 *In:* Effects of Watershed Development and Management on Aquatic Ecosystems. L. A. Roesner (ed.). New York: American Society of Civil Engineers.

Karr, J. R. 1995. Protecting aquatic ecosystems: clean water is not enough. Pp. 7–13 *In:* Biological Assessment and Criteria: Tools for Water Resource Planning and Decision Making. W. S. Davis and T. P. Thomas (eds.). Boca Raton, FL: CRC Press, Inc.

Karr, J., and E. Chu. 1999. Restoring Life in Running Waters: Better Biological Monitoring. Island Press. ISBN 1559636742, 9781559636742. 206 pages

Karr, J. R., and C. O. Yoder. 2004. Biological assessment and criteria improve total maximum daily load decision making. Journal of Environmental Engineering 130(6):594–604.

Kennedy, A. J., D. S. Cherry1, and R. J. Currie. 2004. Evaluation of ecologically relevant bioassays for a lotic system impacted by a coal-mine effluent, using Isonychia. Environmental Monitoring and Assessment 95(1–3):37–55.

Khan, B. I., H. M. Solo-Gabriele, T. G. Townsend, and Y. Cai. 2006. Release of arsenic to the environment from CCA-treated wood. 1. Leaching and speciation during service. Environmental Science and Technology 40:988–993.

King, R. M. Baker, D. Whigham, D. Weller and K. Hurd. 2005. Spatial considerations for linking watershed land cover to ecological indicators in streams. Ecological Applications 15:137–153.

Kline, M. 2002. Storm water official plans to send $2,000 bill to sealing company. The Tennessean (Nashville). October 2.

Konrad, C. P., and D. B. Booth. 2005. Ecological significance of hydrologic changes in urban streams. Pp. 157–177 *In:* Effects of Urbanization on Stream Ecosystems. L. R. Brown, R. H. Gray, R. M. Hughes, and M. R.

Meador (eds.). American Fisheries Society, Symposium 47.

Konrad, C. P., D. B. Booth, and S. J. Burges. 2005. Effects of urban development in the Puget Lowland, Washington, on interannual streamflow patterns: consequences for channel form and streambed disturbance. Water
Resources Research 41(7):1–15.

Krajeswki, W. F., and J. A. Smith. 2002. Radar hydrology: rainfall estimation.
Advances in Water Resources 25:1387–1394.

Krause, C.W., B. Lockard, T. J. Newcomb, D. Kibler, V. Lohani, and D. J. Orth.
2004. Predicting influences of urban development on thermal habitat in a
warm water stream. Journal of the American Water Resources Association
40:1645–1658.

Kunkel, K. E., K, Amdsager, and D. R. Easterling. 1999. Long-term trends in
extreme precipitation events over the coterminous United States and Canada. Journal of Climate 12:2515–2527.

LAC DPW (Los Angeles County, Department of Public Works). 1985. Plan for
Flood Control and Water Conservation, Maps 2, 3, 7, 16, and 17. LAC
DPW, Los Angeles, CA.

Lee, J., and J. Heaney. 2003. Estimation of urban imperviousness and its impacts on storm water systems. Journal of Water Resources Planning and
Management 129(5):419–426.

Leopold, L., and Maddock. 1953. The hydraulic geometry of stream channels
and some physiographic implications. USGS Professional Paper 252.
Washington, DC: U.S. Department of the Interior.

Leopold, L. B. 1968. Hydrology for Urban Planning—A Guidebook on the
Hydrologic Effects of Urban Land Use. USGS Circular 554. Washington,
DC: USGS.

Leopold, L. B. 1973. River channel change with time—an example.
Geological Society of America Bulletin 84:1845–1860.

Lerberg, S., F. Holland, and D. Singer. 2000. Responses of tidal creek macro-
benthic communities to the effects of watershed development. Estuaries
23(6):838–853.

Line, D., and N. White. 2007. Effect of development on runoff and pollutant
export. Water Environment Research 75(2):184–194.

Lorantas, R. M., and D. P. Kristine. 2004. Annual monitoring of young of the
year abundance of smallmouth bass in rivers in Pennsylvania: relationships
between abundance, river characteristics, and spring fishing. Pennsylvania
Fish & Boat Commission, Bellefonte.

Lowrance, R., L. S. Altier, J. D. Newbold, R. R. Schnabel, P. M. Groffman, J.
M. Denver, D. L. Correll, J. W. Gilliam, J. L. Robinson, R. B. Brinsfield, K.
W. Staver, W. Lucas, and A. H. Todd. 1997. Water quality functions of riparian forest buffers in Chesapeake Bay watersheds. Environmental Management 21:687–712.

Lu, R., R. P. Turco, K. Stolzenbach, S. K. Friendlander, C. Xiong, K. Schiff,
and L. Tiefenthaler. 2003. Dry deposition of airborne trace metals on the
Los Angeles Basin and adjacent coastal waters. Journal of Geophysical Re-

search 102(D2):4074.

Mack, J. J., and M. Micacchion. 2007. An Ecological and Functional Assessment of Urban Wetlands in Central Ohio. Volume 1: Condition of Urban Wetlands Using Rapid (Level 2) and Intensive (Level 3) Assessment Methods. Ohio EPA Technical Report WET/2007-3A. Columbus, OH: Ohio Environmental Protection Agency, Wetland Ecology Group, Division of Surface Water.

MacRae, C. R. 1997. Experience from morphological research on Canadian streams: is control of the two-year frequency runoff event the best basis for stream channel protection? *In:* Proceedings of the Effects of Watershed Developments and Management on Aquatic Ecosystems Conference, Snowbird, UT, August 4–9, 1996. L. A. Roesner (ed.). New York: ASCE.

Madsen, T., and E. Figdor. 2007. When It Rains, It Pours: Global Warming and the Rising Frequency of Extreme Precipitation in the United States. Penn Environment Research & Policy Center, Philadelphia, PA. Available at *http://www.pennenvironment.org.*

Maestre, A., Pitt, R. E., and Derek Williamson. 2004. Nonparametric statistical tests comparing first flush with composite samples from the NPDES Phase 1 municipal stormwater monitoring data. Stormwater and Urban Water Systems Modeling. Pp. 317–338 *In:* Models and Applications to Urban Water Systems, Vol. 12. W. James (ed.). Guelph, Ontario: CHI.

Maestre, A., and R. Pitt. 2005. The National Stormwater Quality Database, Version 1.1, A Compilation and Analysis of NPDES Stormwater Monitoring Information. Final Draft Report. Washington, DC: EPA Office of Water.

Maestre, A., and R. Pitt. 2006. Identification of significant factors affecting stormwater quality using the National Stormwater Quality Database. Pp. 287–326 *In:* Stormwater and Urban Water Systems Modeling, Monograph 14. W. James, K. N. Irvine, E. A. McBean, and R. E. Pitt (eds.). Guelph, Ontario: CHI.

Mahler, B. J., P. C. VanMetre, T. J. Bashara, J. T. Wilson, and D. A. Johns. 2005. Parking lot sealcoat: an unrecognized source of urban polycyclic aromatic hydrocarbons. Environmental Science and Technology 39:5560–5566.

Mallin, M., S. Ensign, M. McIver, C. Shank, and P. Fowler. 2001. Demographic, landscape and meteorological factors controlling the microbial pollution of coastal waters. Hydrobiologia 460:185–193.

Mallin, M., D. Parsons, V. Johnson, M. McIver, and H. Covan. 2004. Nutrient limitation and algal blooms in urbanizing tidal creeks. Journal of Experimental Marine Biology and Ecology 298:211–231.

Masterson, J. P., and R. T. Bannerman. 1994. Impacts of stormwater runoff on urban streams in Milwaukee County, Wisconsin. In National Symposium on Water Quality, American Water Resources Association.

Maxted, J. R. 1999. The effectiveness of retention basins to protect aquatic life and physical habitat in three regions of the United States. Pp. 215–222 *In:*

Proceedings of the Comprehensive Stormwater and Aquatic Ecosystem Management Conference, Auckland, New Zealand, February 1999.

May, C. W. 1996. Assessment of Cumulative Effects of Urbanization on Small Streams in the Puget Sound Lowland Ecoregion: Implications for Salmonid Resource Management. Ph.D. dissertation, University of Washington, Seattle, WA.

May, C. W., R. R. Horner, J. R. Karr, B. W. Mar, and E. B. Welch. 1997. Effects of urbanization on small streams in the Puget Sound Lowland Ecoregion. Watershed Protection Techniques 2(4):483–494.

McBride, M., and D. Booth. 2005. Urban impacts on physical stream condition: effects on spatial scale, connectivity, and longitudinal trends. Journal of the American Water Resources Association 6:565–580.

McClain, M. E., E. W. Boyer, C. L. Dent, S. E. Gergel, N. B. Grimm, P. M. Groffman, S. C. Hart, J. W. Harvey, C. A. Johnston, E. Mayorga, W. H. McDowell, and G. Pinay. 2003. Biogeochemical hot spots and hot moments at the interface of terrestrial and aquatic ecosystems. Ecosystems 6:301–312.

McPherson, T. N., S. J. Burian, H. J. Turin, M. K. Stenstrom, and I. H. Suffet. 2002. Comparison of the pollutant loads in dry and wet weather runoff in a southern California urban watershed. Water Science and Technology 45:255–261.

Meador, M. R., J. F. Coles, and H. Zappia. 2005. Fish assemblage responses to urban intensity gradients in contrasting metropolitan areas: Birmingham, Alabama and Boston, Massachusetts. American Fisheries Society Symposium 47:409–423.

Meierdiercks, K. J., J. A. Smith, A. J. Miller, and M. L. Baeck. 2004. The urban drainage network and its control on extreme floods. Transactions of the American Geophysical Union 85(47), Abstract H11F-0370.

Meyer, J. L. 1994. Sources of carbon for the microbial loop. Microbial Ecology 28:195–199.

Meyer, J. L., and J. B. Wallace. 2001. Lost linkages and lotic ecology: rediscovering small streams. Pp. 295–317 *In:* Proceedings of the 41st Symposium of the British Ecological Society jointly sponsored by the Ecological Society of America. M. C. Press, N. J. Huntly, and S. Levin (eds.). Orlando, FL: Blackwell Science.

Meyer, J. L., G. C. Poole, and K. L. Jones. 2005a. Buried alive: potential consequences of burying headwater streams in drainage pipes. Proceedings of the 2005 Georgia Water Resources Conference, held April 25–27, 2005, at the University of Georgia. K. J. Hatcher (ed.). Athens, GA: Institute of Ecology, The University of Georgia.

Meyer, J. L., M. J. Paul, and W. K. Taulbee. 2005b. Stream ecosystem function in urbanizing landscapes. Journal of the North American Benthological Society 24(3):602–612.

Meyer, S. C. 2005. Analysis of base flow trends in urban streams, northeastern Illinois, USA. Hydrogeology Journal 13(5–6):871–885.

Miller, W., and A. J. Boulton. 2005. Managing and rehabilitating ecosystem processes in regional urban streams in Australia. Hydrobiologia 552.

Miltner, R. J., and E. T. Rankin. 1998. Primary nutrients and the biotic integrity of rivers and streams. Freshwater Biology 40(1):145–158.

Miltner, R. J., D. White, and C. O. Yoder. 2004. The biotic integrity of streams in urban and suburbanizing landscapes. Landscape and Urban Planning 69:87–100.

MNCPPC (Maryland National Capital Park and Planning Commission). 2000. Stream Condition Cumulative Impact Models for the Potomac Subregion. Silver Spring, MD.

Moore, A. A., and M. A. Palmer. 2005. Invertebrate biodiversity in agricultural and urban headwater streams: Implications for conservation and management. Ecological Applications 15(4):1169–1177.

Morisawa, M., and E. LaFlure. 1982. Hydraulic geometry, stream equilibrium and urbanization. Pp. 333350 *In:* Adjustments of the Fluvial System. D. D. Rhodes and G. P. Williams (eds.). London: Allen and Unwin.

Morley, S., and J. Karr. 2002. Assessing and restoring the health of urban streams in the Puget Sound Basin. Conservation Biology 16(6):1498–1509.

Morquecho, R. 2005. Pollutant Associations with Particulates in Stormwater. PhD dissertation. University of Alabama.

Morrisey, D. J., R. B. Williamson, L. Van Dam, and D. J. Lee. 2000. Stormwater-contamination of urban estuaries. 2. Testing a predictive model of the build-up of heavy metals in sediments. Estuaries 23(1):67–79.

Moscrip, A. L., and D. R. Montgomery. 1997. Urbanization, flood frequency, and salmon abundance in Puget Lowland streams. Journal of the American Water Resources Association 33:1289–1297.

Nadeau, T. L., and M. C. Rains. 2007. Hydrological connectivity of headwaters to downstream waters: how science can inform policy. Journal of the American Water Resources Association 43:118–133.

Nanson, G. C., and R. W. Young. 1981. Downstream reduction of rural channel size with contrasting urban effects in small coastal streams of southeastern Australia. Journal of Hydrology 52:239–255.

Neller, R. J. 1988. A comparison of channel erosion in small urban and rural catchments, Armidale, New South Wales. Earth Surface Processes and Landforms 13:1–7.

Nelson, E. J., and D. B. Booth. 2002. Sediment budget of a mixed-land use, urbanizing watershed. Journal of Hydrology 264:51–68.

Nelson, K. C., and M. A. Palmer. 2007. Stream temperature surges under urbanization and climate change: Data, models, and responses. Journal of the American Water Resources Association 43(2):440–452.

Ng, E., and P. C. Miller. 1980. Soil Moisture Relations in the Southern California Chaparral. Ecology 61(1):98-107.

Novotny, V., and G. Chesters. 1981. Handbook of Nonpoint Pollution. New York: VanNostrand Reinhold.

Novotny, V., et al. 1986. Effect of Pollution from Snow and Ice on Quality of

Water from Drainage Basins. Technical report, Marquette University.

NRC (National Research Council). 1984. Wastewater characteristics. Chapter 2 in Ocean Disposal Systems for Sewage Sludge and Effluent. Washington, DC: National Academy Press.

NRC. 2001. Assessing the TMDL Approach to Water Quality Management. Washington, DC: National Academies Press.

NRC. 2002. Riparian Areas: Functions and Strategies for Management. Washington, DC: National Academy Press.

NRC. 2004. Assessing the National Streamflow Information Program. Washington, DC: National Academy Press.

Oberts, G. L. 1994. Influence of snowmelt dynamics on stormwater runoff quality. Watershed Protection Techniques 1(2).

Odemerho, F. O. 1992. Limited downstream response of stream channel size to urbanization in a humid tropical basin. Professional Geographer 44:332–339.

Ohio EPA (Ohio Environmental Protection Agency). 1992. Biological and Habitat Investigation of Greater Cincinnati Area Streams (Hamilton and Clermont Counties, Ohio). Ohio EPA Tech. Rept. 1992-1-1. Columbus, OH: Division of Water Quality Planning and Assessment.

Ohio EPA. 2002. Field Evaluation Manual for Ohio's Primary Headwater Habitat Streams, Final Version 1.0. Columbus, OH: Ohio Environmental Protection Agency, Division of Surface Water.

Oliver, B., J. B. Milne, and N. LaBarre. 1974. Chloride and lead in urban snow. Journal of Water Pollution Control Federation 46(4).

Paul, M. J., and J. L. Meyer. 2001. Streams in the urban landscape. Annual Review of Ecology and Systematics 32:333–365.

Perez-Rivas, M. 2000. Asphalt sealer kills 1,000 fish in Montgomery. The Washington Post (Washington, DC), July 29.

Pierstorff, B. W., and P. L. Bishop. 1980. Water pollution from snow removal operations. Journal of the Environmental Engineering Division 106(2):377-388.

Pitt, R. 1987. Small Storm Urban Flow and Particulate Washoff Contributions to Outfall Discharges. Ph.D. Dissertation. Department of Civil and Environmental Engineering, University of Wisconsin–Madison.

Pitt, R., and J. McLean. 1986. Toronto Area Watershed Management Strategy Study. Humber River Pilot Watershed Project. Toronto, Ontario: Ontario Ministry of the Environment.

Pitt, R., and J. Voorhees. 1995. Source loading and management model (SLAMM). Pp. 225–243 *In:* National Conference on Urban Runoff Management: Enhancing Urban Watershed Management at the Local, County, and State Levels, March 30–April 2, 1993. EPA/625/R-95/003. Cincinnati, OH: EPA Center for Environmental Research Information.

Pitt, R., and J. Voorhees. 2002. SLAMM, the Source Loading and Management Model. Pp. 79–101 *In:* Wet-Weather Flow in the Urban Watershed: Technology and Management. R. Field and D. Sullivan (eds.). Boca Raton, FL:

Lewis Publishers.

Pitt, R., M. Lalor, R. Field, D. D. Adrian, and D. Barbe'. 1993. Investigation of Inappropriate Pollutant Entries into Storm Drainage Systems. EPA/600/R-92/238. Cincinnati, OH: EPA Office of Research and Development.

Pitt, R. E., R. Field, M. Lalor, and M. Brown. 1995. Urban stormwater toxic pollutants: assessment, sources, and treatability. Water Environment Research 67(3):260–275.

Pitt, R., B. Robertson, P. Barron, A. Ayyoubi, and S. Clark. 1999. Stormwater Treatment at Critical Areas: The Multi-Chambered Treatment Train (MCTT). EPA/600/R-99/017. Cincinnati, OH: EPA Wet Weather Flow Management Program, National Risk Management Research Laboratory.

Pitt, R., R. Bannerman, S. Clark, and D. Williamson. 2005a. Sources of pollutants in urban areas (Part 1)—Older monitoring projects. Pp. 465–484 and 507–530 *In:* Effective Modeling of Urban Water Systems, Monograph 13. W. James, K. N. Irvine, E. A. McBean, and R. E. Pitt (eds.). Guelph, Ontario: CHI.

Pitt, R., R. Bannerman, S. Clark, and D. Williamson. 2005b. Sources of pollutants in urban areas (part 2)—recent sheetflow monitoring results. Pp. 485–530 *In:* Effective Modeling of Urban Water Systems, Monograph 13. W. James, K. N. Irvine, E. A. McBean, and R. E. Pitt (eds.). Guelph, Ontario: CHI.

Pitt, R., D. Williamson, and J. Voorhees. 2005c. Review of historical street dust and dirt accumulation and washoff data. Pp. 203–246 *In:* Effective Modeling of Urban Water Systems, Monograph 13. W. James, K. N. Irvine, E. A. McBean, and R. E. Pitt (eds.). Guelph, Ontario: CHI.

Pitt, R., S. Clark, and D. Lake. 2007. Construction Site Erosion and Sediment Controls: Planning, Design, and Performance. Lancaster, PA: DEStech Publications.

Pitt, R., A. Maestre, H. Hyche, and N. Togawa. 2008. The updated National Stormwater Quality Database (NSQD), Version 3. Conference CD. 2008 Water Environment Federation Technical Exposition and Conference, Chicago, October 2008.

Poff, N. L., J. D. Allan, M. B. Bain, J. R. Karr, K. L. Prestegaard, B. D. Richter, R. E. Sparks, and J. C. Stromberg. 1997. The natural flow regime. Bioscience 47(11):769–784.

Poff, N. L., B. P. Bledsoe, and C. O. Cuhaciyan. 2006. Hydrologic variation with land use across the contiguous United States: geomorphic and ecological consequences for stream ecosystems. Geomorphology 79(3–4):264–285.

Pomeroy, C. A., L. A. Roesner, J. C. Coleman II, and E. Rankin. 2008. Protocols for Studying Wet Weather Impacts and Urbanization Patterns. Report No. 03WSM3. Alexandria, VA: Water Environment Research Foundation.

Porcella, D. B., and D. L. Sorenson. 1980. Characteristics of Non-point Source Urban Runoff and Its Effects on Stream Ecosystems. EPA-600/3-80-032. Washington, DC: EPA.

Power, M. 1997. Assessing the effects of environmental stressors on fish populations. Aquatic Toxicology 39(2):151–169.

Prince Georges County, Maryland. 2000. Low-Impact Development Design Strategies. EPA 841-B-00-003.

Pruss, A. 1998. Review of epidemiological studies on health effects from exposure to recreational water. International Journal of Epidemiology 27(1):1–9.

Purcell, A. H., D. W. Bressler, M. J. Paul, M. T. Barbour, E. T. Rankin, J. L. Carter, and V. H. Resh. 2009. Assessment Tools for Urban Catchments: Developing Biological Indicators. Journal of the American Water Resources Association.

Ramcheck, J. M., and R. L. Crunkilton. 1995. Toxicity Evaluation of Urban Stormwater Runoff in Lincoln Creek, Milwaukee, Wisconsin. College of Natural Resources, University of Wisconsin, Stevens Point, WI.

Ricciardi, A., and R. Kipp. 2008. Predicting the number of ecologically harmful exotic species in an aquatic system. Diversity and Distributions 14:374–380.

Richter, B. D., J. V. Baumgartner, J. Powell, and D. P. Braun. 1996. A method for assessing hydrologic alteration within ecosystems. Conservation Biology 10(1):163–174.

Richter, B. D., J. V. Baumgartner, R. Wigington, and D. P. Braun. 1997a. How much water does a river need? Freshwater Biology 37(1):231–249.

Richter, B. D., D. P. Braun, M. A. Mendelson, and L. L. Master. 1997b. Threats to imperiled freshwater fauna. Conservation Biology 11:1081–1093.

Richter, B. D., R. Mathews, and R. Wigington. 2003. Ecologically sustainable water management: Managing river flows for ecological integrity. Ecological Applications 13(1):206–224.

Rieman, B., D. Lee, J. McIntyre, K. Ovetton, and R. Thurow. 1993. Consideration of Extinction Risks for Salmonids. Fish Habitat Relationships Technical Bulletin Number 14. USDA Forest Service, Intermountain Research Station.

Riley, S. P. D., G. T. Busteed, L. B. Kats, T. L. Vandergon, L. F. S. Lee, R. G. Dagit, J. L. Kerby, R. N. Fisher, and R. M. Sauvajot. 2005. Effects of urbanization on the distribution and abundance of amphibians and invasive species in Southern California streams. Conservation Biology 19(6):1894–1907.

Rodríguez-Iturbe, I., A. Rinaldo, R. Rigon, R. L. Bras, A. Marani, and E. J. Ijjasz-Vásquez. 1992. Energy dissipation, runoff production, and the 3-dimensional structure of river basins. Water Resources Research 28(4):1095–1103.

Roesner, L. A., and B. P. Bledsoe. 2003. Physical effects of wet weather flows on aquatic habitats: present knowledge and research needs. Water Environment Research Foundation Rept. No. 00-WSM-4.

Rogge, W. F., L. M. Hildemann, M. A. Mazurek, G. R. Cass, and B. R. T. Simoneit. 1997. Sources of fine organic aerosol. 7. Hot asphalt roofing tar pot fumes. Environmental Science and Technology 31(10):2726–2730.

Rose, S., and N. E. Peters. 2001. Effects of urbanization on streamflow in the Atlanta area (Georgia, USA): a comparative hydrological approach. Hydrological Processes 18:1441–1457.

Rosselot, K. S. 2006a. Copper Released from Brake Lining Wear in the San Francisco Bay Area, Final Report, Brake Pad Partnership, January.

Rosselot, K. S. 2006b. Copper Released from Non-Brake Sources in the San Francisco Bay Area, Final Report, Brake Pad Partnership, January.

Roy, A., B. Freeman, and M. Freeman. 2006. Riparian influences on stream fish assemblage structure on urbanizing streams. Landscape Ecology 22(3):385–402.

Roy, A. H., A. D. Rosemond, M. J. Paul, D. S. Leigh, and J. B. Wallace. 2003. Stream macroinvertebrate response to catchment urbanization (Georgia, U.S.A.). Freshwater Biology 48:329–346.

Roy, A. H., M. C. Freeman, B. J. Freeman, S. J. Wenger, W. E. Ensign, and J. L. Meyer. 2005. Investigating hydrologic alteration as a mechanism of fish assemblage shifts in urbanizing streams. Journal of the North American Benthological Society 24(3):656–678.

Royer, T. V., M. T. Monaghan, and G. W. Minshall. 1999. Processing of native and exotic leaf litter in two Idaho (U.S.A.) streams. Hydrobiologia 400:123–128.

Sabin, L.D., J.H. Lim, K.D. Stolzenbach, and K.C. Schiff. 2005. Contributions of trace metals from atmospheric deposition to stormwater runoff in a small impervious urban catchment. Water Research 39(16):3929-3937.

Salaita, G. N., and P. H. Tate. 1998. Spectroscopic and microscopic charaterization of Portland cement based unleached and leached solidified waste. Applied Surface Science 133(1–2):33–46.

Sandahl, J. F., D. H. Baldwin, J. J. Jenkins, and N. Scholz. 2007. A sensory system at the interface between urban stormwater runoff and salmon survival. Environmental Science and Technology 41:2998–3004.

Schmid-Araya, J. M. 2000. Invertebrate recolonization patterns in the hyporheic zone of a gravel stream. Limnology and Oceanography 45(4):1000–1005.

Scholz, J. G., and D. B. Booth. 2001. Monitoring urban streams: strategies and protocols for humid-region lowland systems. Environmental Monitoring and Assessment 71:143–164.

Schueler, T. 1983. Atmospheric loading sources. Chapter IV in Urban Runoff in the Washington Metropolitan Area, Final Report. Washington, DC.

Schueler, T. R. 1987. Controlling Urban Runoff: A Practical Manual for Planning and Designing Urban BMPs. Department of Environmental Programs, Metropolitan Washington Council of Governments, Water Resources Planning Board.

Schueler, T. 1994. The importance of imperviousness. Watershed Protection Techniques 1(1):1–15.

Schueler, T. 2004. An integrated framework to restore small urban watersheds. Urban Subwatershed Restoration Manual Series. Ellicott City, MD: Center for Watershed Protection.

Schwarz, G. E., A. B. Hoos, R. B. Alexander, and R. A. Smith. 2006. The SPARROW Surface Water-Quality Model—Theory, Application, and User Documentation: U.S. Geological Survey Techniques and Methods, Book 6, Section B, Chapter 3. Available at *http://pubs.usgs.gov/tm/2006/tm6b3*. Last accessed August 6, 2007.

SEWRPC (Southeastern Wisconsin Planning Commission). 1978. Sources of Water Pollution in Southeastern Wisconsin: 1975. Technical Rept. No. 21. Waukesha, WI.

Sharp, J. M., L. N. Christian, B. Garcia-Fresca, S. A. Pierce, and T. J. Wiles. 2006. Changing recharge and hydrogeology in an urbanizing area— Example of Austin, Texas, USA. Philadelphia Annual Meeting (October 22–25, 2006), Geological Society of America. Abstracts with Programs 38(7):289.

Shields, C.A., L. E. Band, N. Law, P. M. Groffman, S. S. Kaushal, K. Savvas, G. T. Fisher, K. T. Belt. 2008. Streamflow distribution of non-point source nitrogen export from urban-rural catchments in the Chesapeake Bay watershed. Water Resources Research 44. W09416, doi:10.1029/2007 WR006360.

Shreve, R. L. 1966. Statistical law of stream numbers. Journal of Geology 74:17–37.

Shreve, R. L. 1967. Infinite topologically *random* channel networks. Journal of Geology 75:179–186.

Shreve, R. L. 1969. Stream lengths and basin areas in topologically random channel networks. Journal of Geology 77:397–414.

Simmons, D. L., and R. J. Reynolds. 1982. Effects of urbanization on base flow of selected South-Shore streams, Long Island, New York. Journal of the American Water Resources Association 18(5):797–805.

Simon, A., and C. R. Hupp. 1986. Channel evolution in modified Tennessee streams. Pp. 71–82 *In:* Proceedings of the Fourth Federal Interagency Sedimentation Conference, March, Las Vegas, NV, Vol. 2.

Simon, T. P., and J. Lyons. 1995. Application of the Index of Biotic Integrity to evaluate water resource integrity in freshwater ecosystems. Pp. 245–262 *In:* Biological Assessment and Criteria: Tools for Water Resource Planning and Decision Making. W. S. Davis and T. P. Simon (eds.). CRC Press.

Simon, A., M. Doyle, M. Kondolf, F. D. Shields, Jr., B. Rhoads, and M. McPhillips. 2007. Critical evaluation of how the Rosgen classification and associated "natural channel design" methods fail to integrate and quantify fluvial processes and channel response. Journal of the American Water Resources Association 43(5):1117–1131.

Smart, J. S. 1968. Statistical properties of stream lengths. Water Resources Research 4:1001–1014.

SMBRP (Santa Monica Bay Restoration Project). 1994. Characterization Study of the Santa Monica Bay Restoration Plan. Santa Monica Bay Restoration Project, Monterey Park, CA.

Smith, S. B., D. R. P. Reader, P. C. Baumann, S. R. Nelson, J. A. Adams, K. A. Smith, M. M. Powers, P. L. Hudson, A. J. Rosolofson, M. Rowan, D. Peterson, V. S. Blazer, J. T. Hickey, and K. Karwowski. 2003. Lake Erie Ecological Investigation; Summary of findings: Part 1; Sediments, Invertebrate Communities, and Fish Communities: Part 2; Indicators, Anomalies, Histopathology, and Ecological Risk Assessment. USGS, Mimeo.

Smock, L. A., and C. M. MacGregor. 1988. Impact of the America chestnut blight on aquatic shredding macroinvertebrates. Journal of the North American Benthological Society 7:212–221.

Sprague, L. A., D. A. Harned, D. W. Hall, L. H. Nowell, N. J. Bauch, and K. D. Richards. 2007. Response of stream chemistry during base flow to gradients of urbanization in selected locations across the conterminous United States, 2002–04. U.S. Geological Survey Scientific Investigations Report 2007–5083, 132 pp.

Stein, E. D., and L. L. Tiefenthaler. 2005. Dry-weather metals and bacteria loading in an arid, urban watershed: Ballona Creek, California. Water, Air, Soil Pollution 164(1–4):367–382.

Stoddard, J. L., D. P. Larsen, C. P. Hawkins, R. K. Johnson, and R. H. Norris. 2006. Setting expectations for the ecological condition of streams: the concept of reference condition. Ecological Applications 16(4):1267–1276.

Stolzenbach, K. D., S. K. Friedlander, R. Turco, A. Winer, R. Lu, C. Xiong, J.-H. Lim, L. Sabin, K. Schiff, and L. Tiefenthaler. 2007. Atmospheric Deposition as a Source of Contaminants in Stormwater Runoff. Presentation to NRC Committee on Reducing Stormwater Discharge Contributions to Water Pollution, May 2, 2007, Austin, TX.

Stone, B. 2004. Paving over paradise: how land use regulations promote residential imperviousness. Landscape and Urban Planning (69):101–113.

Strahler, A. N. 1957. Quantitative analysis of watershed geomorphology. Transactions of the American Geophysical Union (38):913–920.

Strahler, A. N. 1964. Quantitative geomorphology of drainage basins and channel networks. Pp. 4-39–4-76 *In:* Handbook of Applied Hydrology. Ven Te Chow (ed.). New York: McGraw-Hill.

Strayer, D. L., J. A. Downing, W. R. Haag, T. L. King, J. B. Layzer, T. J. Newton, and S. Nichols. 2004. Changing Perspectives on Pearly Mussels, North America's Most Imperiled Animals. BioScience 54(5):429–439.

Stribling, J. B., E. W. Leppo, J. D. Cummins, J. Galli, S. Meigs, L. Coffman, M. Cheng. 2001. Relating instream biological condition to BMPs in watershed. Pp. 287–304 *In:* Linking Stormwater BMP Designs and Performance to Receiving Water Impact Mitigation. B. R. Urbonas (ed.). Proceedings of an Engineering Foundation Conference, Snowmass Village, CO, August 19–24, 2001. United Engineering Foundation, Environmental and Water Resources Institute of ASCE.

Suter, G. 2006. Ecological risk assessment and ecological epidemiology for contaminated sites. Human and Ecological Risk Assessment 12(1):31–38.

Suter, G. W., II. 1993. Ecological Risk Assessment. Boca Raton, FL: Lewis Publishers.

Sutherland, R. 2000. Methods for estimating the effective impervious area of urban watersheds. Pp. 193–195 *In:* The Practice of Watershed Protection. T. R. Schueler and H. K. Holland (eds.). Ellicott City, MD: Center for Watershed Protection.

Sweeney, B., W. Bott, T. Jackson, K. Kaplan, L. Newbold, J. Standley, L. Hession, and R. Horowitz. 2004. Riparian deforestation, stream narrowing and loss of stream ecosystem services. Proceedings of the National Academy of Sciences 101:14132–14137.

Thomson, J. D., G. Weiblen, B. A. Thomson, S. Alfaro, and P. Legendre. 1996. Untangling multiple factors in spatial distributions: lilies, gophers, and rocks. Ecology 77(6):1698–1715.

Tobiason, S. 2004. Stormwater metals removal by media filtration: field assessment case study. Watershed 2004 Conference Proceedings, CD-ROM. Alexandria, VA: Water Environment Federation.

Trimble, S. W. 1997. Contribution of stream channel erosion to sediment yield from an urbanizing watershed. Science 278:1442–1444.

Urban, M., D. Skelly, D. Burchsted, W. Price, and S. Lowry. 2006. Stream communities across a rural-urban landscape gradient. Diversity and Distributions 12:337–350.

U.S. Department of Agriculture (USDA). 2000. Summary Report: 1997 National Resources Inventory (revised December 2000). Natural Resources Conservation Service, Washington, DC, and Statistical Laboratory, Iowa State University, Ames, Iowa, 89 pages.

USGS (U. S. Geological Survey). 1998. A New Evaluation of the USGS Streamgaging Network: A Report to Congress dated November 30, 1998. Reston, VA: U.S. Geological Survey.

USGS. 1999. Analyzing Land Use Change in Urban Environments. USGS Fact Sheet 188-99.

Vogel, R. M., J. R. Stedinger, and R. P. Hooper. 2003. Discharge indices for water quality loads. Water Resources Research 39(10), doi:10.1029/2002WR001872.

Wallinder, I. O., B. Bahar, C. Leygraf, and J. Tidblad. 2007. Modelling and mapping of copper runoff for Europe. Journal of Environmental Monitoring 9:66–73.

Walsh, C., K. Waller, J. Gehling, and R. MacNally. 2007. Riverine invertebrate assemblages are degraded more by catchment urbanization than riparian deforestation. Freshwater Biology 52(3):574–587.

Walsh, C. J., A. H. Roy, J. W. Feminella, P. D. Cottingham, P. M. Groffman, and R. P. Morgan. 2005. The urban stream syndrome: current knowledge and the search for a cure. Journal of the North American Benthological Society 24(3):706–723.

Walters, D. M., M. C. Freeman, D. S. Leigh, B. J. Freeman, and C. M. Pringle.

2005. Urbanization effects on fishes and habitat quality in a southern pied-mont river basin. Pp. 69–85 *In:* Effects of Urbanization on Stream Ecosys-tems. L. R. Brown, R. M. Hughes, R. Gray, and M. R. Meador (eds.). Be-thesda, MD: American Fisheries Society.

Wang, L., and P. Kanehl. 2003. Influences of watershed urbanization and in-stream habitat on macroinvertebrates in cold water streams. Journal of the American Water Resources Association 39(5):1181–1196.

Wang, L., J. Lyons, P. Rasmussen, P. Simons, T. Wiley, and P. Stewart. 2003. Watershed, reach, and riparian influences on stream fish assemblages in the Northern Lakes and Forest Ecoregion. Canadian Journal of Fisheries and Aquatic Science 60:491–505.

Warren, M., and M. G. Pardew. 1998. Road crossings as barriers to small-stream fish movement. Transactions of the American Fishery Society 127:637–644.

Waters, T. F. 1995. Sediment in Streams—Sources, Biological Effects and Control. American Fisheries Society Monograph 7. Bethesda, MD: Ameri-can Fisheries Society.

Webster, J. R., and D. J. D'Angelo. 1997. A regional analysis of the physical characteristics of streams. Journal of the North American Benthological Society 16(1):87–95.

Wells, C. 1995. Skinny streets and one-sided sidewalks: a strategy for not pav-ing paradise. Watershed Protection Techniques 1(3):135–137.

Whitlow, J. R., and K. J. Gregory. 1989. Changes in urban stream channels in Zimbabwe. Regulated Rivers Research and Management 4:27–42.

Wiebusch, B., M. Ozaki, H. Watanabe, and C. F. Seyfried. 1998. Assessment of leaching tests on construction material made of incinerator ash (sewage sludge): investigations in Japan and Germany. Water Science and Technol-ogy 38(7):195–205.

Willett, G. 1980. Urban erosion. In National Conference on Urban Erosion and Sediment Control; Institutions and Technology. EPA 905/9-80-002. Wash-ington, DC: EPA.

Williams, J. E., J. E. Johnson, D. A. Hendrickson, S. Contreras-Balderas, J. D. Williams, M. Navarro Mendoza, D. E. McAllister, J. E. Deacon. 1989. Fishes of North America endangered, threatened, or of special concern: 1989. Fisheries 14:2–20.

Winter, T. 2007. The role of groundwater in generating streamflow in headwa-ter areas in maintaining baseflow. Journal of American Water Resources Association 43(1):15–25.

Wolman, M. G., and W. P. Miller. 1960. Magnitude and frequency of forces in geomorphic processes. Journal of Geology 68:54–74.

Wolman, M. G., and A. Schick. 1967. Effects of construction on fluvial sedi-ment, urban and suburban areas of Maryland. Water Resources Research 3:451–464.

Wright, T., J. Tomlinson, T. Schueler, and K. Cappiella. 2007. Direct and indi-rect impacts of urbanization on wetland quality. Wetlands and Watersheds

Article 1. Washington, DC: EPA Office of Wetlands, Oceans, and Watersheds, and Ellicott City, MD: Center for Watershed Protection.

Xiao, Q. F. 1998. Rainfall Interception of Urban Forest. Ph.D. Dissertation.
University of California, Davis.

Yee, D., and A. Franz. 2005. Castro Valley Atmospheric Deposition Monitoring. PowerPoint presentation, Brake Pad Partnership Stakeholder Conference.

4
Monitoring and Modeling

As part of its statement of task, the committee was asked to consider several aspects of stormwater monitoring, including how useful the activity is, what should be monitored and when and where, and how benchmarks should be established. As noted in Chapter 2, the stormwater monitoring requirements under the U.S. Environmental Protection Agency (EPA) stormwater program are variable and generally sparse, which has led to considerable skepticism about their usefulness. This chapter first considers the value of the data collected over the years by municipalities and makes suggestions for improvement. It then does the same for industrial stormwater monitoring, which has lagged behind the municipal separate storm sewer system (MS4) program both in requirements and implementation.

It should be noted upfront that this chapter does not discuss the fine details of MS4 and industrial monitoring that pertain to regulatory compliance—questions such as should the average end of pipe concentrations meet water quality standards, how many exceedances should be allowed per year, or should effluent concentrations be compared to acute or chronic criteria. Individual benchmarks and effluent limits for specific chemicals emanating from specific industries are not provided. The current state of MS4 and industrial stormwater monitoring and the paucity of high quality data are such that it is premature and in many cases impossible to make such determinations. Rather, the chapter suggests both *how* to monitor an individual industry and *how* to determine benchmarks and effluent limits for industrial categories. It suggests how monitoring requirements should be tailored to accommodate the risk level of an individual industrial discharger. Finally, it makes numerous technical suggestions for improving the monitoring of MS4s, building on the data already submitted and analyzed as part of the National Stormwater Quality Database. Policy recommendations about the monitoring of both industries and MS4s are found in Chapter 6.

This chapter's emphasis on monitoring of stormwater should not be interpreted as a disinterest in other types of monitoring, such as biomonitoring of receiving waters, precipitation measurements, or determination of land cover. Indeed, these latter activities are extremely important (they are introduced in the preceding chapter) and they underpin the new permitting program proposed in Chapter 6 (especially biological monitoring). Stormwater management would benefit most substantially from a well-balanced monitoring program that encompasses chemical, biological, and physical parameters from outfalls to receiving waters. Currently, however, decisions about stormwater management are usually made with incomplete information; for example, there are continued recommendations by many that street cleaning will solve a municipality's problems, even when the municipality does not have any information on the sources of the material being removed.

A second charge to the committee was to define the elements of a "protocol" to link pollutants in stormwater discharges to ambient water quality criteria. As described in Chapter 3, many processes connect sources of pollution to an effect observed in a downstream receiving water. More and more, these processes can be represented in watershed models, which are the key to linking stormwater sources to effects observed in receiving waters. The latter half of the chapter explores the current capability of models to make such links, including simple models, statistical and conceptual models, and more involved mechanistic models. At the present time, associating a single discharger with degraded in-stream conditions is generally not possible because of the state of both modeling and monitoring of stormwater.

MONITORING OF MS4s

EPA's regulations for stormwater monitoring of MS4s is very limited, in that only the application requirements are stated [see 40 CFR § 122.26(d)]. The regulations require the MS4 program to identify five to ten stormwater discharge outfalls and to collect representative stormwater data for conventional and priority toxic pollutants from three representative storm events using both grab and composite sampling methods. Each sampled storm event must have a rainfall of at least 0.1 inch, must be preceded by at least 72 hours of a dry period, and the rain event must be within 50 percent of the average or median of the per storm volume and duration for the region. While the measurement of flow is not specifically required, an MS4 must make estimates of the event mean concentrations (EMCs) for pollutants discharged from all outfalls to surface waters, and in order to determine EMCs, flow needs to be measured or calculated.

Other than these requirements, the exact type of MS4 monitoring that is to be conducted during the permit term is left to the discretion of the permitting authority. EPA has not issued any guidance on what would be considered an adequate MS4 monitoring program for permitting authorities to evaluate compliance. Some guidance for MS4 monitoring based on desired management questions has been developed locally (for example, see the SCCWRP Technical Report No. 419, SMC 2004, Model Monitoring Program for MS4s in Southern California).

In the absence of national guidance from EPA, the MS4 monitoring programs for Phase I MS4s vary widely in structure and objectives, and Phase II MS4 programs largely do not perform any monitoring at all. The types of monitoring typically contained in Phase I MS4 permits include the (1) wet weather outfall screening and monitoring to characterize stormwater flows, (2) dry weather outfall screening and monitoring under illicit discharge detection and elimination programs, (3) biological monitoring to determine storm water impacts, (4) ambient water quality monitoring to characterize water quality conditions, and (5) stormwater control measure (SCM) effectiveness monitoring.

The Nationwide Stormwater Quality Database

Stormwater monitoring data collected by a portion of Phase I MS4s has been evaluated for years by the University of Alabama and the Center for Watershed Protection and compiled in a database called the Nationwide Stormwater Quality Database (NSQD). These data were collected in order to describe the characteristics of stormwater on a national level, to provide guidance for future sampling needs, and to enhance local stormwater management activities in areas with limited data. The MS4 monitoring data collected over the past ten years from more than 200 municipalities throughout the country have great potential in characterizing the quality of stormwater runoff and comparing it against historical benchmarks. Version 3 of the NSQD is available online at: http://unix.eng.ua.edu/~rpitt/Research/ms4/mainms4.shtml. It contains data from more than 8,500 events and 100 municipalities throughout the country. About 5,800 events are associated with homogeneous land uses, while the remainder are for mixed land uses.

The general approach to data collection was to contact EPA regional offices to obtain state contacts for the MS4 data, then the individual municipalities with Phase I permits were targeted for data collection. Selected outfall data from the International BMP Database were also included in NSQD version 3, eliminating any source area and any treated stormwater samples. Some of the older National Urban Runoff Program (NURP) (EPA, 1983) data were also included in the NSQD, along with some data from specialized U.S. Geological Survey (USGS) stormwater monitoring activities in order to better represent nationwide conditions and additional land uses. Because there were multiple sources of information, quality assurance and quality control reviews were very important to verify the correctness of data added to the database, and to ensure that no duplicate entries were added.

The NSQD includes sampling location information such as city, state, land use, drainage area, and EPA Rain Zone, as well as date, season, and rain depth. The constituents commonly measured for in stormwater include total suspended solids (TSS), 5-day biological oxygen demand (BOD$_5$), chemical oxygen demand (COD), total phosphorus (TP), total Kjeldahl nitrogen (TKN), nitrite plus nitrate (NO$_2$+NO$_3$), total copper (Cu), total lead (Pb), and total zinc (Zn). Less information is available for many other constituents (including filterable heavy metals and bacteria). Figure 4-1 is a map showing the EPA Rain Zones in the United States, along with the locations of the communities contributing to the NSQD, version 3. Table 4-1 shows the number of samples for each land use and for each Rain Zone. This table does not show the number of mixed land-use site samples. Rain Zones 8 and 9 have very few samples, and institutional and open-space areas are poorly represented. However, residential, commercial, industrial, and freeway data are plentiful, except for the few Rain Zones noted above.

Land use has an important impact on the quality of stormwater. For example, the concentrations of heavy metals are higher for industrial land-use areas

TABLE 4-1 Number of Samples per Land Use and EPA Rain Zone

Single Land Use	1	2	3	4	5	6	7	8	9	Total
Commercial	234	484	131	66	42	37	64	0	22	1080
Freeways	0	241	14	0	262	189	28	0	0	734
Industrial	100	327	90	51	83	74	146	0	22	893
Institutional	9	46	0	0	0	0	0	0	0	55
Open Space	68	37	0	18	0	2	0	0	0	125
Residential	294	1470	290	122	105	32	532	7	81	2933
Total	705	2605	525	257	492	334	770	7	125	5820

Note: there are no mixed-use sites in this table. SOURCE: National Stormwater Quality Database.

due to manufacturing processes and other activities that generate these materials. Fecal coliform concentrations are relatively high for residential and mixed residential land uses, and nitrate concentrations are higher for the freeway land use. Open-space land-use areas show consistently low concentrations for the constituents examined. Seasons could also be a factor in the variation of nutrient concentrations in stormwater due to seasonal uses of fertilizers and leaf drop occurring during the fall season. Most studies also report lower bacteria concentrations in the winter than in the summer. Lead concentrations in stormwater have also significantly decreased since the elimination of lead in gasoline (see Figure 2-6). Most of the statistical tests used are multivariate statistical evaluations that compare different constituent concentrations with land use and geographical location. More detailed discussions of the earlier NSQD results are found in various references, including Maestre et al. (2004, 2005) and Pitt et al. (2003, 2004).

How to use the NSQD to Calculate Representative EMC Values

EMC values were initially used during the NURP to describe typical concentrations of pollutants in stormwater for different monitoring locations and land uses. An EMC is intended to represent the average concentration for a single monitored event, usually based on flow-weighted composite sampling. It can also be calculated from discrete samples taken during an event if flow data are also available. Many individual subsamples should be taken throughout most of the event to calculate the EMC for that event. Being an overall average value, an EMC does not represent possible extremes that may occur during an event.

The NSQD includes individual EMC values from about 8,500 separate events. Stormwater managers typically want a representative single value for a land use for their area. As such, they typically evaluate a series of individual

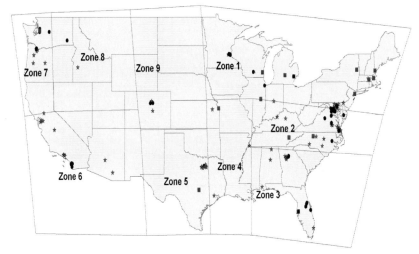

Database Representation

- BMP
- NURP
- USGS
- MS4

FIGURE 4-1 Sampling Locations for Data Contained in the National Stormwater Quality Database, version 3.

storm EMC values for conditions similar to those representing their site of concern. With the NSQD in a spreadsheet form, it is relatively simple to extract suitable events representing the desired conditions. However, the individual EMC values will likely have a large variability. Maestre and Pitt (2006) reviewed the NSQD data to better explain the variability according to different site and sampling conditions (land use, geographical location, season, rain depth, amount of impervious area, sampling methods, antecedent dry period, etc.). The most common significant factor was land use, with some geographical and fewer seasonal effects observed. As with the original NURP data, EMCs in the NSQD are usually expressed using medians and coefficients of variation to reflect uncertainty, assuming lognormal distributions of the EMC values. Figure 4-2 shows several lognormal probability plots for a few constituents from the NSQD. Probability plots shown as straight lines indicate that the concentrations can be represented by lognormal distributions (see Box 4-1).

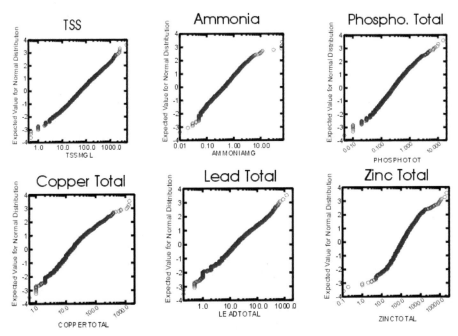

FIGURE 4-2 Lognormal probability plots of stormwater quality data for selected constituents (pooled data from NSQD version 1.1).

Fitting a known distribution is important as it helps indicate the proper statistical tests that may be conducted. Using the *median* EMC value in load calculations, without considering the data variability, will result in smaller mass loads compared to actual monitored conditions. This is due to the medians underrepresenting the larger concentrations that are expected to occur. The use of *average* EMC values will represent the larger values better, although they will still not represent the variability likely to exist. If all of the variability cannot be further explained adequately (such as being affected by rain depth), which would be highly unlikely, then a set of random calculations (such as that obtained using Monte Carlo procedures) reflecting the described probability distribution of the constituents would be the best method to use when calculating loads.

Municipal Monitoring Issues

As described in Chapter 2, typical MS4 monitoring requirements involve sampling during several events per year at the most common land uses in the area. Obviously, a few samples will not result in very useful data due to

---

BOX 4-1
Probability Distributions of Stormwater Data

The coefficient of variation (COV) values for many constituents in the NSQD range from unusually low values of about 0.1 (for pH) to highs between 1 and 2. One objective of a data analysis procedure is to categorize the data into separate stratifications, each having small variations in the observed concentrations. The only stratification usually applied is for land use. However, further analyses indicated many differences by geographical area and some differences by season. When separated into appropriate stratifications, the COV values are reduced, ranging between about 0.5 to 1.0. With a reasonable confidence of 95 percent (α= 0.05) and power of 80 percent (β= 0.20), and a suitable allowable error goal of 25 percent, the number of samples needed to characterize these conditions would there-fore range from about 25 to 50 (Burton and Pitt, 2002). In a continuing monitoring program (such as the Phase I stormwater National Pollutant Discharge Elimination System [NPDES] permit monitoring effort) characterization data will improve over time as more samples are obtained, even with only a few samples collected each year from each site.

Stormwater managers have generally accepted the assumption of lognormality of stormwater constituent concentrations between the 5th and 95th percentiles. Based on this assumption, it is common to use the log-transformed EMC values to evaluate differences between land-use categories and other characteristics. Statistical inference methods, such as estimation and tests of hypothesis, and analysis of variance, require statistical informa-tion about the distribution of the EMC values to evaluate these differences. The use of the log-transformed data usually includes the location and scale parameter, but a lower-bound parameter is usually neglected.

Maestre et al. (2005) conducted statistical tests using NSQD data to evaluate the log-normality assumptions of selected common constituents. It was found in almost all cases that the log-transformed data followed a straight line between the 5th and 95th percentile, as illustrated in Figure 4-3 for total dissolved solids (TDS) in residential areas.

For many statistical tests focusing on the central tendency (such as for determining the concentrations that are to be used for mass balance calculations), this may be a suit-able fit. As an example, the model WinSLAMM (Pitt, 1986; Pitt and Voorhees, 1995) uses a Monte Carlo component to describe the likely variability of stormwater source flow pollut-ant concentrations using either lognormal or normal probability distributions for each con-stituent. However, if the most extreme values are of importance, such as when dealing with the influence of many non-detectable values on the predicted concentrations, or de-termining the frequency of observations exceeding a numerical standard, a better descrip-tion of the extreme values may be important.

The NSQD contains many factors for each sampled event that likely affect the ob-served concentrations. These include such factors as seasons, geographical zones, and rain intensities. These factors may affect the shape of the probability distribution. The only way to evaluate the required number of samples in each category is by using the power of the test, where power is the probability that the test statistic will lead to a rejection of the null hypothesis (Gibbons and Chakraborti, 2003).

In the NSQD, most of the data were from residential land uses. The Kolmogorov-Smirnov test was used to indicate if the cumulative empirical probability distribution of the residential stormwater constituents can be adequately represented with a lognormal distri-bution. The number of collected samples was sufficient to detect if the empirical distribu-

continues next page

BOX 4-1 Continued

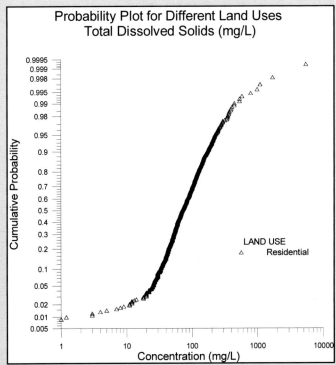

FIGURE 4-3 Probability plot of total dissolved solids in residential land uses (NSQD version 1.1 data).

tion was located inside an interval of width 0.1 above and below the estimated cumulative probability distribution. If the interval was reduced to 0.05, the power varies between 40 and 65 percent. Another factor that must be considered is the importance of relatively small errors in the selected distribution and the problems of false-negative determinations. It may not be practical to collect as many data observations as needed when the distributions are close. Therefore, it is important to understand what types of further statistical and analysis problems may be caused by having fewer samples than optimal. For example, Figure 4-4 (total phosphorus in residential areas) shows that most of the data fall along the straight line (indicating a lognormal fit), with fewer than 10 observations (out of 933) in the tails being outside of the obvious path of the line, or a false-negative rate of about 0.01 (1 percent).

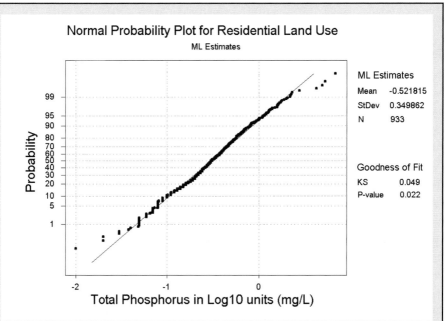

FIGURE 4-4 Normality test for total phosphorus in residential land uses using the NSQD.

Further analyses to compare the constituent concentration distributions to other common probability distributions (normal, lognormal, gamma, and exponential) were also conducted for all land uses by Maestre et al. (2004). Most of the stormwater constituents can be assumed to follow a lognormal distribution with little error. The use of a third parameter in the estimated lognormal distribution may be needed, depending on the number of samples. When the number of samples is large per category (approximately more than 400 samples) the maximum likelihood and the two-parameter lognormal distribution better fit the empirical distribution. For large sample sizes, the L-moments method usually unacceptably truncates the distribution in the lower tail. However, when the sample size is more moderate per category (approximately between 100 and 400 samples), the three-parameter lognormal method, estimated by L-moments, better fits the empirical distribution. When the sample size is small (less than 100 samples, as is common for most stormwater programs), the use of the third parameter does not improve the fit with the empirical distribution and the common two-parameter lognormal distribution produces a better fit than the other two methods. The use of the lognormal distribution also has an advantage over the other distribution types because it can be easily transformed to a normal distribution and the data can then be correctly examined using a wide variety of statistical tests.

the variability of stormwater characteristics. However, during the period of a five-year permit with three samples per year, about 15 events would be sampled for each land use. While still insufficient for many analyses, this number of data points likely allows the confidence limits to be reasonably calculated for the average conditions. When many sites of the same land use are monitored for a region, substantial data may be collected during a permit cycle. This was the premise of the NSQD where MS4 data were collected for many locations throughout the country. These data were evaluated and various findings made. The following comments are partially based on these analyses, along with additional data sources.

Sampling Technique and Compositing

There are a variety of methods for collecting and compositing stormwater samples that can result in different values for the EMC. The first distinction is the mode of sample collection, either as grab samples or automatic sampling. Obviously, grab sampling is limited by the speed and accuracy of the individuals doing the sampling, and it is personnel intensive. It is for this reason that about 80 percent of the NSQD samples are collected using automatic samplers. Manual sampling has been observed to result in slightly lower TSS concentrations compared to automatic sampling procedures. This may occur, for example, if the manual sampling team arrives after the start of runoff and therefore misses an elevated first flush (if it exists for the site), resulting in reduced EMCs.

A second important concept is how and whether the samples are combined following collection. With *time-based discrete sampling,* samplers (people or machines) are programmed to take an aliquot after a set period of time (usually in the range of every 15 minutes) and each aliquot is put into a separate bottle (usually 1 liter). Each bottle is processed separately, so this method can have high laboratory costs. This is the only method, however, that will characterize the changes in pollutant concentrations during the event. *Time-based composite sampling* refers to samplers being programmed to take an aliquot after a set period of time (as short as every 3 minutes), but then the aliquots are combined into one container prior to analysis (compositing). All parts of the event receive equal weight with this method, but the large number of aliquots can produce a reasonably accurate composite concentration. Finally, *flow-weighted composite sampling* refers to samplers being programmed to collect an aliquot (usually 1 liter) for a set volume of discharge. Thus, more samples are collected during the peak of the hydrograph than toward the trailing edge of the hydrograph. All of the aliquots are composited into one container, so the concentration for the event is weighted by flow.

Most communities calculate their EMC values using flow-weighted composite sample analyses for more accurate mass discharge estimates compared to time-based compositing. This is especially important for areas with a first flush of very short duration, because time-composited samples may overly emphasize these higher flows. An automatic sampler with flow-weighted samples, in conjunction with a bed-load sampler, is likely the most accurate sampling method, but only if the sampler can obtain a representative sample at the location (such as sampling at a cascading location, or using an automated depth-integrated sampler) (Clark et al., 2008).

Time- and flow-weighted composite options have been evaluated in residential, commercial, and industrial land uses in EPA Rain Zone 2 and in industrial land uses in EPA Rain Zone 3 for the NSQD data. No significant differences were observed for BOD_5 concentrations using either of the compositing schemes for any of the four categories. TSS and total lead median concentrations in EPA Rain Zone 2 were two to five times higher in concentration when time-based compositing was used instead of flow-based compositing. Nutrients in EPA Rain Zone 2 collected in residential, commercial, and industrial areas showed no significant differences using either compositing method. The only exceptions were for ammonia in residential and commercial land-use areas and total phosphorus in residential areas where time-based composite samples had higher concentrations. Metals were higher when time-based compositing was used in residential and commercial land-use areas. No differences were observed in industrial land-use areas, except for lead. Again, in most cases, mass discharges are of the most importance in order to show compliance with TMDL requirements. Flow-weighted sampling is the most accurate method to obtain these values (assuming sufficient numbers of subsamples are obtained). However, if receiving water effects are associated with short-duration high concentrations, then discrete samples need to be collected and analyzed, with no compositing of the samples during the event. Of course, this is vastly more costly and fewer events are usually monitored if discrete sampling is conducted.

Numbers of Data Observations Needed

The biggest issue associated with most monitoring programs is the number of data points needed. In many cases, insufficient data are collected to address the objectives of the monitoring program with a reasonable amount of confidence and power. Burton and Pitt (2002) present much guidance in determining the amount of data that should be collected. A basic equation that can be used to estimate the number of samples to characterize a set of conditions is as follows:

$$n = [COV(Z_{1-\alpha} + Z_{1-\beta})/(error)]^2$$

where:

n = number of samples needed.

α = false-positive rate ($1-\alpha$ is the degree of confidence; a value of α of 0.05 is usually considered statistically significant, corresponding to a $1-\alpha$ degree of confidence of 0.95, or 95%).

β = false-negative rate ($1-\beta$ is the power; if used, a value of β of 0.2 is common, but it is frequently and improperly ignored, corresponding to a β of 0.5).

$Z_{1-\alpha}$ = Z score (associated with area under a normal curve) corresponding to $1-\alpha$; if α is 0.05 (95% degree of confidence), then the corresponding $Z_{1-\alpha}$ score is 1.645 (from standard statistical tables).

$Z_{1-\beta}$ = Z score corresponding to $1-\beta$ value; if β is 0.2 (power of 80%), then the corresponding $Z_{1-\beta}$ score is 0.85 (from standard statistical tables); however, if power is ignored and β is 0.5, then the corresponding $Z_{1-\beta}$ score is 0.

error = allowable error, as a fraction of the true value of the mean.

COV = coefficient of variation (sometimes noted as CV), the standard deviation divided by the mean (dataset assumed to be normally distributed).

Figures 4-5 and 4-6 can be used to estimate the sampling effort, based on the expected variability of the constituent being monitored, the allowable error in the calculated mean value, and the associated confidence and power. Figure 4-5 can be used for a single sampling point that is being monitored for basic characterization information, while Figure 4-6 is used for paired sampling when two locations are being compared. Confidence and power are needed to control the likelihood of false negatives and false positives. The sample needs increase dramatically as the difference between datasets becomes small when comparing two conditions with a paired analysis, as shown in Figure 4-6 (above and below an outfall, influent vs. effluent, etc.). Typically, being able to detect a difference of at least about 25 percent (requiring about 50 sample pairs with typical sample variabilities) is a reasonable objective for most stormwater projects. This is especially important when monitoring programs attempt to distinguish test and control conditions associated with SCMs. It is easy to confirm significant differences between influent and effluent conditions at wet detention ponds, as they have relatively high removal rates. Less effective controls are much more difficult to verify, as the sampling program requirements become very expensive.

First-Flush Effects

First flush refers to an assumed elevated load of pollutants discharged in the beginning of a runoff event. The first-flush effect has been observed more often in small catchments than in large catchments (Thompson et al., 1995, cited by WEF and ASCE, 1998). Indeed, in large catchments (>162 ha, 400 acres), the

Number of Samples Required
(alpha = 0.05, beta = 0.20)

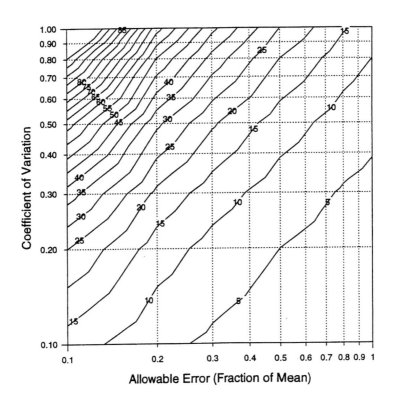

FIGURE 4-5 Number of samples to characterize median (power of 80% and confidence of 95%). SOURCE: Reprinted, with permission, from, Burton and Pitt (2002). Copyright 2002 by CRC Press.

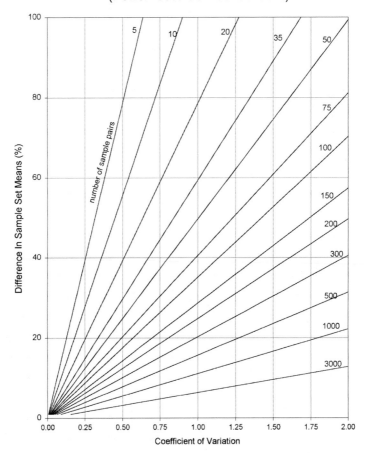

FIGURE 4-6 Number of paired samples needed to distinguish between two sets of observations (power 80% and confidence of 95%). SOURCE: Reprinted, with permission from, Burton and Pitt (2002). Copyright 2002 by CRC Press.

highest concentrations are usually observed at the times of flow peak (Brown et al., 1995; Soeur et al., 1995). Adams and Papa (2000) and Deletic (1998) both concluded that the presence of a first flush depends on numerous site and rainfall characteristics.

Figure 4-7 is a plot of monitoring data from the Villanova first-flush study (Batroney, 2008) showing the flows, rainfall, TSS concentration, TDS concentration, and TDS and TSS event mean concentrations for the inflow to an infiltration trench. Because of the first-flush effect, a grab sample early in the storm would have over-predicted the TSS event mean concentration of the site, and a later sample would have under-predicted this same value, although for TDS the results would have been similar.

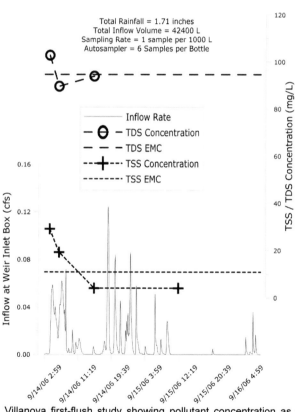

FIGURE 4-7 Villanova first-flush study showing pollutant concentration as a function of inflow rainfall volume. This study collected runoff leaving the top floor of a parking garage. Samples were taken of the runoff in one-quarter-inch increments, up to an inch of rain, and then every inch thereafter. The plot of TSS concentration versus rainfall increment shows a strong first flush for this storm, while the TDS concentration does not. SOURCE: Reprinted, with permission, Batroney (2008). Copyright 2008 by T. Thomas Batroney.

Figure 4-8 shows data for a short-duration, high-intensity rain in Tuscaloosa, Alabama, that had rain intensities as great a 6 inches per hour for a 10-minute period. The drainage area was a 0.4-ha paved parking lot with some landscaping along the edges. The turbidity plot shows a strong first flush for this event, and the particle size distributions indicate larger particles at the beginning of the event, then becoming smaller as the event progresses, and then larger near the end. Most of the other pollutants analyzed had similar first-flush patterns like the turbidity, with the notable exception of bacteria. Both *E. coli* and enterococci concentrations started off moderately low, but then increased substantially near the end of the rain. Several rains have been monitored at this site so far, and most show a similar pattern with decreasing turbidity and increasing bacteria as the rain continues.

Sample collection conducted for some of the NPDES MS4 Phase I permits required both a grab and a composite sample for each event. A grab sample was to be taken during the first 30 minutes of discharge to capture the first flush, and a flow-weighted composite sample was to be taken for the entire time of discharge (every 15 to 20 minutes for at least three hours or until the event ended). Maestre et al. (2004) examined about 400 paired sets of 30-minute and 3-hour samples from the NSQD, as shown in Table 4-2. Generally, a statistically significant first flush is associated with a median concentration ratio of about 1.4 or greater (the exceptions are where the number of samples in a specific category is much smaller). The largest ratios observed were about 2.5, indicating that for these conditions the first 30-minute flush sample concentrations are about 2.5 times greater than the composite sample concentrations. More of the larger ratios are found for the commercial and institutional land-use categories, where larger paved areas are likely to be found. The smallest ratios are associated with the residential, industrial, and open-space land uses—locations where there may be larger areas of unpaved surfaces.

The data in Table 4-2 were from North Carolina (76.2 percent), Alabama (3.1 percent), Kentucky (13.9 percent), and Kansas (6.7 percent) because most other states' stormwater permits did not require this sampling strategy. The NSQD investigation of first-flush conditions for these data locations indicated that a first-flush effect was not present for all the land-use categories and certainly not for all constituents. Commercial and residential areas were more likely to show this phenomenon, especially if the peak rainfall occurred near the beginning of the event. It is expected that this effect will more likely occur in a watershed with a high level of imperviousness, but even so, the data indicated first flushes for less than 50 percent of the samples for the most impervious areas. This reduced frequency of observed first flushes in areas most likely to have first flushes is probably associated with the varying rain conditions during the different events, including composite samples that did not represent the complete runoff duration.

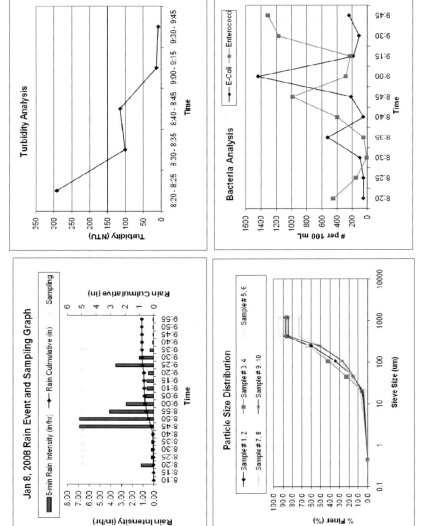

FIGURE 4-8 Pollutant variations during rain period (0.4-ha drainage area, mostly paved parking with small fringe turf area, Tuscaloosa, Alabama). SOURCE: Robert Pitt, University of Alabama.

TABLE 4-2 Significant First Flush Ratios (First Flush to Composite Median Concentration)

Parameter	Commercial				Industrial				Institutional			
	n	sc	R	ratio	n	sc	R	ratio	n	sc	R	ratio
Turbidity, NTU	11	11	=	1.32			X				X	
COD, mg/L	91	91	≠	2.29	84	84	≠	1.43	18	18	≠	2.73
TSS, mg/L	90	90	≠	1.85	83	83	=	0.97	18	18	≠	2.12
Fecal coliform, col/100mL	12	12	=	0.87			X				X	
TKN, mg/L	93	86	≠	1.71	77	76	≠	1.35			X	
Phosphorus total, mg/L	89	77	≠	1.44	84	71	=	1.42	17	17	=	1.24
Copper, total, µg/L	92	82	≠	1.62	84	76	≠	1.24	18	7	=	0.94
Lead, total, µg/L	89	83	≠	1.65	84	71	≠	1.41	18	13	≠	2.28
Zinc, total, µg/L	90	90	≠	1.93	83	83	≠	1.54	18	18	≠	2.48

Parameter	Open Space				Residential				All Combined			
	n	sc	R	ratio	n	sc	R	ratio	n	sc	R	ratio
Turbidity, NTU			X		12	12	=	1.24	26	26	=	1.26
COD, mg/L	28	28	=	0.67	140	140	≠	1.63	363	363	≠	1.71
TSS, mg/L	32	32	=	0.95	144	144	≠	1.84	372	372	≠	1.60
Fecal coliform, col/100mL			X		10	9	=	0.98	22	21	=	1.21
TKN, mg/L	32	14	=	1.28	131	123	≠	1.65	335	301	≠	1.60
Phosphorus, total, mg/L	32	20	=	1.05	140	128	≠	1.46	363	313	≠	1.45
Copper, total, µg/L	30	22	=	0.78	144	108	≠	1.33	368	295	≠	1.33
Lead, total, µg/L	31	16	=	0.90	140	93	≠	1.48	364	278	≠	1.50
Zinc, total, µg/L	21	21	=	1.25	136	136	≠	1.58	350	350	≠	1.59

Note: n, number of total possible events; sc, number of selected events with detected values; R, result; X, not enough data; =, not enough evidence to conclude that median values are different; ≠, median values are different. "Ratio" is the ratio of the first flush to the full-period sample concentrations.
SOURCE: NSQD, as reported by Maestre et al. (2004).

Groups of constituents showed different behaviors for different land uses. All the heavy metals evaluated showed higher concentrations at the beginning of the event in the commercial land-use category. Similarly, all the nutrients showed higher initial concentrations in residential land-use areas, except for total nitrogen and orthophosphorus. This phenomenon was not found in the bacterial analyses. None of the land uses showed a higher population of bacteria at the beginning of the event.

The general conclusion from these data is that, in areas having low and generally even-intensity rains, first-flush observations are more common, especially in small and mostly paved areas. As an area increases in size, multiple routing pathways tend to blend the water, and runoff from the more distant locations reaches the outfall later in the event. SCMs located at outfalls in areas having low levels of impervious cover should be selected and sized to treat the complete event, if possible. Preferential treatment of first flushes may only be justified for small impervious areas, but even then, care needs to be taken to prevent undersizing and missing substantial fractions of the event.

Seasonal first flushes refer to larger portions of the annual runoff and pollutant discharges occurring during a short rain season. Seasonal first flushes may be observed in more arid locations where seasonal rainfalls are predominant. As an example, central and southern California can have dry conditions for extended periods, with the initial rains of the season occurring in the late fall. These rains can be quite large and, since they occur after prolonged dry periods, may carry substantial portions of the annual stormwater pollutant load. This is especially pronounced if later winter rains are more mild in intensity and frequent. For these areas, certain types of seasonally applied SCMs may be effective. As an example, extensive street, channel, and inlet cleaning in the late summer and early fall could be used to remove large quantities of debris and leaves from the streets before the first heavy rains occur. Other seasonal maintenance operations benefiting stormwater quality should also be scheduled before these initial rains.

Rain Depth Effects

An issue related to first flushes pertains to the effects of rain depth on stormwater quality. The NSQD contains much rainfall data along with runoff data for most areas of the country. Figure 4-9 contains scatter plots showing concentrations plotted against rain depth for some NSQD data. Although many might assume a correlation between concentrations and rain depth, in fact there are no obvious trends of concentration associated with rain depth. Rainfall energy determines erosion and wash-off of particulates, but sufficient runoff volume is needed to carry the particulate pollutants to the outfalls. Different travel times from different locations in the drainage areas results in these materials arriving at different times, plus periods of high rainfall intensity (that increase pollutant wash-off and movement) occur randomly throughout the storm. The

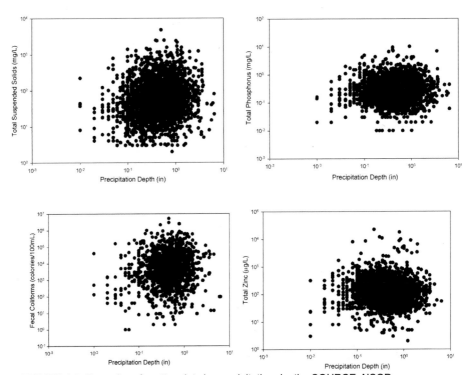

FIGURE 4-9 Examples of scatter plots by precipitation depth. SOURCE: NSQD.

resulting outfall stormwater concentration patterns for a large area having various surfaces is therefore complex and rain depth is just one of the factors involved.

Reported Monitoring Problems

A number of monitoring problems were described in the local Phase I community MS4 annual monitoring reports that were summarized as part of assembling the NSQD. About 58 percent of the communities described monitoring problems. Problems were mostly associated with obtaining reliable data for the targeted events. These problems increased costs because equipment failures had to be corrected and sampling excursions had to be rescheduled. One of the basic sampling requirements was to collect three samples every year for each of the land-use stations. These samples were to be collected at least one month apart during storm events having at least 0.1-inch rains, and with at least 72 hours from the previous 0.1-inch storm event. It was also required (when feasible) that the variance in the duration of the event and the total rainfall not exceed the median rainfall for the area. About 47 percent of the communities reported problems meeting these requirements. In many areas of the country, it was difficult to have three storm events per year with these characteristics. Furthermore, the complete range of site conditions needs to be represented in the data-collection effort; focusing only on a narrow range of conditions limits the representativeness of the data.

The second most frequent problem, reported by 26 percent of the communities, concerned backwater tidal influences during sampling, or that the outfall became submerged during the event. In other cases, it was observed that there was flow under the pipe (flowing outside of the pipe, in the backfill material, likely groundwater), or sometimes there was no flow at all. These circumstances all caused contamination of the collected samples, which had to be discarded, and prevented accurate flow monitoring. Greater care is obviously needed when locating sampling locations to eliminate these problems.

About 12 percent of the communities described errors related to malfunctions of the sampling equipment. When reported, the equipment failures were due to incompatibility between the software and the equipment, clogging of the rain gauges, and obstruction in the sampling or bubbler lines. Memory losses in the equipment recording data were also periodically reported. Other reported problems were associated with lighting, false starts of the automatic sampler before the runoff started, and operator error due to misinterpretation of the equipment configuration manual.

The reported problems suggest that the following changes should be made. First, the rain gauges need to be placed close to the monitored watersheds. Large watersheds cannot be represented with a single rain gauge at the monitoring station. In all cases, a standard rain gauge needs to supplement a tipping bucket rain gauge, and at least three rain gauges should be used in the research watersheds. Second, flow-monitoring instrumentation also needs to be used at all water quality monitoring stations. The lack of flow data greatly hinders the value of the chemical data. Third, monitoring needs to cover the complete storm duration. Automatic samplers need to be properly programmed and maintained to handle very short to very long events. It is unlikely that manual samplers were able to initiate sampling near the beginning of the events, unless they were deployed in anticipation of an event later in the day. A more cost-effective and reliable option would be to have semi-permanent monitoring stations at the various locations with sampling equipment installed in anticipation of a monitored event. Most monitoring agencies operated three to five land-use stations at one time. This number of samplers, and flow equipment, could have been deployed in anticipation of an acceptable event and would not need to be continuously installed in the field at all sampling locations.

Non-Detected Analyses

Left-censored data involve observations that are reported as below the limits of detection, whereas right-censored data involve above-range observations. Unfortunately, many important stormwater measurements (such as for filtered heavy metals) have large fractions of undetected values. These incomplete data greatly hinder many statistical tests. To estimate the problems associated with censored values, it is important to identify the probability distributions of the data in the dataset and the level of censoring. As discussed previously, most of the constituents in the NSQD follow a lognormal distribution. When the frequencies of the censored observations were lower than 5 percent, the means, standard deviations, and COVs were almost identical to the values obtained when the censored observations were replaced by half of the detection limit. As the percentage of nondetected values increases, replacing the censored observation by half of the detection limit instead of estimating them using Cohen's maximum likelihood method produced lower means and larger standard deviations. Replacing the censored observations by half of the detection limit is not recommended for levels of censoring larger than 15 percent. Because the Cohen method uses the detected observations to estimate the nondetected values, it is not very accurate, and therefore not recommended, when the percentage of censored observations is larger than 40 percent (Burton and Pitt, 2002). In this case, summaries should only be presented for the detected observations, with clear notations stating the level of nondetected observations.

The best method to eliminate problems associated with left-censored data is to use an appropriate analytical method. By keeping the nondetectable level below 5 percent, there are many fewer statistical analysis problems and the value of the datasets can be fully realized. Table 4-3 summarizes the recommended minimum detection limits for various stormwater constituents to obtain manageable nondetection frequencies (< 5 percent), based on the NSQD data observations. Some of the open-space stormwater measurements (lead, and oil and grease, for example) would likely have greater than 5 percent nondetections, even with the detection limits shown. The detection limits for filtered heavy metals should also be substantially less than shown on this table.

Seasonal Effects

Another factor that some believe may affect stormwater quality is the season when the sample was obtained. If the few samples collected for a single site were all collected in the same season, the results may not be representative of the whole year. The NPDES sampling protocols were designed to minimize this effect by requiring the three samples per year to be separated by at least one month. The few samples still could be collected within a single season, but not within the same week. Seasonal variations for residential fecal coliform data are shown in Figure 4-10 for NSQD data for all residential areas. These data were

TABLE 4-3 Suggested Analytical Detection Limits for Stormwater Monitoring Programs to Obtain Less Than 5 Percent Nondetections

Parameter	Residential, Commercial, Industrial, Freeway		Open Space	
Conductivity	20	µS/cm	20	µS/cm
Hardness	10	mg/L	10	mg/L
Oil and grease	0.5	mg/L	0.5	mg/L
TDS	10	mg/L	10	mg/L
TSS	5	mg/L	1	mg/L
BOD_5	2	mg/L	1	mg/L
COD	10	mg/L	5	mg/L
Ammonia	0.05	mg/L	0.01	mg/L
$NO_2 + NO_3$	0.1	mg/L	0.05	mg/L
TKN	0.2	mg/L	0.2	mg/L
Dissolved P	0.02	mg/L	0.01	mg/L
Total P	0.05	mg/L	0.02	mg/L
Total Cu	2	µg/L	2	µg/L
Total Pb	3	µg/L (residential µg/L)	1	µg/L
Total Ni	2	µg/L	1	µg/L
Total Zn	20	µg/L (residential 10 µg/L)	5	µg/L

SOURCE: Maestre and Pitt (2005).

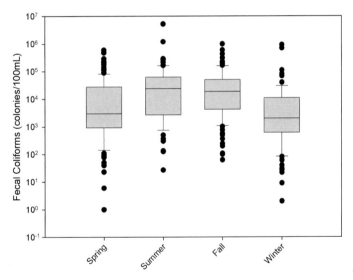

FIGURE 4-10 Fecal coliform concentrations in stormwater by season. SOURCE: NSQD.

the only significant differences in concentration by season for any constituent measured. The bacteria levels are lowest during the winter season and highest during the summer and fall (a similar conclusion was obtained during the NURP data evaluations).

Recommendations for MS4 Monitoring Activities

The NSQD is an important tool for the analysis of stormwater discharges at outfalls. About a fourth of the total existing information from the NPDES Phase I program is included in the database. Most of the statistical analyses in this research were performed for residential, commercial, and industrial land uses in EPA Rain Zone 2 (the area of emphasis according to the terms of the EPA-funded research). Many more data are available from other stormwater permit holders that are not included in this database. Acquiring these additional data for inclusion in the NSQD is a recommended and cost-effective activity and should be accomplished as additional data are also being obtained from ongoing monitoring projects.

The use of automatic samplers, coupled with bed-load samplers, is preferred over manual sampling procedures. In addition, flow monitoring and on-site rainfall monitoring need to be included as part of all stormwater characterization monitoring. The additional information associated with flow and rainfall data will greatly enhance the usefulness of the much more expensive water quality monitoring. Flow monitoring must also be correctly conducted, with adequate verification and correct base-flow subtraction methods applied. A related issue frequently mentioned by the monitoring agencies is the lack of on-site precipitation information for many of the sites. Using regional rainfall data from locations distant from the monitoring location is likely to be a major source of error when rainfall factors are being investigated.

Many of the stormwater permits only required monitoring during the first three hours of the rain event. This may have influenced the EMCs if the rain event continued much beyond this time. Flow-weighted composite monitoring should continue for the complete rain duration. Monitoring only three events per year from each monitoring location requires many years before statistically adequate numbers of observations are obtained. In addition, it is much more difficult to ensure that such a small fraction of the total number of annual events is representative. Also, there is minimal value in obtaining continued data from an area after sufficient information is obtained. It is recommended that a more concentrated monitoring program be conducted for a two- or three-year period, with a total of about 30 events monitored for each site, covering a wide range of rain conditions. Periodic checks can be made in future years, such as repeating concentrated monitoring every 10 years or so (and for only 15 events during the follow-up surveys).

Finally, better watershed area descriptions, especially accurate drainage-area delineations, are needed for all monitored sites. While the data contained in the NSQD are extremely useful, future monitoring information obtained as part of the stormwater permit program would be greatly enhanced with these additional considerations.

MONITORING OF INDUSTRIES INCLUDING CONSTRUCTION

The various industrial stormwater monitoring requirements of the EPA Stormwater Program have come under considerable scrutiny since the program's inception. Input to the committee at its first meeting conveyed the strong sense that monitoring as it is being done is nearly useless, is burdensome, and produces data that are not being utilized. The requirements consist of the following. All industrial sectors covered under the Multi-Sector General Permit (MSGP) must conduct visual monitoring four times a year. This visual monitoring is performed by collecting a grab sample within the first hour of stormwater discharge and observing its characteristics qualitatively (except for construction activities—see below). A subset of MSGP industries are required to perform analytical monitoring for benchmark pollutant parameters (see Table 2-5) four times in year 2 of permit coverage and again in year 4 if benchmarks are exceeded in year 2. A benchmark sample is collected as a grab sample within the first hour of stormwater discharge after a rainfall event of 0.1 inch or greater and with an interceding dry period of at least 72 hours. An even smaller subset of MSGP industries that are subject to numerical effluent guidelines under 40 C.F.R. must, in addition, collect grab samples of their stormwater discharge after every discharge event and analyze it for specific pollutant parameters as specified in the effluent guidelines (see Table 2-6). There is no monitoring requirement for stormwater discharges from construction activity in the Construction General Permit. There is only an elective requirement that the construction site be visually inspected within 24 hours after the end of a storm event that is 0.5 inch or greater, if inspections are not performed weekly.

EPA selected the benchmark analytical parameters for industry subsectors to monitor using data submitted by industrial groups in 1993 as part of their group applications. The industrial groups were required to sample a minimum of 10 percent of facilities within an industry group for pH, TSS, BOD_5, oil and grease, COD, TKN, nitrate plus nitrite nitrogen, and total phosphorous. Each sampling facility within a group collected a minimum of one grab sample within the first 30 minutes of discharge and one flow-weighted composite sample. Other nonconventional pollutants such as fecal coliform bacteria, iron, and cobalt were analyzed only if the industry group expected it to be present. Similarly, toxic pollutants such as lead, copper, and zinc were not sampled but rather self-identified only if expected to be present in the stormwater discharge. As a result of the self-directed nature of these exercises, the data submitted with the group applications were often incomplete, inconsistent, and not representative of the potential risk posed by the stormwater discharge to human health and aquatic life. EPA has not conducted or funded independent investigations and has relied solely on the data submitted by industry groups to determine which pollutant parameters are appropriate for the analytical monitoring of an industry subsector. Thus, there are glaring deficiencies; for example, the only benchmark parameter for asphalt paving and roofing materials is TSS, even though current science shows that the most harmful pollutants in stormwater discharges from the asphalt manufacturing industry are polycyclic aromatic hydrocarbons (compare Table 2-5 with Mahler et al., 2005).

Aside from the suitability of benchmark parameters is the fact the too few samples are collected to sufficiently characterize the variability of pollutant concentrations associated with industrial facilities within a sector. This is discussed in detail in Box 4-2, which describes one of the few efforts to collect and analyze data from the benchmark monitoring of industries done in Southern California. EPA has not requested a nationwide effort to compile these data, as was done for the MS4 program, although this could potentially lead to average effluent concentrations by industrial sector that could be used for a variety of purposes, including more considerate regulations. Finally, the compliance monitoring that is presently being conducted under the MSGP is of limited usefulness because it is being done to comply with effluent guidelines that have not been updated to reflect the best available technology relevant to pollutants of most concern. All of these factors have led to an industrial stormwater monitoring program that is not very useful for the purposes of reducing stormwater pollution from industries or informing operators on which harmful pollutants to expect from their sites.

Industrial-Area Monitoring Issues

Monitoring at industrial sites has some unique issues that must be overcome. The most important aspect for any monitoring program is understanding and specifying the objectives of the monitoring program and developing and following a detained experimental design to allow these objectives to be met. The following discussion is organized around the reasons why monitoring at industrial sites may be conducted.

Regional Monitoring of Many Facilities

An important monitoring objective would be regional monitoring to calibrate and verify stormwater quality models, to randomly verify compliance at facilities not normally requiring monitoring, and to establish benchmarks for compliance. As shown in Box 4-2, haphazard monitoring throughout an area would require a very large effort, and would still likely result in large errors in the expected data. It is recommended that a regional stormwater authority coordinate regional monitoring as part of the MS4 monitoring requirements, possibly

BOX 4-2
The Plight of Industrial Stormwater Data

Unlike the data collected by municipalities and stored in the NSQD, the benchmark monitoring data collected by permitted industries are not compiled or analyzed on a national basis. However, there has been at least one attempt to compile these data on a more local basis. California required that industrial facilities submit their benchmark monitoring data over a nine-year period, and it was subsequently analyzed by Michael Stenstrom and colleagues at UCLA (Stenstrom and Lee, 2005; Lee et al., 2007). The collected data were for such parameters as pH, turbidity, specific conductance, oil and grease (or total organic carbon), and several metals. There are more than 6,000 industries covered under the California general permit, each of which was to have collected two grab samples per year for a limited number of parameters. Whether these data were collected each year and for each industry was highly variable.

The analysis of the data from Los Angeles and Ventura counties revealed that stormwater monitoring data are not similar to the types of data that the environmental engineering field is used to collecting, in particular wastewater data. Indeed, as shown in Figure 4-11, stormwater data are many orders of magnitude more variable than drinking water and wastewater data. The coefficients of variation for municipal and industrial stormwater were almost two orders of magnitude higher than for drinking water and wastewater, with the industrial stormwater data being particularly variable. This variability comes from various sources, including intrinsic variability given the episodic nature of storm events, analytical methods that are more variable when applied to stormwater, and sampling technique problems and error.

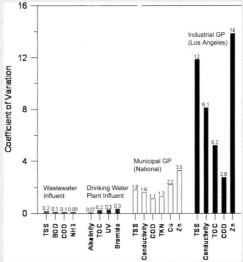

FIGURE 4-11 A comparison of data from four sources: wastewater influent, drinking water plant effluent, municipal stormwater, and industrial stormwater. SOURCE: Reprinted, with permission, from Stenstrom (2007). Copyright 2007 by Michael K. Stenstrom.

continues next page

BOX 4-2 Continued

This enormous variability means that it is extremely difficult to make meaningful statements. For example, it was impossible, using different analyses, to correlate certain chemical pollutants with certain industries. Furthermore, although the data revealed that there are exceedances of benchmark values for certain parameters (Al, Cu, Fe, Pb, and Zn in particular), the data are not of sufficient quantity or quality to identify problem polluters. Finally, there were also large numbers of outliers (that is, samples whose concentrations were well above the 75th percentile range).

Because of these large coefficients of variation, greater numbers of samples are needed to be able to say there is a significant difference between samples. As shown in Figure 4-12 using COD and a 50 percent difference in means as an example, one would need six data points to tell the difference between two wastewater influents, 80 data points if one had municipal stormwater data, and around 1,000 data points for industrial stormwater. These numbers obviously eclipse what is required under all states' MSGPs.

For drinking water treatment, monitoring is done to ensure the quality of the product, while for wastewater, there is a permit that requires the plant to meet a specific quality of water. Unlike these other areas of water resources, there are few incentives that might compel an industry to increase its frequency of stormwater monitoring. As a result, industries are less invested in the process and rarely have the expertise needed to carry out self-monitoring.

Permitted industries are not required to sample flow. However, Stenstrom and colleagues used Los Angeles rainfall data (see Figure 4-13) as a surrogate for flow and demonstrated that there is a seasonal first-flush phenomenon occurring in early fall. That is, samples taken after a prolonged dry spell will have higher pollutant concentrations. There are always high concentrations of contaminants during the first rainfall because contaminants have had time to accumulate since the previous rainfall. This is important because EPA asks the industrial permittees to collect data from the first rainfall, such that they may end up overestimating the mass emissions for the year. Furthermore, it shows that numeric limits for grab samples would be risky because the measured data are highly affected by the timing of the storm.

The controversy about numeric limits for industrial stormwater dischargers has existed for more than ten years in California. A recent expert panel concluded that in some cases, numeric limits are appropriate (for construction, but not for municipalities). Stenstrom's recommendations are that industrial monitoring should be either ended or upgraded (for competent industries). If upgraded, it should include more types of monitored parameters, a sampling method with a lower coefficient of variation, real-time monitoring as opposed to grab samples, more quality assurance/quality control, and web-based reporting. A fee-based program with a subset of randomly selected industries may be better than requiring every industry to sample. Stenstrom and Lee (2005) suggest who might do this monitoring if the industry does not have the necessary trained personnel. There is concern that the California water boards are too understaffed to administer such programs and respond to high emitters.

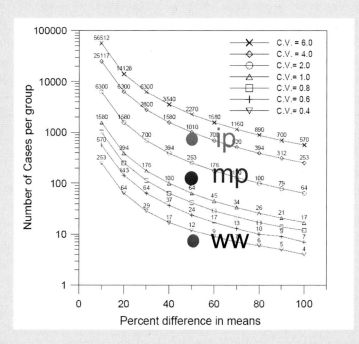

FIGURE 4-12 Number of cases needed to detect a certain percentage difference in the means, using COD as an example. SOURCE: Reprinted, with permission, from Stenstrom (2007). Copyright 2007 by Michael K. Stenstrom.

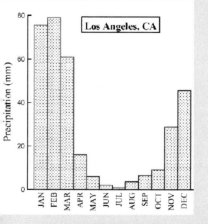

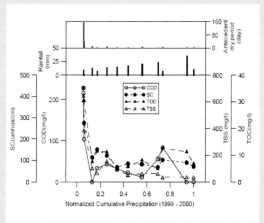

FIGURE 4-13 Annual precipitation in Los Angeles (left) and seasonal first flushes of various contaminants (right). SOURCE: Reprinted, with permission, from Stenstrom (2007). Copyright 2007 by Michael K. Stenstrom.

SOURCES: Stenstrom and Lee (2005), Lee et al. (2007), Stenstrom (2007).

even at the state level covering several Phase I municipalities. A coordinated effort would be most cost-effective with the results compiled for a specific objective. The general steps in this effort would include the following.

(1) Compiling available regional stormwater quality data and comparing the available data to the needs (such as calibration of a regional model; verifying compliance of facilities not requiring monitoring; and establishing regional benchmarks). This may include expanding the NSQD for the region to include all of the collected data, plus examination of data collected as part of other specialized monitoring activities. These objectives will result in different data needs, so it is critical that the uses of the data are identified before sampling plans are established.

(2) Identifying monitoring opportunities as part of other on-going activities that can be expanded to also meet data gaps for these specific objectives. It is important to understand the time frame for the monitoring and ensure that it will meet the needs. As an example, current NPDES stormwater monitoring only requires a few events to be sampled per year at a facility. It may take many years before sufficient data are obtained unless the monitoring effort is accelerated.

(3) Preparing an experimental design that identifies the magnitude of the needed data, considering the allowable errors in the results, and carrying out the sampling program. Different types of data may have varying data quality objectives, depending on their use. It may be possible to truncate some of the monitoring when a sufficient understanding is obtained.

A regionally calibrated and verified model can be used to review development plans and proposed SCMs for new facilities. When suitably integrated with receiving-water modeling tools, a stormwater model can also be used to develop discharge objectives and numeric discharge limits that are expected to meet regulatory requirements. Eventually, it may be possible to couple watershed stormwater models with regional receiving water assessments and beneficial use studies. Haphazard monitoring of a few events each year will be very difficult to correlate with regional receiving water objectives, while a calibrated and verified watershed model, along with receiving water assessments, will result in a much more useful tool and understanding of the local problems.

Regional monitoring can also be targeted to categories of industries that were previously determined to be of low priority. This monitoring activity would randomly target a specific number of these facilities for monitoring to verify the assumption that they are of low priority and are still carrying out the minimum management practices. This activity would also quantify the discharges from these facilities and the performance of the minimum controls. If the discharges are excessive when compared to the initial assumptions, or the management practices being used are not adequate, then corrective actions would be instigated. A single category of specific industries could be selected for any one year, and a team from the regional stormwater management authority could randomly select and monitor a subset of these facilities. An efficient experimental design would need to be developed based on expected conditions, but it is expected that from 10 to 15 such facilities would be monitored for at least a year in a large metropolitan area that has a Phase I stormwater permit, or even state-wide.

Regional monitoring is also necessary to more accurately establish benchmarks for numeric permits. Geographical location, along with land use, is normally an important factor affecting stormwater quality. Receiving water impacts and desired beneficial uses also vary greatly for different locations. It is therefore obvious that compliance benchmarks also be established that consider these regional differences. This could be a single statewide effort if the state agency has the permit authority and if the state has minimal receiving water and stormwater variations. However, in most cases, significant variations occur throughout the state and separate monitoring activities would be needed for each region. In the simplest case, probability distributions of stormwater discharge quality can be developed for different discharge categories and the benchmarks would be associated with a specific probability value. In some cases, an overall distribution may be appropriate, and only the sites having concentrations greater than the benchmark value would need to have additional treatment. In all cases, a basic level of stormwater management should be expected for all sites, but the benchmark values would identify sites where additional controls are necessary. The random monitoring of sites not requiring extensive monitoring could be used to identify and adjust the basic levels of control needed for all categories of stormwater dischargers.

Identification of Critical Source Areas Associated with Specific Industrial Operations

The objective of this monitoring activity would be to identify and characterize critical source areas for specific industries of concern. If critical source areas can be identified, targeted control or treatment can be much more effective than relying only on outfall monitoring. Many of the treatment strategies for industrial sites involve pollution prevention, ranging from covering material or product storage areas to coating galvanized metal. Other treatment strategies involve the use of highly effective treatment devices targeting a small area, such as filters used to treat zinc in roof runoff or lamella plate separators for pretreatment of storage yard runoff before wet pond treatment. Knowledge of the characteristics of the runoff from the different areas at a facility is needed in order to select and design the appropriate treatment methods.

Box 4-3 is a case study of one such group monitoring effort—for a segment of the telecommunications industry targeting a specific maintenance practice. Instead of having each telecommunication company throughout the country conduct a detailed monitoring program for individual stormwater permits associated with maintenance efforts, many of the companies joined together under an

BOX 4-3
Monitoring to Support a General Stormwater Group Permit
Application for the Telecommunications Industry

This monitoring program was conducted to support a group permit application for the telecommunications industry, specifically to cover maintenance operations associated with pumping water out of communications manholes that is then discharged into the storm drainage system. Under federal and state environmental statues, the generator (owner or operator) is responsible for determining if the discharged water needs treatment. The work performed under this project covered characterization, prevention, and treatment methods of water found in manholes.

The objective of this project was to develop a test method to quickly evaluate water in manholes and then to recommend on-site treatment and preventative methods. To meet the telecommunication industry needs, the evaluating tests of water found in manholes need to be simple, quick, inexpensive, field applicable, and accurate indicators of contaminated conditions. The on-site treatment methods must be cost-effective and quickly reduce the concentrations of the contaminant of concern to acceptable levels before the water from manholes is discharged, to result in a safe environment for workers.

A sampling effort was conducted by Pitt et al. (1998) to characterize the quality of the water and sediment found in manholes. More than 700 water samples and 300 sediment samples were analyzed over a three-year period, representing major land-use, age, season, and geographical factors from throughout the United States. The samples were analyzed for a wide range of common and toxic constituents. The statistical procedures identified specific relationships between these main factor categories and other manhole characteristics. Part of the project was to evaluate many field analytical methods. Finally, research was also conducted to examine possible water treatment methods for water being pumped from telecommunication manholes.

Summary of Sampling Effort and Strategy

The objective of the monitoring program was to characterize telecommunication manhole water and sediment. Important variables affecting the quality of these materials were also determined. A stratified random sampling design was followed, with the data organized in a full 2^4 factorial design, with repeated sampling of the same manholes for each season. The goal for the minimum number of samples per strata was ten. This sampling effort enabled the determination of errors associated with the results, which was expected to be less than 25 percent. In addition, this level of effort enabled comparison tests to be made outside of the factorial design. Table 4-4 lists the constituents that were evaluated for each of the sample types.

The immense amount of data collected during this project and the adherence to the original experimental design enabled a comprehensive statistical evaluation of the data. Several steps in data analysis were performed, including:

• exploratory data analyses (mainly probability plots and grouped box plots),
• simple correlation analyses (mainly Pearson correlation matrices and associated scatter plots),
• complex correlation analyses (mainly cluster and principal component analyses, plus Kurskal-Wallis comparison tests), and

• model building (based on complete 2^4 factorial analyses of the most important factors).

The toxicity screening tests (using the Azur Microtox® method) conducted on both un-filtered and filtered water samples from telecommunication manholes indicated a wide range of toxicity, with no obvious trends for season, land use, or age. About 60 percent of the samples were not considered toxic (less than an I25 light reduction of 20 percent, the light reduction associated with phosphorescent bacteria after a 25-minute exposure to undi-luted samples), about 20 percent were considered moderately toxic, while about 10 percent were considered toxic (light reductions of greater than 40 percent), and 10 percent were considered highly toxic (light reductions of greater than 60 percent). Surprisingly, samples from residential areas generally had greater toxicities than samples from commercial and industrial areas. Samples from newer areas were also more toxic than those from older areas. Further statistical tests of the data indicated that the high toxicity levels were likely associated with periodic high concentrations of salt (in areas using de-icing salt), heavy metals (especially filterable zinc, with high values found in most areas), and pesticides (associated with newer residential areas).

TABLE 4-4 Constituents Examined in Water and Sediment from Telecommunication Manholes

Constituent	Unfiltered Water	Filtered Water	Sediment
Solids, volatile solids, COD, Cu, Pb, and Zn	X	X	X
Turbidity, color, and toxicity (Microtox screening method)	X	X	
pH, conductivity, hardness, phosphate, nitrate, ammonia, boron, fluoride, potassium, and detergents	X		
Odor, color, and texture			X
E. coli, enterococci, particle size, and chromium	Selected		
Metal scan (ICP)			Selected
PAHs, phenols (GC/MSD), and pesticides	X	Selected	Selected

SOURCE: Modified from Pitt et al. (1998).

Concentrations of copper, lead, and zinc were evaluated in almost all of the water samples, and some filtered samples were also analyzed for chromium. From 470 to 548 samples (75 to 100 percent of all unfiltered samples analyzed) had detectable concentra-tions of these metals. Filterable lead concentrations in the water were as high as 160 µg/L, while total lead concentrations were as high as 810 µg/L. Zinc values in filtered and unfil-tered samples were as high as about 3,500 µg/L. Some of the copper concentrations were also high in both filtered and unfiltered samples (as high as 1,400 µg/L). Chromium con-centrations as high as 45 µg/L were also detected.

continues next page

BOX 4-3 Continued

About 300 sediment samples were analyzed and reviewed for heavy metals. An ICP/MS was used to obtain a broad range of metals with good detection limits. The following list shows the median observed concentrations for some of the constituents found in the sediments (expressed as milligrams of the constituent per kilogram of dry sediment):

Aluminum	14,000 mg/kg
COD	85,000 mg/kg
Chromium	<10 mg/kg
Copper	100 mg/kg
Lead	200 mg/kg
Strontium	35 mg/kg
Zinc	1,330 mg/kg

Geographical area had the largest effect on the data observations, while land use, season, and age influenced many fewer parameters. The most obvious relationship was found for high dissolved solids and conductivity associated with winter samples from snowmelt areas. The high winter concentrations slowly decreased with time, with the lowest concentrations noted in the fall. Another important observation was the common association between zinc and toxicity. Residential-area samples generally had larger zinc concentrations than the samples from commercial and industrial areas. Samples from the newest areas also had higher zinc concentrations compared to samples from older areas. No overall patterns were observed for zinc concentrations in sediment samples obtained from manholes. Other constituents (especially nutrients and pesticides) were also found to have higher concentrations in water collected from manholes in newer residential areas. Very few organic toxicants were found in the water samples, but sediment sample organic toxicant concentrations appeared to be well correlated to sediment texture and color. About 10 to 25 percent of the sediment samples had relatively large concentrations of organics. Bacteria analyses indicated some relatively high bacteria counts in a small percentage of the samples. Bacteria were found in lower amounts during sampling periods that were extremely hot or extremely cold. Pacific Northwest samples also had the lowest bacteria counts.

The data were used to develop and test predictive equations based on site conditions. These models were shown to be valid for most of the data, but the highest concentrations were not well predicted. Therefore, special comparisons of many site conditions were made for the manholes having water with the highest concentrations of critical constituents for comparison to the other locations. It was interesting to note that about half of the problem manholes were repeated samples from the same sites (after complete pumping), but at different seasons, indicating continuous problems and not discrete incidents. In addition, the problem manholes were found for all areas of the country and for most rain conditions. Water clarity and color, along with sediment texture, were found to be significant factors associated with the high concentrations of other constituents, while land use was also noted as a significant factor. These factors can be used to help identify problem manholes, but the rates of false positives and false negatives were found to be high. Therefore, these screening criteria can be used to identify more likely problematic manholes, but other methods (such as confirmation chemical analyses) are also needed to identify those that could not be identified using these simpler methods.

continues next page

BOX 4-3 Continued

The field analytical test methods worked reasonably well, but had much higher detection limits than advertised, limiting their usefulness. Due to the complexity and time needs for many of these on-site analyses, it is usually more effective to analyze samples at a central facility. For scheduled maintenance operations, a crew could arrive at the site before the maintenance time to collect samples and have them analyzed before the maintenance crew arrives. For emergency repairs, it is possible to pump the collected water into a tank truck for later analyses, treatment, and disposal.

The treatment scenario developed and tested is relatively rapid and cheap and can be used for all operations, irrespective of screening analyses. Chemical addition (using ferric chloride) to the standing water in the manhole was found to reduce problematic levels of almost all constituents to low levels. Slow pumping from the water surface over about a 15- to 30-minute period, with the discharged water then treated in 20-μm cartridge filters, allows the manhole to be entered and the repairs made relatively rapidly, with the water safely discharged. The remaining several inches of water in the bottom of the manhole, along with the sediment, can be removed at a later time for proper disposal.

SOURCE: Pitt et al. (1998).

---

industrial trade group to coordinate the monitoring and to apply for a group permit. This was a significant effort that was conducted over several years and involved the participation of many regional facilities throughout the nation. This coordinated effort spread the cost over these different participants, and also allowed significant amounts of data to be collected, control practices to be evaluated, and the development of screening methods that allow emergency maintenance operations of the telecommunication system to proceed in a timely manner. The experimental design of this monitoring program allowed an efficient examination of factors affecting stormwater discharges from these operations. This enabled the efficient implementation of effective control programs that targeted specific site and operational characteristics. Although the total cost for this monitoring program was high, it was much less costly than if each individual company had conducted their own monitoring. In addition, this group effort resulted in much more useful information for the industry as a whole.

Outfall Monitoring at a Single Industrial Facility for Permit Compliance and to Demonstrate Effectiveness of Control Practices

Sampling at an individual facility results in outfall data that can be compared to pre-control conditions and numeric standards. There are many guidance documents and reports available describing how to monitor stormwater at an outfall. Two comprehensive sources that describe stormwater monitoring procedures include the handbook written by Burton and Pitt (2002) and a recent guidance report prepared by Shaver et al. (2007). There are a number of basic components that need to be included for an outfall characterization monitoring effort, many which have been described in this report. These include the following:

- rainfall monitoring in the drainage area (rate and depth, at least at two locations).
- flow monitoring at the outfall (calibrated with known flow or using dye dilution methods).
- flow-weighted composite sampler, with sampler modified to accommodate a wide range of rain events.
- recommended use of water quality sonde to obtain high-resolution and continuous measurements of such parameters as turbidity, conductivity, pH, oxidation reduction potential, dissolved oxygen (DO), and temperature.
- preparation of adequate experimental design that quantifies the needed sampling effort to meet the data quality objectives (adequate numbers of samples in all rain categories and seasons).
- selection of constituents that meet monitoring objectives. In addition, the analytical methods must be appropriately selected to minimize "nondetected" values.
- monitoring station maintenance must also be conducted appropriately to ensure reliable sample collection. Sampling plan must also consider sample retrieval, sample preparation and processing, and delivery to the analytical laboratory to meet quality control requirements.

Burton and Pitt (2002) describe these monitoring components in detail, along with many other monitoring elements of potential interest (e.g., receiving water biological, physical, and chemical monitoring, including sediment and habitat studies), and include many case studies addressing these components, along with basic statistical analyses and interpretation of the collected data. Box 4-4 provides a detailed example of industrial stormwater monitoring at individual sites in Wisconsin.

In general, monitoring of industries should be tailored to their stormwater pollution potential, considering receiving water uses and problems. There are a number of site survey methods that have been developed to rank industry by risk that mostly rely on visual inspections and information readily available from regional agencies. The Center for Watershed Protection developed a hot-spot investigation procedure that is included in the Urban Subwatershed Restoration Manual No. 11 (Wright et al., 2005). This site survey reconnaissance method ranks each site according to its likely stormwater pollutant discharge potential. A detailed field sheet is used when surveying each site to assist with the visual inspections. Cross and Duke (2008) developed a methodology, described in greater detail in Chapter 6, to visually assess industrial facilities based on the level of activities exposed to stormwater. They devised four categories—Category A, no activities exposed to stormwater; Category B, low intensity; Category C, medium intensity; and Category D, high intensity—and tested this

BOX 4-4
Wisconsin's Monitoring of Industrial Stormwater

The State of Wisconsin also uses a site assessment method to rank industrial operations into three tiers, mostly based on their standard industrial codes. This system groups facilities by industry and how likely they are to contaminate stormwater. The general permits differ in monitoring requirements, inspection frequency, plan development requirements, and the annual permit fee. The Tier 1 general permit covers the facilities that are considered "heavy" industries, such as paper manufacturing, chemical manufacturing, petroleum refining, ship building/repair, and bulk storage of coal, minerals, and ores. The monitoring required of these facilities is presented in this box. The Tier 2 general permit covers facilities that are considered "light" industries and includes such sites as furniture manufacturing, printing, warehousing, and textiles. Facilities with no discharge of contaminated stormwater are in the Tier 3 category and include sites that have no outdoor storage of materials or waste products.

In accordance with the Wisconsin MSGP, Tier 1 industries are required to perform an annual chemical stormwater sampling at each outfall for those residual pollutants listed in the industry's stormwater pollution prevention plan. The one runoff event selected for sampling must occur between March and November and the rainfall depth must be at least 0.1 inch. At least 72 hours must separate the sampled event and the previous rainfall of 0.1 inch. The concentration of the pollutant must represent a composite of at least three grab samples collected in the first 30 minutes of the runoff event. There is concern about the value of collecting so few samples from just one storm each year.

To evaluate how well this sampling protocol characterizes pollutant concentrations in industrial runoff, the Wisconsin Department of Natural Resources partnered with the USGS to collect stormwater samples from three Tier 2 industrial sites (Roa-Espinosa and Bannerman, 1994). Seven runoff events were monitored at each site, and the samples were collected using five different sampling methods, including (1) flow-weighted composites, (2) time-based discrete samples, (3) time-based composites, (4) a composite of discrete samples from first 30 minutes, and (5) time-based composite sheet flow samples. The first three methods have been described previously. For the composite of discrete samples from the first 30 minutes, the sampler is programmed to take an aliquot after a set period of time (usually every 5 minutes) and the aliquots are combined into one container. The sampler stops collecting samples after 30 minutes. For many sites the samples are collected manually, so there is a high probability the sample does not represent the first 30 minutes of the event. For the time-based composite sheet flow samples, a sheet flow sampler is programmed to take an aliquot of sheet flow after a set period of time (usually about every 5 to 15 minutes). All the aliquots are deposited in one bottle beneath the surface of the ground. All of the parts of the hydrograph receive equal weight in the final concentration, but the larger number of aliquots makes for a reasonably accurate composite concentration. This method is unique in that it can be placed near the source of concern. Automatic samplers were used for the first four methods, while sheet flow samplers designed by the USGS were used for the fifth method (Bannerman et al., 1993). Samples were collected during the entire event. All the automatic samplers had to be installed at a location with concentrated flow, such as an outfall pipe, while the sheet flow samplers could be installed in the pavement near a potential source, such as a material storage area.

continues next page

BOX 4-4 Continued

The time-based discrete, time-based composite, first-30-minute composite, and sheet flow samples were analyzed for COD, total recoverable copper, total recoverable lead, total recoverable zinc, TSS, total solids, and hardness. In addition to these constituents, the flow-weighted composite samples were analyzed for antimony, arsenic, beryllium, chromium, ammonia-N, nitrate plus nitrite, TKN, and TP. All the analysis was done at the State Laboratory of Hygiene in Madison, Wisconsin, and the data are stored in the USGS's QWDATA database.

The number of samples collected during a runoff event varied greatly among the five types of sampling. By design, the median number of samples collected for the first 30 minutes was three. Limits on the funds available for laboratory cost limited the time-based discrete sampling to about six per storm. Since they are not restricted by laboratory cost, the composites can be based on more sub-samples during a storm. Thus, the median numbers of sub-samples collected for the flow-weighted composite and time-based composite were 13 and 24, respectively. The time-based composite sheet flow sample could not document the number of samples it collected, but it was set to collect a sample every few minutes.

To judge the accuracy of the sampling methods, one method had to be selected as the most representative of the concentration and load affecting the receiving water. Because a relatively large number of samples are collected and the timing of the sampling is weighted by volume, the flow-weighted composite concentrations were used as the best representation of the quality of the industrial runoff. Concentrations in water samples collected by the time-based composite method compared very well to those collected by the flow-weighted composite method, especially if the time-based composite resulted in 20 sub-samples or more. This was not true for the discrete sampling method, because many fewer sub-samples were used to represent changes across the hydrograph. The time-based composite sheet flow sampler produced concentrations slightly higher than the time-based composite samplers collecting water in the concentrated flow. Concentrations from the sheet flow sampler are probably not diluted by other source areas such as the roof.

Concentrations of total recoverable zinc and TSS collected in the first 30 minutes of the event were usually two to three times higher than the flow-weighted composite samples. For many of the events, the highest concentration of these constituents occurred in the first 10 minutes of the event. Although the concentrations might be higher in the first part of the event, the earlier parts of the event might only represent one third or less of the total runoff volume. Thus, using the concentrations from the first 30 minutes of the event could greatly overestimate the constituent load from the site.

Along with accuracy, the selection of an appropriate sampling method must consider cost and the criteria for installing the sampling equipment. To measure flow, the site must have a location where the flow is concentrated, such as a pipe or well-defined channel, and the runoff is just coming from the site. Out of 474 sites evaluated for this project, only 14 met the criteria for an accurate flow measurement. A few more sites might be suitable for using an automatic sampler without flow measurements, but the number of sites would still be limited. Sheet flow samplers can be used on most sites, since they are simply installed in the pavement near the source of concern.

For each sampling method, approximate costs were determined including equipment, installation of equipment, and the analysis of one sample (Table 4-5). Collecting the samples and processing the data should also be included, but they were not because this cost is highly variable. Flow-weighted composite and time-based discrete sampling had the highest cost. Flow measurements made the composite sampling more expensive, while the laboratory cost of analyzing six discrete samples increased the cost of the time-based

TABLE 4-5 Cost of Using Different Sampling Methods in 1993 Dollars

Method	Estimated Cost for Equipment, Installation, and Analysis of One Sample
Flow-weighted composite	$16,052
Time-based discrete	$22,682
Time-based composite	$5,920
First-30-minutes (automatic sampler)	$6,000
First-30-minutes (grab sample)	$1,800[1]
Time-based composite sheet flow sampler	$2,889

[1]Cost of laboratory analysis only. SOURCE: Reprinted, with permission, from Roa-Espinosa and Bannerman (1994). Copyright 1994 by the American Society of Civil Engineers.

discrete method. It should be noted that hand grab samples could be used to collect the discrete samples in the first 30 minutes at lower cost, although this depends strongly on the skill of the person collecting the sample. The sheet flow sampler could be the most cost effective approach to sampling an industrial site.

A determination must be made of how many runoff events should be sampled in order to accurately characterize a site's water quality. As shown in Table 4-6, representing a site with the results from one storm can be very misleading. Concentrations in Table 4-6 were collected by the flow-weighted composite method. The geometric means of EMCs from five or more events were very different than the lowest or highest concentration observed for the set of storms. The chances of observing an extreme value by sampling just one event is increased by selecting a sampling method designed to collect a limited number of sub-samples, such as the first-30-minutes method. Too few storms were monitored in this project to properly evaluate the variability in the EMCs, but sufficient changes occur between the zinc and TSS geometric means in Table 4-6 to suggest that a compliance monitoring schedule should include a minimum of five events be sampled each year.

To overcome the high COV observed for municipal stormwater data collected in Wisconsin, EMCs should be determined for about 40 events (Selbig and Bannerman, 2007; Horwatich et al., 2008). The 40 event mean concentrations would probably represent the long-range distribution of rainfall depths, and there would be sufficient data available to perform some trend analysis, such as evaluating the benefits of an SCM implemented at an industrial site. Monitoring 40 events each year, however, would be too costly for an annual compliance monitoring schedule for each industrial site.

Results from this project indicate that the stormwater monitoring required at industrial sites cannot adequately characterize the quality of runoff from an industrial site. Only collecting samples from the first 30 minutes of a storm is probably an overestimate of the concentration, and a load calculated from this concentration would exaggerate the impact of the site on the receiving waters. Time- and flow-based composite sampling would be much better methods for monitoring a site if there are locations to operate an automatic sampler. For sites without such a location, the time-based composite sheet flow sampler offers the best results at the least cost. Given all the variability in concentrations between runoff events, the annual monitoring schedule for any site should include sampling multiple storms.

continues next page

BOX 4-4 Continued

TABLE 4-6 Effects of Including a Different Number of Events in the Geometric Mean Calculation for Zinc and TSS[a]

Number of Events	Total Recoverable Zinc	Total Suspended Solids
AC Rochester		
1 (Lowest Concentration)	57	8
1 (Highest Concentration)	150	84
3	76	24
5	91	36
PPG Industries		
1 (Lowest Concentration)	140	32
1 (Highest Concentration)	330	49
3	153	57
6	186	53
Warman International		
1 (Lowest Concentration)	68	17
1 (Highest Concentration)	140	56
3	67	15
5	81	26
7	74	19

[a]Samples were collected using the flow-weighted composite method. SOURCE: Reprinted, with permission, from Roa-Espinosa and Bannerman (1994). Copyright 1994 by the American Society of Civil Engineers.

scheme by examining many southern Florida industrial facilities. About 25 percent of the facilities surveyed that were officially included in the stormwater permit program had no stormwater exposure (Category A), but very few had submitted the necessary application to qualify for an exception under the "no exposure" rule. Slightly more than half of the of the surveyed facilities were included in the "no exposure" and "low exposure" categories, obviously deserving less attention compared to the higher impact categories.

Recommendations for Industrial Stormwater Monitoring

Suitable industrial monitoring programs can be implemented for different categories of industrial activities. The following is one such suggestion, based on the likely risks associated with stormwater discharges from each type of facility.

No Exposure to Industrial Activities and Other Low-Risk Industrial Operations

For sites having limited stormwater exposure to industrial operations, such as no outdoor storage of materials or waste products, basic monitoring would not normally be conducted. However, roof runoff (especially if galvanized metals are used) and large parking areas need to be addressed under basic stormwater regulations dealing with these common sources of contaminants and the large amounts of runoff that may be produced. Simple SCM guidance manuals can be used to select and size any needed controls for these sites, based on the areas of concern at the facility. For these facilities, simple visual inspections with no monitoring requirements may be appropriate to ensure compliance with the basic stormwater regulations. A regionally calibrated stormwater quality model can be used to evaluate these basic stormwater conditions and to calculate the expected benefits of control measures. Periodic random monitoring of sites in this category should be conducted to verify the small magnitude of discharges from these sites and the performance of SCMs.

Medium-Risk Industrial Operations

For "medium-intensity" industry facilities, site inspections and modeling should be supplemented with suitable outfall monitoring to ensure compliance. As noted in Box 4-2, there can be a tremendous amount of variability in industrial runoff characteristics. However, the dataset described in that example was a compilation of data from many different types of facilities, with no separation by industrial type. Even different facilities in a single industrial group may have highly variable runoff characteristics. However, a single facility has much less variability, and reasonable monitoring strategies can be developed for compliance purposes. As noted in Box 4-4, about 40 samples were expected to be needed for each site in that example. With typical permit periods of five years, this would require that less than ten samples per year (more than the three samples per year currently obtained at many locations) be collected in order to determine the EMC for the site for comparison to allowable discharge conditions. Obviously, the actual number of samples needed is dependent on the variability of the runoff characteristics and the allowable error, as described elsewhere. After about 10 to 15 storms have been monitored for a site, it would be possible to better estimate the total number of samples actually needed based on the data quality objectives. If the monitoring during the permit period indicated excessive stormwater discharges, then the SCMs are obviously not adequate and would need improvement. The permit for the next five-year period could then be modified to reflect the need for more stringent controls, and suitable fines accessed if the facility was not in compliance. It is recommended that absolute compliance not be expected in the industrial permits, but that appropriate benchmarks be established that allow a small fraction of the monitored events to exceed the goals. This is similar to discharge permit requirements for combined sewers, and for air quality regulations, where a certain number of excessive periods are allowed per year.

High-Risk Industrial Facilities

For "high-risk" industrial sites of the most critical nature, especially if non-compliance may cause significant human and environmental health problems, visual inspections and site modeling should be used in conjunction with monitoring of each event during the permit period. Because of the potential danger associated with noncompliance, the most stringent and robust controls would be required, and frequent monitoring would be needed to ensure compliance. If noncompliance was noted, immediate action would be needed to improve the discharge conditions. This is similar to industrial and municipal NPDES monitoring requirements for point sources.

MODELING TO LINKING SOURCES OF POLLUTION TO EFFECTS IN RECEIVING WATERS

Stormwater permitting is designed to regulate dischargers, develop information, and reduce the level of stormwater pollutants and impact on receiving waterbodies. An important assumption is that the level of understanding of the stormwater system, through a combination of monitoring and modeling, is sufficient to associate stormwater discharges with receiving waterbody impacts. Impairment of waterbodies can occur for a variety of physical, chemical, and biological reasons, often with a complex combination of causes. The ambient water quality of a receiving waterbody, which may result in a determination of impairment, is itself a function of the total mass loading of pollutant; dilution with stream discharge or standing waterbody volume; the capacity of the aquatic ecosystem to assimilate, transform, or disperse the pollutant; and transport out of the waterbody. In addition to the chemical and physical attributes of the water, impairment may also be characterized by degraded biologic structure or geomorphic form of the waterbody (e.g., channel incision in urban areas). Interactions between multiple pollutant loadings, long turnover and residence times, saturation effects, and cascading feedbacks with biological communities complicate the apparent response of waterbodies to pollutant discharge. This is particularly important when considering cumulative watershed effects, in which interactions between stressors and long-term alteration of watershed conditions may contribute to threshold responses of a waterbody to continued loading or alteration. Under these conditions, simple "loading-response" relations are often elusive and require consideration of historical and local watershed conditions.

As an example, pollutant loading at high stream flow or into strong tidally flushed systems may be advected downstream or into the coastal ocean without building up significant concentrations, while pollutant loading at low flow may not be effectively transported and dispersed and may build up to harmful concentrations. In the former case the pollutant may be rapidly transported out of the local waterbody, but may impact a more distant, downstream system. In addition, certain pollutants, such as inorganic nitrogen, may be discharged into surface waters and subsequently transformed and removed from the water column into vegetation or outgassed (e.g., volatilized or denitrified) into the atmosphere under certain ecosystem conditions. Sediment and other pollutants may be stored for long time periods in alluvial or lacustrine deposits, and then remobilized long after the initial loading into a stream reach or standing waterbody in response to extreme climate events, land-use change, reservoir management, or even reductions in the pollutant concentrations in the water column. Consequently, long lags may exist between the actual discharge of the sediment (and any pollutants adsorbed or otherwise stored within the deposits) and their contribution to waterbody impairment. Therefore, understanding the fate of pollutants, particularly nonconservative forms, may require consideration of the full ecosystem cycling and transport of the material over long time periods.

Impairment of waterbodies can be assessed on the basis of biological indicators, as discussed in Chapter 2. As organisms and communities respond to multiple stressors, it is not always clear what the direct or indirect effects of any specific pollutant discharge is, or how that may be exacerbated by correlated or interacting activity in the watershed. The association of specific types of impairment with surrounding land use implicitly accounts for these interactions but does not provide a mechanistic understanding of the linkage sufficient to specify effective remedial activity. However, much progress has been made in determining toxic effects of certain contaminants on different aquatic species assemblages (see, e.g., Shaver et al., 2007) and on quantifying impacts of land use on flow duration curves, EMCs, and loading rates for a number of pollutants (Maestre and Pitt, 2005). For the latter effort, it has been shown that there is large variability within land-use categories, both as a function of specific SCMs and of innate differences due to historical legacies, climate, and hydrogeology.

A protocol linking pollutants in stormwater discharges to ambient water quality criteria should be based on conservation of mass, in which the major inputs, outputs, transformations, and stores of the pollutant can be quantified. Indeed, these are the components of hydrologic and watershed models used to simulate the fate and transport of stormwater and its pollutants. SCMs that improve ambient water quality criteria are designed to act on one or more of these mass balance terms. A number of these measures act to reduce the magnitude of a stormwater source (e.g., porous pavement), while others are designed to absorb or dissipate a pollutant within a hydrologic flowpath downstream from a source (e.g., rain garden, detention pond, stream restoration). The latter requires some consideration of the flowpath from the source to the receiving waterbody. Therefore, determining the major sources, sinks, and transformations of the pollutant should be the first step in this procedure. For a number of pollutants there may be very few potential sources, while for others there may be multiple significant sources. The spatial diversity of these sources and sinks may also range from uniform distribution to "hot spot" patterns that are difficult to detect and quantify. Many stormwater models work effectively with sources, but are not structured to follow the transport or transformation of pollutants from source to waterbody along hydrologic flowpaths.

Figure 4-14 shows the drainage area of Jordan Lake, an important regional

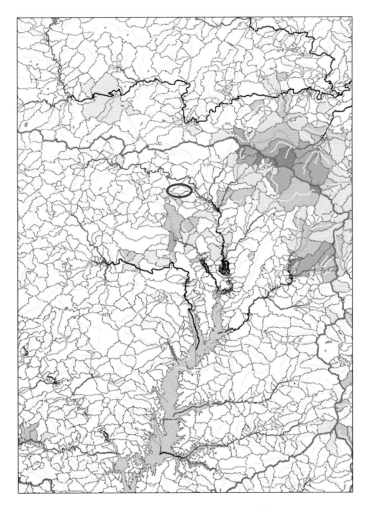

FIGURE 4-14 The drainage area to Jordan Lake, a major drinking water reservoir in the Triangle area of North Carolina, is under nutrient-sensitive rules, requiring reductions in total nitrogen and phosphorus. Drainage flowlines and catchment areas are from NHDplus, and are shaded according to their percentage of industrial and commercial land cover from the NLCD. The area outlined in black is a small urban catchment, detailed in Figure 4-15, and comprised of a wooded central region, surrounded by residential and institutional land use. SOURCE: Data from the NHD[+].

drinking water source in the Triangle area of North Carolina. Catchment areas are shaded to relate the percentage of industrial and commercial land cover, according to the National Land Cover Database (NLCD). Figure 4-15 shows a small tributary within the Jordan Lake watershed in Chapel Hill (outlined in Figure 4-14) with a high-resolution image of all impervious surfaces overlain on the topographically defined surface flowpath network. Each of the distributed sources of stormwater is routed through a flowpath consisting of other pervious and impervious segments, within which additions, abstractions, and transformations of water and pollutants occur depending on weather, hydrologic, and eco-system conditions. The cumulative delivery and impact of all stormwater sources include the transformations occurring along the flowpaths, which could include specific SCMs such as detention or infiltration facilities or simply infil-tration or transformations in riparian areas or low-order streams. The riparian area may be bypassed depending on stormwater concentration or piping, and it may have various levels of effectiveness on reducing pollutants depending on geomorphic, ecosystem, and hydrologic conditions. The ability of a stormwater model to capture these types of effects is a key property influencing its ability to associate a stormwater source with a waterbody outcome.

FIGURE 4-15 A small urban catchment in the Lake Jordan watershed of North Carolina with distributed sources of impervious surface (buildings and roads) stormwater arranged within the full surface drainage flowpath system. Stormwater from each source is routed down surface and subsurface flowpaths to the nearest tributary and out the drainage network, with additions and abstractions of water and pollutants along each flowpath segment. SOURCE: Data from the NHD[+].

This section discusses the fundamentals of stormwater modeling and the capabilities of commonly used models. Much of this information is captured in a summary table at the end of the section (Table 4-7). The models included are the following:

- The Rational Method, or $Q = C*I*A$, where Q is the peak discharge for small urban catchments, A is the catchment area, I is the rainfall intensity, and C is a rainfall-runoff coefficient.
- The Simple Method, which classifies stormwater generation and impact regimes by the percent impervious cover
- TR-20 and TR-55
- The Generalized Watershed Loading Function (GWLF)
- Program for Predicting Polluting Particle Passage through Pits, Puddles, and Ponds (P8)
- Model for Urban Stormwater Improvement Conceptualization (MUSIC)
- Stormwater Management Model (SWMM)
- Source Loading and Management Model (WinSLAMM)
- Soil and Water Assessment Tool (SWAT)
- Hydrologic Simulation Program–Fortran (HSPF)
- Western Washington Hydrologic Model
- Chesapeake Bay Watershed Model (CBWM)

Fundamentals of Stormwater Models

Stormwater models are designed to evaluate the impacts of a stormwater discharge on a receiving waterbody. In order to do this, the model must have the capability of describing the nature of the source term (volumes, constituents), transport and transformation to the receiving waterbody, and physical, chemical, and biological interaction with the receiving water body and ecosystem. No model can mechanistically reproduce all of these interactions because of current limitations in available data, incomplete understanding of all processes, and large uncertainties in model and data components. Computer resources, while rapidly advancing, still limit the complexity of certain applications, especially as spatial data become increasingly available and it is tempting to model at ever-increasing resolution and comprehensiveness. Therefore, models must make a set of simplifying assumptions, emphasizing more reliable and available data, while attempting to retain critical processes, feedbacks, and interactions. Models are typically developed for a variety of applications, ranging from hydraulic design for small urban catchments to urban and rural pollutant loading at a range of watershed scales.

An evaluation of the current state of stormwater modeling should say much about our ability to link pollutant sources with effects in receiving waters. Both stormwater models and models supporting the evaluation of SCM design and effectiveness are based on simulating a mass budget of water and specific pollutants. The detail of mass flux, transformation, and storage terms vary depending on the scale and purpose of the application, level of knowledge regarding the primary processes, and available data. In many cases, mechanisms of transformation may be either poorly understood or may be dependent on detailed interactions. As an example, nitrogen-cycle transformations are sensitive to very short temporal and spatial conditions, termed "hot spots" and "hot moments" relative to hydrologic flowpaths and moisture conditions (McClain et al., 2003).

Stormwater runoff production and routing are common components of these models. All models include an approach to estimate the production of stormwater runoff from one or more zones in the watershed, although runoff routing from the location(s) of runoff production to a point or waterbody is not always included explicitly. Major divisions between approaches are found in the representation of the watershed "geography" in terms of patterns and heterogeneity, and in runoff production and routing. Some stormwater models do not consider the effects of routing from a runoff source to a local waterbody directly, but may attempt to reproduce net impacts at larger scales through the use of unit hydrograph theory to estimate peak flows, and delivery ratios or stormwater control efficiency factors to estimate export to a waterbody.

There are a number of different approaches and paradigms used in stormwater models that include varying degrees of watershed physical, biological, and chemical process detail, as well as spatial and temporal resolution and the representation of uncertainty in model estimates. A number of researchers have written about the nature of watershed models (e.g., Beven, 2001; Pitt and Vorhees, 2002). At present, many hydrologic and stormwater models have become so complex, with multiple choices for different components, that standard descriptions apply only to specific components of the models. The following discussion is generalized; most models fit the descriptions only to certain degrees or only under specific conditions in which they are operated.

Lumped Versus Distributed Approaches

Central to the design of watershed models is the concept of a "control volume," which is a unit within which material and energy contents and balances are defined, with boundaries across which material and energy transport occurs. Control volumes can range from multiple subsurface layers and vegetation canopy layers bounded in three dimensions to a full watershed. Lumped models ignore or average spatial heterogeneity and patterns of watershed conditions, representing all control volumes, and the stores, sources, and sinks of water and pollutants in a vertically linked set of conceptual components, such as surface interception, unsaturated and saturated subsurface zones, and a single stream or river reach. For example, SWAT or HSPF are conceptually lumped at the scale of subwatersheds (e.g., the level of geography in Figure 4-14) and do not show any spatial patterns at higher resolutions (e.g., Figure 4-15) than these units. While multiple land-use/soil combinations may be represented, these models do not represent the connectivity of the land segments (e.g., which land segments drain into which land segments) and assume all unique land segment types drain directly to a stream.

Distributed models include some scheme to represent spatial heterogeneity of the watershed environment pertinent to stormwater generation, including land cover, soils, topography, meteorological inputs, and stream reach properties distributed through a set of linked control volumes. Control volumes representing land elements, including vertically linked surface and subsurface stores, are connected by a representation of water and pollutant lateral routing through a network of flowpaths that may be predefined or set by the dynamics of surface, soil, and saturated zone water storage. The land elements may be grid cells in a regular lattice, or irregular elements (e.g., triangles) with the pattern adapted to variations in land surface characteristics or hydraulic gradients.

A number of models are intermediate between lumped and distributed, with approaches such as lumping at the subwatershed scale, incorporating statistical distributions of land element types within subwatersheds but without explicit pattern representation, or lumping some variables and processes (such as groundwater storage and flux), while including distributed representation of topography and land cover. Thus, within the model SLAMM (Pitt and Vorhees, 2002), the catchment is described in sufficient detail to summarize the breakdown of different drainage sequences. As an example, roof area will be broken down to the proportion that drains to pervious areas and to directly connected impervious areas. An important distinction is that there is no routing of the output of one land element into another, such that there is no drainage sequence that may significantly modify the stormwater runoff from its source to the stream. Implicitly, all land elements drain directly into a stream, although a loss rate or delivery ratio can be specified.

The choice of a more lumped or distributed model is often dependent on available data and overall complexity of the model. Simpler, lumped models may be preferred in the absence of sufficient data to effectively parameterize a distributed approach, or for simplicity and computational speed. However, fully lumped models may be limited in their ability to represent spatial dependency, such as the development and dynamics of riparian zones, or the effects of SCM patterns and placement. As there is typically an irreducible level of spatial heterogeneity in land surface characteristics down to very small levels below the resolution of individual flow elements, we note that all models lump at some scale (Beven, 2000).

Mechanistic Versus Conceptual Process Representation

Mechanistic, or process-based, approaches attempt to reproduce key stormwater transport and transformation processes with more physically, chemically, or biologically based detail, while conceptual models represent fluxes between stores and transformations with aggregate, simplified mathematical forms. No operational models are built purely from first principles, so the distinction between mechanistic and conceptual process basis is one of degree.

The level of sampling necessary to support detailed mechanistic models, as well as remaining uncertainty in physicochemical processes active in heterogeneous environments typically limits the application of first-principle methods. The development or application of more mechanistic approaches is currently limited by available measurements, which require both time and resources to adequately carry out. Unfortunately, modeling and monitoring have often been mutually exclusive in terms of budgets, although it is necessary for both to be carefully planned and integrated. A new generation of sensors and a more rigorous and formal sampling protocol for existing methods will be necessary to advance beyond the current practice.

At present, most operational hydrologic and transport models are based on a strong set of simplifying assumptions regarding active processes and/or the spatial variation of sources, sinks, and stores in the watershed. Runoff production can be computed by a range of more mechanistic to more conceptual or empirical methods. More mechanistic methods include estimation of infiltration capacities based on soil hydraulic properties and moisture conditions, excess runoff production, and hydraulic routing over land surfaces into and through a stream-channel network. More conceptual approaches use a National Resources Conservation Service (NRCS) curve number approach (see Box 4-5) and unit hydrograph methods to estimate runoff volume and time of concentration. Pollutant concentrations or loads are often estimated on the basis of look-up tables using land use or land cover. Land use- or land cover-specific EMC or unit area loading for pollutants can be developed directly from monitoring data or from local, regional, or national databases. The NSQD statistically summarizes the results of a large number of stormwater monitoring projects (as discussed previously in this chapter). The effects of SCM performance (typically percent removal) can be estimated from similar databases (e.g., www.bmpdatabase.org). A set of models, such as SWAT, incorporate fairly detailed descriptions of nutrient cycling as an alternative to using EMC, requiring more detailed inputs of soil, crop, and management information. Unfortunately, the detailed biogeochemistry of this and similar models is typically not matched by the hydrology, which remains lumped at individual Hydrologic Response Unit (HRU) levels using NRCS curve number methods, although options exist to incorporate more mechanistic infiltration excess runoff.

BOX 4-5
NRCS Technical Release 55

NRCS methods to estimate runoff volumes and flows have been popular since the early 1950s (Rallison, 1980). Fundamentally they can be broken into the separation of runoff from the rainfall volume (Curve Number Method), the pattern of runoff over time (dimensionless unit hydrograph), and their application within computer simulation models. In the late 1970s these components were packaged together in a desktop hydrology method known as Technical Release 55 (TR-55). TR-55 became the primary model used by the majority of stormwater designers, and there is considerable confusion over the terms used to describe what aspects of the NRCS methods are in use.

The NRCS Curve Number Method was first derived in the 1950s for prediction of runoff from ungauged agricultural areas. It relates two summation ratios, that of runoff to rainfall and that of moisture retained to maximum potential retention. Two statistically based relations were developed to drive the ratio, the first of which is based on a "curve number" which depicts the soil type, land cover, and initial moisture content. The second or initial abstraction is defined as the volume of losses that occur prior to the initiation of runoff, and is also related to the curve number. Data were used to derive curve numbers for each soil type and cover as shown in Figure 4-16 (Rallison, 1980).

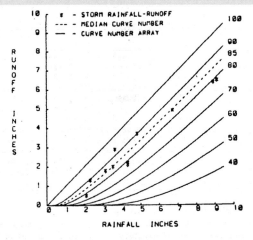

FIGURE 4-16 Development of curve number from collected data. SOURCE: Reprinted, with permission, from Rallison (1980). Copyright 1980 by the American Society of Civil Engineers.

The Curve Number method is a very practical method that gives "average" runoff results from a watershed and is used in many models (WIN TR-55, TR-20, SWMM, GWLF, HEC-HMS, etc.). Caution has to be exercised when using it for smaller urbanizing storm events. For example, past practice was to average curve numbers for developments for pavement and grass based on percent imperviousness. While this works well for large storms, for smaller storms it gives erroneous answers through violation of the initial abstraction relationship. Current state manuals (MDE, 2000; PaDEP, 2006) do not allow paved- and unpaved-area curve numbers to be averaged. When applied to continuous simulation models (such as in SWMM or GWLF), it requires an additional method to recover the capacity to remove runoff because the soil capacity to infiltrate water is restored over time.

The NRCS Dimensionless Unit Hydrograph has also evolved over many years and simply creates a temporal pattern from the runoff generated from the curve number method. This transformation is based upon the time of concentration, defined as the length of time the water takes to travel from the top to the bottom of the watershed. The dimensionless curve ensures that conservation of mass is maintained. The main purpose of this method is to estimate how long it takes the runoff generated by the curve number to run off the land and produce discharge at the watershed outlet.

The NRCS curve number and dimensionless unit hydrograph were first incorporated in the Soil Conservation Service (SCS) TR-20 hydrologic computer model developed in the 1960s. As most stormwater professionals did not have access to mainframes, SCS put together TR-55, which created a hand or calculator method to apply the curve number and dimensionless unit hydrograph. In order to create this hand method, many runs were generated using TR-20 to develop patterns for different times of concentration. The difficulty with using the original TR-55 in the modern era is that the simplifications to the hydrograph development do not allow the benefits of SCMs to be easily accounted for.

The use of the term TR-55 has been equated with the curve number method; this has created confusion, especially when it is included in municipal code. Further clouding the issue, there are two types of TR-55 computer models available. One is based on the original, outdated, simplified hand method, and the other (Win TR-55) returns to the more appropriate application of the curve number and dimensionless hydrograph methods. In either case, the focus of these models is on single event hydrology and cannot easily incorporate or demonstrate the benefits of the wide range of structural and nonstructural SCMs. Note that the curve number and dimensionless unit hydrograph methods are incorporated in many continuous flow models, including SWMM and GWLF, as the basis of runoff generation and runoff timing.

Deterministic Versus Stochastic Methods

Deterministic models are fully determined by their equation sets, initial and boundary conditions, and forcing meteorology. There are no components that include random variation. In a stochastic model, at least one parameter or variable is drawn from a probability distribution function such that the same model set-up (initial and boundary conditions, meteorology, parameter sets) will have randomly varying results. The advantage of the latter approach is the ability to generate statistical variability of outcomes, reflecting uncertainty in parameters, processes, or any other component. In fact, any deterministic model can be operated in a stochastic manner by sampling parameter values from specified probability distributions.

It is recognized that information on the probability distribution of input parameters may be scarce. For situations with limited information on parameter values, one option is to assume a uniform distribution that brackets a range of values of the parameter reported in the literature. This would at least be a start in considering the impacts of the variability of model inputs on outputs. A thorough discussion on methods for incorporating uncertainty analysis into model evaluation is provided in Chapter 14 of Ramaswami et al. (2005). It should be noted that the ability to generate probability distribution information on stormwater outcomes requires a potentially large number of model runs, which may be difficult for detailed mechanistic and distributed models that have large computational loads.

Continuous Versus Event-Based Approaches

Another division between modeling approaches is the time domain of the simulation. Event-based models limit simulation time domains to a storm event, covering the time of rainfall and runoff generation and routing. Initial conditions need to be estimated on the basis of antecedent moisture or precipitation conditions. For catchments in which runoff is dominated by impervious surfaces, this is a reasonable approach. In landscapes dominated by variable source area runoff dynamics in which runoff is generated from areas that actively expand and contract on the basis of soil moisture conditions, a fuller accounting of the soil moisture budget is required. Furthermore, event-based modeling is inappropriate for water quality purposes because it will not reproduce the full distribution of receiving water problems. Continuous models include simulation of a full time domain composed of storm and inter-storm periods, thus tracking soil moisture budgets up to and including storm events.

Outfall Models

After beneficial use impairments are recognized, cause-and-effect relationships need to be established and restorative discharge goals need to be developed. Models are commonly used to calculate the expected discharges for different outfalls affecting the receiving water in a community. All of the models shown in Table 4-7 can calculate outfall discharge quantities, although some may only give expected average annual discharge. Models calculate these discharges using a variety of processes, but all use an urban hydrology component to determine the runoff quantity and various methods to calculate the quality of the runoff. The runoff quantity is multiplied by the pollutant concentration in the outfall to obtain the mass discharges of the different pollutants. The outfall mass discharge from the various outfalls in the area can then be compared to identify the most significant outfalls that should be targeted for control.

The most common hydrology "engines" in simple stormwater models are the NRCS curve number method or a simple volumetric runoff coefficient—R_v, the ratio of runoff to rainfall—for either single rainfall events or the total annual rainfall depth. Runoff quality in the simple models is usually calculated based on published EMCs for similar land uses in the same geographical area. More complex models may use build-up and wash-off of pollutants from impervious surfaces in a time series or they may derive pollutant concentrations from more detailed biogeochemical cycling mechanisms, including atmospheric deposition and other inputs (e.g., fertilizer). Some models use a combination of these processes depending on the area considered, and others offer choices to the model user. Again, these processes all need local calibration and verification to reduce the likely uncertainty associated with the resultant calculated discharge conditions.

Source Area

When the outfalls are ranked according to their discharges of the pollutants of importance, further detailed modeling can be conducted to identify sources of the significant pollutants within the outfall drainage area. Lumped parameter models cannot be used, as the model parameters vary within the drainage area according to the different source areas. Distributed area models can be used to calculate contributions from different source areas within the watershed area. This information can then be used to rank the land uses and source area contributions. In-stream responses can be calculated if the land-area models are linked to appropriate receiving-water models.

Need for Coupling Models

As urban areas become increasingly extensive and heterogeneous, including a gradient of dense urban to forest and agricultural areas, linkage and coupling of models to develop feedback and interactions (e.g., impacts of urban runoff hydraulics with stream scour and sedimentation, mixed with agricultural nutrient and sediment production on receiving waterbodies) is a critical area that requires more development. In general, stormwater models were designed to track and predict discharges from sources by surface water flowpaths into receiving waterbodies, such that infiltration was considered to be a loss (or retention) of water and its constituents. To fully evaluate catchment-scale impacts of urbanization on receiving waterbodies, the infiltration term needs to be considered a source term for the groundwater, and a groundwater component or model needs to be coupled to complete the surface–subsurface hydrologic interactions and loadings to the waterbody.

Finally, each of the models may or may not incorporate explicit consideration of SCM performance based on design, implementation and location within the catchment. As discussed in the next chapter, SCM models can range from simple efficiency factors (0–1 multipliers on source discharge) to more detailed treatment of physical, chemical, and biological transport and transformations.

Linking to Receiving-Water Models

Specific problems for urban receiving waters need to be identified through comprehensive field monitoring and modeling. Monitoring can identify current problems and may identify the stressors of importance (see Burton and Pitt [2002] for tools to evaluate receiving water impairments). However, monitoring cannot predict conditions that do not yet exist and for other periods of time that are not represented at the time of monitoring. Modeling is therefore needed to gain a more comprehensive understanding of the problem. In small-scale totally urbanized systems, less complex receiving-water models are needed. However, as the watershed becomes more complex and larger with multiple land uses, the receiving-water models also need to become more complex. Complex receiving-water models need to include transport and transformations of the pollutants of concern, for example. Examples of models shown on the comparison table that include receiving-water processes are MUSIC and HSPF. Other models (such as WinSLAMM) provide direct data links to external receiving-water models. Calibration and verification of important receiving-water processes that are to be implemented in a model can be very expensive and time consuming, and still result in substantial uncertainty.

Model Calibration and Verification

Calibration is the process where model parameters are adjusted to minimize the difference between model output and field measurements, with an aim of keeping model parameters within a range of values reported in the literature. Model *verification*, similar to model validation, is used to mean comparison between calibrated model results using part of a data set as input and results from application of the calibrated model using a second (independent) part of the data set as input. Oreskes et al. (1994) present the viewpoint that no model can really be verified; at best, verification should be taken to mean that a model is consistent with a physical system under a given set of comparison data. This is not synonymous with saying that the model can reliably represent the real system under any set of conditions. In general, the water quantity aspects of stormwater modeling are easier to calibrate and verify than the water quality aspects, in part because there are more water quantity data available and because chemical transformations are more complex to simulate. A thorough discussion of the broad topic of model evaluation is provided by several excellent texts on this subject, including Schnoor (1996) and Ramaswami et al. (2005).

Models in Practice Today

Table 4-7 presents a set of models used for stormwater evaluation that range in complexity from first-generation stormwater models making use of simple empirical land cover/runoff and loading relations to more detailed and information-demanding models. The columns in Table 4-7 provide an abbreviated description of some of the attributes of these models—common usage, typical application scales, the degree of model complexity, some data requirements (for the hydrologic component), whether the model addresses groundwater, and whether the model has the ability to simulate SCMs. Models capable of simulating a water quality component require EMC data, with some models also having a simple build-up/wash-off approach to water quality simulation (e.g., SWMM, WinSLAMM, and MUSIC) and others simulating more complex geochemistry (e.g., SWAT and HSPF). The set of columns in Table 4-7 is not meant to be exhaustive in describing the models, which is why websites are provided for comprehensive model descriptions and data requirements.

In addition to the models listed in Table 4-7, a representative set of emerging research models that are not specifically designed for stormwater, but may offer some advantages for specific uses, are also described below. In general, it is important that models that integrate hydrologic, hydraulic, meteorologic, water quality, and biologic processes maintain balance in their treatment of process details. Both model design and data collection should proceed in concert and should be geared toward evaluating and diagnosing the consistency of model or coupled model predictions and the uncertainty attached to each component and the integrated modeling system. The models should be used in a manner that produces both best estimates of stormwater discharge impacts on receiving waterbodies, as well as the level of uncertainty in the predictions.

The Rational Method is a highly simplified model widely used to estimate peak flows for in sizing storm sewer pipes and other low level drainage pathways. The method assumes a constant rainfall rate (intensity), such that the runoff rate will increase until the time at which all of the drainage area contributes to flow at its outlet (termed the *time of concentration*). The product of the drainage area and rainfall intensity is considered to be the input flow rate to the drainage area under consideration; the ratio of the input flow rate to an outflow discharge rate is termed the runoff coefficient. Runoff coefficients for a variety of land surface types and slopes have been compiled in standard tables (see e.g., Chow et al., 1988). The outflow is determined by multiplying inflow (rainfall intensity times drainage area) by the runoff coefficient for the land-surface type. As pointed out by Chow et al. (1988), this method is often criticized owing to its simplified approach, so its use is limited to stormwater inlet and piping designs.

The Simple Method estimates stormwater pollutant loads for urban areas, and it is most valuable for assessing and comparing the relative stormwater pollutant load changes of different land use and stormwater management scenarios. It requires a modest amount of information, including the subwatershed drainage area and impervious cover, stormwater pollutant concentrations (as defined by the EMC), and annual precipitation. The subwatershed can be broken up into specific land uses, such that annual pollutant loads are calculated for each type of land use. Stormwater pollutant concentrations are usually estimated from local or regional data, or from national data sources. The Simple Method estimates pollutant loads for chemical constituents as a product of annual runoff volume and pollutant concentration, as $L = 0.226\,R \times C \times A$, where L = annual load (lbs), R = annual runoff (inches), C = pollutant concentration (mg/l), and A = area (acres).

Of slightly increased complexity are those models initially developed decades ago by the Soil Conservation Service, now the NRCS of the U.S. Department of Agriculture (USDA). NRCS Technical Releases (TR) 20 and 55 are widely used in many municipalities, despite the availability of more rigorous, updated stormwater models. Box 4-5 provides an overview of the NRCS TR-55 assumptions and approaches.

A number of watershed models that are used for stormwater assessment are lumped, conceptual forms, with varying levels of process simplification and spatial patterns aggregated at the subwatershed level, with aspatial statistical distribution of land types as described above. The GWLF model (Haith and Shoemaker, 1987) is an example of this type of approach, using simple land use-based EMC with NRCS curve number estimates of runoff within a watershed context. GWLF is a continuous model with simplified upper- and lower-zone subsurface water stores, and a simple linear aquifer to deliver groundwater flow. EMCs are assigned or calibrated for subsurface and surface flow delivery, while sediment erosion and delivery are computed with the use of the Universal Soil Loss Equation and delivery coefficients. The methods are easily linked to a Geographical Information System (GIS), which provides land-use composition at the subwatershed level and develops estimates of runoff and loading that are typically used to estimate annual loading. AVGWLF links GWLF with Arc-View and is used as a planning- or screening-level tool. A recent example of AVGWLF for nutrient loading linked to a simple stream network nutrient decay model for the development of a TMDL for a North Carolina water supply area is given in Box 4-6.

BOX 4-6
The B. Everett Jordan Lake GWLF Watershed Model Development

Jordan Lake is a regionally important water supply reservoir at the base of the 1,686-square-mile Haw watershed in North Carolina (see Figure 4-17). It is considered a nutrient-sensitive waterbody. Officials are now in the process of implementing watershed goals to reduce nitrogen and phosphorus, with the reduction goals differentiated by geographic location within the basin. In support of the development of these rules as part of a TMDL effort, the North Carolina Division of Water Quality commissioned a water quality modeling study (Tetra Tech, 2003). The modeling effort was needed to support the evaluation of nutrient reduction strategies in different parts of the watershed relative to Jordan Lake, which requires both a model of nutrient loading, as well as river transport and transformation. Given data and resource restrictions, a more detailed model was not considered feasible. As GWLF does not support nutrient transformations in the stream network, the model was used in conjunction with a method to decay nutrient source loading by river transport distance to the lake. A spreadsheet model was designed to take as input GWLF estimates of seasonal loads for 14-digit hydrologic unit code (HUC) subbasins of the Haw, and to reduce the loads by river miles between the subwatershed and Jordan Lake. The GWLF loading model was calibrated to observations in small subwatersheds within the Haw using HRUs developed from soil and NLCD land classes, updated with additional information from county GIS parcel databases and the 2000 Census. This information was used to estimate subwatershed impervious surface cover, fertilizer inputs, runoff curve numbers, soil water capacity, and vegetation cover to adjust evapotranspiration rates. Wastewater disposal (sewer or septic) was estimated on the basis of urban service boundaries. GWLF was used to provide loading estimates, using limited information on soil and groundwater nutrient concentrations, and calibrated delivery ratios. In-stream loss was based on a first-order exponential decay function of river travel time to Jordan Lake, with the decay coefficient generated by estimates of residence time in the river network, and upstream/downstream nutrient loads following non-linear regression methods used in SPARROW (Alexander et al., 2000). Further adjustments based on impoundment trapping of sediment and associated nutrient loads were carried out for larger reservoirs in the Haw. The results provided estimates of both loading and transport efficiency to Jordan Lake, with estimates of relative effectiveness of sectoral loading reductions in different parts of the watershed.

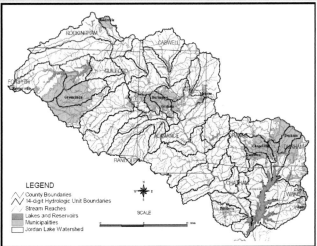

FIGURE 4-17 14 digit HUCs draining to Jordan Lake in the Haw River watershed of North Carolina. SOURCE: Tetra Tech (2003).

P8 (Program for Predicting Polluting Particle Passage through Pits, Puddles, and Ponds) is a curve number-based model for predicting the generation and transport of stormwater runoff pollutants in urban watersheds, originally developed to help design and evaluate nutrient control in wet detention ponds (Palmstrom and Walker, 1990; http://wwwalker.net/p8/). Continuous water-balance and mass-balance calculations are performed and consist of the following elements: watersheds, devices, particle classes, and water quality components. Continuous simulations use hourly rainfall and daily air temperature time series. The model was initially calibrated to predict runoff quality typical of that measured under NURP (EPA, 1983). SCMs in P8 include detention ponds (wet, dry, extended), infiltration basins, swales, and buffer strips. Groundwater and baseflows are also included in the model using linear reservoir processes.

MUSIC is a part of the Catchment Modelling Toolkit (www.toolkit.net.au) developed by the Cooperative Research Center for Catchment Hydrology in Australia (Wong et al., 2001). The model concentrates on the quality and quantity of urban stormwater, including detailed accounting of multiple SCMs acting within a treatment train and life-cycle costing. It employs a simplified rainfall–runoff model (Chiew and McMahon, 1997) based on impervious area and two moisture stores (shallow and deep). TSS, total nitrogen, and total phosphorus are based on EMCs, sampled from lognormal distributions. The model does not contain detailed hydraulics required for routing or sizing of SCMs, and it is designed as a planning tool.

EPA's SWMM has the capability of simulating water quantity and quality for a single storm event or for continuous runoff. The model is commonly used to design and evaluate storm, sanitary, and combined sewer systems. SWMM accounts for hydrologic processes that produce runoff from urban areas, including time-varying rainfall, evaporation, snow accumulation and melting, depression storage, infiltration into soil, percolation to groundwater, interflow between groundwater and the drainage system, and nonlinear reservoir routing of overland flow. Spatial variability is modeled by dividing a study area into a collection of smaller, homogeneous subcatchment areas, each containing its own fraction of pervious and impervious sub-areas. Overland flow can be routed between sub-areas, between subcatchments, or between entry points of a drainage system. SWMM can also be used to estimate the production of pollutant loads associated with runoff for a number of user-defined water quality constituents. Transport processes include dry-weather pollutant buildup over different land uses, pollutant wash-off from specific land uses, direct contribution of rainfall deposition, and the action of such SCMs as street cleaning, source control, and treatment in storage units, among others.

Watershed models such as SWAT (Arnold et al., 1998) or HSPF (Bicknell et al., 1997, 2005) have components based on similar land-use runoff and loading factors, but also incorporate options to utilize detailed descriptions of interception, infiltration, runoff, routing, and biogeochemical transformations. Both models are based on hydrologic models that were developed prior to the availability of detailed digital spatial information on watershed form and use conceptual control volumes that are not spatially linked. HRUs are based on land use, soils, and vegetation (and crop) type, among other characteristics, and are considered uniformly distributed through a subbasin. Within each HRU, simplified representations of soil upper and lower zones, or unsaturated and saturated components, are vertically integrated with a conceptual groundwater storage-release component. There is no land surface routing and all runoff from a land element is considered to reach the river reach, with some delivery ratio if appropriate for sediment and other constituents. Like GWLF, the models are typically not designed to estimate loadings from individual dischargers, but are used to help guide and develop TMDL for watersheds. SWAT and HSPF are integrated within the EPA BASINS system (http://www.epa.gov/waterscience/basins) with GIS tools designed to use available spatial data to set up and parameterize simulations for watersheds within the United States. Examples of combining one of these models, typically designed for larger-scale applications (such as the area shown in Figure 4-14) with more site-specific models such as SLAMM or SWMM, are given in Box 4-7.

BOX 4-7
Using SWAT and WinSLAMM to Predict Phosphorus Loads
in the Rock River Basin, Wisconsin

Wisconsin Administrative Code NR 217 states that wastewater treatment facilities in Wisconsin must achieve an effluent concentration of 1 mg/L for phosphorus. Alternative limits are allowed if it can be demonstrated that achieving the 1 mg/L limit will not "result in an environmentally significant improvement in water quality" (NR 217.04(2)(b)1). In response to NR 217, a group of municipal wastewater treatment facilities formed the Rock River Partnership (RRP) to assess water quality management issues (Kirsch, 2000). The RRP and the Wisconsin Department of Natural Resources funded a study to seek water quality solutions across all media, and not just pursue additional reductions from point sources. A significant portion of the study required a modeling effort to determine the magnitude of various nutrient sources and determine potential reductions through the implementation of global SCMs.

The Rock River Basin covers approximately 9,530 square kilometers and lies within the glaciated portion of south central and eastern Wisconsin (Figure 4-18). The Rock River and its numerous tributaries thread their way through this landscape that spreads over 10 counties inhabited by more than 750,000 residents. There are 40 permitted municipalities in the watershed, representing 4 percent of the land area, and they are served by 57 sewage treatment plants. Urban centers include Madison, Janesville, and Beloit as well as smaller cities such as Waupun, Watertown, Oconomowoc, Jefferson, and Beaver Dam. Although the basin is experiencing rapid growth, it is still largely rural in character with agriculture using nearly 75 percent of the land area. Crops range from continuous corn and corn–soybean rotations in the south to a mix of dairy, feeder operations, and cash cropping in the north. The basin enjoys a healthy economy with a good balance of agricultural, industrial, and service businesses.

continues next page

BOX 4-7 Continued

The focus of the modeling was to construct an intermediate-level macroscale model to better quantify phosphorus loads from point and nonpoint sources throughout the basin. The three goals of the modeling effort were to (1) estimate the average annual phosphorus load, (2) estimate the relative contribution of phosphorus loads from both nonpoint (urban and agricultural) and point sources, and (3) estimate changes in average annual phosphorus loads from the application of global SCMs and point source controls.

SWAT was selected for the agricultural analysis and WinSLAMM was selected to develop phosphorus loads for the urban areas. WinSLAMM was selected to make estimates of stormwater loads, because it is already calibrated in Wisconsin for stormwater volumes and pollutant concentrations. Outputs of phosphorus loads from WinSLAMM were used as input to SWAT. One output of SWAT was a total nonpoint phosphorus load based on agricultural loads calculated in SWAT and stormwater loads estimated by WinSLAMM.

SWAT was calibrated with data from 23 USGS gauging stations in the Rock River Basin. Hydrology was balanced first on a yearly basis looking at average annual totals, then monthly to verify snowfall and snowmelt routines, and then daily. Daily calibration was conducted to check crop growth, evapotranspiration, and daily peak flows. Crop yields predicted by SWAT were calibrated to those published in the USDA Agricultural Statistics.

Under current land-use and management conditions, the model predicted an average annual load of approximately 1,680,000 pounds of total phosphorus for the basin with 41 percent from point sources and 59 percent from nonpoint sources. Less than 10 percent of the annual phosphorus load is generated by the urban areas in the watershed. Evaluation of various SCM scenarios shows that with implementation of NR 217 (applicable point source effluent at 1 mg/L) and improvement in tillage practices and nutrient management practices, total phosphorus can be reduced across the basin by approximately 40 percent. It is important to note that the nonpoint management practices that were analyzed were limited to two options: modifications in tillage practices, and adoption of recommended nutrient application rates. No other management practices (i.e., urban controls, riparian buffer strips, etc.) were simulated. Urban controls were not included because the urban areas contributed a relatively small percentage of the total phosphorus load. Thus, loadings depicted by SWAT under these management scenarios do not necessarily represent the lowest attainable loads. Results suggest that a combination of point and nonpoint controls will be required to attain significant phosphorus reductions.

The CBWM is a detailed watershed model that is extended from HSPF as a base, but includes additional components to incorporate stormwater controls at the land segment level. HSPF is operated for a number of subbasins, and each subbasin model includes different land segments based on land cover and soil units as aspatial, lumped distribution functions, but also includes representation of SCMs and (large) stream routing. Model implementation at the scale of the full Chesapeake Bay watershed requires fairly coarse-grained land partitioning. A threshold of 100 cfs mean annual flow is used to represent streams and rivers, and the one-to-one mapping of land segment to river reach produces large, heterogeneous land segments as the basic runoff-producing zones. SCMs are implemented either at the field or runoff production unit as distinct land segment types in terms of management or land cover, or as "edge-of-field" reductions of runoff or pollutant loads. The latter are assigned as static efficiency factors irrespective of flow conditions or season, with all SCMs within a land segment integrated into a single weighted efficiency value.

FIGURE 4-18 Rock River Basin, Wisconsin. SOURCE: Reprinted, with permission, from Kirsch (2000). Copyright 2000 by American Society for Biological and Agricultural Engineers.

SLAMM is designed for complex, urban catchments and is used as a planning tool to assess both stormwater and pollutant runoff production and the capability of specific stormwater control strategies to reduce stormwater discharges from urban sources. It is specifically designed to capture the most significant distributed and sequential drainage effects of variable source areas in urban catchments (Pitt and Vorhees, 2002) and is based on detailed descriptions of the catchment composition, including both type and relative position (drainage sequence) of land elements. The model is dependent on high-resolution classification or description of the catchment that has become increasingly available in urban areas over the past two decades, and comprehensive field assessment of runoff and pollutant loading from different urban land elements. SLAMM uses continuous simulation for some aspects, such as the build up of street pollutant loads between storms, while using event-based simulation for runoff. The description of build-up and wash-off is a critical component in urban stormwater models applied to areas with substantial impervious surfaces and is a good example of the need to match detailed and rigorous field sampling in order to adequately describe and represent dominant processes. Details of measurement and model representation for build-up and wash-off of contaminants are given in Box 4-8.

Potential New Applications of Coupled Distributed Models

The advent of high-resolution digital topographic and land-cover data over the past two decades has fueled a significant shift in runoff modeling towards "spatially explicit" simulations that distinguish and connect runoff producing elements in a detailed flow routing network. While models developed prior to the availability of high-resolution data or based on older paradigms developed in the absence of this information required spatial and conceptual lumping of control volumes, more recently developed distributed models may contain control volumes linked in multiple vertical layers (soil and aquifer elements) and laterally from a drainage divide to the stream, including stream-channel and riparian segments. A set of models has been developed and applied to stormwater generation using this paradigm that can be applied at the scale of residential neighborhoods, resolving land cover and topography at the parcel level. These models also vary in terms of their emphasis, with some models better representing coupled surface water–groundwater interactions, water, carbon and nutrient cycling, or land–atmosphere interactions. Boyer et al. (2006) have recently reviewed a set of hydrologic and ecosystem models in terms of their ability to simulate sources, transport, and transformation of nitrogen within terrestrial and aquatic ecosystems. Data and information requirements are typically high, and the level of process specificity may outstrip the available information necessary to parameterize the integrated models. However, an emphasis is placed on providing mechanistic linkage and feedbacks between important surface, subsurface, atmospheric, and ecosystem components. Examples of these models include the Distributed Hydrology Soil Vegetation model (DHSVM, Wigmosta et al., 1994); the Regional Hydro-Ecologic Simulation System (RHESSys, Band et al., 1993; Tague and Band, 2004); ParFlow-Common Land Model (CLM, Maxwell and Miller, 2007); the Penn State Integrated Hydrologic Model (PIHM, Qu and Duffy, 2007); the Soil Moisture Distribution and Routing (SMDR) model (Easton et al., 2007); and that of Xiao et al. (2007).

One advantage of integrating surface and subsurface flow systems within any of these model structures is the ability to incorporate different SCMs by specifying characteristics of specific locations within the flow element networks linked to the subsurface drainage. Examples can include alteration of surface detention storage and release curves to simulate detention ponds, or soil depth, texture, vegetation, and drainage release for rainfall gardens. The advantage of this approach is the tight coupling of these SCM features with the connected surface and subsurface drainage systems, allowing the direct incorporation of the SCM as sink or source terms within the flowpath network. Burgess et al. (1998) effectively demonstrated that suburban lawns can become the major

BOX 4-8
Build-up and Wash-off of Contaminants from Impervious Surfaces

The accumulation and wash-off of street particulates have been studied for many years (Sartor and Boyd, 1972; Pitt, 1979, 1985, 1987) and are important considerations in many stormwater models, such as SWMM, HSPF, and SLAMM, that require information pertaining to the movement of pollutants over land surfaces. Accumulation rates are usually obtained through trial and error during calibration, with little, if any, actual direct measurements. Furthermore, those direct measurements that have been made are often misapplied in modeling applications, resulting in unreasonable model predictions.

Historically, streets have been considered the most important directly connected impervious surface. Therefore, much early research was directed toward measuring the processes on these surfaces. Although it was eventually realized that other surfaces can also be significant pollutant sources (see Pitt et al., 2005a,b for reviews), additional research to study accumulation and wash-off for these other areas has not been conducted, such that the following discussion is focused on street dirt accumulation and wash-off.

Accumulation of Particulates on Street Surfaces

The permanent storage component of street surface particulates is a function of street texture and condition and is the quantity of street dust and dirt that cannot be removed naturally by rain or wind, or by street cleaning equipment. It is literally trapped in the texture of the street. The street dirt loading at any time is this initial permanent loading plus the accumulation amount corresponding to the exposure period, minus the resuspended material removal by wind and traffic-induced turbulence.

One of the first research studies to attempt to measure street dirt accumulation was conducted by Sartor and Boyd (1972). Field investigations were conducted between 1969 and 1971 in several cities throughout the United States and in residential, commercial, and industrial land-use areas. Figure 4-19 is a plot of the 26 test area measurements collected from different cities, but separated by the three land uses. The data are the accumulated solids loading plotted against the number of days since the street had been cleaned by the municipal street cleaning operation or a "significant" rain. There is a large amount of variability. The street cleaning and this rain were both assumed to remove all of the street dirt; hence, the curves were all forced through zero loading at zero days.

A more thorough study was conducted in San Jose, California by Pitt (1979), during which the measured street dirt loading for a smooth street was also found to be a function of time. As shown in Figure 4-20, both accumulation rates and increases in particle size of the street dirt increase as time between street cleaning lengthens. However, it is also evident that there is a substantial residual loading on the streets immediately after the street cleaning, which differs substantially from the assumption of Sartor and Boyd that rains reduce street dirt to zero.

The San Jose study also investigated the role of different street textures, which resulted in very different street dirt loadings. Although the accumulation and deposition rates are quite similar, the initial loading values (the permanent storage values) are very different, with greater amounts of street dirt trapped by the coarser (oil and screens) pavement. Street cleaning and rains are not able to remove this residual material. The early, uncorrected Sartor and Boyd accumulation rates that ignored the initial loading values were almost ten times the corrected values that had reasonable "initial loads."

continues next page

BOX 4-8 Continued

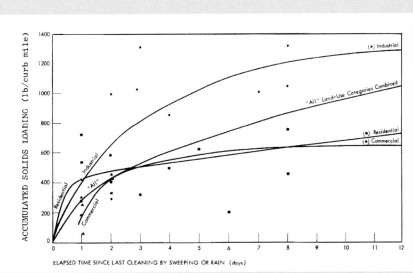

FIGURE 4-19   Accumulation curves developed during early street cleaning research. SOURCE: Sartor and Boyd (1972).

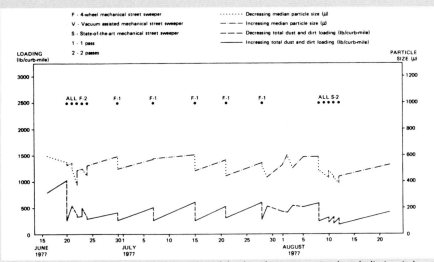

FIGURE 4-20 Street dirt accumulation and particle size changes on good asphalt streets in San Jose, California. SOURCE: Pitt (1979).

Finally, it was found that, at very long accumulation periods relative to the rain frequency, the wind losses (fugitive dust) may approximate the deposition rate, resulting in very little increases in loading. In Bellevue, Washington, with inter-event rain periods averaging about three days, steady loadings were observed after about one week (Pitt, 1985). However, in Castro Valley, California, the rain inter-event periods were much longer (ranging from about 20 to 100 days), and steady loadings were never observed (Pitt and Shawley, 1982).

Taking many studies into account (Sartor and Boyd 1972—corrected; Pitt, 1979, 1983, 1985; Pitt and Shawley, 1982; Pitt and Sutherland, 1982; Pitt and McLean, 1986), the most important factors affecting the initial loading and maximum loading values have been found to be street texture and street condition, and not land use. When data from many locations are studied, it is apparent that smooth streets have substantially less loadings at any accumulation period compared to rough streets for the same land use. Very long accumulation periods relative to the rain frequency result in high street dirt loadings. However, during these conditions the wind losses of street dirt (as fugitive dust) may approximate the deposition rate, resulting in relatively constant street dirt loadings.

Wash-off of Street Surface Pollutants

Wash-off of particulates from impervious surfaces is dependent on the available supply of particulates on the surface that can be removed by rains, the rain energy available to loosen the material, and the capacity of the runoff to transport the loosened material. Observations of particulate wash-off during controlled tests have resulted in empirical wash-off models. The earliest controlled street dirt wash-off experiments were conducted by Sartor and Boyd (1972) to estimate the percentage of the available particulates on the streets that would wash off during rains of different magnitudes. Sartor and Boyd fitted their data to an exponential curve, as shown in Figure 4-21 (accumulative wash-off curves for several particle sizes). The empirical equation that they developed, $N = N_o\, e^{-kR}$, is only sensitive to the total rain depth up to the time of interest and the initial street dirt loading.

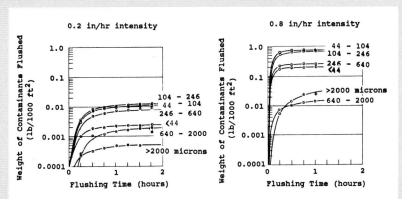

FIGURE 4-21   Street dirt wash-off during high-intensity rain tests.   SOURCE: Sartor and Boyd (1972).

continues next page

BOX 4-8 Continued

There are several problems with this approach. First, these figures did not show the total street dirt loading that was present before the wash-off tests. Most modelers have assumed that the asymptotic maximum shown was the total "before-rain" street dirt loading; that is, the N_o factor has been assumed to be the total initial street loading, when in fact it is only the portion of the total street load available for wash-off (the maximum asymptotic wash-off load observed during the wash-off tests). The actual total street dirt loadings were several times greater than the maximum wash-off amounts observed. STORM and SWMM now use an availability factor (A) for particulate residue as a calibration procedure in order to reduce the wash-off quantity for different rain intensities (Novotny and Chesters, 1981). Second, the proportionality constant, k, was found by Sartor and Boyd to be slightly dependent on street texture and condition, but was independent of rain intensity and particle size. The value of this constant is usually taken as 0.18/mm, assuming that 90 percent of the particulates will be washed from a paved surface in one hour during a 13 mm/h rain. However, Alley (1981) fitted this model to watershed outfall runoff data and found that the constant varied for different storms and pollutants for a single study area. Novotny examined "before" and "after" rain-event street particulate loading data using the Milwaukee NURP stormwater data (Bannerman et al., 1983) and found almost a three-fold difference between the proportionality constant value for fine (<45 μm) and medium-sized particles (100 to 250 μm). Jewell et al. (1980) also found large variations in outfall "fitted" values for different rains compared to the typical default value. They stressed the need to have local calibration data before using the exponential wash-off equation, as the default values can be very misleading. The exponential wash-off equation for impervious areas is justified, but wash-off coefficients for each pollutant would improve its accuracy. The current SWMM5 version discourages the use of accumulation and wash-off functions due to lack of data, and the misinterpretation of available data.

It turns out that particle dislodgement and transport characteristics at impervious areas can be directly measured using relatively simple wash-off tests. The Bellevue, Washington, urban runoff project (Pitt, 1985) included about 50 pairs of street dirt loading observations close to the beginnings and ends of rains to determine the differences in loadings that may have been caused by the rains. The observations were affected by rains falling directly on the streets, along with flows and particulates originating from non-street areas. When all the data were considered together, the net loading difference was about 10 to 13 g/curb-m removed, which amounted to a street dirt load reduction of about 15 percent. Large reductions in street dirt loadings for the small particles were observed during these Bellevue rains. Most of the weight of solid material in the runoff was concentrated in fine particle sizes (<63 μm). Very few wash-off particles greater than 1,000 μm were found; in fact, street dirt loadings increased for the largest sizes, presumably due to settled erosion materials. Urban runoff outfall particle size analyses in Bellevue (Pitt, 1985) resulted in a median particle size of about 50 μm; similar results were obtained in the Milwaukee NURP study (Bannerman et al., 1983). The results make sense because the rain energy needed to remove larger particles is much greater than for small particles.

In order to clarify street dirt wash-off, Pitt (1987) conducted numerous controlled wash-off tests on city streets in Toronto. The experimental factors examined included rain intensity, street texture, and street dirt loading. The differences between available and total street dirt loads were also related to the experimental factors. The runoff flow quantities were also carefully monitored to determine the magnitude of initial and total rain water losses on impervious surfaces. The test setup was designed and tested to best represent actual rainfall conditions, such as rain intensities (3 mm/h) and peak rain intensities (12 mm/h). The kinetic energies of the "rains" during these tests were therefore comparable to actual rains under investigation. Figure 4-22 shows the asymptotic wash-off values observed in the tests, along with the measured total street dirt loadings. The maximum asymptotic values are the "available" street dirt loadings (N_o). As can be seen, the measured

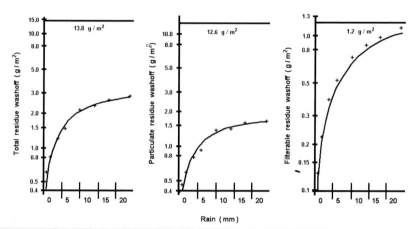

Rain (mm)

FIGURE 4-22 Wash-off plots for high rain intensity, dirty street, and smooth street test, showing the total street dirt loading. SOURCE: Pitt (1987).

total loadings are several times larger than these "available" loading values. For example, the asymptotic available total solids value for the high-intensity rain–dirty street–smooth street test was about 3 g/m^2 while the total load on the street for this test was about 14 g/ m^2, or about five times the available load. The differences between available and total loadings for the other tests were even greater, with the total loads typically about ten times greater than the available loads. The total loading and available loading values for dissolved solids were quite close, indicating almost complete wash-off of the very small particles.

The availability factor (the ratio of the available loading, N_0, to the total loading) depended on the rain intensity and the street roughness, such that wash-off was more efficient for the higher rain energy and smoother pavement tests. The worst case was for a low rain intensity and rough street, where only about 4.5 percent of the street dirt would be washed from the pavement. In contrast, the high rain intensities on the smooth streets were more than four times more efficient in removing street dirt (20 percent removal).

A final important consideration in calculating wash-off of street dirt during rains is the carrying capacity of the flowing water to transport sediment. If the calculated wash-off is greater than the carrying capacity (such as would occur for relatively heavy street dirt loads and low to moderate rain intensities), then the carrying capacity is limiting. For high rain intensities, the carrying capacity is likely sufficient to transport most or all of the wash-off material. Figure 4-23 shows the maximum wash-off amounts (g/m^2) for the different tests conducted on smooth streets plotted against the rain intensity (mm/h) used for the tests (data from Sartor and Boyd, 1972, and Pitt, 1987). Wash-off limitations for rough streets would be more restrictive.

continues next page

BOX 4-8 Continued

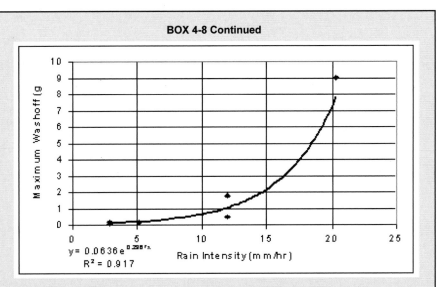

FIGURE 4-23 Maximum wash-off capacity for smooth streets (based on measurements of Sartor and Boyd, 1972; Pitt, 1987). If the predicted wash-off, using the previous "standard" wash-off equations, is smaller than the values shown in this figure, then those values can be used directly. However, if the predicted wash-off is greater than the values shown in this figure, then the values in the figure should be used.

Accumulation and Wash-off Summary

This discussion summarized street particulate wash-off observations obtained during special wash-off tests, along with associated street dirt accumulation measurements. The objectives of these tests were to identify the significant rain and street factors affecting particulate wash-off and to develop appropriate wash-off models. The controlled wash-off experiments identified important relationships between "available" and "total" particulate loadings and the significant effects of the test variables on the wash-off model parameters. Past modeling efforts have typically ignored or misused this relationship to inaccurately predict the importance of street particulate wash-off. The available loadings were almost completely washed off streets during rains of about 25 mm (as previously assumed). However, the fraction of the total loading that was available was at most only 20 percent of the total loading, and averaged only 10 percent, with resultant actual wash-offs of only about 9 percent of the total loadings.

In many model applications, total initial loading values (as usually measured during field studies) are used in conjunction with model parameters as the available loadings, resulting in predicted wash-off values that are many times larger than observed. This has the effect of incorrectly assuming greater pollutant contributions originating from streets and less from other areas during rains. This in turn results in inaccurate estimates of the effectiveness of different source area urban runoff controls. Although streets can be important sources of runoff and stormwater pollutants, their significance varies greatly depending on the land use and rainfall pattern. They are much more important sources in areas having relatively mild rains (e.g., the Pacific Northwest), where contaminants from other potential sources are not effectively transported to the storm drainage system.

source of stormwater in seasonally wet conditions (Seattle), while Cuo et al. (2008) have explored the modification of DHSVM to include detention SCMs. Xiao et al. (2007) explicitly integrated and evaluated parcel scale SCM design and efficiency into their model. Wang et al. (2008) integrated a canopy interception model with a semi-distributed subsurface moisture scheme (TOP-MODEL) to evaluate the effectiveness of urban tree canopy interception on stormwater production, utilizing a detailed spatial dataset of urban tree cover. Band et al. (2001) and Law (2003) coupled a water-, carbon-, and nitrogen-cycling model to a distributed water routing system modified from DHSVM to simulate nitrogen cycling and export in a high-spatial-resolution representation of forested and suburban catchments. While these models have the potential to directly link stormwater generation with specific dischargers, the challenge of scaling to larger watersheds remains. SMDR (Easton et al., 2007) has recently been used to integrate rural and urban stormwater production, including dissolved phosphorus source and transport in New York State.

Alternatives to mass budget-based models include fully statistical approaches such as simple regressions based on watershed land use and population (e.g., Boyer et al., 2002); nonlinear regression using detailed watershed spatial data and observed loads to estimate retention parameters and loading of nutrients, sediment, and other pollutants (e.g., Smith et al., 1997; Brakebill and Preston, 1999; Schwarz et al., 2006); and Bayesian chain models (e.g., Reckhow and Chapra, 1999; Borsuk et al., 2001). These models have the advantage of being data-based, and therefore capable of assimilating observations as they become available to update water quality probabilities, but also lack a process basis that might support management intervention. A major debate exists within the literature as to the relative advantages of detailed process-based models that may not have inadequate information for parameterization, and the more empirical, data-based approaches.

Limitations in Extending Stormwater Models to Biological Impacts

The mass budget approach may be successful in developing the physical and chemical characteristics of the receiving waterbody in terms of the flow (or stage) duration curve, the distribution of concentrations over time, and the integrated pollutant storage and flux (load) terms. However, the biological status of the waterbody requires a link between the physical and chemical conditions, primary productivity, and trophic system interactions. Progressing from aquatic ecosystem productivity to trophic systems includes increasingly complex ecological processes such as competition, herbivory, predation, and migration. To date, mechanistic linkage between flow path hydraulics, biogeochemistry, and the ecological structure of the aquatic environment has not been developed. Instead, habitat suitability for different communities is identified through empirical sampling and analysis, with the implicit assumption that, as relative

TABLE 4-7 Example Mathematical Models That Have Been or Can Be Used in Stormwater Modeling

Model	Common Use	Typical Scale	Complexity	Data Requirements	Ground-water	SCM	Reference
Rational Method	Urban hydraulic design—peak flow	Small	Simple	Land cover, rainfall intensity, T_c	None	None	Standard hydrology text
Simple Method	Urban annual runoff, loads	Small to medium	Simple	Impervious surface cover, land use, annual rainfall	None	None	http://www.stormwa tercenter.net/monit oring%20and%20a ssessment/simple %20meth/simple.ht m
TR-20 TR-55	Rural/urban runoff production for simple stormwater models, hydraulic design	Small to medium	Simple to medium	Land use, soil texture, T_c	None	Pond sizing for hydraulic benefits and others through CN modification	http://www.wsi.nrcs .usda.gov/products/ W2Q/H&H/Tools_ Models
GWLF	Rural/urban runoff, pollutant loading	Medium to watershed	Simple to medium	Land use, soil texture, precipitation time series	Simple linear reservoir	Runoff reduction with CN modification	Haith and Shoemaker (1987) http://www.avgwlf.p su.edu/overview.ht m
P8	Urban runoff, pollutant loading	Small to large	Simple to medium	Land use, soil texture, precipitation time series, SCM type and sizing	Simple linear reservoir	Runoff reduction with CN modification, ponds (evaluation and sizing), infiltration, street cleaning	Palmstrom and Walker (1990) http://www.wwwalk er.net/p8/
MUSIC	Urban runoff, pollutant loading, hydraulic design, simple receiving water	Small to large	Medium to complex	Land use, soil texture, precipitation/PET? time series, drainage system details, SCM type and sizing	Simple linear reservoir	Comprehensive evaluation of SCM systems	Wong (2000) (proprietary) http://www.toolkit.n et.au/cgi-bin/WebObjects/too lkit.woa/wa/product Details?productID= 1000000

Model	Purpose	Scale	Complexity	Input data	Simple linear reservoir?	SCM	Reference/URL
SWMM	Urban runoff, pollutant loading, hydraulic design	Small to large	Medium to complex	Land use, soil texture, meteorological time series, drainage system details, SCM type and sizing	Simple linear reservoir?	Infiltration practices, ponds, street cleaning	http://www.epa.gov/ednnrmrl/models/swmm
PCSWMM	Same as above	Same as above	Same as above	Same as above	Same as above	Enhanced SCM compared to SWMM	(proprietary) http://www.computationalhydraulics.com/Software/PCSWMM.NET
WinSLAMM	Urban runoff, pollutant loads	Small to large	Intermediate	Land cover, land use, development characteristics, soil texture, compaction, rainfall event time series, monthly PET, monthly water evaporation, SCM type and sizing	Mounding under infiltration controls	Comprehensive evaluation of SCM systems	(proprietary) http://www.winslamm.com/prod01.htm
SWAT	Rural runoff, loading	Medium to watershed	Intermediate	Land cover/land use, soil texture, precipitation, temperature, humidity, solar radiation time or PET series	Simple subbasin reservoir	Impoundments, agricultural conservation practices, nutrient management, buffers	http://www.epa.gov/waterscience/BASINS/bsnsdocs.html#swat
HSPF	Comprehensive watershed evaluation, receiving water dynamics	Medium to watershed	Complex	Land cover/land use, soil texture, precipitation, temperature, humidity, solar radiation or PET time series	Subbasin reservoir	Infiltration, ponds	Bicknell et al. (2005) http://www.epa.gov/ceampubl/swater/hspf/index.htm http://www.epa.gov/waterscience/BASINS/bsnsdocs.html#hspf
WWHM	HSPF engine with regional modifications,	Puget Sound	Complex	Same as above	Same as above	Enhanced infiltration, ponds (from HSPF)	http://www.ecy.wa.gov/programs/wq/stormwater/wwhm_training/index.html.
CBWM	HSPF engine with regional modifications,	Chesapeake Bay Watershed	Complex	Same as above	Same as above	Enhanced infiltration, ponds (from HSPF)	http://www.chesapeakebay.net/phase5.htm

Note: CN, curve number

habitat suitability changes, transitions will occur between species or assemblages. These methods may work well at the base of the trophic system (algae, phytoplankton) and for specific conditions such as DO limitations on fish communities, but the impacts of low to moderate concentrations of pollutants on aquatic ecosystems may still be poorly understood. A critical assumption in these and similar models (e.g., ecological community change resulting from physical changes to the watershed or climate) is the substitution of space for time. More detailed understanding of the mechanisms leading to a shift in ecological communities and interactions with the physical environment is necessary to develop models of transient change, stability of the shifts, and feedback to the biophysical environment.

Given these limitations, it should be noted that statistical databases on species tolerance to a range of aquatic conditions have been compiled that will allow the development of habitat suitability mapping as a mechanism for (1) targeting ecosystem restoration, (2) determining vulnerable sites (for use in application of the Endangered Species Act), and (3) assessing aquatic ecosystem impairment and "best use" relative to reference sites.

***

Stormwater models have been developed to meet a range of objectives, including small-scale hydraulic design (e.g., siting and sizing a detention pond), estimation of potential contributions of stormwater pollutants from different land covers and locations using empirically generated EMC, and large watershed hydrology and gross pollutant loading. The ability to associate a given discharger with a particular waterbody impairment is limited by the scale and complexity of watersheds (i.e., there maybe multiple discharge interactions); by the ability of a model to accurately reproduce the distribution function of discharge events and their cumulative impacts (as opposed to focusing only on design storms of specific return periods); and by the availability of monitoring data of sufficient number and design to characterize basic processes (e.g., build-up/wash-off), to parameterize the models, and to validate model predictions.

In smaller urban catchments with few dominant dischargers and significant impervious area, current modeling capabilities may be sufficient to associate the cumulative impact of discharge to waterbody impairment. However, many impaired waterbodies have larger, more heterogeneous stormwater sources, with impacts that are complex functions of current and past conditions. The level of sampling that would be necessary to support linked model calibration and verification using current measurement technologies is both time-consuming and expensive. In order to develop a more consistent capability to support stormwater permitting needs, there should be increased investment in improving model paradigms, especially the practice and methods of model linkage as described above, and in stormwater monitoring. The latter may require investment in a new generation of sensors that can sample at temporal resolutions that can adjust to characterize low flow and the dynamics of storm flow, but are sufficiently inexpensive and autonomous to be deployed in multiple locations from distributed sources to receiving waterbodies of interest. Finally, as urban areas extend to encompass progressively lower-density development, the interactions of surface water and groundwater become more critical to the cumulative impact of stormwater on impaired waterbodies.

EPA needs to ensure continuous support and development of their water quality models and spatial data infrastructure. Beyond this, a set of distributed watershed models has been developed that can resolve the location and position of parcels within hydrologic flow fields; these are being modified for use as urban stormwater models. These models avoid the pitfalls of lumping, but they require much greater volumes of spatial data, provided by current remote sensing technology (e.g., lidar, airborne digital optical and infrared sensors) as well as the emerging set of in-stream sensor systems. While these methods are not yet operational or widespread, they should be further investigated and tested for their capabilities to support stormwater management.

CONCLUSIONS AND RECOMMENDATIONS

This chapter addresses what might be the two weakest areas of the stormwater program—monitoring and modeling of stormwater. The MS4 and particularly the industrial stormwater monitoring programs suffer from (1) a paucity of data, (2) inconsistent sampling techniques, (3) a lack of analyses of available data and guidance on how permittees should be using the data to improve stormwater management decisions, and (4) requirements that are difficult to relate to the compliance of individual dischargers. The current state of stormwater modeling is similarly limited. Stormwater modeling has not evolved enough to consistently say whether a particular discharger can be linked to a specific waterbody impairment, although there are many correlative studies showing how parameters co-vary in important but complex and poorly understood ways (see Chapter 3). Some quantitative predictions can be made, particularly those that are based on well-supported causal relationships of a variable that responds to changes in a relatively simple driver (e.g., modeling how a runoff hydrograph or pollutant loading change in response to increased impervious land cover). However, in almost all cases, the uncertainty in the modeling and the data, the scale of the problems, and the presence of multiple stressors in a watershed make it difficult to assign to any given source a specific contribution to water quality impairment. More detailed conclusions and recommendations about monitoring and modeling are given below.

Because of a ten-year effort to collect and analyze monitoring data from MS4s nationwide, the quality of stormwater from urbanized areas is well characterized. These results come from many thousands of storm events, systematically compiled and widely accessible; they form a robust dataset of utility to theoreticians and practitioners alike. These data make it possible to accurately estimate the EMC of many pollutants. Additional data are available from other stormwater permit holders that were not originally included in the database and from ongoing projects, and these should be acquired to augment the database and improve its value in stormwater management decision-making.

Industry should monitor the quality of stormwater discharges from certain critical industrial sectors in a more sophisticated manner, so that permitting authorities can better establish benchmarks and technology-based effluent guidelines. Many of the benchmark monitoring requirements and effluent guidelines for certain industrial subsectors are based on inaccurate and old information. Furthermore, there has been no nationwide compilation and analysis of industrial benchmark data, as has occurred for MS4 monitoring data, to better understand typical stormwater concentrations of pollutants from various industries. The absence of accurate benchmarks and effluent guidelines for critical industrial sectors discharging stormwater may explain the lack of enforcement by permitting authorities, as compared to the vigorous enforcement within the wastewater discharge program.

Industrial monitoring should be targeted to those sites having the greatest risk associated with their stormwater discharges. Many industrial sites have no or limited exposure to runoff and should not be required to undertake extensive monitoring. Visual inspections should be made, and basic controls should be implemented at these areas. Medium-risk industrial sites should conduct monitoring so that a sufficient number of storms are measured over the life of the permit for comparison to regional benchmarks. Again, visual inspections and basic controls are needed for these sites, along with specialized controls to minimize discharges of the critical pollutants. Stormwater from high-risk industrial sites needs to be continuously monitored, similar to current point source monitoring practices. The use of a regionally calibrated stormwater model and random monitoring of the lower-risk areas will likely require additional monitoring.

Continuous, flow-weighted sampling methods should replace the traditional collection of stormwater data using grab samples. Data obtained from too few grab samples are highly variable, particularly for industrial monitoring programs, and subject to greater uncertainly because of experimenter error and poor data-collection practices. In order to use stormwater data for decision making in a scientifically defensible fashion, grab sampling should be abandoned as a credible stormwater sampling approach for virtually all applications. It should be replaced by more accurate and frequent continuous sampling methods that are flow weighted. Flow-weighted composite monitoring should continue for the duration of the rain event. Emerging sensor systems that provide high temporal resolution and real-time estimates for specific pollutants should be further investigated, with the aim of providing lower costs and more extensive monitoring systems to sample both streamflow and constituent loads.

Flow monitoring and on-site rainfall monitoring need to be included as part of stormwater characterization monitoring. The additional information associated with flow and rainfall data greatly enhance the usefulness of the much more expensive water quality monitoring. Flow monitoring should also be correctly conducted, with adequate verification and correct base-flow subtraction methods applied. Using regional rainfall data from locations distant from the monitoring location is likely to be a major source of error when rainfall factors are being investigated. The measurement, quality assurance, and maintenance of long-term precipitation records are both vital and nontrivial to stormwater management.

Whether a first flush of contaminants occurs at the start of a rainfall event depends on the intensity of rainfall, the land use, and the specific pollutant. First flushes are more common for smaller sites with greater imperviousness and thus tend to be associated with more intense land uses such as commercial areas. Even though a site may have a first flush of a constituent of concern, it is still important that any SCM be designed to treat as much of the runoff from the site as possible. In many situations, elevated discharges may occur later in an event associated with delayed periods of peak rainfall intensity.

Stormwater runoff in arid and semi-arid climates demonstrates a seasonal first-flush effect (i.e., the dirtiest storms are the first storms of the season). In these cases, it is important that SCMs are able to adequately handle these flows. As an example, early spring rains mixed with snowmelt may occur during periods when wet detention ponds are still frozen, hindering their performance. The first fall rains in the southwestern regions of the United States may occur after extended periods of dry weather. Some SCMs, such as street cleaning targeting leaf removal, may be more effective before these rains than at other times of the year.

Watershed models are useful tools for predicting downstream impacts from urbanization and designing mitigation to reduce those impacts, but they are incomplete in scope and typically do not offer definitive causal links between polluted discharges and downstream degradation. Every model simulates only a subset of the multiple interconnections between physical, chemical, and biological processes found in any watershed, and they all use a grossly simplified representation of the true spatial and temporal variability of a watershed. To speak of a "comprehensive watershed model" is thus an oxymoron, because the science of stormwater is not sufficiently far advanced to determine causality between all sources, resulting stressors, and their physical, chemical, and biological responses. Thus, it is not yet possible to create a protocol that mechanistically links stormwater dischargers to the quality of receiving waters. The utility of models with more modest goals, however, can still be high—as long as the questions being addressed by the model are in fact relevant and important to the functioning of the watershed to which that model is being applied, and sufficient data are available to calibrate the model for the processes included therein.

EPA needs to ensure that the modeling and monitoring capabilities of the nation are continued and enhanced to avoid losing momentum in understanding and eliminating stormwater pollutant discharges. There is a need to extend, develop, and support current modeling capabilities, emphasizing (1) the impacts of flow energy, sediment transport, contaminated sediment, and acute and chronic toxicity on biological systems in receiving waterbodies; (2) more mechanistic representation (physical, chemical, biological) of SCMs; and (3) coupling between a set of functionally specific models to promote the linkage of source, transport and transformation, and receiving water impacts of stormwater discharges. Stormwater models have typically not incorporated interactions with groundwater and have treated infiltration and recharge of groundwater as a loss term with minimal consideration of groundwater contamination or transport to receiving waterbodies. Emerging distributed modeling paradigms that simulate interactions of surface and subsurface flowpaths provide promising tools that should be further developed and tested for applications in stormwater analysis.

REFERENCES

Adams, B., and F. Papa. 2000. Urban Stormwater Management Planning with Analytical Probabilistic Methods. New York: John Wiley and Sons.

Alexander, R. B., R. A. Smith, and G. E. Schwarz. 2000. Effect of stream channel size on the delivery of nitrogen to the Gulf of Mexico. Nature 403:758–761.

Alley, W. M. 1981. Estimation of impervious-area washoff parameters. Water Resources Research 17(4):1161–1166.

Arnold, J. G., R. Srinivasan, R. S. Muttiah, and J. R. Williams. 1998. Large area hydrologic modeling and assessment Part I: Model development. Journal of the American Water Resources Association 34(1):73e89.

Band, L. E., P. Patterson, R. R. Nemani, and S. W. Running. 1993. Forest ecosystem processes at the watershed scale: 2. Adding hillslope hydrology. Agricultural and Forest Meteorology 63:93–126.

Band, L. E., C. L. Tague, P. Groffman, and K. Belt. 2001. Forest ecosystem processes at the watershed scale: hydrological and ecological controls of nitrogen export. Hydrological Processes 15:2013–2028.

Bannerman, R., K. M. Baun, P. E. Bohn, and D. A. Graczyk. 1983. Evaluation of Urban Nonpoint Source Pollution Management in Milwaukee County, Wisconsin. PB 84-114164. Chicago: EPA.

Bannerman, R. T., D. W. Owens, R. B. Dodds, and N. J. Hornewer. 1993. Sources of pollutants in Wisconsin stormwater. Water Science and Technology 28(3–5):241–259.

Batroney, T. 2008. The Implications of the First Flush Phenomenon on Infiltration BMP Design. Master's Thesis. Water Resources and Environmental Engineering, Villanova University, Villanova, PA, May.

Beven, K. J. 2000. On the future of distributed modelling in hydrology. Hydrological Processes 14:3183–3184.

Beven, K. J. 2001. Dalton Medal Lecture: How far can we go in distributed hydrological modelling? Hydrology and Earth System Sciences 5(1):1–12.

Bicknell, B. R., J. C. Imhoff, J. L. Kittle, Jr., A. S. Donigian, Jr., and R. C. Johanson. 1997. Hydrological Simulation Program—Fortran, User's manual for version 11. EPA/600/R-97/080. Athens, GA: EPA National Exposure Research Laboratory.

Bicknell, B. R., J. C. Imhoff, J. L. Kittle, Jr., T. H. Jobes, and A. S. Donigian, Jr. 2005. HSPF Version 12.2 User's Manual. AQUA TERRA Consultants, Mountain View, CA. In cooperation with Office of Surface Water, Water Resources Discipline, U.S. Geological Survey, Reston, VA, and National Exposure Research Laboratory Office of Research and Development, EPA, Athens, GA.

Borsuk, M. E., D. Higdon, C. A. Stow, and K. H. Reckhow. 2001. A Bayesian hierarchical model to predict benthic oxygen demand from organic matter loading in estuaries and coastal zones. Ecological Modeling 143:165–181.

Boyer, E. W., C. L. Goodale, N. A. Jaworski, and R.W. Howarth. 2002. Anthropogenic nitrogen sources and relationships to riverine nitrogen export in the northeastern USA. Biogeochemistry 57:137–169.

Boyer, E. W., R. B. Alexander, W. J. Parton, C. Li, K. Butterbach-Bahl, S. D. Donner, R. W. Skaggs, and S. J. Del Grosso. 2006. Modeling denitrification in terrestrial and aquatic ecosystems at regional scales. Ecological Applications 16:2123–2142.

Brakebill, J. W., and S. E. Preston. 1999. Digital Data Used to Relate Nutrient Inputs to Water Quality in the Chesapeake Bay Watershed, Version 1.0. U.S. Geological Survey Water Open-File Report 99-60.

Brown, T., W. Burd, J. Lewis, and G. Chang. 1995. Methods and procedures in stormwater data collection. *In:* Stormwater NPDES Related Monitoring Needs. H. C. Torno (ed.). Reston, VA: Engineering Foundation and ASCE.

Burgess, S. J., M. S. Wigmosta, and J. M. Meena. 1998. Hydrological effects of land-use change in a zero-order catchment. Journal of Hydrological Engineering ASCE 3:86–97.

Burton, G. A., Jr., and R. Pitt. 2002. Stormwater Effects Handbook: A Tool Box for Watershed Managers, Scientists, and Engineers. Boca Raton, FL: CRC Press, 911 pp.

Chiew, F. H. S., and T. A. McMahon. 1997. Modelling daily runoff and pollutant load from urban catchments. Water (AWWA Journal) 24:16–17.

Chow, V. T., D. R. Maidment, and L. W. Mays. 1988. Applied Hydrology. McGraw Hill.

Clark, S. E., C. Y. S. Siu, R. E. Pitt, C. D. Roenning, and D. P. Treese. 2008. Peristaltic pump autosamplers for solids measurement in stormwater runoff.

Water Environment Research 80. doi:10.2175/106143008X325737.

Cross, L. M., and L. D. Duke. 2008. Regulating industrial stormwater: state permits, municipal implementation, and a protocol for prioritization. Journal of the American Water Resources Association 44(1):86–106.

Cuo, L., D. P. Lettenmaier, B. V. Mattheussen, P. Storck, and M. Wiley. 2008. Hydrologic prediction for urban watersheds with the Distributed Hydrology-Soil-Vegetation-Model. Hydrological Processes, DOI: 10.1002/hyp.7023

Deletic, A. 1998. The first flush load of urban surface runoff. Water Research 32(8):2462–2470.

Easton, Z. M., P. Gérard-Marchant, M. T. Walter, A. M. Petrovic, and T. S. Steenhuis. 2007. Hydrologic assessment of an urban variable source watershed in the northeast United States. Water Resources Research 43:W03413, doi:10.1029/2006WR005076.

EPA (U.S. Environmental Protection Agency). 1983. Results of the Nationwide Urban Runoff Program. PB 84-185552. Washington, DC: Water Planning Division.

Gibbons, J., and S. Chakraborti. 2003. Nonparametric Statistical Inference, 4th edition. New York: Marcel Dekker, 645 pp.

Haith, D. A., and L. L. Shoemaker. 1987. Generalized watershed loading functions for stream flow nutrients. Water Resources Bulletin 23(3):471–478.

Horwatich, J., R. Bannerman, and R. Pearson. 2008. Effectiveness of Hydrodynamic Settling Device and a Stormwater Filtration Device in Milwaukee, Wisconsin. U.S. Geological Survey Investigative Report. Middleton, WI: USGS.

Jewell, T. K., D. D. Adrian, and D. W. Hosmer. 1980. Analysis of stormwater pollutant washoff estimation techniques. International Symposium on Urban Storm Runoff. University of Kentucky, Lexington, KY.

Kirsch, K. J. 2000. Predicting Sediment and Phosphorus Loads in the Rock River Basin Using SWAT. Presented at the ASAE International Meeting, Paper 002175. July 9–12. St. Joseph, MI: ASAE.

Law, N. L. 2003. Sources and Pathways of Nitrate in Urbanizing Watersheds. Ph.D. Dissertation Proposal, Department of Geography, University of North Carolina at Chapel Hill.

Lee, H., X. Swamikannu, D. Radulescu, S. Kim, and M. K. Stenstrom. 2007. Design of stormwater monitoring programs. Journal of Water Research, doi:10.1016/j.watres.2007.05.016.

Maestre, A., and R. Pitt. 2005. The National Stormwater Quality Database, Version 1.1. A Compilation and Analysis of NPDES Stormwater Monitoring Information. Washington, DC: EPA Office of Water.

Maestre, A., and R. Pitt. 2006. Identification of significant factors affecting stormwater quality using the National Stormwater Quality Database. Pp. 287–326 *In:* Stormwater and Urban Water Systems Modeling, Monograph 14. W. James, K. N. Irvine, E. A. McBean, and R. E. Pitt (eds.). Guelph, Ontario: CHI.

Maestre, A., R. E. Pitt, and D. Williamson. 2004. Nonparametric statistical tests comparing first flush with composite samples from the NPDES Phase 1 municipal stormwater monitoring data. Stormwater and urban water systems modeling. Pp. 317–338 *In:* Models and Applications to Urban Water Systems, Vol. 12. W. James (ed.). Guelph, Ontario: CHI.

Maestre, A., R. Pitt, S. R. Durrans, and S. Chakraborti. 2005. Stormwater quality descriptions using the three parameter lognormal distribution. Stormwater and urban water systems modeling. *In:* Models and More for Urban Water Systems, Monograph 13. W. James, K. N. Irvine, E. A. McBean, and R. E. Pitt (eds.). Guelph, Ontario: CHI.

Mahler, B. J., P. C. VanMetre, T. J. Bashara, J. T. Wilson, and D. A. Johns. 2005. Parking lot sealcoat: An unrecognized source of urban polycyclic aromatic hydrocarbons. Environmental Science and Technology 39:5560–5566.

Maxwell, R. H., and M. Miller. 2005. Development of a coupled land surface and groundwater model. Journal Hydrometeorology 6:233–247.

McClain, M. E., E. W. Boyer, C. L. Dent, S. E. Gergel, N. B. Grimm, P. M. Groffman, S. C. Hart, J. W. Harvey, C. A. Johnston, E. Mayorga, W. H. McDowell, and G. Pinay. 2003. Biogeochemical hot spots and hot moments at the interface of terrestrial and aquatic ecosystems . Ecosystems 6:301–312.

MDE (Maryland Department of the Environment). 2000. 2000 Maryland Stormwater Design Manual, Volumes I & II. Prepared by the Center for Watershed Protection and the Maryland Department of the Environment, Water Management Administration, Baltimore, MD.

Novotny, V., and G. Chesters. 1981. Handbook of Nonpoint Pollution Sources and Management. New York: Van Nostrand Reinhold.

Oreskes, N., Shrader-Frechette K., Belitz, K. 1994. Verification, Validation, and Confirmation of Numerical Models in the Earth Sciences. Science, New Series, 263(5147):641-646.

PaDEP (Pennsylvania Department of Environmental Protection). 2006. Pennsylvania Stormwater Best Management Practices Manual. Harrisburg, PA: Pennsylvania Department of Environmental Protection.

Palmstrom, N., and W. Walker. 1990. The P8 Urban Catchment Model for Evaluating Nonpoint Source Controls at the Local Level. Enhancing States' Lake Management Programs, USEPA.

Pitt, R. 1979. Demonstration of Nonpoint Pollution Abatement Through Improved Street Cleaning Practices. EPA-600/2-79-161 Cincinnati, OH: EPA Office of Research and Development.

Pitt, R. 1983. Urban Bacteria Sources and Control in the Lower Rideau River Watershed. Ottawa, Ontario: Ontario Ministry of the Environment.

Pitt, R. 1985. Characterizing and Controlling Urban Runoff through Street and Sewerage Cleaning. EPA/600/S2-85/038. PB 85-186500. Cincinnati, OH: EPA. Storm and Combined Sewer Program. Risk Reduction Engineering Laboratory.

Pitt, R. 1986. The incorporation of urban runoff controls in the Wisconsin Priority Watershed Program. Pp. 290–313 *In:* Advanced Topics in Urban Runoff Research. B. Urbonas and L. A. Roesner (eds.). New York: Engineering Foundation and ASCE.

Pitt, R. 1987. Small Storm Urban Flow and Particulate Washoff Contributions to Outfall Discharges. Ph.D. Dissertation. Department of Civil and Environmental Engineering, University of Wisconsin–Madison.

Pitt, R., and J. McLean. 1986. Toronto Area Watershed Management Strategy Study. Humber River Pilot Watershed Project. Toronto, Ontario: Ontario Ministry of the Environment.

Pitt, R., and G. Shawley. 1982. A Demonstration of Nonpoint Source Pollution Management on Castro Valley Creek. Alameda County Flood Control and Water Conservation District (Hayward, CA) for the Nationwide Urban Runoff Program. Washington, DC: EPA, Water Planning Division.

Pitt, R., and R. Sutherland. 1982. Washoe County Urban Stormwater Management Program. Reno, NV: Washoe Council of Governments.

Pitt, R., and J. Voorhees. 1995. Source loading and management model (SLAMM). Pp. 225–243 *In:* Seminar Publication: National Conference on Urban Runoff Management: Enhancing Urban Watershed Management at the Local, County, and State Levels. March 30–April 2, 1993. EPA/625/R-95/003. Cincinnati, OH: EPA Center for Environmental Research Information.

Pitt, R., and J. Voorhees. 2002. SLAMM, the Source Loading and Management Model. Pp. 103–139 *In:* Wet-Weather Flow in the Urban Watershed. R. Field and D. Sullivan (eds.). Boca Raton, FL: CRC Press.

Pitt, R., A. Maestre, H. Hyche, and N. Togawa. 2008. The updated National Stormwater Quality Database (NSQD), Version 3. Conference CD. 2008 Water Environment Federation Technical Exposition and Conference, Chicago, IL.

Pitt, R. S., Clark, J. Lantrip, and J. Day. 1998. Telecommunication Manhole Water and Sediment Study; Vol. 1: Evaluation of Field Test Kits, 483 pp.; Vol. 2: Water and Sediment Characteristics, 1290 pp.; Vol. 3: Discharge Evaluation Report, 218 pp.; Vol. 4: Treatment of Pumped Water, 104 pp. Special Report SR-3841. Morriston, NJ: Bellcore, Inc.

Pitt, R., A. Maestre, and R. Morquecho. 2003. Evaluation of NPDES Phase I municipal stormwater monitoring data. *In:* National Conference on Urban Stormwater: Enhancing the Programs at the Local Level. EPA/625/R-03/003.

Pitt, R. E., A. Maestre, R. Morquecho, and D. Williamson. 2004. Collection and examination of a municipal separate storm sewer system database. Stormwater and Urban Water Systems Modeling. Pp. 257–294 *In:* Models and Applications to Urban Water Systems, Vol. 12. W. James (eds.). Guelph, Ontario: CHI.

Pitt, R., R. Bannerman, S. Clark, and D. Williamson. 2005a. Sources of pollutants in urban areas (Part 1)—Older monitoring projects. Pp. 465–484 and

507–530 *In:* Effective Modeling of Urban Water Systems, Monograph 13. W. James, K. N. Irvine, E. A. McBean, and R. E. Pitt (eds.). Guelph, Ontario: CHI.

Pitt, R., R. Bannerman, S. Clark, and D. Williamson. 2005b. Sources of pollutants in urban areas (Part 2)—Recent sheetflow monitoring results. Pp. 485–530 *In:* Effective Modeling of Urban Water Systems, Monograph 13. W. James, K. N. Irvine, E. A. McBean, and R. E. Pitt (eds.). Guelph, Ontario: CHI.

Qu, Y., and C. J. Duffy. 2007. A semidiscrete finite volume formulation for multiprocess watershed simulation. Water Resources Research 43:W08419, doi:10.1029/2006WR005752.

Rallison, R. K. 1980. Origin and evolution of the SCS runoff equation. Pp. 912–924 *In:* Proceedings of Symposium on Watershed Management, Boise, Idaho. New York: American Society of Civil Engineers.

Ramaswami, A., J. B. Milford, and M. J. Small. 2005. Integrated Environmental Modeling: Pollutant Transport, Fate, and Risk in the Environment. New York: John Wiley and Sons.

Reckhow, K. H., and S. C. Chapra. 1999. Modeling excessive nutrient loading in the environment. Environmental Pollution 100:197–207.

Roa-Espinosa, A., and R. Bannerman. 1994. Monitoring BMP effectiveness at industrial sites. Pp. 467–486 *In:* Proceedings of Engineering Foundation Conference, Stormwater NPDES Related Monitoring Needs. H. C. Torno (ed.).

Sartor, J., and G. Boyd. 1972. Water Pollution Aspects of Street Surface Contaminants. EPA-R2-72-081.

Schnoor, J. L. 1996. Environmental Modeling: Fate and Transport of Pollutants in Water, Air, and Soil. New York: John Wiley and Sons.

Schwarz, G. E., A. B. Hoos, R. B. Alexander, and R. A. Smith. 2006. The SPARROW surface water-quality model: theory, application, and user documentation. U.S. Geological Survey Techniques and Methods Report, Book 6, Chapter B3.

Selbig, W. R., and R. T. Bannerman. 2007. Evaluation of Street Sweeping as a Stormwater-Quality-Management Tool in Three Residential Basins in Madison, Wisconsin. USGS Scientific Investigations Report 2007-5156, 103 pp.

Shaver, E., R. Horner, J. Skupien, C. May, and G. Ridley. 2007. Fundamentals of Urban Runoff Management: Technical and Institutional Issues, 2nd edition. Madison, WI: EPA and North American Lake Management Society, 237 pp.

Smith, R. A., G. E. Schwarz, and R. B. Alexander. 1997. Regional interpretation of water-quality monitoring data. Water Resources Research 33(12):2781–2798.

Soeur, C., J. Hubka, and G. Chang. 1995. Methods for assessing urban storm water pollution. Pp. 558 *In:* Stormwater NPDES Related Monitoring Needs. H. C. Torno (ed.). Reston, VA: Engineering Foundation and

ASCE.

Stenstrom, M. 2007. The Future of Industrial Stormwater Monitoring or, What Are Some Pitfalls and Ways We Can Improve Monitoring, for Industries and Others as Well. Presentation to the NRC Committee on Reducing Stormwater Discharge Contributions to Water Pollution. December 17, Irvine, CA.

Stenstrom, M. K., and H. Lee. 2005. Final Report. Industrial Stormwater Monitoring Program. Existing Statewide Permit Utility and Proposed Modifications.

Tague, C. L., and L. E. Band. 2004. RHESSys: Regional Hydro-Ecologic Simulation System—An object-oriented approach to spatially distributed modeling of carbon, water, and nutrient cycling. Earth Interactions 8:1–42.

TetraTech. 2003. B. Everett Jordan Lake TMDL Watershed Model Development. NC DWQ Contract No. EW030318, Project Number 1-2.

Wang, J., T. A. Endrenyi, and D. J. Nowak. 2008. Mechanistic simulation of tree canopy effects in an urban water balance model. Hydrological Processes 44:75–85.

WEF and ASCE. 1998. Urban Runoff Quality Management. WEF Manual Practice No. 23. ASCE Manual and Report on Engineering Practice No. 87. Reston, VA.

Wigmosta, M. S., L. W. Vail, and D. A. Lettenmaier. 1994. A distributed hydrology-vegetation model for complex terrain. Water Resources Research 30:1665–1679.

Wong, T. H. F. 2000. Improving Urban Stormwater Quality–From Theory to Implementation. Journal of the Australian Water Association 27(6):28–31.

Wong, T. H. F., H. P. Duncan, T. D. Fletcher, G. A. Jenkins, and J. R. Coleman. 2001. A Unified Approach to Modeling Urban Stormwater Treatment. Paper presented at the Second South Pacific Stormwater Conference, Auckland, June 27–29, pp. 319–327.

Wright, T., C. Swann, K. Cappiella, and T. Schueler. 2005. Urban Subwatershed Restoration Manual No. 11: Unified Subwatershed and Site Reconnaissance: A User's Manual (Version 2.0). Ellicott City, MD: Center for Watershed Protection.

Xiao, Q., E. G. McPherson, J. R. Simpson, S. L. Ustin and Mo. 2007. Hydrologic processes at the urban residential scale. Hydrological Processes 21:2174-2188.

5
Stormwater Management Approaches

A fundamental component of the U.S. Environmental Protection Agency's (EPA) Stormwater Program, for municipalities as well as industries and construction, is the creation of stormwater pollution prevention plans. These plans invariably document the stormwater control measures that will be used to prevent the permittee's stormwater discharges from degrading local waterbodies. Thus, a consideration of these measures—their effectiveness in meeting different goals, their cost, and how they are coordinated with one another—is central to any evaluation of the Stormwater Program. This report uses the term stormwater control measure (SCM) instead of the term best management practice (BMP) because the latter is poorly defined and not specific to the field of stormwater.

The committee's statement of task asks for an evaluation of the relationship between different levels of stormwater pollution prevention plan implementation and in-stream water quality. As discussed in the last two chapters, the state of the science has yet to reveal the mechanistic links that would allow for a full assessment of that relationship. However, enough is known to design systems of SCMs, on a site scale or local watershed scale, to lessen many of the effects of urbanization. Also, for many regulated entities the current approach to stormwater management consists of choosing one or more SCMs from a preapproved list. Both of these facts argue for the more comprehensive discussion of SCMs found in this chapter, including information on their characteristics, applicability, goals, effectiveness, and cost. In addition, a multitude of case studies illustrate the use of SCMs in specific settings and demonstrate that a particular SCM can have a measurable positive effect on water quality or a biological metric. The discussion of SCMs is organized along the gradient from the rooftop to the stream. Thus, pollutant and runoff prevention are discussed first, followed by runoff reduction and finally pollutant reduction.

HISTORICAL PERSPECTIVE ON
STORMWATER CONTROL MEASURES

Over the centuries, SCMs have met different needs for cities around the world. Cities in the Mesopotamian Empire during the second millennium BC had practices for flood control, to convey waste, and to store rain water for household and irrigation uses (Manor, 1966) (see Figure 5-1). Today, SCMs are considered a vital part of managing flooding and drainage problems in a city. What is relatively new is an emphasis on using the practices to remove pollutants from stormwater and selecting practices capable of providing groundwater recharge. These recent expectations for SCMs are not readily accepted and require an increased commitment to the proper design and maintenance of the practices.

With the help of a method for estimating peak flows (the Rational Method, see Chapter 4), the modern urban drainage system came into being soon after World War II. This generally consisted of a system of catch basins and pipes to prevent flooding and drainage problems by efficiently delivering runoff water to the nearest waterbody. However, it was soon realized that delivering the water too quickly caused severe downstream flooding and bank erosion in the receiving water. To prevent bank erosion and provide more space for flood waters, some stream channels were enlarged and lined with concrete (see Figure 5-2). But while hardening and enlarging natural channels is a cost-effective solution to erosion and flooding, the modified channel increases downstream peak flows and it does not provide habitat to support a healthy aquatic ecosystem.

FIGURE 5-1 Cistern tank, Kamiros, Rhodes (ancient Greece, 7th century BC). SOURCE: Robert Pitt, University of Alabama.

FIGURE 5-2 Concrete channel in Lincoln Creek, Milwaukee, Wisconsin. SOURCE: Roger Bannerman, Wisconsin Department of Natural Resources.

Some way was needed to control the quantity of water reaching the end of pipes during a runoff event, and on-site detention (Figure 5-3) became the standard for accomplishing this. Ordinances started appearing in the early 1970s, requiring developers to reduce the peaks of different size storms, such as the 10-year, 24-hour storm. The ordinances were usually intended to prevent future problems with peak flows by requiring the installation of flow control structures, such as detention basins, in new developments. Detention basins can control peak flows directly below the point of discharge and at the property boundary. However, when designed on a site-by-site basis without taking other basins into account, they can lead to downstream flooding problems because volume is not reduced (McCuen, 1979; Ferguson, 1991; Traver and Chadderton, 1992; EPA, 2005d). In addition, out of concerns for clogging, openings in the outlet structure of most basins are generally too large to hold back flows from smaller, more frequent storms. Furthermore, low-flow channels have been constructed or the basins have been graded to move the runoff through the structure without delay to prevent wet areas and to make it easier to mow and maintain the detention basin.

Because of the limitations of on-site detention, infiltration of urban runoff to control its volume has become a recent goal of stormwater management. Without stormwater infiltration, municipalities in wetter regions of the country can expect drops in local groundwater levels, declining stream base flows (Wang et al., 2003a), and flows diminished or stopped altogether from springs feeding wetlands and lakes (Leopold, 1968; Ferguson, 1994).

FIGURE 5-3 On-site detention. SOURCE: Tom Schueler, Chesapeake Stormwater Network, Inc.

The need to provide volume control marked the beginning of low-impact development (LID) and conservation design (Arendt, 1996; Prince George's County, 2000), which were founded on the seminal work of landscape architect Ian McHarg and associates decades earlier (McHarg and Sutton, 1975; McHarg and Steiner, 1998). The goal of LID is to allow for development of a site while maintaining as much of its natural hydrology as possible, such as infiltration, frequency and volume of discharges, and groundwater recharge. This is accomplished with infiltration practices, functional grading, open channels, disconnection of impervious areas, and the use of fewer impervious surfaces. Much of the LID focus is to manage the stormwater as close as possible to its source—that is, on each individual lot rather than conveying the runoff to a larger regional SCM. Individual practices include rain gardens (see Figure 5-4), disconnected roof drains, porous pavement, narrower streets, and grass swales. In some cases, LID site plans still have to include a method for passing the larger storms safely, such as a regional infiltration or detention basin or by increasing the capacity of grass swales.

Infiltration has been practiced in a few scattered locations for a long time. For example, on Long Island, New York, infiltration basins were built starting in 1930 to reduce the need for a storm sewer system and to recharge the aquifer, which was the only source of drinking water (Ferguson, 1998). The Cities of Fresno, California, and El Paso, Texas, which faced rapidly dropping groundwater tables, began comprehensive infiltration efforts in the 1960s and 1970s. In the 1980s Maryland took the lead on the east coast by creating an ambitious

FIGURE 5-4 Rain Garden in Madison, Wisconsin. SOURCE: Roger Bannerman, Wisconsin Department of Natural Resources.

statewide infiltration program. The number of states embracing elements of LID, especially infiltration, has increased during the 1990s and into the new century and includes California, Florida, Minnesota, New Jersey, Vermont, Washington, and Wisconsin.

Evidence gathered in the 1970s and 1980s suggested that pollutants be added to the list of things needing control in stormwater (EPA, 1983). Damages caused by elevated flows, such as stream habitat destruction and floods, were relatively easy to document with something as simple as photographs. Documentation of elevated concentrations of conventional pollutants and potentially toxic pollutants, however, required intensive collection of water quality samples during runoff events. Samples collected from storm sewer pipes and urban streams in the Menomonee River watershed in the late 1970s clearly showed the concentrations of many pollutants, such as heavy metals and sediment, were elevated in urban runoff (Bannerman et al., 1979). Levels of heavy metals were especially high in industrial-site runoff, and construction-site erosion was calculated to be a large source of sediment in the watershed. This study was followed by the National Urban Runoff Program, which added more evidence about the high levels of some pollutants found in urban runoff (Athayde et al., 1983; Bannerman et al., 1983).

***

With new development rapidly adding to the environmental impacts of existing urban areas, the need to develop good stormwater management programs is more urgent than ever. For a variety of reasons, the greatest potential for stormwater management to reduce the footprint of urbanization is in the suburbs. These areas are experiencing the fastest rates of growth, they are more amenable to stormwater management because buildings and infrastructure are not yet in place, and costs for stormwater management can be borne by the developer rather than by taxpayers. Indeed, most structural SCMs are applied to new development rather than existing urban areas. Many of the most innovative stormwater programs around the country are found in the suburbs of large cities such as Seattle, Austin, and Washington, D.C. When stormwater management in ultra-urban areas is required, it entails the retrofitting of detention basins and other flow control structures or the introduction of innovative below-ground structures characterized by greater technical constraints and higher costs, most of which are charged to local taxpayers.

Current-day SCMs represent a radical departure from past practices, which focused on dealing with extreme flood events via large detention basins designed to reduce peak flows at the downstream property line. As defined in this chapter, SCMs now include practices intended to meet broad watershed goals of protecting the biology and geomorphology of receiving waters in addition to flood peak protection. The term encompasses such diverse actions as using more conventional practices like basins and wetland to installing stream buffers, reducing impervious surfaces, and educating the public.

REVIEW OF STORMWATER CONTROL MEASURES

Stormwater control measures refer to what is defined by EPA (1999) as "a technique, measure, or structural control that is used for a given set of conditions to manage the quantity and improve the quality of stormwater runoff in the most cost-effective manner." SCMs are designed to mitigate the changes to both the quantity and quality of stormwater runoff that are caused by urbanization. Some SCMs are engineered or constructed facilities, such as a stormwater wetland or infiltration basin, that reduce pollutant loading and modify volumes and flow. Other SCMs are preventative, including such activities as education and better site design to limit the generation of stormwater runoff or pollutants.

Stormwater Management Goals

It is impossible to discuss SCMs without first considering the goals that they are expected to meet. A broadly stated goal for stormwater management is to reduce pollutant loads to waterbodies and maintain, as much as possible, the natural hydrology of a watershed. On a practical level, these goals must be made specific to the region of concern and embedded in the strategy for that region. Depending on the designated uses of the receiving waters, climate, geomorphology, and historical development, a given area may be more or less sensitive to both pollutants and hydrologic modifications. For example, goals for groundwater recharge might be higher in an area with sandy soils as compared to one with mostly clayey soils; watersheds in the coastal zone may not require hydrologic controls. Ideally, the goals of stormwater management should be linked to the water quality standards for a given state's receiving waters. However, because of the substantial knowledge gap about the effect of a particular stormwater discharge on a particular receiving water (see Chapter 3 conclusions), surrogate goals are often used by state stormwater programs in lieu of water quality standards. Examples include credit systems, mandating the use of specific SCMs, or achieving stormwater volume reduction. Credit systems might be used for practices that are known to be productive but are difficult to quantify, such as planting trees. Specific SCMs might be assumed to remove a percent of pollutants, for example 85 percent removal of total suspended solids (TSS) within a stormwater wetland. Reducing the volume of runoff from impervious surfaces (e.g., using an infiltration device) might be assumed to capture the first flush of pollutants during a storm event. Before discussing specific state goals, it is worth understanding the broader context in which goals are set.

Trade-offs Between Stormwater Control Goals and Costs

The potentially substantial costs of implementing SCMs raise a number of fundamental social choices concerning land-use decisions, designated uses, and priority setting for urban waters. To illustrate some of these choices, consider a hypothetical urban watershed with three possible land-cover scenarios: 25, 50, and 75 percent impervious surface. A number of different beneficial uses could be selected for the streams in this watershed. At a minimum, the goal may be to establish low-level standards to protect public health and safety. To achieve this, sufficient and appropriate SCMs might be applied to protect residents from flooding and achieve water quality conditions consistent with secondary human contact. Alternatively, the designated use could be to achieve the physical, chemical, and/or biological conditions sufficient to provide exceptional aquatic habitat (e.g., a high-quality recreational fishery). The physical, biological, and chemical conditions supportive of this use might be similar to a reference stream located in a much less disturbed watershed. Achieving this particular designated use would require substantially greater resources and effort than achieving a secondary human contact use. Intermediate designated uses could also be imagined, including improving ambient water quality conditions that would make the water safe for full-body emersion (primary human contact) or habitat conditions for more tolerant aquatic species.

Figure 5-5 sketches what the marginal (incremental) SCM costs (opportunity costs) might be to achieve different designated uses given different amounts of impervious surface in the watershed. The horizontal axis orders potential

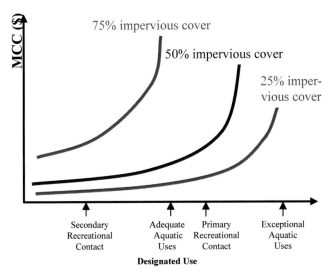

FIGURE 5-5 Cost of achieving designated uses in a hypothetical urban watershed. MCC is the marginal control cost, which represents the incremental costs to achieve successive expansion of designated uses through SCMs. The curves are constructed on the assumption that the lowest cost combination of SCMs would be implemented at each point on the curve.

designated uses in terms of least difficult to most difficult to achieve. The three conceptual curves represent the SCM costs under three different impervious surface scenarios. The relative positions of the cost curves indicate that achieving any specific designated use will be more costly in situations with a higher percentage of the watershed in impervious cover. All cost curves are upward sloping, reflecting the fact that incremental improvements in designated uses will be increasingly costly to achieve. The cost curves are purely conceptual, but nonetheless might reasonably reflect the relative costs and direction of change associated with achieving specific designated uses in different watershed conditions.

The locations of the cost curves suggest that in certain circumstances not all designated uses can be achieved or can be achieved only at an extremely high cost. For example, the attainment of exceptional aquatic uses may be unachievable in areas with 50 percent impervious surface even with maximum application of SCMs. In this illustration, the cost of achieving even secondary human contact use is high for areas with 75 percent impervious surfaces. In such highly urbanized settings, achievement of only adequate levels of aquatic uses could be exceedingly high and strain the limits of what is technically achievable. Finally, the existing and likely expected future land-use conditions have significant implications for what is achievable and at what cost. Clearly land-use decisions have an impact on the cost and whether a use can be achieved, and thus they need to be included in the decision process. The trade-off between costs and achieving specific designated uses can change substantially given different development patterns.

The purpose of Figure 5-5 is not to identify the precise location of the cost curves or to identify thresholds for achieving specific designated uses. Rather, these concepts are used to illustrate some fundamental trade-offs that confront public and private investment and regulatory decisions concerning stormwater management. The general relationships shown in Figure 5-5 suggest the need for establishing priorities for investments in stormwater management and controls, and connecting land usage and watershed goals. Setting overly ambitious or costly goals for urban streams may result in the perverse consequence of causing more waters to fail to meet designated uses. For example, consider efforts to secure ambitious designated uses in highly developed areas or in an area slated for future high-density development. Regulatory requirements and investments to limit stormwater quantity and quality through open-space requirements, areas set aside for infiltration and water detention, and strict application of maximum extent practicable controls have the effect of both increasing development costs and diminishing land available for residential and commercial properties. Policies designed to achieve exceedingly costly or infeasible designated uses in urban or urbanizing areas could have the net consequence of shifting development (and associated impervious surface) out into neighboring areas and watersheds. The end result might be minimal improvements in "within-watershed" ambient conditions but a decrease in designated uses (more impairments) elsewhere. In such a case, it might be sound water quality policy to accept higher levels of impervious surface in targeted locations, more stormwater-related impacts, and less ambitious designated uses in urban watersheds in order to preserve and protect designated uses in other watersheds.

Setting unrealistic or unachievable water quality objectives in urban areas can also pose political risks for stormwater management. The cost and difficulty of achieving ambitious water quality standards for urban stream goals may be understood by program managers but pursued nonetheless in efforts to demonstrate public commitment to achieving high-quality urban waters. Yet, promising what cannot be realistically achieved may act to undermine public support for urban stormwater programs. Increasing costs without significant observable improvements in ambient water conditions or achievement of water quality standards could ultimately reduce public commitment to the program. Thus, there are risks of "setting the bar" too high, or not coordinating land use and designated stream uses.

The cost of setting the bar too low can also be significant. Stormwater requirements that result in ineffective stormwater management will not achieve or maintain the desired water uses and can result in impairments. Loss of property, degraded waters, and failed infrastructure are tangible costs to the public (Johnston et al., 2006). Streambank rehabilitation costs can be severe, and loss of confidence in the ability to meet stormwater goals can result.

The above should not be construed as an argument for or against devoting resources to SCMs; rather, such decisions should be made with an open and transparent acknowledgment and understanding of the costs and consequences involved in those decisions.

Common State Stormwater Goals

Most states do not and have never had an overriding water quality objective in their stormwater program, but rather have used engineering criteria for SCM performance to guide stormwater management. These criteria can be loosely categorized as:

- Erosion and sedimentation control,
- Recharge/base flow,
- Water quality,
- Channel protection, and
- Flooding events.

The SCMs used to address these goals work by minimizing or eliminating increases in stormwater runoff volume, peak flows, and/or the pollutant load carried by stormwater.

The criteria chosen by any given state usually integrate state, federal, and regional laws and regulations. Areas of differing climates may emphasize one goal over another, and the levels of control may vary drastically. Contrast a desert region where rainwater harvesting is extremely important versus a coastal region subject to hurricanes. Some areas like Seattle have frequent smaller volume rainfalls—the direct opposite of Austin, Texas—such that small volume controls would be much more effective in Seattle than Austin. Regional geology (karst) or the presence of Brownfields may affect the chosen criteria as well.

The committee's survey of State Stormwater Programs (Appendix C) reflects a wide variation in program goals as reflected in the criteria found in their SCM manuals. Some states have no specific criteria because they do not produce SCM manuals, while others have manuals that address every category of criteria from flooding events to groundwater recharge. Some states rely upon EPA or other states' or transportation agencies' manuals. In general, soil and erosion control criteria are the most common and often exist in the absence of any other state criteria. This wide variation reflects the difficulties that states face in keeping up with rapidly changing information about SCM design and performance.

The criteria are ordered below (after the section on erosion and sediment control) according to the size of the storm they address, from smallest to most extreme. The criteria can be expressed in a variety of ways, from a simple requirement to control a certain volume of rainfall or runoff (expressed as a depth) to the size of a design storm to more esoteric requirements, such as limiting the time that flow can be above a certain threshold. The volumes of rainfall or run-off are based on statistics of a region's daily rainfall, and they approximate one another as the percentage of impervious cover increases. Design storms for lar-ger events that address channel protection and flooding are usually based on extreme event statistics and tend to represent a temporal pattern of rainfall over a set period, usually a day. Finally, it should be noted that the categories are not mutually exclusive; for example, recharge of groundwater may enhance water quality via pollutant removal during the infiltration process.

Erosion and Sedimentation Control. This criterion refers to the preven-tion of erosion and sedimentation of sites during construction and is focused at the site level. Criteria usually include a barrier plan to prevent sedimentation from leaving the site (e.g., silt fences), practices to minimize the potential ero-sion (phased construction), and facilities to capture and remove sediment from the runoff (detention). Because these measures are considered temporary, smaller extreme events are designated as the design storm than what typically would be used if flood control were the goal.

Recharge/Base Flow. This criterion is focused on sustaining the precon-struction hydrology of a site as it relates to base flow and recharge of groundwa-ter supplies. It may also include consideration of water usage of the property owners and return through septic tanks and tile fields. The criterion, expressed as a volume requirement, is usually to capture around 0.5 to 1.0 inch of runoff from impervious surfaces depending on the climate and soil type of the region. (For this range of rainfall, very little runoff occurs from grass or forested areas, which is why runoff from impervious surfaces is used as the criterion.)

Water Quality. Criteria for water quality are the most widespread, and are usually crafted as specific percent removal for pollutants in stormwater dis-charge. Generally, a water quality criterion is based on a set volume of storm-water being treated by the SCM. The size of the storm can run from the first inch of rainfall off impervious surfaces to the runoff from the one-year, 24-hour extreme storm event. It should be noted that the term "water quality" covers a wide range of groundwater and surface water pollutants, including water tem-perature and emerging contaminants.

Many of the water quality criteria are surrogates for more meaningful pa-rameters that are difficult to quantify or cannot be quantified, or they reflect situations where the science is not developed enough to set more explicit goals. For example, the Wisconsin state requirement of an 80 percent reduction in TSS in stormwater discharge does not apply to receiving waters themselves. How-ever, it presumes that there will be some water quality benefits in receiving wa-ters; that is, phosphorus and fecal coliform might be captured by the TSS re-quirement. Similarly water quality criteria may be expressed as credits for good practices, such as using LID, street sweeping, or stream buffers.

Channel Protection. This criterion refers to protecting channels from accelerated erosion during storm events due to the increased runoff. It is tied to either the presumed "channel-forming event"—what geomorphologists once believed was the storm size that created the channel due to erosion and deposition—or to the minimum flow that accomplishes any degree of sediment transport. It is generally defined as somewhere between the one- and five-year, 24-hour storm event or a discharge level typically exceeded once to several times per year. Some states require a reduction in runoff volume for these events to match preconstruction levels. Others may require that the average annual duration of flows that are large enough to erode the streambank be held the same on an annual basis under pre- and postdevelopment conditions.

It is not uncommon to find states where a channel protection goal will be written poorly, such that it does not actually prevent channel widening. For example, MacRae (1997) presented a review of the common "zero runoff increase" discharge criterion, which is commonly met by using ponds designed to detain the two-year, 24-hour storm. MacRae showed that stream bed and bank erosion occur during much lower events, namely mid-depth flows that generally occur more than once a year, not just during bank-full conditions (approximated by the two-year event). This finding is entirely consistent with the well-established geomorphological literature (e.g., Pickup and Warner, 1976; Andrews, 1984; Carling, 1988; Sidle, 1988). During monitoring near Toronto, MacRae found that the duration of the geomorphically significant predevelopment mid-bankfull flows increased by more than four-fold after 34 percent of the basin had been urbanized. The channel had responded by increasing in cross-sectional area by as much as three times in some areas, and was still expanding.

Flooding Events. This criterion addresses public safety and the protection of property and is applicable to storm events that exceed the channel capacity. The 10- through the 100-year storm is generally used as the standard. Volume-reduction SCMs can aid or meet this criterion depending on the density of development, but usually assistance is needed in the form of detention SCMs. In some areas, it may be necessary to reduce the peak flow to below preconstruction levels in order to avoid the combined effects of increased volume, altered timing, and a changed hydrograph. It should be noted that some states do not consider the larger storms (100-year) to be a stormwater issue and have separate flood control requirements.

Each state develops a framework of goals, and the corresponding SCMs used to meet them, which will depend on the scale and focus of the stormwater management strategy. A few states have opted to express stormwater goals within the context of watershed plans for regions of the state. However, the setting of goals on a watershed basis is time-consuming and requires study of the watersheds in question. The more common approach has been to set generic or minimal controls for a region that are not based on a watershed plan. This has been done in Maryland, Wisconsin (see Box 5-1), and Pennsylvania (see Box 5-2). This strategy has the advantage of more rapid implementation of

---

BOX 5-1
Wisconsin Statewide Goal of TSS Reduction for Stormwater Management

To measure the success of stormwater management, Wisconsin has statewide goals for sediment and flow (Wisconsin DNR, 2002). A lot is known about the impacts of sediment on receiving waters, and any reduction is thought to be beneficial. Flow can be a good indicator of other factors; for example, reducing peak flows will prevent bank erosion.

Developing areas in Wisconsin are required to reduce the annual TSS load by 80 percent compared to no controls (Wisconsin DNR, 2002). Two flow-rated requirements for developing areas are in the administrative rules. One is that the site must maintain the peak flow for the two-year, 24-hour rainfall event. Second, the annual infiltration volume for postdevelopment must be within 90 percent of the predevelopment volumes for residential land uses; the number for non-residential is 60 percent. Both of these flow control goals are thought to also have water quality benefits.

The goal for existing urban areas is an annual reduction in TSS loads. Municipalities must reduce their annual TSS loads by 20 percent, compared to no controls, by 2008. This number is increased to 40 percent by 2013. All of these goals were partially selected to be reasonable based on cost and technical feasibility.

BOX 5-2
Volume-Based Stormwater Goals in Pennsylvania

Pennsylvania has developed a stormwater *Best Management Practices* manual to support the Commonwealth's Storm Water Management Act. This manual and an accompanying sample ordinance advocates two methods for stormwater control based on volume, termed Control Guidance (CG) 1 and 2. The first (CG-1) requires that the runoff volume be maintained at the two-year, 24-hour storm level (which corresponds to approximately 3.5 inches of rainfall in this region) through infiltration, evapotranspiration, or reuse. This criterion addresses recharge/base flow, water quality, and channel protection, as well as helping to meet flooding requirements.

The second method (CG-2) requires capture and removal of the first inch of runoff from paved areas, with infiltration strongly recommended to address recharge and water quality issues. Additionally, to meet channel protection criteria, the second inch is required to be held for 24 hours, which should reduce the channel-forming flows. (This is an unusual criterion in that it is expressed as what an SCM can accomplish, not as the flow that the channel can handle.) Peak flows for larger events are required to be at preconstruction levels or less if the need is established by a watershed plan. These criteria are the starting point for watershed or regional plans, to reduce the effort of plan development. Some credits are available for tree planting, and other nonstructural practices are advocated for dissolved solids mitigation. See http://www.dep.state.pa.us/dep/deputate/watermgt/wc/subjects/stormwatermanagement/default.htm.

some SCMs because watershed management plans are not required. In order to be applicable to all watersheds in the state, the goals must target common pollutants or flow modification factors where the processes are well known. It must also be possible for these goals to be stated in National Pollutant Discharge Elimination System (NPDES) permits. Many states have selected TSS reduction, volume reduction, and peak flow control as generic goals. A generic goal is not usually based on potentially toxic pollutants, such as heavy metals, due to the complexity of their interaction in the environment, the dependence on the existing baseline conditions, and the need for more understanding on what are acceptable levels. The difficulty with the generic approach is that specific watershed issues are not addressed, and the beneficial uses of waters are not guaranteed.

One potential drawback of a strategy based on a generic goal coupled to the permit process is that the implementation of the goal is usually on a site-by-site basis, especially for developing areas. Generic goals may be appropriate for certain ubiquitous watershed processes and are clearly better than having no goals at all. However, they do not incorporate the effects of differences in past development and any unique watershed characteristics; they should be considered just a good starting point for setting watershed-based goals.

Role of SCMs in Achieving Stormwater Management Goals

One important fundamental change in SCM design philosophy has come about because of the recent understanding of the roles of smaller storms and of impervious surfaces. This is demonstrated by Box 3-4, which shows that for the Milwaukee area more than 50 percent of the rainfall by volume occurs in storms that have a depth of less then 0.75 inch. If extreme events are the only design criteria for SCMs, the vast majority of the annual rainfall will go untreated or uncontrolled, as it is smaller than the minimum extreme event. This relationship is not the same in all regions. For example, in Austin, Texas, the total yearly rainfall is smaller than in Milwaukee, but a large part of the volume occurs during larger storm events, with long dry periods in between.

The upshot is that the design strategy for stormwater management, including drainage systems and SCMs, should take a region's rainfall and associated runoff conditions into account. For example, an SCM chosen to capture the majority of the suspended solids, recharge the baseflow, reduce streambank erosion, and reduce downstream flooding in Pennsylvania or Seattle (which have moderate and regular rainfall) would likely not be as effective in Texas, where storms are infrequent and larger. In some areas, a reduction in runoff volume may not be sufficient to control streambank erosion and flooding, such that a second SCM like an extended detention stormwater wetland may be needed to meet management goals.

Finally, as discussed in greater detail in a subsequent section, SCMs are most effective from the perspective of both efficiency and cost when stormwater management is incorporated in the early planning stages of a community. Retrofitting existing development with SCMs is much more technically difficult and costly because the space may not be available, other infrastructure is already installed, or utilities may interfere. Furthermore, if the property is on private land or dedicated as an easement to a homeowners association, there may be regulatory limitations to what can be done. Because of these barriers, retrofitting existing urban areas often depends on engineered or manufactured SCMs, which are more expensive in both construction and operation.

Stormwater Control Measures

SCMs reduce or mitigate the generation of stormwater runoff and associated pollutants. These practices include both "structural" or engineered devices as well as more "nonstructural measures" such as land-use planning, site design, land conservation, education, and stewardship practices. Structural practices may be defined as any facility constructed to mitigate the adverse impacts of stormwater and urban runoff pollution. Nonstructural practices, which tend to be longer-term and lower-maintenance solutions, can greatly reduce the need for or increase the effectiveness of structural SCMs. For example, product substitution and land-use planning may be key to the successful implementation of an infiltration SCM. Preserving wooded areas and reducing street widths can allow the size of detention basins in the area to be reduced.

Table 5-1 presents the expansive list of SCMs that are described in this chapter. For most of the SCMs, each listed item represents a class of related practices, with individual methods discussed in greater detail later in the chapter. There are nearly 20 different broad categories of SCMs that can be applied, often in combination, to treat the quality and quantity of stormwater runoff. A primary difference among the SCMs relates to which stage of the development cycle they are applied, where in the watershed they are installed, and who is responsible for implementing them.

The development cycle extends from broad planning and zoning to site design, construction, occupancy, retrofitting, and redevelopment. As can be seen, SCMs are applied throughout the entire cycle. The scale at which the SCM is applied also varies considerably. While many SCMs are installed at individual sites as part of development or redevelopment applications, many are also applied at the scale of the stream corridor or the watershed or to existing municipal stormwater infrastructure. The final column in Table 5-1 suggests who would implement the SCM. In general, the responsibility for implementing SCMs primarily resides with developers and local stormwater agencies, but planning agencies, landowners, existing industry, regulatory agencies, and municipal separate storm sewer system (MS4) permittees can also be responsible for implementing many key SCMs.

In Table 5-1, the SCMs are ordered in such a way as to mimic natural systems as rain travels from the roof to the stream through combined application of

TABLE 5-1 Summary of Stormwater Control Measures—When, Where, and Who

Stormwater Control Measure	When	Where	Who
Product Substitution	Continuous	National, state, regional	Regulatory agencies
Watershed and Land-Use Planning	Planning stage	Watershed	Local planning agencies
Conservation of Natural Areas	Site and watershed planning stage	Site, watershed	Developer, local planning agency
Impervious Cover Minimization	Site planning stage	Site	Developer, local review authority
Earthwork Minimization	Grading plan	Site	Developer, local review authority
Erosion and Sediment Control	Construction	Site	Developer, local review authority
Reforestation and Soil Conservation	Site planning and construction	Site	Developer, local review authority
Pollution Prevention SCMs for Stormwater Hotspots	Post-construction or retrofit	Site	Operators and local and state permitting agencies
Runoff Volume Reduction— Rainwater harvesting	Post-construction or retrofit	Rooftop	Developer, local planning agency and review authority
Runoff Volume Reduction— Vegetated	Post-construction or retrofit	Site	Developer, local planning agency and review authority
Runoff Volume Reduction— Subsurface	Post-construction or retrofit	Site	Developer, local planning agency and review authority
Peak Reduction and Runoff Treatment	Post-construction or retrofit	Site	Developer, local planning agency and review authority
Runoff Treatment	Post-construction or retrofit	Site	Developer, local planning agency and review authority
Aquatic Buffers and Managed Floodplains	Planning, construction and post-construction	Stream corridor	Developer, local planning agency and review authority, landowners
Stream Rehabilitation	Postdevelopment	Stream corridor	Local planning agency and review authority

continues next page

TABLE 5-1 Continued

Stormwater Control Measure	When	Where	Who
Municipal Housekeeping	Postdevelopment	Streets and stormwater infrastructure	MS4 Permittee
Illicit Discharge Detection and Elimination	Postdevelopment	Stormwater infrastructure	MS4 Permittee
Stormwater Education	Postdevelopment	Stormwater infrastructure	MS4 Permittee
Residential Stewardship	Postdevelopment	Stormwater infrastructure	MS4 Permittee

Note: Nonstructural SCMs are in italics.

a series of practices throughout the entire development site. This order is upheld throughout the chapter, with the implication that no SCM should be chosen without first considering those that precede it on the list.

Given that there are 20 different SCM groups and a much larger number of individual design variations or practices within each group, it is difficult to authoritatively define the specific performance or effectiveness of SCMs. In addition, our understanding of their performance is rapidly changing to reflect new research, testing, field experience, and maintenance history. The translation of these new data into design and implementation guidance is accelerating as well. What is possible is to describe their basic hydrologic and water quality objectives and make a general comparative assessment of what is known about their design, performance, and maintenance as of mid-2008. This broad technology assessment is provided in Table 5-2, which reflects the committee's collective understanding about the SCMs from three broad perspectives:

- Is widely accepted design or implementation guidance available for the SCM and has it been widely disseminated to the user community?
- Have enough research studies been published to accurately characterize the expected hydrologic or pollutant removal performance of the SCM in most regions of the country?
- Is there enough experience with the SCM to adequately define the type and scope of maintenance needed to ensure its longevity over several decades?

Affirmative answers to these three questions are needed to be able to reliably quantify or model the ability of the SCM, which is an important element in defining whether the SCM can be linked to improvements in receiving water quality. As will be discussed in subsequent sections of this chapter, there are many SCMs for which there is only a limited understanding, particularly those that are nonstructural in nature.

The columns in Table 5-2 summarize several important factors about each SCM, including the ability of the SCM to meet hydrologic control objectives and water quality objectives, the availability of design guidance, the availability of performance studies, and whether there are maintenance protocols. The hydrologic control objectives range from complete prevention of stormwater flow to reduction in runoff volume and reduction in peak flows. The column on water quality objectives describes whether the SCM can prevent the generation of, or remove, contaminants of concern in stormwater.

The availability of design guidance tends to be greatest for the structural practices. Some but not all nonstructural practices are of recent origin, and communities lack available design guidance to include them as an integral element of local stormwater solutions. Where design guidance is available, it may not yet have been disseminated to the full population of Phase II MS4 communities.

TABLE 5-2 Current Understanding of Stormwater Control Measure Capabilities

SCM	Hydrologic Control Objectives	Water Quality Objectives	Available Design Guidance	Performance Studies Available	Defined Maintenance Protocols
Product Substitution	NA	Prevention	NA	Limited	NA
Watershed and Land-Use Planning	All objectives	Prevention	Available	Limited	Yes
Conservation of Natural Areas	Prevention	Prevention	Available	None	Yes
Impervious Cover Minimization	Prevention and reduction	Prevention	Available	Limited	No
Earthwork Minimization	Prevention	Prevention	Emerging	Limited	Yes
Erosion and Sediment Control	Prevention and reduction	Prevention and removal	Available	Limited	Yes
Reforestation and Soil Conservation	Prevention and reduction	Prevention and removal	Emerging	None	No
Pollution Prevention SCMs for Hotspots	NA	Prevention	Emerging	Very few	No
Runoff Volume Reduction— Rainwater harvesting	Reduction	NA	Emerging	Limited	Yes
Runoff Volume Reduction— Vegetated (Green Roofs, Bioretention, Bioinfiltration, Bioswales)	Reduction and some peak attenuation	Removal	Available	Limited	Emerging
Runoff Volume Reduction— Subsurface (Infiltration Trenches, Pervious Pavements)	Reduction and some peak attenuation	Removal	Available	Limited	Yes

continues next page

TABLE 5-2 Continued

SCM	Hydrologic Control Objectives	Water Quality Objectives	Available Design Guidance	Performance Studies Available	Defined Maintenance Protocols
Peak Reduction and Runoff Treatment (Stormwater Wetlands, Dry/Wet Ponds)	Peak attenuation	Removal	Available	Adequate	Yes
Runoff Treatment (Sand Filters, Manufactured Devices)	None	Removal	Emerging	Adequate— sand filters Limited— manufactured devices	Yes
Aquatic Buffers and Managed Floodplains	NA	Prevention and removal	Available	Very few	Emerging
Stream Rehabilitation	NA	Prevention and removal	Emerging	Limited	Unknown
Municipal Housekeeping (Street Sweeping/ Storm-Drain Cleanouts)	NA	Removal	Emerging	Limited	Emerging
Illicit Discharge Detection/ Elimination	NA	Prevention and removal	Available	Very few	No
Stormwater Education	Prevention	Prevention	Available	Very few	Emerging
Residential Stewardship	Prevention	Prevention	Emerging	Very few	No

Note: Nonstructural SCMs are in italics.

Key:		
Hydrologic Objective	Water Quality Objective	Available Design Guidance?
Prevention: Prevents generation of runoff **Reduction:** Reduces volume of runoff **Treatment:** Delays runoff delivery only **Peak Attenuation**: Reduction of peak flows through detention	**Prevention**: Prevents generation, accumulation, or wash-off of pollutants and/or reduces runoff volume **Removal**: Reduces pollutant concentrations in runoff by physical, chemical, or biological means	**Available:** Basic design or implementation guidance is available in most areas of the country are readily available **Emerging:** Design guidance is still under development, is missing in many parts of the country, or requires more performance data
Performance Data Available?	Defined Maintenance Protocol?	Notes:
Very Few: Handful of studies, not enough data to generalize about SCM performance **Limited:** Numerous studies have been done, but results are variable or inconsistent **Adequate**: Enough studies have been done to adequately define performance	**No:** Extremely limited understanding of procedures to maintain SCM in the future **Emerging:** Still learning about how to maintain the SCM **Yes**: Solid understanding of maintenance for future SCM needs	**NA:** Not applicable for the SCM

The column on the availability of performance data is divided into those SCMs where enough studies have been done to adequately define performance, those SCMs where limited work has been done and the results are variable, and those SCMs where only a handful of studies are available. A large and growing number of performance studies are available that report the efficiencies of structural SCMs in reducing flows and pollutant loading (Strecker et al., 2004; ASCE, 2007; Schueler et al., 2007; Selbig and Bannerman, 2008). Many of these are compiled in the Center for Watershed Protection's National Pollutant Removal Performance Database for Stormwater Treatment Practices (http://www.cwp.org/Resource_Library/Center_Docs/SW/bmpwriteup_092 007_v3.pdf), in the International Stormwater BMP Database (http://www.bmp-database.org/Docs/Performance%20Summary%20June%202008.pdf), and by the Water Environment Research Foundation (WERF, 2008). In cases where there is incomplete understanding of their performance, often information can be gleaned from other fields including agronomy, forestry, petroleum exploration, and sanitary engineering. Current research suggests that it is not a question if whether structural SCMs "work" but more of a question of to what degree and with what longevity (Heasom et al., 2006; Davis et al., 2008; Emerson and Traver, 2008). There is considerably less known about the performance of nonstructural practices for stormwater treatment, partly because their application has been uneven around the country and it remains fairly low in comparison to structural stormwater practices.

Finally, defined maintenance protocols for SCMs can be nonexistent, emerging, or fully available. SCMs differ widely in the extent to which they can be considered permanent solutions. For those SCMs that work on the individual site scale on private property, such as rain gardens, local stormwater managers may be reluctant to adopt such practices due to concerns about their ability to enforce private landowners to conduct maintenance over time. Similarly, those SCMs that involve local government decisions (such as education, residential stewardship practices, zoning, or street sweeping) may be less attractive because governments are likely to change over time.

The following sections contain more detailed information about the individual SCMs listed in Tables 5-1 and 5-2, including the operating unit processes, the pollutants treated, the typical performance for both runoff and pollutant reduction, the strengths and weaknesses, maintenance and inspection requirements, and the largest sources of variability and uncertainty.

Product Substitution

Product substitution refers to the classic pollution prevention approach of reducing the emissions of pollutants available for future wash-off into stormwater runoff. The most notable example is the introduction of unleaded gasoline, which resulted in an order-of-magnitude reduction of lead levels in stormwater runoff in a decade (Pitt et al., 2004a,b). Similar reductions are expected with the phase-out of methyl tert-butyl ether (MTBE) additives in gasoline. Other examples of product substitution are the ban on coal-tar sealants during parking lot renovation that has reduced PAH runoff (Van Metre et al., 2006), phosphorus-free fertilizers that have measurably reduced phosphorus runoff to Minnesota lakes (Barten and Johnson, 2007), the painting of galvanized metal surfaces, and alternative rooftop surfaces (Clark et al., 2005). Given the importance of coal power plant emissions in the atmospheric deposition of nitrogen and mercury, it is possible that future emissions reductions for such plants may result in lower stormwater runoff concentrations for these two pollutants.

The level of control afforded by product substitution is quite high if major reductions in emissions or deposition can be achieved. The difficulty is that these reductions require action in another environmental regulatory arena, such as air quality, hazardous waste, or pesticide regulations, which may not see stormwater quality as a core part of their mission.

Watershed and Land-Use Planning

Communities can address stormwater problems by making land-use decisions that change the location or quantity of impervious cover created by new development. This can be accomplished through zoning, watershed plans, comprehensive land-use plans, or Smart Growth incentives.

The unit process that is managed is the amount of impervious cover, which is strongly related to various residential and commercial zoning categories (Cappiella and Brown, 2000). Numerous techniques exist to forecast future watershed impervious cover and its probable impact on the quality of aquatic resources (see the discussion of the Impervious Cover Model in Chapter 3; CWP, 1998a; MD DNR, 2005). Using these techniques and simple or complex simulation models, planners can estimate stormwater flows and pollutant loads through the watershed planning process and alter the location or intensity of development to reduce them.

The level of control that can be achieved by watershed and land-use planning is theoretically high, but relatively few communities have aggressively exercised it. The most common application of downzoning has been applied to watersheds that drain to drinking water reservoirs (Kitchell, 2002). The strength of this practice is that it has the potential to directly address the underlying causes of the stormwater problem rather than just treating its numerous symptoms. The weakness is that local decisions on zoning and Smart Growth are reversible and often driven by other community concerns such as economic development, adequate infrastructure, and transportation. In addition, powerful consumer and market forces often have promoted low-density sprawl development. Communities that use watershed-based zoning often require a compelling local environmental goal, since state and federal regulatory authorities have traditionally been extremely reluctant to interfere with the local land-use and zoning powers.

Conservation of Natural Areas

Natural-area conservation protects natural features and environmental re-sources that help maintain the predevelopment hydrology of a site by reducing runoff, promoting infiltration, and preventing soil erosion. Natural areas are protected by a permanent conservation easement prescribing allowable uses and activities on the parcel and preventing future development. Examples include any areas of undisturbed vegetation preserved at the development site, including forests, wetlands, native grasslands, floodplains and riparian areas, zero-order stream channels, spring and seeps, ridge tops or steep slopes, and stream, wet-land, or shoreline buffers. In general, conservation should maximize contiguous area and avoid habitat fragmentation.

While natural areas are conserved at many development sites, most of these requirements are prompted by other local, state, and federal habitat protections, and are not explicitly designed or intended to provide runoff reduction and stormwater treatment. To date, there are virtually no data to quantify the runoff reduction and/or pollutant removal capability of specific types of natural area conservation, or the ability to explicitly link them to site design.

Impervious Cover Reduction

A variety of practices, some of which fall under the broader term "better site design," can be used to minimize the creation of new impervious cover and dis-connect or make more permeable the hard surfaces that are needed (Nichols et al., 1997; Richman, 1997; CWP, 1998a). A list of some common impervious cover reduction practices for both residential and commercial areas is provided below.

Elements of Better Site Design: Single-Family Residential
- o   Maximum residential street width
- o   Maximum street right-of-way width
- o   Swales and other stormwater practices can be located within the right-of-way
- o   Maximum cul-de-sac radius with a bioretention island in the center
- o   Alternative turnaround options such as hammerheads are acceptable if they reduce impervious cover
- o   Narrow sidewalks on one side of the street (or move pedestrian path-ways away from the street entirely)
- o   Disconnect rooftops from the storm-drain systems
- o   Minimize driveway length and width and utilize permeable surfaces
- o   Allow for cluster or open-space designs that reduce lot size or setbacks in exchange for conservation of natural areas
- o   Permeable pavement in parking areas, driveways, sidewalks, walkways, and patios

Elements of Better Site Design: Multi-Family Residential and Commercial
- o Design buildings and parking to have multiple levels
- o Store rooftop runoff in green roofs, foundation planters, bioretention areas, or cisterns
- o Reduce parking lot size by reducing parking demand ratios and stall dimensions
- o Use landscaping areas, tree pits, and planters for stormwater treatment
- o Use permeable pavement over parking areas, plazas, and courtyards

CWP (1998a) recommends minimum or maximum geometric dimensions for subdivisions, individual lots, streets, sidewalks, cul-de-sacs, and parking lots that minimize the generation of needless impervious cover, based on a national roundtable of fire safety, planning, transportation and zoning experts. Specific changes in local development codes can be made using these criteria, but it is often important to engage as many municipal agencies that are involved in development as possible in order to gain consensus on code changes.

At the present time, there is little research available to define the runoff reduction benefits of these practices. However, modeling studies consistently show a 10 to 45 percent reduction in runoff compared to conventional development (CWP, 1998b,c, 2002). Several monitoring studies have documented a major reduction in stormwater runoff from development sites that employ various forms of impervious cover reduction and LID in the United States and Australia (Coombes et al., 2000; Philips et al., 2003; Cheng et al., 2005) compared to those that do not.

Unfortunately, better site design has been slowly adopted by local planners, developers, designers, and public works officials. For example, although the project pictured in Figure 5-6 has been very successful in terms of controlling stormwater, the better-site-design principles used have not been widely adopted in the Seattle area. Existing local development codes may discourage or even prohibit the application of environmental site design practices, and many engineers and plan reviewers are hesitant to embrace them. Impervious cover reduction must be incorporated at the earliest stage of site layout and design to be effective, but outdated development codes in many communities can greatly restrict the scope of impervious cover reduction (see Chapter 2). Finally, the performance and longevity of impervious cover reduction are dependent on the infiltration capability of local soils, the intensity of development, and the future management actions of landowners.

Earthwork Minimization

This source control measure seeks to limit the degree of clearing and grading on a development site in order to prevent soil compaction, conserve soils, prevent erosion from steep slopes, and protect zero-order streams. This is accomplished by (1) identifying key soils, drainage features, and slopes to protect

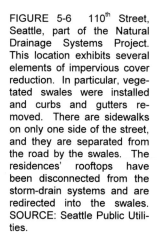

FIGURE 5-6   110th Street, Seattle, part of the Natural Drainage Systems Project. This location exhibits several elements of impervious cover reduction. In particular, vegetated swales were installed and curbs and gutters removed. There are sidewalks on only one side of the street, and they are separated from the road by the swales. The residences' rooftops have been disconnected from the storm-drain systems and are redirected into the swales. SOURCE: Seattle Public Utilities.

and then (2) establishing a limit of disturbance where construction equipment is excluded. This element is an important, but often under-utilized component of local erosion and sediment control plans.

Numerous researchers have documented the impact of mass grading, clearing, and the passage of construction equipment on the compaction of soils, as measured by increase in bulk density, declines in soil permeability, and increases in the runoff coefficient (Lichter and Lindsey, 1994; Legg et al., 1996; Schueler, 2001a,b; Gregory et al., 2006). Another goal of earthwork minimization is to protect zero-order streams, which are channels with defined banks that emanate from a hollow or ravine with convergent contour lines (Gomi et al., 2002). They represent the uppermost definable channels that possess temporary or intermittent flow. Functioning zero-order channels provide major watershed functions, including groundwater recharge and discharge (Schollen et al., 2006; Winter, 2007), important nutrient storage and transformation functions (Bernot and Dodds, 2005; Groffman et al., 2005), storage and retention of eroded hillslope sediments (Meyers, 2003), and delivery of leaf inputs and large woody debris. Compared to high-order network streams, zero-order streams are disproportionately disturbed by mass grading, enclosure, or channelization (Gomi et al., 2002; Meyer, 2003).

The practice of earthwork minimization is not widely applied across the country. This is partly due to the limited performance data available to quantify its benefits, and the absence of local or national design guidance or performance benchmarks for the practice.

Erosion and Sediment Control

Erosion and sediment control predates much of the NPDES stormwater permitting program. It consists of the temporary installation and operation of a series of structural and nonstructural practices throughout the entire construction process to minimize soil erosion and prevent off-site delivery of sediment. Because construction is expected to last for a finite and short period of time, the design standards are usually smaller and thus riskier (25-year versus the 100-year storm). By phasing construction, thereby limiting the exposure of bare earth at any one time, the risk to the environment is reduced significantly.

The basic practices include clearing limits, dikes, berms, temporary buffers, protection of drainage-ways, soil stabilization through hydroseeding or mulching, perimeter controls, and various types of sediment traps and basins. All plans have some component that requires filtration of runoff crossing construction areas to prevent sediment from leaving the site. This usually requires a sediment collection system including, but not limited to, conventional settling ponds and advanced sediment collection devices such as polymer-assisted sedimentation and advanced sand filtration. Silt fences are commonly specified to filter distributed flows, and they require maintenance and replacement after storms as shown in Figure 5-7. Filter systems are added to inlets until the streets are paved and the surrounding area has a cover of vegetation (Figure 5-8). Sedimentation basins (Figure 5-9) are constructed to filter out sediments through rock filters, or are equipped with floating skimmers or chemical treatment to settle out pollutants. Other common erosion and sediment control measures include temporary seeding and rock or rigged entrances to construction sites to remove dirt from vehicle tires (see Figure 5-10).

Control of the runoff's erosive potential is a critical element. Most erosion and sediment control manuals provide design guidance on the capacity and ability of swales to handle runoff without eroding, on the design of flow paths to transport runoff at non-erosive velocities, and on the dissipation of energy at pipe outlets. Examples include rock energy dissipaters, level spreaders (see Figure 5-11), and other devices.

Box 5-3 provides a comprehensive list of recommended construction SCMs. The reader is directed to reviews by Brown and Caraco (1997) and Shaver et al. (2007) for more information. Although erosion and sediment control practices are temporary, they require constant operation and maintenance during the complicated sequence of construction and after major storm events. It is exceptionally important to ensure that practices are frequently inspected and repaired and that sediments are cleaned out. Erosion and sediment control are

FIGURE 5-7 A functioning silt fence (top) and an improperly maintained silt fence (bottom). SOURCES: Top, EPA NPDES Menu of BMPs (available at http://cfpub.epa.gov/npdes/storm-water/menuofbmps/index.cfm? action=factsheet_results& view=specific&bmp=56) and, bottom, Robert Traver, Villanova University.

FIGURE 5-8 Sediment filter left in place after construction. SOURCE: Robert Traver, Villanova University.

FIGURE 5-9   Sediment basin.   SOURCE: EPA NPDES Menu of BMPs (available at http://cfpub.epa.gov/npdes/stormwater/menuofbmps/index.cfm?action=factsheet_results& vew=specific&bmp=56).

FIGURE 5-10 Rumble strips to remove dirt from vehicle tires.   SOURCE: Laura Ehlers, National Research Council.

FIGURE 5-11 Level spreader. SOURCE: Robert Traver, Villanova University.

BOX 5-3
Recommended Construction Stormwater Control Measures

1. As the top priority, emphasize construction management SCMs as follows:
   • Maintain existing vegetation cover, if it exists, as long as possible.
   • Perform ground-disturbing work in the season with smaller risk of erosion, and work off disturbed ground in the higher risk season.
   • Limit ground disturbance to the amount that can be effectively controlled in the event of rain.
   • Use natural depressions and planning excavation to drain runoff internally and isolate areas of potential sediment and other pollutant generation from draining off the site, so long as safe in large storms.
   • Schedule and coordinate rough grading, finish grading, and erosion control application to be completed in the shortest possible time overall and with the shortest possible lag between these work activities.

2. Stabilize with cover appropriate to site conditions, season, and future work plans. For example:
   • Rapidly stabilize disturbed areas that could drain off the site, and that will not be worked again, with permanent vegetation supplemented with highly effective temporary erosion controls until achievement of at least 90 percent vegetative soil cover.
   • Rapidly stabilize disturbed areas that could drain off the site, and that will not be worked again for more than three days, with highly effective temporary erosion controls.
   • If at least 0.1 inch of rain is predicted with a probability of 40 percent or more, before rain falls stabilize or isolate disturbed areas that could drain off the site, and that are being actively worked or will be within three days, with measures that will prevent or minimize transport of sediment off the property.

continues next page

BOX 5-3 Continued

3. As backup for cases where all of the above measures are used to the maximum extent possible but sediments still could be released from the site, consider the need for sediment collection systems including, but not limited to, conventional settling ponds and advanced sediment collection devices such as polymer-assisted sedimentation and advanced sand filtration.

4. Specify emergency stabilization and/or runoff collection (e.g., using temporary depressions) procedures for areas of active work when rain is forecast.

5. If runoff can enter storm drains, use a perimeter control strategy as backup where some soil exposure will still occur, even with the best possible erosion control (above measures) or when there is discharge to a sensitive waterbody.

6. Specify flow control SCMs to prevent or minimize to the extent possible:
• Flow of relatively clean off-site water over bare soil or potentially contaminated areas;
• Flow of relatively clean intercepted groundwater over bare soil or potentially contaminated areas;
• High velocities of flow over relatively steep and/or long slopes, in excess of what erosion control coverings can withstand; and
• Erosion of channels by concentrated flows, by using channel lining, velocity control, or both.

7. Specify stabilization of construction entrance and exit areas, provision of a nearby tire and chassis wash for dirty vehicles leaving the site with a wash water sediment trap, and a sweeping plan.

8. Specify construction road stabilization.

9. Specify wind erosion control.

10. Prevent contact between rainfall or runoff and potentially polluting construction materials, processes, wastes, and vehicle and equipment fluids by such measures as enclosures, covers, and containments, as well as berming to direct runoff.

---

widely applied in many communities, and most states have some level of design guidance or standards and specifications. Nonetheless, few communities have quantified the effectiveness of a series of construction SCMs applied to an individual site, nor have they clearly defined performance benchmarks for individual practices or their collective effect at the site. In general, there has been little monitoring in the past few decades to characterize the performance of construction SCMs, although a few notable studies have been recently published (e.g., Line and White, 2007). Box 5-4 describes the effectiveness of filter fences and filter fences plus grass buffers to reduce sediment loadings from construction activities and the resulting biological impacts.

BOX 5-4
Receiving Water Impacts Associated with Construction Site Discharges

The following is a summary of a recent research project that investigated in-stream biological conditions downstream of construction sites having varying levels of erosion controls (none, the use of filter fences, and filter fences plus grass buffers) for comparison. The project title is *Studies to Evaluate the Effectiveness of Current BMPs in Controlling Stormwater Discharges from Small Construction Sites* and was conducted for the Alabama Water Resources Research Institute, Project 2001AL4121B, by Drs. Robert Angus, Ken Marion, and Melinda Lalor of the University of Alabama at Birmingham. The initial phase of the project, described below, was completed in 2002 (Angus et al., 2002). While this case study is felt to be representative of many sites across the United States, there are other examples of where silt fences have been observed to be more effective (e.g., Barrett et al., 1998).

Methods

This study was conducted in the upper Cahaba River watershed in north central Alabama, near Birmingham. The study areas had the following characteristics. (1) Topography and soil types representative of the upland physiographic regions in the Southeast (i.e., southern Appalachian and foothill areas); thus, findings from this study should be relevant to a large portion of the Southeast. (2) The rainfall amounts and intensities in this region are representative of many areas of the Southeast and (3) the expanding suburbs of the Birmingham metropolitan area are rapidly encroaching upon the upper Cahaba River and its tributaries. Stormwater runoff samples were manually collected from sheet flows above silt fences, and from points below the fence within the vegetated buffer. Water was sampled during "intense" (≥1 inch/hour) rain events. The runoff samples were analyzed for turbidity, particle size distribution (using a Coulter Counter Multi-Sizer IIe), and total solids (dissolved solids plus suspended/non-filterable solids). Sampling was only carried out on sites with properly installed and well-maintained silt fences, located immediately upgrade from areas with good vegetative cover.

Six tributary or upper mainstream sites were studied to investigate the effects of sedimentation from construction sites on both habitat quality and the biological "health" of the aquatic ecosystem (using benthic macroinvertebrates and fish). EPA's Revision to Rapid Bioassessment Protocols for Use in Streams and Rivers was used to assess the habitat quality at the study sites. Each site was assessed in the spring to evaluate immediate effects of the sediment, and again during the following late summer or early fall to evaluate delayed effects.

Results

Effectiveness of Silt Fences. Silt fences were found to be better than no control measures at all, but not substantially. The mean counts of small particles (<5 μm) below the silt fences were about 50 percent less than that from areas with no erosion control measures, even though the fences appeared to be properly installed and in good order. However, the variabilities were large and the difference between the means was not statistically significant. For every variable measured, the mean values of samples taken below silt fences were significantly higher ($p < 0.001$) than samples collected from undisturbed vegetated control sites collected nearby and at the same time. These data therefore indicate that silt fences are only marginally effective at reducing soil particulates in runoff water.

Effectiveness of Filter Fences with Vegetated Buffers. Runoff samples were also collected immediately below filter fences, and below filter fences after flow over buffers having 5, 10, and 15 feet of dense (intact) vegetation. Mean total solids in samples collected below silt fences and a 15-foot-wide vegetated buffer zone were about 20 percent lower, on average, than those samples collected only below the silt fence. The installation of filter fences above an intact, good vegetated buffer removes sediment from construction site runoff more effectively than with the use of filter fences alone.

Biological Metrics Sensitive to Sedimentation Effects (Fish). Analysis of the fish biota indicates that various metrics used to evaluate the biological integrity of the fish community also are affected by highly sedimented streams. As shown in Figure 5-12, the overall composition of the population, as quantified by the Index of Biotic Integrity (IBI) is lower; the proportion and biomass of darters, a disturbance-sensitive group, is lower; the proportion and biomass of sunfish is higher; the Shannon-Weiner diversity index is lower; and the number of disturbance-tolerant species is higher as mean sediment depth increases.

Benthic Macroinvertebrates. A number of stream benthic macroinvertebrate community characteristics were also found to be sensitive to sedimentation. Metrics based on these characteristics differ greatly between sediment-impacted and control sites (Figure 5-13). Some of the metrics that appear to reflect sediment-associated stresses include the Hilsenhoff Biotic Index (HBI), a variation of the EPT index (percent EPT minus Baetis), and the Sorensen Index of Similarity to a reference site. The HBI is a weighted mean tolerance value; high HBI values indicate sites dominated by disturbance-tolerant macroinvertebrate taxa. The EPT% index is the percent of the collection represented by organisms in the generally disturbance-sensitive orders *Ephemeroptera, Plecoptera,* and *Trichoptera.* Specimens of the genus *Baetis* were not included in the index as they are relatively disturbance-tolerant. The HBI and the EPT indices also show positive correlations to several other measures of disturbance, such as percent of the watershed altered by development.

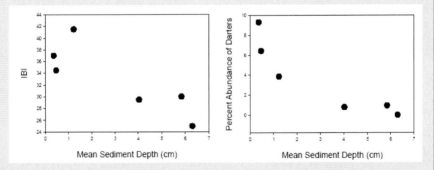

FIGURE 5-12 Association between two fish metrics and amount of stream sediment. NOTE: The IBI is based on numerous characteristics of the fish population. The percent relative abundance of darters is the percentage of darters to all the fish collected at a site. SOURCE: Angus et al. (2002).

continues next page

BOX 5-4 Continued

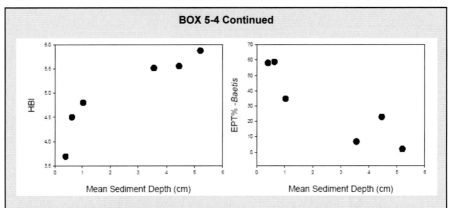

FIGURE 5-13 Associations between two macroinvertebrate metrics and the amount of stream sediment. SOURCE: Angus et al. (2002).

Reforestation and Soil Compost Amendments

This set of practices seeks to improve the quality of native vegetation and soils present at the site. Depending on the ecoregion, this may involve forest, prairie, or chapparal plantings, tilling, and amending compacted soils to improve their hydrologic properties.

The goal is to maintain as much predevelopment hydrologic function at a development site as possible by retaining canopy interception, duff/soil layer interception, evapotranspiration, and surface infiltration. The basic methods to implement this practice are described in Cappiella et al. (2006), Pitt et al. (2005), Chollak and Rosenfeld (1998), and Balusek (2003).

At this time, there are few monitoring data to assess the degree to which land reforestation or soil amendments can improve the quality of stormwater runoff at a particular development site, apart from the presumptive watershed research that has shown that forests with undisturbed soils have very low rates of surface runoff and extremely low levels of pollutants in runoff (Singer and Rust, 1975; Johnson et al., 2000; Chang, 2006). More data are needed on the hydrologic properties of urban forests and soils whose ecological functions are stressed or degraded by the urbanization process (Pouyat et al., 1995, 2007).

Pollution Prevention SCMs for Stormwater Hotspots

Certain classes of municipal and industrial operations are required to maintain a series of pollution prevention practices to prevent or minimize contact of pollutants with rainfall and runoff. Pollution prevention practices involve a wide range of operational practices at a site related to vehicle repairs, fueling, washing and storage, loading and unloading areas, outdoor storage of materials, spill prevention and response, building repair and maintenance, landscape and turf management, and other activities that can introduce pollutants into the stormwater system (CWP, 2005). Training of personnel at the affected area is needed to ensure that industrial and municipal managers and employees understand and implement the correct stormwater pollution prevention practices needed for their site or operation.

Examples of municipal operations that may need pollution prevention plans include public works yards, landfills, wastewater treatment plants, recycling and solid waste transfer stations, maintenance depots, school bus and fleet storage and maintenance areas, public golf courses, and ongoing highway maintenance operations. The major industrial categories that require stormwater pollution prevention plans were described in Table 2-3. Both industrial and municipal operations must develop a detailed stormwater pollution prevention plan, train employees, and submit reports to regulators. Compliance has been a significant issue with this program in the past, particularly for small businesses (Duke and Augustenberg, 2006; Cross and Duke, 2008) Recently filed investigations of stormwater hotspots indicate many of these operations are not fully implementing their stormwater pollution prevention plans, and a recent GAO report (2007) indicates that state inspections and enforcement actions are extremely rare.

The goal of pollution prevention is to prevent contact of rainfall or stormwater runoff with pollutants, and it is an important element of the post-construction stormwater plan. However, with the exception of a few industries such as auto salvage yards (Swamikannu, 1994), basic research is lacking on how much greater event mean concentrations are at municipal and industrial stormwater hotspots compared to other urban land uses. In addition, little is presently known about whether aggressive implementation of stormwater pollution prevention plans actually can reduce stormwater pollutant concentrations at hot spots.

Runoff Volume Reduction—Rainwater Harvesting

A primary goal of stormwater management is to reduce the volume of runoff from impervious surfaces. There are several classes of SCMs that can achieve this goal, including rainwater harvesting systems, vegetated SCMs that evapotranspirate part of the volume, and infiltration SCMs. For all of these measures, the amount of runoff volume to be captured depends on watershed goals, site conditions including climate, upstream nonstructural practices employed, and whether the chosen SCM is the sole management measure or part of a treatment train. Generally, runoff-volume-reduction SCMs are designed to handle at least the first flush from impervious surfaces (1 inch of rainfall). In Pennsylvania, control of the 24-hour, two-year storm volume (about 8 cm) is considered the standard necessary to protect stream-channel geomorphology, while base flow recharge and the first flush can be addressed by capturing a much smaller volume of rain (1–3 cm). Where both goals must be met, the designer is permitted to either oversize the volume reduction device to control the larger volume, or build a smaller device and use it in series with an extended detention basin to protect the stream geomorphology (PaDEP, 2006). Some designers have reported that in areas with medium to lower percentage impervious surfaces they are able to control up to the 100-year storm by enlarging runoff-volume-reduction SCMs and using the entire site. In retrofit situations, capture amounts as small as 1 cm are a distinct improvement. It should be noted that there are important, although indirect, water quality benefits of all runoff-volume-reduction SCMs—(1) the reduction in runoff will reduce streambank erosion downstream and the concomitant increases in sediment load, and (2) volume reductions lead to pollutant load reductions, even if pollutant concentrations in stormwater are not decreased.

Rainwater harvesting systems refer to use of captured runoff from roof tops in rain barrels, tanks, or cisterns (Figures 5-14 and 5-15). This SCM treats runoff as a resource and is one of the few SCMs that can provide a tangible economic benefit through the reduction of treated water usage. Rainwater harvesting systems have substantial potential as retrofits via the use of rain barrels or cisterns that can replace lawn or garden sprinkling systems. Use of this SCM to provide gray water within buildings (e.g., for toilet flushing) is considerably more complicated due to the need to construct new plumbing and obtain the necessary permits.

FIGURE 5-14 Rainwater harvesting tanks at a Starbucks in Austin, Texas. SOURCE: Laura Ehlers, National Research Council.

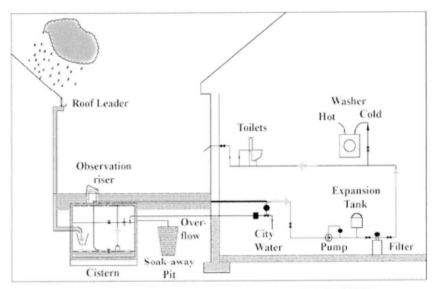

FIGURE 5-15 A Schematic of rainwater harvesting. SOURCE: PaDEP (2006).

The greatest challenge with these systems is the need to use the stored water and avoid full tanks, since these cannot be responsive in the event of a storm. That is, these SCMs are effective only if the captured runoff can be regularly used for some grey water usage, like car washing, toilet flushing, or irrigation systems (golf courses, landscaping, nurseries). In some areas it might be possible to use the water for drinking, showering, or washing, but treatment to potable water quality would be required. Sizing of the required storage is dependent on the climate patterns, the amount of impervious cover, and the frequency of water use. Areas with frequent rainfall events require less storage as long as the water is used regularly, while areas with cold weather will not be able to utilize the systems for irrigation in the winter and thus require larger storage.

One substantial advantage of these systems is their ability to reduce water costs for the user and the ability to share needs. An example of this interaction is the Pelican Hill development in Irvine, California, where excess runoff from the streets and houses is collected in enormous cisterns and used for watering of a nearby golf course. Furthermore, compared to other SCMs, the construction of rainwater harvesting facilities provide a long-term benefit with minimal maintenance cost, although they do require an upfront investment for piping and storage tanks.

Coombes et al. (2000) found that rainwater harvesting achieved a 60 to 90 percent reduction in runoff volume; in general, few studies have been conducted to determine the performance of these SCMs. It should be noted that rainwater harvesting systems do collect airborne deposition and acid rain.

Runoff Volume Reduction—Vegetated

A large and very promising class of SCMs includes those that use infiltration and evapotranspiration via vegetation to reduce the volume of runoff. These SCMs also directly address water quality of both surface water and groundwater by reducing streambank erosion, capturing suspended solids, and removing other pollutants from stormwater during filtration through the soil (although the extent to which pollutants are removed depends on the specific pollutant and the local soil chemistry). Depending on their design, these SCMs can also reduce peak flows and recharge groundwater (if they infiltrate). These SCMs can often be added as retrofits to developed areas by installing them into existing lawns, rights of way, or traffic islands. They can add beauty and property value.

Flow volume is addressed by this SCM group by first capturing runoff, creating a temporary holding area, and then removing the stored volume through infiltration and evapotranspiration. Examples include bioswales, bioretention, rain gardens, green roofs, and bioinfiltration. Swales refer to grassy areas on the side of the road that convey drainage. These were first designed to move runoff away from paved areas, but can now be designed to achieve a certain contact time with runoff so as to promote infiltration and pollutant removal (see Figure 5-16). Bioretention generally refers to a constructed sand filter with soil and vegetation growing on top to which stormwater runoff from impervious surfaces is directed (Figure 5-17). The original rain garden or bioretention facilities were constructed with a fabric at the bottom of the prepared soil to prevent infiltration and instead had a low-level outflow at the bottom. Green roofs (Figure 5-18) are very similar to bioretention SCMs. They tend to be populated with a light expanded shale-type soil and succulent plants chosen to survive wet and dry periods. Finally, bioinfiltration is similar to bioretention but is better engineered to achieve greater infiltration (Figure 5-19). All of these devices are usually at the upper end of a treatment train and designed for smaller storms, which minimizes their footprint and allows for incorporation within existing infrastructure (such as traffic control devices and median strips). This allows for distributed treatment of the smaller volumes and distributed volume reduction.

These SCMs work by capturing water in a vegetated area, which then infiltrates into the soil below. They are primarily designed to use plant material and soil to evapotranspirate the runoff over several days. A shallow depth of ponding is required, since the inflows may exceed the possible infiltration ability of the native soil. This ponding is maintained above an engineered sandy soil mixture and is a surface-controlled process (Hillel, 1998). Early in the storm, the soil moisture potential creates a suction process that helps draw water into the SCM. This then changes to a steady rate that is "practically equal to the saturated hydraulic conductivity" of the subsurface (Hillel, 1998). The hydrologic design goal should be to maximize the volume of water that can be held in the soil, which necessitates consideration of the soil hydraulic conductivity (which varies with temperature), climate, depth to groundwater, and time to drain.

FIGURE 5-16 Vegetated swale. SOURCE: PaDEP (2006).

FIGURE 5-17 Bioretention during a storm event at the University of Maryland. SOURCE: Reprinted, with permission, from Davis et al. (2008). Copyright 2008 by the American Society of Civil Engineers.

FIGURE 5-18 City Hall in the center of Chicago's downtown was retrofitted with a green roof to reduce the heat island effect, remove airborne pollutants, and attenuate stormwater flows as a demonstration of innovative stormwater management in an ultra-urban setting. SOURCE: Courtesy of the Conservation Design Forum.

FIGURE 5-19 Retrofit bioinfiltration at Villanova University immediately following a storm event. SOURCE: Robert Traver, Villanova University.

Usually these devices are designed to empty between 24 and 72 hours after a storm event. In some cases (usually bioretention), these SCMs have an underdrain.

The choice of vegetation is an important part of the design of these SCMs. Many sites where infiltration is desirable have highly sandy soils, and the vegetation has to be able to endure both wet and dry periods. Long root growths are desired to promote infiltration (Barr Engineering Co., 2001), and plants that attract birds can reduce the insect population. Bioretention cells may be wet for longer periods than bioinfiltration sites, requiring different plants. Denser plant-ings or "thorns" may be needed to avoid the destruction caused by humans and animals taking shortcuts through the beds.

The pollutant removal mechanism operating for volume-reduction SCMs are different for each pollutant type, soil type, and volume-reduction mecha-nism. For bioretention and SCMs using infiltration, the sedimentation and filtra-tion of suspended solids in the top layers of the soil are extremely efficient. Several studies have shown that the upper layers of the soil capture metals, par-ticulate nutrients, and carbon (Pitt, 1996; Deschesne et al., 2005; Davis et al., 2008). The removal of dissolved nutrients from stormwater is not as straight-forward. While ammonia is caught by the top organic layer, nitrate is mobile in the soil column. Some bioretention systems have been built to hold water in the soil for longer periods in order to create anaerobic conditions that would pro-mote denitrification (Hunt and Lord, 2006a). Phosphorus removal is related to the amount of phosphorus in the original soil. Some studies have shown that bioretention cells built with agricultural soils increased the amount of phospho-rus released. Chlorides pass through the system unchecked (Ermilio and Traver, 2006), while oils and greases are easily removed by the organic layer. Hunt et al. (2008) have reported in studies in North Carolina that the drying cycle ap-pears to kill off bacteria. Temperature is not usually a concern as most storms do not overflow these devices. Green roofs collect airborne deposition and acid rain and may export nutrients when they overflow. However, this must be tem-pered by the fact that in larger storms, most natural lands would produce nutri-ents.

A group of new research studies from North America and Australia have demonstrated the value of many of these runoff-volume-reduction practices to replicate predevelopment hydrology at the site. The results from 10 recent stud-ies are given in Table 5-3, which shows the runoff reduction capability of biore-tention. As can be seen, the reduction in runoff volume achieved by these prac-tices is impressive—ranging from 20 to 99 percent with a median reduction of about 75 percent. Box 5-5 discusses the excellent performance of the bioswales installed during Seattle's natural drainage systems project (see also Horner et al., 2003; Jefferies, 2004; Stagge, 2006). Bioinfiltration has been less studied, but one field study concluded that close to 30 percent of the storm volume was able to be removed by bioinfiltration (Sharkey, 2006). A very recent case study of bioinfiltration is provided in Box 5-6, which demonstrates that the capture of small storms through these SCMs is extremely effective in areas where the ma-jority of the rainfall falls in smaller storms.

The strengths of vegetated runoff-volume-reduction SCMs include the flexibility to utilize the drainage system as part of the treatment train. For ex-ample, bioswales can replace drainage pipes, green roofs can be installed on buildings, and bioretention can replace parking borders (Figure 5-27), thereby reducing the footprint of the stormwater system. Also, through the use of swales and reducing pipes and inlets, costs can be offset. Vegetated systems are more tolerant of the TSS collected, and their growth cycle maintains pathways for

TABLE 5-3 Volumetric Runoff Reduction Achieved by Bioretention

Bioretention Design	Location	Runoff Reduction	Reference
Infiltration	CT	99%	Dietz and Clausen (2006)
	PA	86%	Ermilio and Traver (2006)
	FL	98%	Rushton (2002)
	AUS	73%	Lloyd et al. (2002)
Underdrain	ONT	40%	Van Seters et al. (2006)
	Model	30%	Perez-Perdini et al. (2005)
	NC	40 to 60%	Smith and Hunt (2007)
	NC	20 to 29%	Sharkey (2006)
	NC	52 to 56%	Hunt et al. (2008)
	MD	52 to 65%	Davis et al. (2008)

BOX 5-5
Bioswale Case Study 110th Street Cascade, Seattle, Washington

A recent example of the ability of SCMs to accomplish a variety of goals was illustrated for water quality swales in Seattle, Washington. As part of its Natural Drainage Systems Project, the City of Seattle retrofitted several blocks of an urban residential neighborhood with curbside vegetated swales. On NW 110th Street, the two-block-long system was

 developed as a cascade, due to the steep slope (6 percent). Twelve stepped, in-series biofilters were installed between properties and the road, each of which contains a storage area and an overflow weir. During rain events, the cells were designed to fill before emptying into the cell downstream. The soils in the bottom of each cell were over one foot thick and consisted of river rocks overlain by a swale mix.   Native plants were chosen to vegetate the sides of the swale.

Extensive flow and water quality sampling occurred during 2003–2006 at the inflow and outflow of the biofilters as well as at references points elsewhere in the neighborhood that are not served by the new SCMs. Perhaps the most profound observation was that almost 50 percent of all rainfall flowing into the cascade was infiltrated, resulting in a corresponding reduction in runoff. Indeed, the cascade discharged measurable flow only during 49 of 235 storm events during the period. Depending on preceding conditions, the cascade was able to retain all of the flow for storms up to 1 inch in magnitude. In addition to the reduction in runoff affected by the swales, they also achieved significant peak flow reduction, as shown in Figure 5-20. Many peak flow rates were entirely dampened, even those where the inflow peak rate was as high as 0.7 cfs.

continues next page

BOX 5-5 Continued

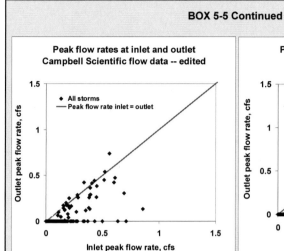
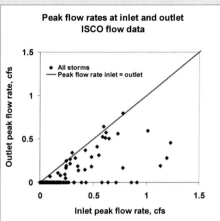

FIGURE 5-20 Peak flow rates at the inlet and outlet of the cascade, as measured by two different devices: Campbell Scientific (left) and ISCO (right). SOURCE: Horner and Chapman (2007).

Water quality data were also extremely encouraging, as shown in Table 5-4. For total suspended solids, influent concentration of 94 mg/L decreased to 29 mg/L at the outlet of the cascade. Similar percent removals were observed for total copper, total phosphorus, total zinc, and total lead (see Table 5-4). Soluble phosphorus concentrations tended to increase from the inflow of the cascade to the outflow.

TABLE 5-4 Typical Outflow Quality from the 110th Street Cascade.

Pollutant	Range (mg/L)
Total Suspended Solids	10–40
Total Nitrogen	0.6–1.4
Total Phosphorus	0.09–0.23
Soluble Reactive Phosphorus	0.02–0.05
Total Copper	0.004–0.008
Dissolved Copper	0.002–0.005
Total Zinc	0.04–0.11
Dissolved Zinc	0.02–0.06
Total Lead	0.002–0.007
Dissolved Lead	<0.001
Motor Oil	0.11–0.33

SOURCE: Horner and Chapman (2007).

Taking both measured concentrations and volume reduction into account, the cascade reduced the mass loadings for the contaminants by 60 percent to greater than 90 percent. As shown in Table 5-5, pollutants associated with sediments were reduced to the greatest extent, while dissolved pollutants were less readily removed.

continues next page

BOX 5-5 Continued

TABLE 5-5 Pollutant Mass Loading Reductions at 110th Street Cascade.

Pollutant	Percent Reduction (90% Confidence Interval)
Total Suspended Solids	84 (72–92)
Total Nitrogen	63 (53–74)
Total Phosphorus	63 (49–74)
Total Copper	83 (77–88)
Dissolved Copper	67 (50–78)
Total Zinc	76 (46–85)
Dissolved Zinc	55 (21–70)
Total Lead	90 (84–94)
Motor Oil	92 (86–97)

SOURCE: Horner and Chapman (2007).

This level of performance was compared to other parts of the neighborhood treated with conventional ditch and pipe systems. The concentrations of almost all pollutants at the outlet of the 100th Cascade was significantly lower than a corresponding outlet at 120th Street. Furthermore, the ability of this SCM to attenuate peak flows and reduce runoff was remarkable.

BOX 5-6
SCM Evaluation Through Monitoring: Villanova Bioinfiltration SCM

The Bioinfiltration Traffic Island located on the campus of Villanova University in Southeastern Pennsylvania is part of the Villanova Urban Stormwater Partnership (VUSP) BMP Demonstration Park (see Figure 5-21). Originally funded through the Pennsylvania Growing Greener Program, and now through the State's 319 nonpoint source monitoring program, the site has been monitored continuously since soon after it was constructed in 2001. This monitoring has lead to a wealth of information about the performance and monitoring needs of infiltration SCMs.

FIGURE 5-21 Villanova Bioinfiltration Traffic Island SCM. SOURCE: Reprinted, with permission, from VUSP. Copyright by Villanova Urban Stormwater Partnership.

The SCM is a retrofit of an existing curb-enclosed traffic island in the parking lot of a university dormitory complex. The original grass area was dug out to approximately six feet. The soil removed during the excavation was then mixed with sand onsite to create a 50 percent sand–soil mixture. This soil mixture was then placed back into the excavation to

continues next page

BOX 5-6 Continued

a depth of approximately four feet, leaving a surface depression that is an average of two feet deep. Care was taken during construction to prevent any compaction of either the soil mixture or the undisturbed soil below. Placement of the mixed soil is shown in Figure 5-22.

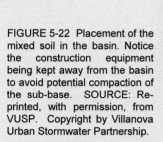

FIGURE 5-22 Placement of the mixed soil in the basin. Notice the construction equipment being kept away from the basin to avoid potential compaction of the sub-base. SOURCE: Reprinted, with permission, from VUSP. Copyright by Villanova Urban Stormwater Partnership.

During construction two curb cuts were created to direct runoff into the SCM. Creation of one of the cuts entailed filling and paving over an existing stormwater inlet to redirect the runoff that previously entered the stormwater drainage system of the parking lot. Another existing inlet was used to collect and redirect runoff into the SCM. Plants were chosen based on their ability to thrive in both extreme wet and dry conditions; the species chosen are commonly found on sand dunes where similar wet/dry conditions may exist.

The contributing watershed is approximately 50,000 square feet and is 52 percent impervious surfaces. The design goal of the SCM was for it to temporarily store the first inch of runoff. The one-inch capture depth is based on an analysis of local historical rainfall data showing that capture of the first inch of each storm would account for approximately 96 percent of the annual rainfall. This capture depth would therefore also account for the majority of the annual pollutant load coming from the drainage area.

Continuous monitoring over multiple years has increased our understanding of how this type of structure operates and its benefits. For example, Heasom et al. (2006) was able to produce a continuous hydrologic flow model of the site based on season. Figure 5-23 shows the variability of the infiltration rate on a seasonal basis, and the relationship between infiltration and temperature (Emerson and Traver, 2008). This work has also shown no statistical change in performance over the five-year monitoring period.

When examining the yearly performance of the site from a surface water standpoint, it is easily shown that on a regular basis approximately 50 to 60 percent of the runoff that reaches the site is removed from the surface waters, and 80 to 85 percent of the rainfall is infiltrated (Figure 5-24).

continues next page

BOX 5-6 Continued

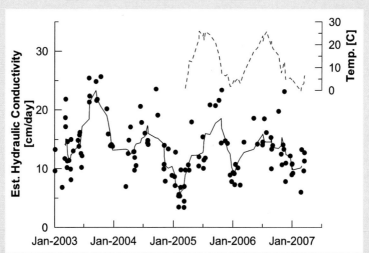

FIGURE 5-23 Seasonal Infiltration Rate. SOURCE: Reprinted, with permission, from Emerson and Traver (2008). Copyright 2008 by Journal of Irrigation and Drainage Engineering.

The performance of the SCM during individual storm events was examined in 2005. Out of 77 rainfall events, overflow was recorded for only seven events. Generally overflow did not occur for rainfalls less than 1.95 inches except for one occasion. As the bowl volume is much less than this value, substantial infiltration must be occurring during the storm event. When one extreme 6-inch storm was recorded (Figure 5-25), it was surprising to note that infiltration occurred all during the storm event, as did some unexpected peak flow reduction. What is even more impressive is to examine the reduction in the duration of flows, which is directly related to downstream channel erosion (Figure 5-26). Clearly the bioinfiltration SCM exceeded its design goals.

Research on this site is currently examining water quality benefits and groundwater interactions. When evaluating the pollutant removal of bioinfiltration, it is critical to consider flow volumes and pollutant levels together. For example, during many of the overflow events, there were higher nutrient levels leaving the SCM than entering due to the plants contained within the SCM. However, when the runoff volume reduction is considered, the total nitrogen and phosphorus removed from the influent is impressive (Davis et al., 2008). Water quality studies of the infiltrated water are still incomplete but generally show some conversion of nitrate to nitrite, and high chlorides from snow melt chemicals moving through the system. Nutrient levels are relatively low in the samples at the 8-foot depth.

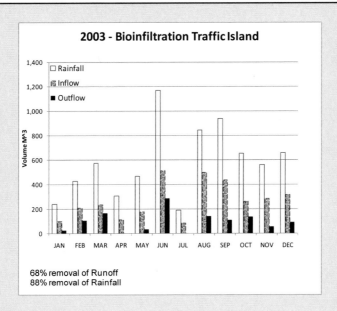

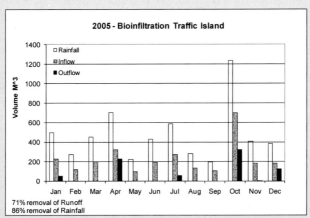

FIGURE 5-24 2003 Performance and 2005 Performance. SOURCE: Reprinted, with permission, from VUSP. Copyright by Villanova Urban Stormwater Partnership.

continues next page

BOX 5-6 Continued

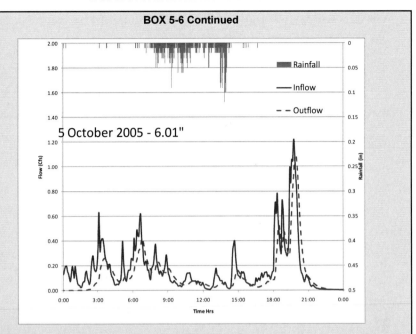

FIGURE 5-25 October 2005 extreme storm event. SOURCE: Reprinted, with permission, from VUSP. Copyright by Villanova Urban Stormwater Partnership.

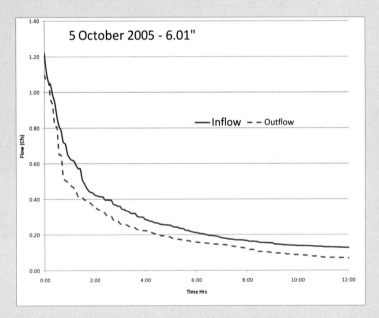

FIGURE 5-26 Flow duration curves, October 2005. SOURCE: Reprinted, with permission, from VUSP. Copyright by Villanova Urban Stormwater Partnership.

FIGURE 5-27 North Carolina Retrofit Bioretention SCMs. SOURCE: Robert Traver, Villanova University.

infiltration and prevents clogging. Freeze–thaw cycles also contribute to pathway maintenance. The aesthetic appeal of vegetated SCMs is also a significant strength.

Weaknesses include the dependence of these SCMs on native soil infiltration and the need to understand groundwater levels and karst geology, particularly for those SCMs designed to infiltrate. For bioinfiltration and bioretention, most failures occur early on and are caused by sedimentation and construction errors that reduce infiltration capacity, such as stripping off the topsoil and compacting the subsurface. Once a good grass cover is established in the contributing area, the danger of sedimentation is reduced. Nonetheless, the need to prevent sediment from overwhelming these structures is critical. The longevity of these SCMs and their vulnerability to toxic spills are a concern (Emerson and Traver, 2008), as is their failure to reduce chlorides. Finally, in areas where the land use is a hot spot, or where the SCM could potentially contaminate the groundwater supply, bioretention, non-infiltrating bioswales, and green roofs may be more suitable than infiltration SCMs.

The role of infiltration SCMs in promoting groundwater recharge deserves additional consideration. Although this is a benefit of infiltration SCMs in regions where groundwater levels are dropping, it may be undesirable in a few limited scenarios. For example, in the arid southwest contributions to base flow from irrigation have turned some dry ephemeral stream systems into perennial streams that support the growth of dense vegetation, which may be less desirable habitat for certain riparian species (like the Arroyo toad in Southern California). Infiltration SCMs could contribute to changing the flow regime in cases such as these. In most urban areas, there is so much impervious cover that it would be difficult to "overinfiltrate." Nonetheless, the use of infiltration SCMs will change local subsurface hydrology, and the ramifications of this—good and bad—should be considered prior to their installation.

Maintenance of vegetated runoff-volume-reduction SCMs is relatively simple. A visit after a rainstorm to check for plant health, to check sediment buildup, and to see if the water is ponded can answer many questions. Maintenance includes trash pickup and seasonal removal of dead grasses and weeds. Sediment removal from pretreatment devices is required. Depending on the pollutant concentrations in the influent, the upper layer of organic matter may need to be removed infrequently to maintain infiltration and to prevent metal and nutrient buildup.

At the site level, the chief factors that lead to uncertainty are the infiltration performance of the soil, particular for the limiting subsoil layer, and how to predict the extent of pollutant removal. Traditional percolation tests are not effective to estimate the infiltration performance; rather, testing hydraulic conductivity is required. Furthermore, the infiltration rate varies depending on temperature and season (Emerson and Traver, 2008). Basing measurements on percent removal of pollutants is extremely misleading, since every site and storm generates different levels of pollutants. The extent of pollutant removal depends on land use, time between storms, seasons, and so forth. These factors should be part of the design philosophy for the site. Finally, it should also be pointed out that climate is a factor determining the effectiveness of some of these SCMs. For example, green roofs are more likely to succeed in areas having smaller, more frequent storms (like the Pacific Northwest) compared to areas subjected to less frequent, more intense storms (like Texas).

Runoff Volume Reduction—Subsurface

Infiltration is the primary runoff-volume-reduction mechanism for subsurface SCMs, such that much of the previous discussion is relevant here. Thus, like vegetated SCMs, these SCMs provide benefits for groundwater recharge, water quality, stream channel protection, peak flow reduction, capture of the suspended solids load, and filtration through the soil (Ferguson, 2002). Because these systems can be built in conjunction with paved surfaces (i.e., they are often buried under parking lots), the amount of water captured, and thus stream protection, may be higher than for vegetated systems. They also have lower land requirements than vegetated systems, which can be an enormous advantage when using these SCMs during retrofitting, as long as the soil is conducive to infiltration.

Similar to vegetated SCMs, this SCM group works primarily by first capturing runoff and then removing the stored volume through infiltration. The temporary holding area is made either of stone or using manufactured vaults. Examples include pervious pavement, infiltration trenches, and seepage pits (see Figures 5-28, 5-29, 5-30, 5-31, and 5-32). As with vegetated SCMs, a shallow

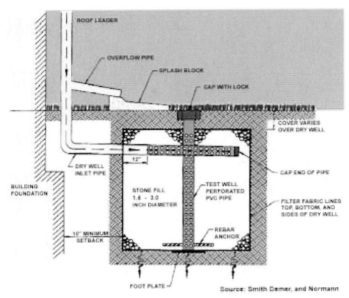

FIGURE 5-28 Schematic of a seepage pit. SOURCE: PaDEP (2006).

FIGURE 5-29 Porous asphalt. SOURCE: PaDEP (2006).

FIGURE 5-30 A retrofitted infiltration trench at Villanova University. SOURCE: Reprinted, with permission, from VUSP. Copyright by VUSP.

FIGURE 5-31 Pervious concrete at Villanova University. SOURCE: Reprinted, with permission from Villanova University. Copyright by VUSP.

FIGURE 5-32 A small office building conversion at the edge of downtown Denver included the replacement of a portion of the site's parking with modular block porous pavement underlain by an 18-inch layer of crushed rock. Rainfall on the porous pavement and roof runoff for most storm events are contained in the reservoir created by the crushed rock. The pavement infiltrates runoff from most storm events for one-third of the impervious area on the half-acre site. SOURCE: Courtesy of Wenk Associates.

depth of ponding is required, since the inflows may exceed the possible infiltration ability of the native soil. In this case, the ponding is maintained within a rock bed under a porous pavement or in an infiltration trench. These devices are usually designed to empty between 24 and 72 hours after the storm event.

The infiltration processes operating for these subsurface SCMs are similar to those for the vegetated devices previously discussed. Thus, much like for vegetated systems, the level of control achieved depends on the infiltration ability of the native soils, the percent of impervious surface area in the contributing watershed, land use contributing to the pollutant loadings, and climate. A large number of recent studies have found that permeable pavement can reduce runoff volume by anywhere from 50 percent (Rushton, 2002; Jefferies, 2004; Bean et al., 2007) to as much as 95 percent or greater (van Seters et al., 2006; Kwiatkowski et al., 2007). Box 5-7 describes the success of a recent retrofitting of asphalt with pervious pavement at Villanova University.

The strengths of subsurface runoff-volume-reduction SCMs are similar to those of their vegetated counterparts. Additional attributes include their ability to be installed under parking areas and to manage larger volumes of rainfall. These SCMs typically have few problems with safety or vector-borne diseases because of their subsurface location and storage capacity, and they can be very aesthetically pleasing. The potential of permeable pavement could be particularly far-reaching if one considers the amount of impervious surface in urban areas that is comprised of roads, driveways, and parking lots.

The weaknesses of these SCMs are also similar to those of vegetated systems, including their dependence on native soil infiltration and the need to understand groundwater levels and karst geology. Simply estimating the soil hydraulic conductivity can have an error rate of an order of magnitude. Specifically for subsurface systems that use geotextiles (not permeable pavement), there is a danger of TSS being compressed against the bottom of the geotextile, preventing infiltration. There are no freeze–thaw cycles or vegetated processes that can reopen pathways, so the control of TSS is even more critical to their life span. In most cases (permeable pavement is an exception), pretreatment is required, except for the cleanest of sources (like a slate roof). Typically, manufactured devices, sediment forebays, or grass strips are part of the design of subsurface SCMs to capture the larger sediment particles.

The maintenance of subsurface runoff-volume-reduction SCMs is relatively simple but critical. If inspection wells are installed, a visit after a rainstorm will check that the volume is captured, and later that it has infiltrated. Porous surfaces should undergo periodic vacuum street sweeping when a sediment source is present. Pretreatment devices require sediment removal. The difficulty with this class of SCMs is that, if a toxic spill occurs or maintenance is not proactive, there are no easy corrective measures other than replacement.

Low-Impact Development. LID refers primarily to the use of small, engineered, on-site stormwater practices to treat the quality and quantity of runoff at its source. It is discussed here because the SCMs that are thought of as LID— particularly vegetated swales, green roofs, permeable pavement, and rain gardens—are all runoff-volume-reduction SCMs. They are designed to capture the first portion of a rainfall event and to treat the runoff from a few hundred square meters of impervious cover.

BOX 5-7
Evaluation Through Monitoring: Villanova Pervious Concrete SCM

Villanova University's Stormwater Research and Demonstration Park is home to a pervious concrete infiltration site (Figure 5-33). The site, formerly a standard asphalt paved area, is located between two dormitories. The area was reconstructed in the summer of 2002 and outfitted with three infiltration beds overlain with pervious concrete. Usage of the site consists primarily of pedestrian traffic with some light automobile traffic. The pervious concrete site is designed to infiltrate small-volume storms (1 to 2 inches). Roof top runoff is directly piped to the rock bed under the concrete. For these smaller events, there is essentially no runoff from the site.

FIGURE 5-33 Villanova University pervious concrete retrofit site. SOURCE: Reprinted, with permission, from VUSP. Copyright by VUSP.

The pervious concrete is outlined with decorative pavers that divide the pervious concrete into three separate sections as seen in Figure 5-33. Underneath these three sections are individual storage beds. Since the site lies on a significant slope it was necessary to create earthen dams that isolate each storage area. At the top of each dam there is an overflow pipe which connects the storage area with the next one downstream. The final storage bed has an overflow that connects to the existing storm sewer. The beds are approximately 4 feet deep and are filled with stone, producing about 40 percent void space within the beds. A geotextile pervious liner was laid down to separate the storage beds from the undisturbed soil below (Figure 5-34). The primary idea was to avoid any upward migration of the in-situ soil, which could possibly reduce the capacity of the beds over time.

continues next page

BOX 5-7 Continued

FIGURE 5-34 Infiltration bed under construction. Pervious concrete has functionality and workability similar to that of regular concrete. However, the pervious concrete mix lacks the sand and other fine particles found in regular concrete. This creates a significant amount of void space which allows water to flow relatively unobstructed through the concrete. This site was the first attempt at creating a pervious concrete SCM in the area, and there were construction and material problems. Since that time the industry has matured, and a second site on campus constructed in 2007 has not had any significant difficulties. SOURCE: Reprinted, with permission, from VUSP. Copyright by VUSP.

Note the runoff from impervious concrete spilling over to the pervious concrete. SOURCE: Robert Traver, Villanova University

Continuous monitoring of the site over a number of years has considerably increased our understanding of infiltration. Similar to the bioinfiltration site (Box 5-6), the infiltration rate of permeable concrete does vary as a function of temperature (Braga et al., 2007; Emerson and Traver, 2008), and the SCM volume reduction is impressive. As shown in Figure 5-35, over 95 percent of the yearly rainfall was infiltrated with minimal overflow. Besides hydrologic plots, water quality plots also show the benefits of permeable concrete (Kwiatkowski et al., 2007). Because over 95 percent of the runoff is infiltrated, well over 95 percent of the pollutant mass is also removed. Figure 5-36 shows the level of copper extracted from lysimeters buried under the rock bed and surrounding grass. The plot is arranged in quartiles, with readings in milligrams per liter. Lysimeter samples from under the surrounding grass and one foot and four feet under the infiltration bed all report almost no copper, compared to samples taken from the port in the rock bed and from the gutters draining the roof tops.

continues next page

BOX 5-7 Continued

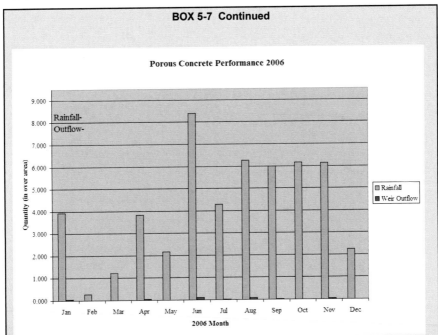

Porous Concrete Performance 2006

FIGURE 5-35 Rainfall and corresponding outflow from the weir of the SCM. SOURCE: Reprinted, with permission, from VUSP. Copyright by VUSP.

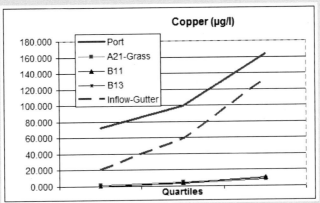

FIGURE 5-36 Copper measured at various locations. The three quartiles correspond to the 25th, 50th, and 75th percentile value of all data collected. A21 is a lysimeter location under the surrounding grass, while B11 and B13 refer to locations that are one foot and four feet under the infiltration bed, respectively. SOURCE: Reprinted, with permission, from VUSP. Copyright by VUSP.

As discussed earlier, several studies have measured the runoff volume reduction of individual LID practices. Fewer studies are available on whether multiple LID practices, when used together, have a cumulative benefit at the neighborhood or catchment scale. Four monitoring studies have clearly documented a major reduction in runoff from developments that employ LID and Better Site Design (see Box 5-8) compared to those that do not. In addition, six studies have documented the runoff reduction benefits of LID at the catchment or watershed scale using a modeling approach (Alexander and Heaney, 2002; Stephens et al., 2002; Holman-Dodds et al., 2003; Coombes, 2004; Hardy et al., 2004; Huber et al., 2006).

Peak Flow Reduction and Runoff Treatment

After efforts are made to prevent the generation of pollutants and to reduce the volume of runoff that reaches stormwater systems, stormwater management focuses on the reduction of peak flows and associated treatment of polluted runoff. The main class of SCMs used to accomplish this is extended detention basins, versions of which have dominated stormwater management for decades. These include a wide variety of ponds and wetlands, including wet ponds (also known as retention basins), dry extended detention ponds (as known as detention basins), and constructed wetlands. By holding a volume of stormwater runoff for an extended period of time, extended detention SCMs can achieve both water quality improvement and reduced peak flows. Generally the goal is to hold the flows for 24 hours at a minimum to maximize the opportunity of settling, adsorption, and transformation of pollutants (based on past pollutant removal studies) (Rea and Traver, 2005). For smaller storm events (one- to two-year storms), this added holding time also greatly reduces the outflows from the SCM to a level that the stream channel can handle. Most wet ponds and stormwater wetlands can hold a "water quality" volume, such that the flows leaving in smaller storms have been held and "treated" for multiple days. Extended detention dry ponds greatly reduce the outflow peaks to achieve the required residence times.

Usually extended detention devices are lower in the treatment train of SCMs, if not at the end. This is both due to their function (they are designed for larger events) and because the required water sources and less permeable soils needed for these SCMs are more likely to be found at the lower areas of the site. Some opportunities exist to naturalize dry ponds or to retrofit wet ponds into stormwater wetlands but it depends on their site configuration and hydrology. Stormwater wetlands are shown in Figures 5-40 and 5-41. A wet pond and a dry extended detention basin are shown in Figures 5-42 and 5-43.

Simple ponds are little more than a hole in the ground, in which stormwater is piped in and out. Dry ponds are meant to be dry between storms, whereas wet ponds have a permanent pool throughout the year. Detention basins reduce peak flows by restricting the outflows and creating a storage area. Depending on the

BOX 5-8
Jordan Cove—An LID Watershed Project

LID refers to the use of a system of small, on-site SCMs to counteract increases in flow and pollution following development and to control smaller runoff events. Although some studies are available that measure the runoff volume reduction of individual LID practices, fewer studies are available on whether multiple LID practices, when used together, have a cumulative benefit at the neighborhood or catchment scale. Of those listed in Table 5-6, Jordan Cove is the most extensively studied, as it was monitored for ten years as part of a paired watershed study that included a site with no SCMs and a site with traditional (detention) SCMs. The watersheds were monitored during calibration, construction, and post-construction periods. The project consisted of 12 lots, and the SCMs used were bioretention, porous pavements, no-mow areas, and education for the homeowners (Figure 5-37).

TABLE 5-6 Review of Recent LID Monitoring Research on a Catchment Scale

Location	Practices	Runoff Reduction
Jordan Cove, USA Dietz and Clausen (2008)	Permeable pavers, bioretention, grass swales, education	84%
Somerset Heights, USA Cheng et al. (2005)	Grass swale, bioretention, and roof-top disconnection	45%
Figtree Place, Australia Coombes et al. (2000)	Rain tanks, infiltration trenches, swales	100%

FIGURE 5-37 Jordan Cove LID subdivision. SOURCE: Reprinted, with permission, from Clausen (2007). Copyright 2007 by John Clausen.

Figure 5-38 (right panel) displays the hydrograph from a post-construction storm comparing the LID, traditional, and control watersheds. Note that the traditional watershed shows the delay and peak reduction from the detention basins, while the LID watershed has almost no runoff. The LID watershed was found to reduce runoff volume by 74 percent by increasing infiltration over preconstruction levels.

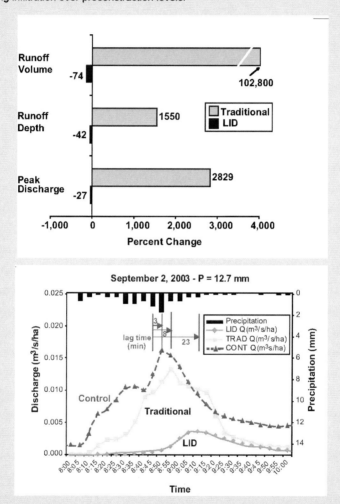

FIGURE 5-38. Significant changes in runoff volume (m³/week), runoff depth (cm/week) and peak discharge (m³/sec/week) after construction was completed (top panel). Hydrograph of all three subdivisions in the project, showing the larger volume and rate of runoff from the traditional and control subdivisions, as compared to the LID (bottom panel). SOURCE: Reprinted, with permission, from Clausen (2007). Copyright 2007 by John Clausen.

continues next page

BOX 5-8 Continued

Comparisons of nutrient and metal concentrations and total export in the surface water shows the value of the LID approach as well as the significance of the reduction in runoff volume. Figure 5-39 shows the changes in pollutant concentration and mass export before and after construction for the traditional and LID subdivisions. Note that concentrations of TSS and nutrients are increased in the LID subdivision (left-hand panel); this is because swales and natural systems are used in place of piping as a "green" drainage system and because only larger storms leave the site. The right-hand panel shows how the large reduction in runoff achieved through infiltration can dramatically reduce the net export of pollutants from the LID watershed.

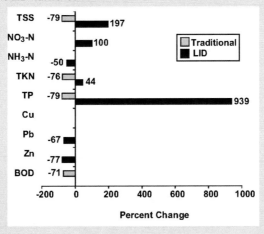

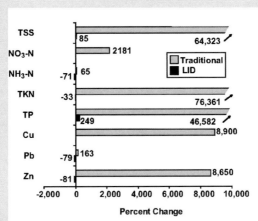

FIGURE 5-39 Significant changes in pollutant concentration, after construction was completed (top). Units are mg/L for NO₃-N, NH₃-N, TKN, TP, and BOD, and µg/L for Cu, Pb, and Zn. Significant changes in mass export (kg/ha/year) after construction was completed (bottom). SOURCE: Reprinted, with permission, from Clausen (2007). Copyright 2007 by John Clausen.

FIGURE 5-40 Constructed wetland. SOURCE: PaDEP (2006).

FIGURE 5-41 Retrofitted stormwater wetland at Villanova University. SOURCE: Re-printed, with permission, from VUSP. Copyright by VUSP.

FIGURE 5-42 Wet pond. SOURCE: PaDEP (2006).

FIGURE 5-43 Dry extended detention pond. SOURCE: PaDEP (2006).

detention time, outflows can be reduced to levels that do not accelerate erosion, that protect the stream channel, and that reduce flooding.

The flow normally enters the structure through a sediment forebay (Figure 5-44), which is included to capture incoming sediment, remove the larger particles through settling, and allow for easier maintenance. Then a meandering path or cell structure is built to "extend" and slow down the flows. The main basin is a large storage area (sometimes over the meandering flow paths). Finally, the runoff exits through an outflow control structure built to retard flow.

Wet ponds, stormwater wetlands, and (to a lesser extent) dry extended detention ponds provide treatment. The first step in treatment is the settling of larger particles in the sediment forebay. Next, for wet ponds a permanent pool of water is maintained so that, for smaller storms, the new flows push out a volume that has had a chance to interact with vegetation and be "treated." This volume is equivalent to an inch of rain over the impervious surfaces in the drainage area. Thus, what exits the SCM during smaller storm events is baseflow contributions and runoff that entered during previous events. For dry extended detention ponds, there is no permanent pool and the outlet is instead greatly restricted. For all of these devices, vegetation is considered crucial to pollutant removal. Indeed, wet ponds are designed with an aquatic bench around the edges to promote contact with plants. The vegetation aids in reduction of flow velocities, provides growth surfaces for microbes, takes up pollutants, and provides filtering (Braskerud, 2001).

The ability of detention structures to achieve a certain level of control is size related—that is, the more peak flow reduction or pollutant removal required, the more volume and surface area are needed in the basin. Because it is not simply the peak flows that are important, but also the duration of the flows that cause damage to the stream channels (McCuen, 1979; Loucks et al., 2005),

FIGURE 5-44 Villanova University sediment forebay. SOURCE: Reprinted, with permission, from VUSP. Copyright by VUSP.

some detention basins are currently sized and installed in series with runoff-volume-reduction SCMs.

The strength of extended detention devices is the opportunity to create habitats or picturesque settings during stormwater management. The weaknesses of these measures include large land requirements, chloride buildup, possible temperature effects, and the creation of habitat for undesirable species in urban areas. There is a perception that these devices promote mosquitoes, but that has not been found to be a problem when a healthy biological habitat is created (Greenway et al., 2003). Another drawback of this class of SCMs is that they often have limited treatment capacity, in that they can reduce pollutants in stormwater only to a certain level. These so-called irreducible effluent concentrations have been documented mainly for ponds and stormwater wetlands, as well as sand filters and grass channels (Schueler, 1998). Finally, it should be noted that either a larger watershed (10–25 acres; CWP, 2004) or a continuous water source is needed to sustain wet ponds and stormwater wetlands.

Maintenance requirements for extended detention basins and wetlands include the removal of built-up sediment from the sediment forebay, harvesting of grasses to remove accumulated nutrients, and repair of berms and structures after storm events. Inspection items relate to the maintenance of the berm and sediment forebay.

While the basic hydrologic function of extended detention devices is well known, their performance on a watershed basis is not. Because they do not significantly reduce runoff volume and are designed on a site-by-site basis using synthetic storm patterns, their exclusive use as a flood reduction strategy at the watershed scale is uncertain (McCuen, 1979; Traver and Chadderton, 1992). Much of this variability is reduced when they are coupled with volume reduction SCMs at the watershed level. Pollutant removal is effected by climate, short-circuiting, and by the schedule of sediment removal and plant harvesting. Extreme events can resuspend captured sediments, thus reintroducing them into the environment. Although there is debate, it seems likely that plants will need to be harvested to accomplish nutrient removal (Reed et al., 1998).

Runoff Treatment

As mentioned above, many SCMs associated with runoff volume reduction and extended detention provide a water quality benefit. There are also some SCMs that focus primarily on water quality with little peak flow or volume effect. Designed for smaller storms, these are usually based on filtration, hydrodynamic separation, or small-scale bioretention systems that drain to a subsequent receiving water or other device. Thus, often these SCMs are used in conjunction with other devices in a treatment train or as retrofits under parking lots. They can be very effective as pretreatment devices when used "higher up" in the watershed than infiltration structures. Finally, in some cases these SCMs are specifically designed to reduce peak flows in addition to providing water quality benefits by introducing elements that make them similar to detention basins; this is particularly the case for sand filters.

The sand filter is relied on as a treatment technology in many regions, particular those where stream geomorphology is less of a concern and thus peak flow control and runoff volume reduction are not the primary goals. These devices can be effective at removing suspended sediments and can extend the longevity and performance of runoff-volume-reduction SCMs. They are also one of the few urban retrofits available, due to the ability to implement them within traditional culvert systems. Figures 5-45 and 5-46 show designs for the Austin sand filter and the Delaware sand filter.

Filters use sand, peat, or compost to remove particulates, similar to the processes used in drinking water plants. Sand filters primarily remove suspended solids and ammonia nitrogen. Biological material such as peat or compost provides adsorption of contaminants such as dissolved metals, hydrocarbons, and other organic chemicals. Hydrodynamic devices use rotational forces to separate the solids from the flow, allowing the solids to settle out of the flow stream. There is a recent class of bioretention-like manufactured devices that combine inlets with planters. In these systems, small volumes are directed to a soil planter area, with larger flows bypassing and continuing down the storm sewer system. In any event, for manufactured items the user needs to look to the manufacturer's published and reviewed data to understand how the device should be applied.

FIGURE 5-45 Austin sand filter. SOURCE: Robert Traver, Villanova University.

FIGURE 5-46 Delaware sand filter. SOURCE: Tom Schueler, Chesapeake Stormwater Network.

The level of control that can be achieved with these SCMs depends entirely on sizing of the device based on the incoming flow and pollutant loads. Each unit has a certified removal rate depending on inflow to the SCM. Also all units have a maximum volume or rate of flow they can treat, such that higher flows are bypassed with no treatment. Thus, the user has to determine what size unit is needed and the number to use based on the area's hydrologic cycle and what criteria are to be met.

With the exception of some types of sand filters, the strengths of water quality SCMs are that they can be placed within existing infrastructure or under parking lots, and thus do not take up land that may be used for other purposes. They make excellent choices for retrofit situations. For filters, there is a wealth of experience from the water treatment community on their operations. For all manufactured devices there are several testing protocols that have been set up to validate the performance of the manufactured devices (the sufficiency of which is discussed in Box 5-9). Weaknesses of these devices include their cost and maintenance requirements. Regular maintenance and inspection at a high level are required to remove captured pollutants, to replace mulch, or to rake and remove the surface layer to prevent clogging. In some cases specialized equipment (vacuum trucks) is required to remove built-up sediment. Although the underground placement of these devices has many benefits, it makes it easy to neglect their maintenance because there are no signs of reduced performance on the surface. Because these devices are manufactured, the unit construction cost is usually higher than for other SCMs. Finally, the numerous testing protocols are confusing and prevent more widespread applications.

The chief uncertainty with these SCMs is due to the lack of certification of some manufactured devices. There is also concern about which pollutants are removed by which class of device. For example, hydrodynamic devices and sand filters do not address dissolved nutrients, and in some cases convert suspended pollutants to their dissolved form. Both issues are related to the false perception that a single SCM must be found that will comprehensively treat stormwater. Such pressures often put vendors in a position of trying to certify that their devices can remove all pollutants. Most often, these devices can serve effectively as part of a treatment train, and should be valued for their incremental contributions to water quality treatment. For example, a filter that removes sediment upstream of a bioinfiltration SCM can greatly prolong the life of the infiltration device.

Aquatic Buffers and Managed Floodplains

Aquatic buffers, sometimes also known as stream buffers or riparian buffers, involve reserving a vegetated zone adjacent to streams, shorelines, or wetlands as part of development regulations or as an ordinance. In most regions of the country, the buffer is managed as forest, although in arid or semi-arid

---

BOX 5-9
Insufficient Testing of Proprietary Stormwater Control Measures

Manufacturers of proprietary SCMs offer a service that can save municipalities time and money. Time is saved by the ability of the manufactures to quickly select a model matching the needs of the site. A city can minimize the cost of buying the product by requiring the different manufacturers to submit bids for the site. All the benefits of the service will have no meaning, however, if the cities cannot trust the performance claims of the different products. Because the United States does not have, at this time, a national program to verify the performance of proprietary SCMs, interested municipalities face a high amount of uncertainty when they select a product. Money could be wasted on products that might have the lowest bid, but do not achieve the water quality goals of the city or state.

The EPA's Environmental Technology Verification (ETV) program was created to facilitate the deployment of innovative or improved environmental technologies through performance verification and dissemination of information. The Wet Weather Flow Technologies Pilot was established as part of the ETV program to verify commercially available technologies used in the abatement and control of urban stormwater runoff, combined sewer overflows, and sanitary sewer overflows. Ten proprietary SCMs were tested under the ETV program (see Figure 5-47), and the results of the monitoring are available on the National Sanitation Foundation International website. Unfortunately, the funding for the ETV program was discontinued before all the stormwater products could be tested. Without a national testing program some states have taken a more regional approach to verifying the performance of proprietary practices, while most states do not have any type of verification or approval program.

The Washington Department of Ecology has supported a testing protocol called Technology Assessment Protocol–Ecology that describes a process for evaluating and reporting on the performance and appropriate uses of emerging SCMs. California, Massachusetts, Maryland, New Jersey, Pennsylvania, and Virginia have sponsored a testing program called Technology Acceptance and Reciprocity Partnership (TARP), and a number of products are being tested in the field. The State of Wisconsin has prepared a draft technical standard (1006) describing methods for predicting the site-specific reduction efficiency of proprietary sedimentation devices. To meet the criteria in the standard the manufacturers can either use a model to predict the performance of the practice or complete a laboratory protocol designed to develop efficiency curves for each product. Although none of these state or federal verification efforts have produced enough information to sufficiently reduce the uncertainty in selection and sizing of proprietary SCMs, many proprietary practices are being installed around the country, because of the perceived advantage of the service being provided by the manufacturers and the sometimes overly optimistic performance claims.

All those involved in stormwater management, including the manufacturers, will have a much better chance of implementing a cost-effective stormwater program in their cities if the barriers to a national testing program for proprietary SCMs are eliminated. Two of the barriers to the ETV program were high cost and the transferability of the results. Also, the ETV testing did not produce results that could be used in developing efficiency curves for the product. A new national testing program could reduce the cost by using laboratory testing instead of field testing. Each manufacturer would only have to do one series of tests in the lab and the results would be applicable to the entire country. The laboratory

continues next page

BOX 5-9 Continued

protocol in the Wisconsin Technical Standard 1006 provides a good example of what should be included to evaluate each practice over a range of particle sizes and flows. These types of laboratory data could also be used to produce efficiency curves for each practice. It would be relatively easy for state and local agencies to review the benefits of each installation if the efficiency curves were incorporated into urban runoff models, such as WinSLAMM or P8.

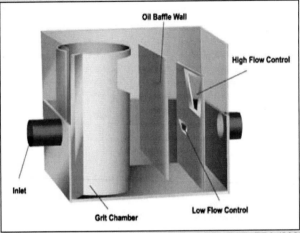

Stormwater 360 Hydrodynamic Separator. SOURCE: EPA (2005c).

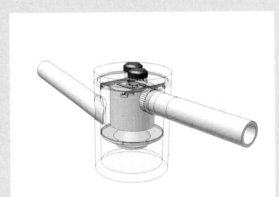

Downstream Defender. SOURCE: Available online at http://epa.gov/Region1/assistance/ ceitts/stormwater/techs/downstreamdefender.html

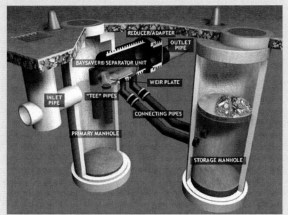

Bay Seperator. SOURCE: EPA (2005a).

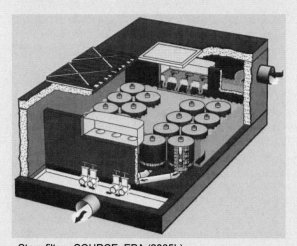

Stormfilter. SOURCE: EPA (2005b).

FIGURE 5-47 Proprietary Manufactured Devices tested by the ETV Program.

regions it may be managed as prairie, chapparal, or other cover. When properly designed, buffers can both reduce runoff volumes and provide water quality treatment to stormwater.

The performance of urban stream buffers cannot be predicted from studies of buffers installed to remove sediment and nutrients from agricultural areas (Lowrance and Sheridan, 2005). Agricultural buffers have been reported to have high sediment and nutrient removal because they intercept sheet flow or shallow groundwater flow in the riparian zone. By contrast, urban stream buffers often receive concentrated surface runoff or may even have a storm-drain pipe that short-circuits the buffer and directly discharges into the stream. Consequently, the pollutant removal capability of urban stream buffers is limited, unless they are specifically designed to distribute and treat stormwater runoff (NRC, 2000). This involves the use of level spreaders, grass filters, and berms to transform concentrated flows into sheet flow (Hathaway and Hunt, 2006). Such designed urban stream buffers have been applied widely in the Neuse River basin to reduce urban stormwater nutrient inputs to this nitrogen-sensitive waterbody.

The primary benefit of buffers is to help maintain aquatic biodiversity within the stream. Numerous researchers have evaluated the relative impact of riparian forest cover and impervious cover on stream geomorphology, aquatic insects, fish assemblages, and various indexes of biotic integrity. As a group, the studies suggest that indicator values for urban stream health increase when riparian forest cover is retained over at least 50 to 75 percent of the length of the upstream network (Goetz et al., 2003; Wang et al., 2003b; McBride and Booth, 2005; Moore and Palmer, 2005). The width of the buffer is also important for enhancing its stream protection benefits, and it ranges from 25 to 200 feet depending on stream order, protection objectives, and community ordinances. At the present time, there are no data to support an optimum width for water quality purposes. The beneficial impact of riparian forest cover is less detectable when watershed impervious cover exceeds 15 percent, at which point degradation by stormwater runoff overwhelms the benefits of the riparian forest (Roy et al., 2005, 2006; Walsh et al., 2007).

Maintenance, inspection, and compliance for buffers can be a problem. In most communities, urban stream buffers are simply a line on a map and are not managed in any significant way after construction is over. As such, urban stream buffers are prone to residential encroachment and clearing, and to colonization by invasive plants. Another important practice is to protect, preserve, or otherwise manage the ultimate 100-year floodplain so that vulnerable property and infrastructure are not damaged during extreme floods. Federal Emergency Management Agency (FEMA), state, and local requirements often restrict or control development on land within the floodway or floodplain. In larger streams, the floodway and aquatic buffer can be integrated together to achieve multiple social objectives.

Stream Rehabilitation

While not traditionally considered an SCM, certain stream rehabilitation practices or approaches can be effective at recreating stream physical habitat and ecosystem function lost during urbanization. When combined with effective SCMs in upland areas, stream rehabilitation practices can be an important component of a larger strategy to address stormwater. From the standpoint of mitigating stormwater impacts, four types of urban stream rehabilitation are common:

• Practices that stabilize streambanks and/or prevent channel incision/enlargement can reduce downstream delivery of sediments and attached nutrients (see Figure 5-48). Although the magnitude of sediment delivery from urban-induced stream-channel enlargement is well documented, there are very few published data to quantify the potential reduction in sediment or nutrients from subsequent channel stabilization.

• Streams can be hydrologically reconnected to their floodplains by building up the profile of incised urban streams using grade controls so that the channel and floodplain interact to a greater degree. Urban stream reaches that have been so rehabilitated have increased nutrient uptake and processing rates, and in particular increased denitrification rates, compared to degraded urban streams prior to treatment (Bukavecas, 2007; Kaushal et al., 2008). This suggests that urban stream rehabilitation may be one of many elements that can be considered to help decrease loads in nutrient-sensitive watersheds.

• Practices that enhance in-stream habitat for aquatic life can improve the expected level of stream biodiversity. However, Konrad (2003) notes that improvement of biological diversity of urban streams should still be considered an experiment, since it is not always clear what hydrologic, water quality, or habitat stressors are limiting. Larson et al. (2001) found that physical habitat improvements can result in no biological improvement at all. In addition, many of the biological processes in urban stream ecosystems remain poorly understood, such as carbon processing and nutrient uptake.

• Some stream rehabilitation practices can indirectly increase stream biodiversity (such as riparian reforestation, which could reduce stream temperatures, and the removal of barriers to fish migration).

It should be noted that the majority of urban stream rehabilitation projects undertaken in the United States are designed for purposes other than mitigating the impacts of stormwater or enhancing stream biodiversity or ecosystem function (Bernhardt et al., 2005). Most stream rehabilitation projects have a much narrower design focus, and are intended to protect threatened infrastructure,

FIGURE 5-48 Three photographs illustrate stream rehabilitation in Denver. The top picture is a creek that has eroded in its bed due to urbanization. The middle picture shows a portion of the stabilized creek immediately after construction. Check structures, which keep the creek from cutting its bed, are visible in the middle distance. The bottom image shows the creek just upstream of one of the check structures two years after stabilization. The thickets of willows established themselves naturally. The only revegetation performed was to seed the area for erosion control. SOURCE: Courtesy of Wenk Associates.

naturalize the stream corridor, achieve a stable channel, or maintain local bank stability (Schueler and Brown, 2004). Improvements in either biological health or the quality of stormwater runoff have rarely been documented.

Unique design models and methods are required for urban streams, compared to their natural or rural counterparts, given the profound changes in hydrologic and sediment regime and stream–floodplain interaction that they experience (Konrad, 2003). While a great deal of design guidance on urban stream rehabilitation has been released in recent years (FISRWG, 2000; Doll and Jennings, 2003; Schueler and Brown, 2004), most of the available guidance has not yet been tailored to produce specific outcomes for stormwater mitigation, such as reduced sediment delivery, increased nutrient processing, or enhanced stream biodiversity. Indeed, several researchers have noted that many urban stream rehabilitation projects fail to achieve even their narrow design objectives, for a wide range of reasons (Bernhardt and Palmer, 2007; Sudduth et al., 2007). This is not surprising given that urban stream rehabilitation is relatively new and rarely addresses the full range of in-stream alteration generated by watershed-scale changes. This shortfall suggests that much more research and testing are needed to ensure urban stream habilitation can meet its promise as an emerging SCM.

Municipal Housekeeping (Street Sweeping and Storm-Drain Cleanouts)

Phase II NPDES stormwater permits specifically require municipal good housekeeping as one of the six minimum management measures for MS4s. Although EPA has not presented definitive guidance on what constitutes "good housekeeping", CWP (2008) outlines ten municipal operations where housekeeping actions can improve the quality of stormwater, including the following:

- municipal hotspot facility management,
- municipal construction project management,
- road maintenance,
- street sweeping,
- storm-drain maintenance,
- stormwater hotline response,
- landscape and park maintenance ,
- SCM maintenance, and
- employee training.

The overarching theme is that good housekeeping practices at municipal operations provide source treatment of pollutants before they enter the storm-drain system. The most frequently applied practices are street sweeping (Figure 5-49) and sediment cleanouts of sumps and storm-drain inlets. Most communities

FIGURE 5-49 Vacuum street sweeper at Villanova University. SOURCE: Robert Traver, Villanova University.

conduct both operations at some frequency for safety and aesthetic reasons, although not specifically for the sake of improving stormwater quality (Law et al., 2008).

Numerous performance monitoring studies have been conducted to evaluate the effect of street sweeping on the concentration of stormwater pollutants in downstream storm-drain pipes (see Pitt, 1979; Bender and Terstriep, 1994; Brinkman and Tobin, 2001; Zarrielo et al., 2002; Chang et al., 2005; USGS, 2005; Law et al., 2008). The basic finding is that regular street sweeping has a low or limited impact on stormwater quality, depending on street conditions, sweeping frequency, sweeper technology, operator training, and on-street parking. Sweeping will always have a limited removal capability because rainfall events frequently wash off pollutants before the sweeper passes through, and only some surfaces are accessible to the sweeper, thus excluding sidewalk, driveways, and landscaped areas. Frequent sweeping (i.e., weekly or monthly) has a moderate capability to remove sediment, trash and debris, coarse solids, and organic matter.

Fewer studies have been conducted on the pollutant removal capability of frequent sediment cleanout of storm-drain inlets, most in regions with arid climates (Lager et al., 1977; Mineart and Singh, 1994; Morgan et al., 2005). These studies have shown some moderate pollutant removal if cleanouts are done on a monthly or quarterly basis. Most communities, however, report that they clean out storm drains on an annual basis or in response to problems or drainage complaints (Law, 2006).

Frequent sweeping and cleanouts conducted on the dirtiest streets and storm drains appear to be the most effective way to include these operations in the stormwater treatment train. However, given the uncertainty associated with the expected pollutant removal for these practices, street sweeping and storm-drain cleanout cannot be relied on as the sole SCMs for an urban area.

Illicit Discharge Detection and Elimination

MS4 communities must develop a program to detect and eliminate illicit discharges to their storm-drain system as a stormwater NPDES permit condition. Illicit discharges can involve illegal cross-connections of sewage or washwater into the storm-drain system or various intermittent or transitory discharges due to spills, leaks, dumping, or other activities that introduce pollutants into the storm-drain system during dry weather. National guidance on the methods to find and fix illicit discharges was developed by Brown et al. (2004). Local illicit discharge detection and elimination (IDDE) programs represent an ongoing and perpetual effort to monitor the network of pipes and ditches to prevent pollution discharges.

The water quality significance of illicit discharges has been difficult to define since they occur episodically in different parts of a municipal storm drain system. Field experience in conducting outfall surveys does indicate that illicit discharges may be present at 2 to 5 percent of all outfalls at any given time. Given that pollutants are being introduced into the receiving water during dry weather, illicit discharges may have an amplified effect on water quality and biological diversity.

Many communities indicate that they employ a citizen hotline to report illicit discharges and other water quality problems (Brown et al., 2004), which sharply increases the number of illicit discharge problems observed.

Stormwater Education

Like IDDE, stormwater education is one of the six minimum management measures that MS4 communities must address in their stormwater NPDES permits. Stormwater education involves municipal efforts to make sure individuals understand how their daily actions can positively or negatively influence water quality and work to change specific behaviors linked to specific pollutants of concern (Schueler, 2001c). Targeted behaviors include lawn fertilization, littering, car fluid recycling, car washing, pesticide use, septic system maintenance, and pet waste pickup. Communities may utilize a wide variety of messages to make the public aware of the behavior and more desirable alternatives through radio, television, newspaper ads, flyers, workshops, or door-to-door outreach. Several communities have performed before-and-after surveys to assess both the penetration rate for these campaigns and their ability to induce changes in actual behaviors. Significant changes in behaviors have been recorded (see Schueler,

2002), although few studies are available to link specific stormwater quality improvements to the educational campaigns (but see Turner, 2005; CASQA, 2007).

Residential Stewardship

This SCM involves municipal programs to enhance residential stewardship to improve stormwater quality. Residents can undertake a wide range of activities and practices that can reduce the volume or quality of runoff produced on their property or in their neighborhood as a whole. This may include installing rain barrels or rain gardens, planting trees, xeriscaping, downspout disconnection, storm-drain marking, household hazardous waste pickups, and yard waste composting (CWP, 2005). This expands on stormwater education in that a municipality provides a convenient delivery service to enable residents to engage in positive watershed behavior. The effectiveness of residential stewardship is enhanced when carrots are provided to encourage the desired behavior, such as subsidies, recognition, discounts, and technical assistance (CWP, 2005). Consequently, communities need to develop a targeted program to educate residents and help them engage in the desired behavior.

SCM Performance Monitoring and Modeling

Stormwater is characterized by widely fluctuating flows. In addition, inflow pollutant concentrations vary over the course of a storm and can be a function of time since the last storm, watershed, size and intensity of rainfall, season, amount of imperviousness, pollutant of interest, and so forth. This variability of the inflow to SCMs along with the very nature of SCMs makes performance monitoring a complex task. Most SCMs are built to manage stormwater, not to enable flow and water quality monitoring. Furthermore, they are incorporated into the collection system and spread throughout developments. Measurement of multiple inflows, outflows, evapotranspiration, and infiltration are simply not feasible for most sites. Many factors, such as temperature and climate, play a role in how well SCMs function. Infiltration rates can vary by an order of magnitude as a function of temperature (Braga et al., 2007; Emerson and Traver, 2008), such that a reading in late summer might be twice that of a winter reading. Determining performance can be further complicated because, e.g., at the start of a storm a detention basin could still be partially full from a previous storm, and removal rates for wetlands are a function of the growing season, not to mention snowmelt events.

Monitoring of SCMs is usually performed for one of two purposes: functionality or more intensive performance monitoring. Monitoring of functionality is primarily to establish that the SCM is functioning as designed. Performance monitoring is focused on determining what level of performance is achieved by the SCM.

Functionality Monitoring

Functionality monitoring, in a broad sense, involves checking to see whether the SCM is functioning and screening it for potential problems. Both the federal and several state industrial and construction stormwater general permits have standard requirements for visual inspections following a major storm event. Visual observations of an SCM by themselves do not provide information on runoff reduction or pollutant removal, but rather only that the device is functioning as designed. Adding some grab samples for laboratory analysis can act as a screening tool to determine if a more complex analysis is required.

The first step of functionality monitoring for any SCM is to examine the physical condition of the device (piping, pervious surfaces, outlet structure, etc.). Visual inspection of sediments, eroded berms, clogged outlets, and other problems are good indications of the SCM's functionality (see Figure 5-50). For infiltration devices, visiting after a storm event will show whether or not the device is functioning. A simple staff gauge (Figure 5-51) or a stilling well in pervious pavement can be used to measure the amount of water-level change over several days to estimate infiltration rates. Minnesota suggests the use of fire equipment or hydrants to fill infiltration sites with a set volume of water to measure the rate of infiltration. For sites that are designed to capture a set volume, for example a green roof, a visit could be coordinated with a rainfall event of the appropriate size to determine whether there is overflow during the event. If so, then clearly further investigation is required.

FIGURE 5-50 Rusted outlet structure. SOURCE: Reprinted, with permission, from Emerson. Copyright by Clay Emerson.

FIGURE 5-51 Staff gauge attached to ultra-sonic sensor after a storm. SOURCE: Reprinted, with permission, from VUSP. Copyright by VUSP.

For extended detention and stormwater wetlands, the depth of water during an event is an indicator of how well the SCM is functioning. Usually high-water marks are easy to determine due to debris or mud marks on the banks or the structures. If the size of the storm event is known, the depths can be compared to what was expected for the structure. Other indicators of problems would include erosion downstream of the SCM, algal blooms, invasive species, poor water clarity, and odor.

For water quality and manufactured devices, visual inspections after a storm event can determine whether the SCM is functioning properly. Standing water over a sand or other media filter 48 hours after a storm is a sign of problems. Odor and lack of flow clarity could be a sign of filter breakthrough or other problems. For manufactured devices, literature about the device should specify inspection and maintenance procedures.

Monitoring of nonstructural SCMs is almost exclusively limited to visual observation due to the difficulty in applying numerical value to their benefits. Visual inspection can identify eroded stream buffers, additional paved areas, or denuded conservation areas (see Figure 5-52).

Performance Monitoring

Performance monitoring is an extremely intensive effort to determine the performance of an SCM over either an individual storm event or over a series of

FIGURE 5-52 Wooded conservation area stripped of trees. Note pile of sawdust. SOURCE: Robert Traver, Villanova University.

storms. It requires integration of flow and water quality data creating both a hydrograph and a polutograph for a storm event as shown in Figure 5-53. The creation of these graphs requires continuous monitoring of the hydrology of the site and multiple water quality samples of the SCM inflow and outflow, the vadose zone, and groundwater. Event mean concentrations can then be determined from these data. There should be clear criteria for the number and type of storms to be sampled and for the conditions preceding a storm. For example, for most SCMs it would be improper to sample a second storm event in series, as the inflow may be free of pollutants and the soil moisture filled, resulting in a poor or negative performance. (Extended detention basins are an exception because the outflow during a storm event may include inflows from previous events.) The size of the sampled storm is also important. If the water quality goal is focused on smaller events, the 100-year storm would not give a proper picture of the performance because the occurrence is so rare that it is not a water quality priority.

For runoff-volume-reduction SCMs, performance monitoring can be extremely difficult because these systems are spread over the project site. The monitoring program must consider multiple-size storms because these SCMs are designed to remove perhaps the first inch of runoff. Therefore, for storms of less than an inch, there is no surface water release, so the treatment is 100

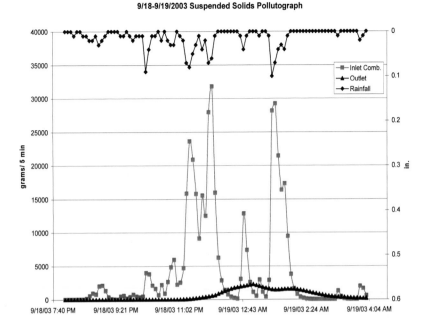

FIGURE 5-53 Example polutograph that displays inflow and outflow TSS during a storm event from the Villanova wetland stormwater SCM. SOURCE: Reprinted, with permission, Rea and Traver (2005). Copyright 2005 by the American Society of Civil Engineers.

percent effective for surface discharges. During larger events, a bioretention SCM or green roof may export pollutants. When viewed over the entire spectrum of storms, these devices are an outstanding success; however, this may not be evident during a hurricane.

Through the use of manufactured weirs (Figure 5-54), it is possible to develop flow-depth criteria based on hydraulic principles for surface flows entering or leaving the SCM. Where this is not practical, various manufacturers have Doppler velocity sensors that, combined with geometry and depth, provide a reasonable continuous record of flow. Measurement of depth within a device can be accomplished through use of pressure transducers, bubblers, float gauges, and ultrasonic sensors. Other common measures would include rainfall and temperature. One advantage of these data recording systems is that they can be connected to water quality probes and automated samplers to provide a flow-weighted sample of the event for subsequent laboratory analysis. Field calibration and monitoring of these systems is required.

Groundwater sampling for infiltration SCMs is a challenge. Although the rate of change in water depth can indicate volume moving into the soil mantle, it is difficult to establish whether this flow is evapotranspirated or ends up as base-flow or deep groundwater input. Sampling in the vadose zone can be established

FIGURE 5-54 Weir flow used to measure flow rate. SOURCE: Robert Traver, Villanova University.

through the use of lysimeters that, through a vacuum, draw out water from the soil matrix. Soil moisture probes can give a rough estimation of the soil moisture content, and weighing lysimeters can establish evapotranspiration rates. Finally groundwater wells can be used to establish the effect of the SCM on the groundwater depth and quality during and after storm events.

Performance monitoring of extended detention SCMs is difficult because the inflows and outflows are variable and may extend over multiple days. Hydrologic monitoring can be accomplished using weirs (Figure 5-54), flow meters, and level detectors. The new generation of temperature, dissolved oxygen, and conductivity probes allows for automated monitoring. (It should be noted that in many cases the conductivity probes are observing chlorides, which are not generally removed by SCMs.) In many cases monitoring of the downstream stream-channel geomorphology and stream habitat may be more useful than performance monitoring when assessing the effect of the SCM.

The performance monitoring of treatment devices is straightforward and involves determining the pollutant mass inflows and outflows. Performance monitoring of manufactured SCMs has been established through several protocols. An example is TARP, used by multiple states (*http://www.dep.state.pa.us/ dep/deputate/pollprev/techservices/tarp/*). This requires the manufacturer to test their units according to a set protocol of lab or field experiments to set performance criteria. Several TARP member and other states have published revised protocols for their use. These and other similar criteria are evolving and the subject of considerable effort by industry organizations that include the American Society of Civil Engineers.

Finally, much needs to be done to determine the performance of nonstructural SCMs, for which little to no monitoring data are available (see Table 5-2). Currently most practitioners expand upon current hydrologic modeling techniques to simulate these techniques. For example, disconnection of impervious surfaces is often modeled by adding the runoff from the roof or parking area as distributed "rainfall" on the pervious area. Experiments and long-term monitoring are needed for these SCMs.

More information on SCM monitoring is available through the International Stormwater BMP Database (*http://www.bmpdatabase.org*).

Modeling of SCM performance

Modeling of SCMs is required to understand their individual performance and their effect on the overall watershed. The dispersed nature of their implementation, the wide variety of possible SCM types and goals, and the wide range of rainfall events they are designed for makes modeling of SCMs extremely challenging. For example, to model multiple SCMs on a single site may require simulation of many hydrologic and environmental processes for each SCM in series. Modeling these effects over large watersheds by simulating each SCM is not only impractical, but the noise in the modeling may make the simulation results suspect. Thus, it is critical to understand the model's purpose, limitations, and applicability.

As discussed in Chapter 4, one approach to simulating SCM performance is through mathematical representation of the unit processes. The large volumes of data needed for process-based models generally restrict their use to smaller-scale modeling. For flow this would start with the hydrograph entering the SCM and include infiltration, evapotranspiration, routing through the system, or whatever flow paths were applicable. The environmental processes that would need to be represented could include settling, adsorption, biological transformation, and soil physics. Currently there are no environmental process models that work across the range of SCMs. Rather, the state of art is to use general removal efficiencies from publications such as the International Stormwater BMP Database (http://www.bmpdatabase.org) and the Center for Watershed Protection's National Pollutant Removal Database (CWP, 2000b, 2007b). Unfortunately, this approach has many limitations. The percent removal used on a site and storm basis does not include storm intensity, period between the storms, land use, temperature, management practices, whether other SCMs are upstream, and so forth. It also should be noted that percent removals are a surface water statistic and do not address groundwater issues or include any biogeochemistry.

Mechanistic simulation of the hydrologic processes within an SCM is much advanced compared to environmental simulation, but from a modeling scale it is still evolving. Indeed, models such as the Prince George's County Decision Support System are greatly improved in that the hydrologic simulation of the SCM includes infiltration, but they still do not incorporate the more rigorous soil physics and groundwater interactions. Some models, such as the Stormwater Management Model (SWMM), have the capability to incorporate mechanistic descriptions of the hydrologic processes occurring inside an SCM.

At larger scales, simulation of SCMs is done primarily using lumped models that do not explicitly represent the unit processes but rather the overall effects. For example, the goal may be to model the removal of 2 cm of rainfall from every storm from bioinfiltration SCMs. Thus, all that would be needed is how many SCMs are present and their configuration and what their capabilities are within your watershed. What is critical for these models is to represent the interrelated processes correctly and to include seasonal effects. Again, the pollutant removal capability of the SCM is represented with removal efficiencies derived from publications.

Regardless of the scale of the model, or the extent to which it is mechanistic or not, nonstructural SCMs are a challenge. Limiting impervious surface or maintenance of forest cover have been modeled because they can be represented as the maintenance of certain land uses. However, aquatic buffers, disconnected impervious surfaces, stormwater education, municipal housekeeping, and most other nonstructural SCMs are problematic. Another challenge from a watershed perspective is determining what volume of pollutants comes from streambank erosion during elevated flows versus from nonpoint source pollution. Most hydrologic models do not include or represent in-stream processes.

In order to move forward with modeling of SCMs, it will be necessary to better understand the unit processes of the different SCMs, and how they differ for hydrology versus transformations. Research is needed to gather performance numbers for the nonstructural SCMs. Until such information is available, it will be virtually impossible to predict that an individual SCM can accomplish a certain level of treatment and thus prevent a nearby receiving water from violating its water quality standard.

DESIGNING SYSTEMS OF STORMWATER CONTROL MEASURES ON A WATERSHED SCALE

Most communities have traditionally relied on stormwater management approaches that result in the design and installation of SCMs on a site-by-site basis. This has created a large number of individual stormwater systems and SCMs that are widely distributed and have become a substantial part of the contemporary urban and suburban landscape. Typically, traditional stormwater infrastructure was designed on a subdivision basis to reduce peak storm flow rates to predevelopment levels for large flood events (> 10-year return period). The problem with the traditional approach is that (1) the majority of storms throughout the year are small and therefore pass through the detention facilities uncontrolled, (2) the criterion of reducing storm flow does not address the need for reducing total storm volume, and (3) the facilities are not designed to work as a system on a watershed scale. In many cases, the site-by-site approach has exacerbated downstream flooding and channel erosion problems as a watershed is gradually built out. For example, McCuen (1979) and Emerson et al. (2005) showed that an unplanned system of site-based SCMs can actually increase flooding on a watershed scale owing to the effect of many facilities discharging into a receiving waterbody in an uncoordinated fashion—causing the very flooding problem the individual basins were built to solve.

With the relatively recent recognition of unacceptable downstream impacts and the regulation of urban stormwater quality has come a rethinking of the design of traditional stormwater systems. It is becoming rapidly understood that stormwater management should occur on a watershed scale to prevent flow control problems from occurring or reducing the chances that they might become worse. In this context, the "watershed scale" refers to the small local watershed to which the individual site drains (i.e., a few square miles within a single municipality). Together, the developer, designer, plan reviewer, owners, and the municipality jointly install and operate a linked and shared system of distributed practices across multiple sites that achieve small watershed objectives. Many metropolitan areas around the country have institutions, such as the Southeast Wisconsin Regional Planning Commission and the Milwaukee Metropolitan Sewage District, that are doing stormwater master planning to reduce flooding, bank erosion, and water quality problems on a watershed scale.

Designing stormwater management on a watershed scale creates the opportunity to evaluate a system of SCMs and maximize overall effectiveness based on multiple criteria, such as the incremental costs to development beyond traditional stormwater infrastructure, the limitations imposed on land area required for site planning, the effectiveness at improving water quality or attenuating discharges, and aesthetics. Because the benefits that accrue with improved water quality are generally not realized by those entities required to implement SCMs, greater value must be created beyond the functional aspects of the facility if there is to be wide acceptance of SCMs as part of the urban landscape. Stormwater systems designed on a watershed basis are more likely to be seen as a multi-functional resource that can contribute to the overall quality of the urban environment. Potential even exists to make the stormwater system a primary component of the civic framework of the community—elements of the public realm that serve to enhance a community's quality of life like public spaces and parks. For example, in central Minneapolis, redevelopment of a 100-acre area called Heritage Park as a mixed-density residential neighborhood was organized around two parks linked by a parkway that served dual functions of recreation and stormwater management.

Key elements of the watershed approach to designing systems of SCMs are discussed in detail below. They include the following:

1.   Forecasting the current and future development types.

2. Forecasting the scale of current and future development.

3. Choosing among on-site, distributed SCMs and larger, consolidated SCMs.

4. Defining stressors of concern.

5. Determining goals for the receiving water.

6. Noting the physical constraints.

7. Developing SCM guidance and performance criteria for the local watershed.

8. Establishing a trading system.

9. Ensuring the safe performance of the drainage network, streams, and floodplains.

10. Establishing community objectives for the publically owned elements of stormwater infrastructure.

11. Establishing a maintenance plan.

Forecasting the Current and Future Development Types

Forecasting the type of current and future development within the local watershed will guide or shape how individual practices and SCMs are generally assembled at each individual site. The development types that are generally thought of include Greenfield development (small and large scales), redevelopment within established communities and on Brownfield sites, and retrofitting of existing urban areas. These development types range roughly from lower density to higher density impervious cover. Box 5-10 explains how the type of development can dictate stormwater management, discussing two main categories—*Greenfield* development and *redevelopment* of existing areas. The former refers to development that changes pristine or agricultural land to urban or suburban land uses, frequently low-density residential housing. Redevelopment refers to changing from an existing urban land use to another, usually of higher density, such as from single-family housing to multi-family housing. Finally, *retrofitting* as used in this report is not a development type but rather the upgrading of stormwater management within an existing land use to meet higher standards.

Table 5-7 shows which SCMs are best suited for Greenfield development (particularly low-density residential), redevelopment of urban areas, and intense industrial redevelopment. The last category is broken out because the suite of SCMs needed is substantially different than for urban redevelopment. Each type of development has a different footprint, impervious cover, open space, land cost, and existing stormwater infrastructure. Consequently, SCMs that are ideally suited for one type of development may be impractical or infeasible for another. One of the main points to be made is that there are more options during Greenfield development than during redevelopment because of existing infrastructure, limited land area, and higher costs in the latter case.

BOX 5-10
Development Types and their Relationship to the Stormwater System

Development falls into two basic types. Greenfield development requires new infrastructure designed according to contemporary design standards for roads, utilities, and related infrastructure. Redevelopment refers to developed areas undergoing land-use change. In contrast to Greenfields, infrastructure in previously developed areas is often in poor condition, was not built to current design standards, and is inadequate for the new land uses proposed. The stormwater management scenarios common to these types of development are described below.

Greenfield Development

At the largest scale, Greenfield development refers to planned communities at the developing edge of metropolitan areas. Communities of this type often vary from several hundred acres to very large projects that encompassed tens of thousands of acres requiring buildout over decades. They often include the trunk or primary stormwater system as well as open stream and river corridors. The most progressive communities of this type incorporate a significant portion of the area to stormwater systems that exist as surface elements. Such stormwater system elements are typically at the subwatershed scale and provide for consolidated conveyance, detention, and water quality treatment. These elements of the infrastructure can be multi-functional in nature, providing for wildlife habitat, trail corridors, and open-space amenities.

Greenfield development can also occur on a small scale—neighborhoods or individual sites within newly developing areas that are served by the secondary public and tertiary stormwater systems. This smaller-scale, incremental expansion of existing urban patterns is a more typical way for cities to grow. A more limited range of SCMs are available on smaller projects of this type, including LID practices.

Redevelopment of Existing Areas

Redevelopment within established communities is typically at the scale of individual sites and occasionally the scale of a small district. The area is usually served by private, on-site systems that convey larger storm events into preexisting stormwater systems that were developed decades ago, either in historic city centers or in "first ring," post-World War II suburbs adjacent to historic city centers. Redevelopment in these areas is typically much denser than the original use. The resulting increase in impervious area, and typically the inadequacy of existing stormwater infrastructure serving the site often results in significant development costs for on-site detention and water quality treatment. Elaborate vaults or related structures, or land area that could be utilized for development, must often be committed to on-site stormwater management to comply with current stormwater regulations.

Brownfields are redevelopments of industrial and often contaminated property at the scale of an individual site, neighborhood, or district. Secondary public systems and private stormwater systems on individual sites typically serve these areas. In many cases, especially in outdated industrial areas, little or no stormwater infrastructure exists, or it is so inadequate as to require replacement. Water quality treatment on contaminated sites may also be necessary. For these reasons, stormwater management in such developments presents special challenges. As an example, the most common methods of remediation of contaminated sites involve capping of contaminated soils or treatment of contaminants in situ, especially where removal of contaminated soils from a site is cost prohibitive. Given that contaminants are still often in place on redeveloped Brownfield sites and must not be disturbed, certain SCMs such as infiltration of stormwater into site soils, or excavation for stormwater piping and other utilities, present special challenges.

TABLE 5-7 Applicability of Stormwater Control Measures by Type of Development

Stormwater Control Measure	Low-Density Greenfield Residential	Urban Redevelopment	Intense Industrial Redevelopment
Product Substitution	○	●	●
Watershed and Land-Use Planning	■	■	○
Conservation of Natural Areas	■	♦	○
Impervious Cover Minimization	■	♦	♦
Earthwork Minimization	■	♦	♦
Erosion and Sediment Control	■	■	■
Reforestation and Soil Conservation	■	●	●
Pollution Prevention SCMs	♦	●	■
Runoff Volume Reduction— Rainwater Harvesting	■	■	●
Runoff Reduction— Vegetated	■	○	●
Runoff Reduction— Subsurface	■	○	♦
Peak Reduction and Runoff Treatment	■	♦	○
Runoff Treatment	●	●	■
Aquatic Buffers and Managed Floodplains	●	♦	○
Stream Rehabilitation	○	♦	♦
Municipal Housekeeping	○	○	NA
IDDE	○	○	○
Stormwater Education	●	●	●
Residential Stewardship	■	●	NA

NOTE: ■, always; ●, often; ○, sometimes; ♦, rarely; NA, not applicable.

Forecasting the Scale of Current and Future Development

The choice of what SCMs to use depends on the area that needs to be serviced. It turns out that some SCMs work best over a few acres, whereas others require several dozen acres or more; some are highly effective only for the smallest sites, while others work best at the stream corridor or subwatershed level. Table 5-1 includes a column that is related the scale at which individual SCMs can be applied ("where" column). The SCMs mainly applied at the site scale include runoff volume reduction—rainwater harvesting, runoff treatment like filtering, and pollution prevention SCMs for hotspots. As one goes up in scale, SCMs like runoff volume reduction—vegetated and subsurface, earthwork minimization, and erosion and sediment control take on more of a role. At the largest scales, watershed and land-use planning, conservation of natural areas, reforestation and soil conservation, peak flow reduction, buffers and managed floodplains, stream rehabilitation, municipal housekeeping, IDDE, stormwater education, and residential stewardship play a more important role. Some SCMs are useful at all scales, such as product substitution and impervious cover minimization.

Choosing Among On-Site, Distributed SCMs and Larger, Consolidated SCMs

There are distinct advantages and disadvantages to consider when choosing to use a system of larger, consolidated SCMs versus smaller-scale, on-site SCMs that go beyond their ability to achieve water quality or urban stream health. Smaller, on-site facilities that serve to meet the requirements for residential, commercial, and office developments tend to be privately owned. Typically, flows are directed to porous landscape detention areas or similar SCMs, such that volume and pollutants in stormwater are removed at or near their source. Quite often, these SCMs are relegated to the perimeter project, incorporated into detention ponds, or, at best, developed as landscape infiltration and parking islands and buffers. On-site infiltration of frequent storm events can also reduce the erosive impacts of stormwater volumes on downstream receiving waters. Maintenance is performed by the individual landowner, which is both an advantage because the responsibility and costs for cleanup of pollutants generated by individual properties are equitably distributed, and a disadvantage because ongoing maintenance incurs a significant expense on the part of individual property owners and enforcement of properties not in compliance with required maintenance is difficult. On the negative side, individual SCMs often require additional land, which increases development costs and can encourage sprawl. Monitoring of thousands of SCMs in perpetuity in a typical city creates a significant ongoing public expense, and special training and staffing may be required to maintain SCM effectiveness (especially for subgrade or in-building vaults used in ultra-urban environments). Finally, given that as much as 30 percent of the urban landscape is comprised of public streets and rights-of-way, there are limited opportunities to treat runoff from streets through individual on-site private SCMs. (Notable exceptions are subsurface runoff-volume-reduction SCMs like permeable pavement that require no additional land and promote full development density within a given land parcel because they use the soil areas below roads and the development site for infiltration.)

In contrast, publicly owned, consolidated SCMs are usually constructed as part of larger Greenfield and infill development projects in areas where there is little or no existing infrastructure. This type of facility—usually an infiltration basin, detention basin, wet/dry pond, or stormwater wetland—tends to be significantly larger, serving multiple individual properties. Ownership is usually by the municipality, but may be a privately managed, quasi-public special district. There must be adequate land available to accommodate the facility and a means of up-front financing to construct the facility. An equitable means of allocating costs for ongoing maintenance must also be identified. However, the advantage of these facilities is that consolidation requires less overall land area, and treatment of public streets and rights-of-way can be addressed. Monitoring and maintenance are typically the responsibility of one organization, allowing for effective ongoing operations to maintain the original function of the facility. If that entity is public, this ensures that the facility will be maintained in perpetuity, allowing for the potential to permanently reduce stormwater volumes and for reduction in the size of downstream stormwater infrastructure. Because consolidated facilities are typically larger than on-site SCMs, mechanized maintenance equipment allows for greater efficiency and lower costs. Finally, consolidated SCMs have great potential for multifunctional uses because wildlife habitat, recreational, and open-space amenities can be integrated to their design. Box 5-11 describes sites of various scales where either consolidated or distributed SCMs were chosen.

Defining Stressors of Concern

The primary pollutants or stressors of concern (and the primary source areas or stormwater hotspots within the watershed likely to produce them) should be carefully defined for the watershed. Although this community decision is made only infrequently, it is critical to ensuring that SCMs are designed to prevent or reduce the maximum load of the pollutants of greatest concern. This choice may be guided by regional water quality priorities (such as nutrient reduction in the Chesapeake Bay or Neuse River watersheds) or may be an outgrowth of the total maximum daily load process where there is known water quality impairment or a listed pollutant. The choice of a pollutant of concern is paramount, since individual SCMs have been shown to have highly variable capabilities to prevent or reduce specific pollutants (see WERF, 2006; ASCE, 2007; CWP, 2007b). In some cases, the capability of SCMs to reduce a specific pollutant may be uncertain or unknown.

BOX 5-11
Examples of Communities Using Consolidated versus Distributed SCMs

Stapleton Airport New Community

This is a mixed-use, mixed-density New Urbanist community that has been under development for the past 15 years on the 4,500-acre former Stapleton Airport site in central Denver. As shown in Figures 5-55 and 5-56, the stormwater system emphasizes surface conveyance and treatment on individual sites, as well as in consolidated regional facilities.

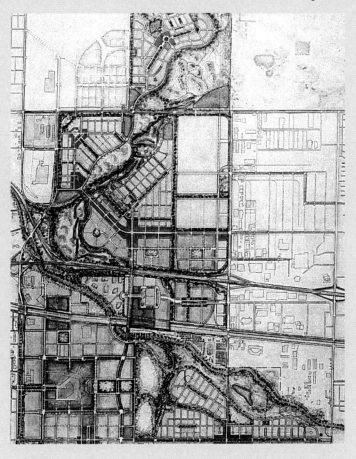

FIGURE 5-55 The community plan, shown on the left, is organized around two day lighted creeks, formerly buried under airport runways, and a series of secondary conveyances which provide recreational open space within neighborhoods. The image above illustrates one of the multi-functional creek corridors. Consolidated stormwater treatment areas and surface conveyances define more traditional park recreation and play areas. SOURCE: Courtesy of the Stapleton Redevelopment Foundation.

FIGURE 5-56 A consolidated treatment area adjacent to one of several neighborhoods that have been constructed as part of the project's build-out. SOURCE: Courtesy of Wenk Associates.

continues next page

BOX 5-11 Continued

Heritage Park Neighborhood Redevelopment

A failed public housing project adjacent to downtown Minneapolis, Minnesota, has been replaced by a mixed-density residential neighborhood. Over 1,200 rental, affordable, and market-rate single- and multi-family housing units have been provided in the 100-acre project area. The neighborhood is organized around two neighborhood parks and a parkway that serve dual functions as neighborhood recreation space and as surface stormwater conveyance and a consolidated treatment system (see Figure 5-57). Water quality treatment is being provided for a combined area of over 660 acres that includes the 100-acre project area and over 500 acres of adjacent neighborhoods. Existing stormwater pipes have been routed through treatment areas with treatment levels ranging from 50 to 85 percent TSS removal, depending on the available land area.

FIGURE 5-57 View of a sediment trap and porous landscape detention area in the central parkway spine of Heritage Park. The sediment trap in the center left of the photo was designed for ease of maintenance access by city crews with standard city maintenance equipment. SOURCE: Courtesy of the SRF Consulting Group, Inc.

The High Point Neighborhood

This Seattle project is the largest example of the city's Natural Drainage Systems Project and it illustrates the incorporation of individual SCMs into street rights-of-way as well as a consolidated facility. The on-site, distributed SCMs in this 600-acre neighborhood are swales, permeable pavement, and disconnected downspouts. A large detention pond services the entire region that is much smaller than it would have been had the other SCMs not been built. Both types of SCMs are shown in Figure 5-58.

FIGURE 5-58 Natural drainage system methods have been applied to a 34-block, 1,600-unit mixed-income housing redevelopment project called High Point. Shown on top, vegetated swales, porous concrete sidewalks, and frontyard rain gardens convey and treat stormwater on-site. Below is the detention pond for the development. SOURCE: top, William Wenk, Wenk Associates, and bottom, Laura Ehlers, National Research Council.

continues next page

BOX 5-11 Continued

Pottsdammer Platz

This project, in the heart of Berlin, Germany, illustrates the potential for stormwater treatment in the densest urban environments by incorporating treatment into building systems and architectural pools that are the centerpiece of a series of urban plazas. As shown in Figure 5-59, on-site, individual SCMs are used to collect stormwater and use it for sanitary purposes.

FIGURE 5-59 As shown to the left and below, stormwater is collected and stored on-site in a series of vaults. Water is circulated through a series of biofiltration areas and used for toilets and other mechanical systems in the building complex. Large storms overflow into an adjacent canal. SOURCE: Reprinted, with permission, from Herbert Dreiseitl, Dieter Grau (2001). Copyright 2001 by Birkhäuser Publishing Ltd.

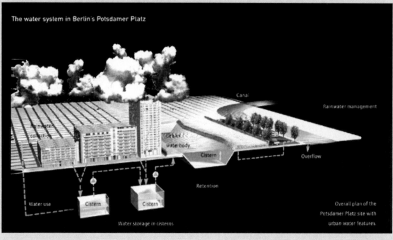

Menomonee Valley Redevelopment, Wisconsin

The 140-acre redevelopment of abandoned railyards illustrates how a Brownfield site within an existing floodplain can be redeveloped using both on-site and consolidated treatment. As shown in Figure 5-60, consolidated treatment is incorporated into park areas which provide recreation for adjacent neighborhoods and serve as a centerpiece for a developing light industrial area that provides jobs to surrounding neighborhoods. Treatment on individual privately owned parcels is limited to the removal of larger sediments and debris only, making more land available for development. The volume of water that, by regulation, must be captured and treated on individual sites is conveyed through a conventional subsurface system for treatment in park areas.

FIGURE 5-60 Illustrations show consolidated treatment areas in proposed parks. The top image illustrates the fair weather condition, the center image the water quality capture volume, and the bottom image the 100-year storm event. Construction was completed in spring 2007. SOURCE: Courtesy of Wenk Associates.

Determining Goals for the Receiving Waters

It is important to set biological and public health goals for the receiving water that are achievable given the ultimate impervious cover intended for the local watershed (see the Impervious Cover Model in Box 3-10). If the receiving water is too sensitive to meet these goals, one should consider adjustments to zoning and development codes to reduce the amount of impervious cover. The biological goals may involve a keystone species, such as salmon or trout, a desired state of biological integrity in a stream, or a maximum level of eutrophication in a lake. In other communities, stormwater goals may be driven by the need to protect a sole-source drinking water supply (e.g., New York watersheds) or to maintain water contact recreation at a beach, lake, or river. Once again, the watershed goals that are selected have a strong influence on the assembly of SCMs needed to meet them, since individual SCMs vary greatly in their ability to achieve different biological or public health outcomes.

Noting the Physical Constraints

The specific physical constraints of the watershed terrain and the development pattern will influence the selection and assembly of SCMs. The application of SCMs must be customized in every watershed to reflect its unique terrain, such as karst, high water tables, low or high slopes, freeze–thaw depth, soil types, and underlying geology. Each SCM has different restrictions or constraints associated with these terrain factors. Consequently, the SCM prescription changes as one moves from one physiographic region to another (e.g., the flat coastal plain, the rolling Piedmont, the ridge and valley, and mountainous headwaters).

Developing SCM Guidance and Performance Criteria for the Local Watershed

Based on the foregoing factors, the community should establish specific sizing, selection, and design requirements for SCMs. These SCM performance criteria may be established in a local, regional, or state stormwater design manual, or by reference in a local watershed plan. The Minnesota Stormwater Steering Committee (MSSC, 2005) provides a good example of how SCM guidance can be customized to protect specific types of receiving waters (e.g., high-quality lakes, trout streams, drinking water reservoirs, and impaired waters). In general, the watershed- or receiving water-based criteria are more specific and detailed than would be found in a regional or statewide stormwater manual. For example, the local stormwater guidance criteria may be more prescriptive with respect to runoff reduction and SCM sizing requirements, outline a preferred sequence for SCMs, and indicate where SCMs should (or should not) be located in the watershed. Like the identification of stressors or pollutants of concerns, this step is rarely taken under current paradigms of stormwater management.

Establishing a Trading System

A stormwater trading or offset system is critical to situations when on-site SCMs are not feasible or desirable in the watershed. Communities may choose to establish some kind of stormwater trading or mitigation system in the event that full compliance is not possible due to physical constraints or because it is more cost effective or equitable to achieve pollutant reduction elsewhere in the local watershed. The most common example is providing an offset fee based on the cost to remove an equivalent amount of pollutants (such as phosphorus in the Maryland Critical Area—MD DNR, 2003). This kind of trading can provide for greater cost equity between low-cost Greenfield sites and higher-cost ultra-urban sites.

Ensuring the Safe and Effective Performance of the Drainage Network, Streams, and Floodplains

The urban water system is not solely designed to manage the quality of run-off. It also must be capable of safely handling flooding from extreme storms to protect life and property. Consequently, communities need to ensure that their stormwater infrastructure can prevent increased flooding caused by development (and possibly exacerbated future climate change). In addition, many SCMs must be designed to safely pass extreme storms when they do occur. This usually requires a watershed approach to stormwater management to ensure that quality and quantity control are integrated together, with an emphasis on the connection and effective use of conveyance channels, streams, riparian buffers, and floodplains.

Establishing Community Objectives for the Publicly Owned Elements of Stormwater Infrastructure

The stormwater infrastructure in a community normally occupies a considerable surface area of the landscape once all the SCMs, drainage easements, buffers, and floodplains are added together. Consequently, communities may require that individual SCM elements are designed to achieve multiple objectives, such as landscaping, parks, recreation, greenways, trails, habitat, sustainability, and other community amenities (as discussed extensively above). In other cases, communities may want to ensure that SCMs do not cause safety or vector problems and that they look attractive. The best way to maximize community benefits is to provide clear guidance in local SCM criteria at the site level and to ensure that local watershed plans provide an overall context for their implementation.

Establishing an Inspection and Maintenance Plan

The long-term performance of any SCM is fundamentally linked to the frequency of inspections and maintenance. As a result, NPDES stormwater permit conditions for industrial, construction, and municipal permittees specify that pollution prevention, construction, and post-construction SCMs be adequately maintained. MS4 communities are also required under NPDES stormwater permits to track, inspect, and ensure the maintenance of the collective system of SCMs and stormwater infrastructure within their jurisdiction. In larger communities, this can involve hundreds or even thousands of individual SCMs located on either public or private property. In these situations, communities need to devise a workable model that will be used to operate, inspect, and maintain the stormwater infrastructure across their local watershed. Communities have the lead responsibility in their MS4 permits to assure that SCMs are maintained properly to ensure their continued function and performance over time. They can elect to assign the responsibility to the public sector, the private sector (e.g., property owners and homeowners association), or a hybrid of the two, but under their MS4 permits they have ultimate responsibility to ensure that SCM maintenance actually occurs. This entails assigning legal and financial responsibilities to the owners of each SCM element in the watershed, as well as maintaining a tracking and enforcement system to ensure compliance.

Summary

Taking all of the elements above into consideration, the emerging goal of stormwater management is to mimic, as much as possible, the hydrological and water quality processes of natural systems as rain travels from the roof to the stream through combined application of a series of practices throughout the entire development site and extending to the stream corridor. The series of SCMs incrementally reduces the volume of stormwater on its way to the stream, thereby reducing the amount of conventional stormwater infrastructure required.

There is no single SCM prescription that can be applied to each kind of development; rather, a combination of interacting practices must be used for full and effective treatment. For a low-density residential Greenfield setting, a combination of SCMs that might be implemented is illustrated in Table 5-8. There are many successful examples of SCMs in this context and at different scales. By contrast, Tables 5-9 and 5-10 outline how the general "roof-to-stream" stormwater approach is adapted for intense industrial operations and urban redevelopment sites, respectively. As can be seen, these development situations require a differ combination of SCMs and practices to address the unique design

TABLE 5-8 From the Roof to the Stream: SCMs in a Residential Greenfield

SCM	What it Is	What it Replaces	How it Works
Land-Use Planning	Early site assessment	Doing SWM design after site layout	Map and plan submitted at earliest stage of development review showing environmental, drainage, and soil features
Conservation of Natural Areas	Maximize forest canopy	Mass clearing	Preservation of priority forests and reforestation of turf areas to intercept rainfall
Earthwork Minimization	Conserve soils and contours	Mass grading and soil compaction	Construction practices to conserve soil structure and only disturb a small site footprint
Impervious Cover Minimization	Better site de-sign	Large streets, lots and cul-de-sacs	Narrower streets, permeable driveways, clustering lots, and other actions to reduce site IC
Runoff Volume Reduction— Rainwater Harvesting	Utilize rooftop runoff	Direct connected roof leaders	A series of practices to capture, disconnect, store, infiltrate, or harvest rooftop runoff
Runoff Volume Reduction— Vegetated	Frontyard bioretention	Positive drainage from roof to road	Grading frontyard to treat roof, lawn, and driveway runoff using shallow bioretention
	Dry swales	Curb/gutter and storm drain pipes	Shallow, well-drained bioretention swales located in the street right-of-way
Peak Reduction and Runoff Treatment	Linear wetlands	Large detention ponds	Long, multi-cell, forested wetlands located in the stormwater conveyance system
Aquatic Buffers and Managed Floodplains	Stream buffer management	Unmanaged stream buffers	Active reforestation of buffers and restoration of degraded streams

Note: SCMs are applied in a series, although all of the above may not be needed at a given residential site. This "roof-to-stream" approach works best for low- to medium-density residential development.

TABLE 5-9 From the Roof to the Outfall: SCMs in an Industrial Context

SCM Category	What it Is	What it Replaces	How it Works
Pollution Prevention	Drainage mapping	No map	Analysis of the locations and connections of the stormwater and wastewater infrastructure from the site
	Hotspot site Investigation	Visual inspection	Systematic assessment of runoff problems and pollution prevention opportunities at the site
	Rooftop management	Uncontrolled rooftop runoff	Use of alternative roof surfaces or coatings to reduce metal runoff, and disconnection of roof runoff for stormwater treatment
	Exterior maintenance practices	Routine plant maintenance	Special practices to reduce discharges during painting, powerwashing, cleaning, sealcoating and sandplasting
	Extending roofs for no exposure	Exposed hotspot operations	Extending covers over susceptible loading/unloading, fueling, outdoor storage, and waste management operations
	Vehicular pollution prevention	Uncontrolled vehicle operations	Pollution prevention practices applied to vehicle repair, washing, fueling, and parking operations
	Outdoor pollution prevention practices	Outdoor materials storage	Prevent rainwater from contact with potential pollutants by covering, secondary containment, or diversion from storm-drain system
	Waste management practices	Exposed dumpster or waste streams	Improved dumpster location, management, and treatment to prevent contact with rainwater or runoff
	Spill control plan and response	No plan	Develop and test response to spills to the storm-drain system, train employees, and have spill control kits available on-site
	Greenscaping	Routine landscape and turf maintenance	Reduce use of pesticides, fertilization, and irrigation in pervious areas, and conversion of turf to forest
	Employee stewardship	Lack of stormwater awareness	Regular ongoing training of employees on stormwater problems and pollution prevention practices
	Site housekeeping and stormwater maintenance	Dirty site and unmaintained infrastructure	Regular sweeping, storm-drain cleanouts, litter pickup, and maintenance of stormwater infrastructure
Runoff Treatment	Stormwater retrofitting	No stormwater treatment	Filtering retrofits to remove pollutants from most severe hotspot areas
IDDE	Outfall analysis	No monitoring	Monitoring of outfall quality to measure effectiveness

Note: While many SCMs are used at each individual industrial site, the exact combination depends on the specific configuration, operations, and footprint of each site.

TABLE 5-10 From the Roof to the Street: SCMs in a Redevelopment Context

SCM Category	What it Is	What it Replaces	How it Works
Impervious Cover Minimization	Site design to prevent pollution	Conventional site design	Designing redevelopment footprint to restore natural area remnants, minimize needless impervious cover, and reduce hotspot potential
Runoff Volume Reduction— Rainwater Harvesting and Vegetated	Treatment on the roof	Traditional rooftops	Use of green rooftops to reduce runoff generated from roof surfaces
	Rooftop runoff treatment	Directly connected roof leaders	Use of rain tanks, cisterns, and rooftop disconnection to capture, store, and treat runoff
	Runoff treatment in landscaping	Traditional landscaping	Use of foundation planters and bioretention areas to treat runoff from parking lots and rooftops
Soil Conservation and Reforestation	Runoff reduction in pervious areas	Impervious or compacted soils	Reducing runoff from compacted soils through tilling and compost amendments, and in some cases, removal of unneeded impervious cover
	Increase urban tree canopy	Turf or landscaping	Providing adequate rooting volume to develop mature tree canopy to intercept rainfall
Runoff Reduction— Subsurface	Increase permeability of impervious cover	Hard asphalt or concrete	Use of permeable pavers, porous concrete, and similar products to decrease runoff generation from parking lots and other hard surfaces.
Runoff Reduction— Vegetated	Runoff treatment in the street	Sidewalks, curb and gutter, and storm drains	Use of expanded tree pits, dry swales and street bioretention cells to further treat runoff in the street or its right-of-way
Runoff Treatment	Underground treatment	Catch basins and storm-drain pipes	Use of underground sand filters and other practices to treat hotspot runoff quality at the site
Municipal Housekeeping	Street cleaning	Unswept streets	Targeted street cleaning on priority streets to remove trash and gross solids
Watershed Planning	Off-site stormwater treatment or mitigation	On-site waivers	Stormwater retrofits or restoration projects elsewhere in the watershed to compensate for stormwater requirements that cannot be met on-site

Note: SCMs are applied in a series, although all of the above may not be needed at a given redevelopment site.

challenges of dense urban environments. The tables are meant to be illustrative of certain situations; other scenarios, such as commercial development, would likely require additional tables.

In summary, a watershed approach for organizing site-based stormwater decisions is generally superior to making site-based decisions in isolation. Communities that adopt the preceding watershed elements not only can maximize the performance of the entire system of SCMs to meet local watershed objectives, but also can maximize other urban functions, reduce total costs, and reduce future maintenance burdens.

COST, FINANCE OPTIONS, AND INCENTIVES

Municipal Stormwater Financing

To be financially sustainable, stormwater programs must develop a stable long-term funding source. The activities common to most municipal stormwater programs (such as education, development design review, inspection, and enforcement) are funded through general tax revenues, most commonly property taxes and sales taxes (NAFSMA, 2006), which is problematic for several reasons. First, stormwater management financed through general tax receipts does not link or attempt to link financial obligation with services received. The absence of such links can reduce the ability of a municipality to adequately plan and meet basic stormwater management obligations. Second, when funded through general tax revenues, stormwater programs must compete with other municipal programs and funding obligations. Finally, in programs funded by general tax revenue, responsibilities for stormwater management tend to be distributed into the work responsibilities of existing and multiple departments (e.g., public works, planning, etc.). One recent survey conducted in the Charles River watershed in Massachusetts found that three-quarters of local stormwater management programs did not have staff dedicated exclusively for stormwater management (Charles River Watershed Association, 2007).

Increasingly, many municipalities are establishing stormwater utilities to manage stormwater (Kaspersen, 2000). Most stormwater utilities are created as a separate organizational entity with a dedicated, self-sustaining source of funding. The typical stormwater utility generates the large majority of revenue through user fees (Florida Stormwater Association, 2003; Black and Veatch, 2005; NAFSMA, 2006). User fees are established and set so as to have a close nexus to the cost of providing the service and, thus, are most commonly based on the amount of impervious surface, frequently measured in terms of equivalent residential unit. For example, an average single-family residence may create 3,000 square feet of impervious surface (roof and driveway area). A per-unit charge is then assigned to this "equivalent runoff unit." To simplify program administration, utilities typically assign a flat rate for residential properties (customer class average) (NAFSMA, 2006). Nonresidential properties are then charged individually based on the total amount of impervious surface (square feet or equivalent runoff units) of the parcel. Fees are sometimes also based on gross area (total area of a parcel) or some combination of gross area and a development intensity measure (Duncan, 2004; NAFSMA, 2006).

Municipalities have the legal authority to create stormwater utilities in most states (Lehner et al., 1999). In addition to creating the utility, a municipality will generally establish the utility rate structure in a separate ordinance. Separating the ordinances allows the municipality flexibility to change the rate structure without revising the ordinance governing the entire utility (Lehner et al., 1999). While municipalities generally have the authority to collect fees, some states have legal restrictions on the ability of local governments to levy taxes (Lehner et al., 1999; NAFSMA, 2006). The legal distinction between a tax and a fee is the most common legal challenge to a stormwater utility. For example, stormwater fees have been subject to litigation in at least 17 states (NAFSMA, 2006). To avoid legal challenges, care must be taken to meet a number of legal tests that distinguish a fee for a specific service and a general tax.

Stormwater utilities typically bill monthly, and fees range widely. A recent survey of U.S. stormwater utilities reported that fees for residential households range from $1 to $14 per month, but a typical residential household rate is in the range of $3 to $6 (Black and Veatch, 2005). Despite the dedicated funding source, the majority of stormwater utilities responding to a recent survey (55 percent) indicated that current funding levels were either inadequate or just adequate to meet their most urgent needs (Black and Veatch, 2005).

Both municipal and state programs can finance administrative programming costs through stormwater permitting fees. Municipal stormwater programs can use separate fees to finance inspection activities. For instance, inspection fees can be charged to cover the costs of ensuring that SCMs are adequately planned, installed, or maintained (Debo and Reese, 2003). Stormwater management programs can also ensure adequate funding for installation and maintenance of SCMs by requiring responsible parties to post financial assurances. Performance bonds, letters of credit, and cash escrow are all examples of financial assurances that require up-front financial payments to ensure that longer-term actions or activities are successfully carried out. North Carolina's model stormwater ordinance recommends that the amount of a maintenance performance security (bond, cash escrow, etc.) be based on the present value of an annuity based on both inspection costs and operation and maintenance costs (Whisnant, 2007).

In addition to fees or taxes, exactions such as impact fees can also be used as a way to finance municipal stormwater infrastructure investments (Debo and Reese, 2003). An impact fee is a one-time charge levied on new development. The fee is based on the costs to finance the infrastructure needed to service the new development. The ability to levy impact fees varies between states. Municipalities that use impact fees are also required to show a close nexus between the size of the fee and the level of benefits provided by the fee; a failure to do so exposes local government to law suits (Keller, 2003). Compared to other funding sources, impact fees also exhibit greater variability in revenue flows because the amount of funds collected is dependent on development growth.

Bonds and grants can supplement the funding sources identified above. Bonds and loans tend to smooth payments over time for large up-front stormwater investments. For example, state and federal loan programs (state revolving funds) provide long-term, low-interest loans to local governments or capital investments (Keller, 2003). In addition, grant opportunities are sometimes available from state and federal sources to help pay for specific elements of local stormwater management programs.

Municipalities require funds to meet federal and state stormwater requirements. Understanding of the municipal costs incurred by implementing stormwater regulations under the Phase I and II stormwater rules, however, is incomplete (GAO, 2007). Of the six minimum measures of a municipal stormwater program (public education, public involvement, illicit discharge detection and elimination, construction site runoff control, post-construction stormwater management, and pollution prevention/good housekeeping—see Chapter 2), a recent study of six California municipalities found that pollution prevention activities (primarily street sweeping) accounted for over 60 percent of all municipal stormwater management costs in these communities (Currier et al., 2005). Annual per-household costs ranged from $18 to $46.

Stormwater Cost Review

Conceptually, the costs of providing SCMs are all opportunity costs (EPA, 2000). Opportunity costs are the value of alternatives (next best) given up by society to achieve a particular outcome. In the case of stormwater control, opportunity costs include direct costs necessary to control and treat runoff such as capital and construction costs and the present value of annual operation and maintenance costs. Initial installation costs should also include the value of foregone opportunities on the land used for stormwater control, typically measured as land acquisition (land price).

Costs also include public and private resources incurred in the administration of the stormwater management program. Private-sector costs might include time and administrative costs associated with permitting programs. Public costs include agency monitoring and enforcement costs.

Opportunity costs also include other values that might be given up as a consequence of stormwater management. For example, the creation of a wet pond in a residential area might be opposed because of perceived safety, aesthetic, or nuisance concerns (undesirable insect or animal species). In this case, the diminished satisfaction of nearby property owners is an opportunity cost associated with the wet pond. On the other hand, if SCMs are considered a neighborhood amenity (e.g., a constructed wetland in a park setting), opportunity costs may decrease. In addition, costs of a given practice may be reduced by reducing costs elsewhere. For example, increasing on-site infiltration rates can reduce off-site storage costs by reducing the volume and slowing the release of runoff.

In general the cost of SCMs is incompletely understood and significant gaps exist in the literature. More systematic research has been conducted on the cost of conventional stormwater SCMs (wet ponds, detention basins, etc.), with less research applied to more recent, smaller-scale, on-site infiltration practices. Cost research is challenging given that stormwater treatment exhibits considerable site-specific variation resulting from different soil, topography, climatic conditions, local economic conditions, and regulatory requirements (Lambe et al., 2005).

The literature on stormwater costs tend to be oriented around construction costs of particular types of SCMs (Wiegand et al., 1986; SWRPC, 1991; Brown and Schueler, 1997; Heaney et al., 2002; Sample et al., 2003; Wossink and Hunt, 2003; Caltrans, 2004; Narayanan and Pitt, 2006; DeWoody, 2007). In many of these studies, construction cost functions are estimated statistically based on a sample of recently installed SCMs and the observed total construction costs. Observed costs are then related statistically to characteristics that influence cost such as practice size. Other studies estimate costs by identifying the individual components of a construction project (pipes, excavation, materials, labor, etc.), estimating unit costs of each component, and then summing all project components. These studies generally find that construction costs decrease on a per-unit basis as the overall size (expressed in volume or drainage area) of the SCM increases (Lambe et al., 2005). These within-practice economies of scale are found across certain SCMs including wet ponds, detention ponds, and constructed wetlands. Several empirical studies, however, failed to find evidence of economies of scale for bioretention practices (Brown and Schueler, 1997; Wossink and Hunt, 2003).

Increasing attention has been paid to small-scale practices, including efforts to increase infiltration and retain water through such means as green roofs, permeable pavements, rain barrels, and rain gardens (under the label of LID). The costs of these practices are less well studied compared to the other stormwater practices identified above. In general, per-unit construction and design costs exceed larger-scale SCMs (Low Impact Development Center, 2007). Higher construction costs, however, may be offset to various degrees by reducing the investments in stormwater conveyance and storage infrastructure (i.e., less storage volume is needed) (CWP, 1998a, 2000a; Low Impact Development Center, 2007). Others have suggested that per-unit costs to reduce runoff may be less for these small-scale distributed practices because of higher infiltration rates and retention rates (MacMullan and Reich, 2007).

Compared to construction costs, less is known about the operation and maintenance costs of SCMs (Wossink and Hunt, 2003; Lambe et al., 2005; MacMullan and Reich, 2007). Most stormwater practices are not maintenance free and can create financial and long-term management obligations for responsible parties (Hager, 2003). Cost-estimation programs and procedures have been developed to estimate operation and maintenance costs as well as construction costs (SWRPC, 1991; Lambe et al., 2005; Narayanan and Pitt, 2006), but examination of observed maintenance costs is less common. Based on estimates from Wossink and Hunt (2003), the total present value of maintenance costs over 20 years can range from 15 to 70 percent of total capital construction costs for wet ponds and constructed wetlands and appear generally consistent with percentages reported in EPA (1999). Operation and maintenance costs were also reported to be a substantial percentage of construction costs of infiltration pits and bioretention areas in Southern California (DeWoody, 2007). Others estimate that over the life of many SCMs, maintenance costs may equal construction costs (CWP, 2000a). In general, maintenance costs tend to decrease as a percentage of total SCM cost as the total size of the SCM increases (Wossink and Hunt, 2003).

Very few quantifiable estimates are available for public and private regulatory compliance costs. Compliance costs could include both initial permitting costs (labor and time delays) of gaining regulatory approval for a particular stormwater design to post-construction compliance costs (administration, inspection monitoring, and enforcement). Compliance monitoring is a particular concern if a stormwater management program relies on widespread use of small-scale distributed on-site practices (Hager, 2003). Unlike larger-scale or regional stormwater facilities that might be located on public lands or on private lands with an active stormwater management plan, a multitude of smaller SCMs would increase monitoring and inspection times by increasing the number of SCMs. Furthermore, municipal governments may be reluctant to undertake enforcement actions against citizens with SCMs located on private land.

Land costs tend to be site specific and exhibit a great deal of spatial variation. Some types of SCMs, such as constructed wetlands, are more land intensive than others. In highly urban areas, land costs may be the single biggest cost outlay of land-intensive SCMs (Wossink and Hunt, 2003).

In general, cost analyses generally find that the cost to treat a given acreage or volume of water is less for regional SCMs than for smaller-scale SCMs (Brown and Schueler, 1997; EPA, 1999; Wossink and Hunt, 2003). For example, considering maintenance, capital construction, and land costs, recent estimates for North Carolina indicate that annual costs for wet ponds and constructed wetlands range between $100 and $3,000 per treated acre (typically less than $1,000). Per-acre annual costs for bioretention and sand filters typically ranged between $300 and $3,500, and between $4,500 and 8,500, respectively. However, if SCMs face space constraints, bioretention areas can become more cost effective. Furthermore, other classes of small, on-site practices, such as grass swales and filter strips, can sometimes be implemented for relatively low cost.

There are exceptions to the general conclusion that larger-scale stormwater practices tend to be less costly on a per-unit basis than more numerous and distributed on-site practices. For instance, in Sun Valley, California, a recent study indicates that installing small distributed practices (infiltration practices, porous pavement, rain gardens) was more cost effective than centralized approaches for a retrofit program (Cutter et al., 2008). In this particular setting, the difference tended to revolve around the high land costs in the urbanized setting. Small-scale practices can be placed on low-valued land or integrated into existing land-scaping, reducing land costs. Centralized stormwater facilities require substantial purchases of high-priced urban properties. Similarly, small distributed practices (porous pavement, green roofs, rain gardens, and constructed wetlands) can also provide a more cost-effective approach to reducing combined sewer overflow (CSO) discharges in a highly urban setting than large structural CSO controls (storage tanks) (Montalto et al., 2007).

SCMs are now a part of most development processes and consequently will increase the cost of the development. Randolph et al. (2006) report on the cost of complying with stormwater and sediment and erosion control regulations for six developments in the Washington, D.C., metropolitan area. These costs include primarily stormwater facility construction and land costs. The findings from these case studies indicate that stormwater and erosion and sediment control comprised about 60 percent of all environmental-related compliance costs for the residential developments studied and added about $5,000 to the average price of a home. Nationwide, stormwater and erosion and sediment controls are estimated to add $1,500 to $9,000 to the cost of a new residential dwelling unit (Randolph et al., 2006).

As a means to control targeted chemical constituents, SCMs may be an expensive control option relative to other control alternatives. For example, nutrients from anthropocentric sources are an increasing water quality concern for many fresh and marine waters. Some states (e.g., Virginia, Maryland, and North Carolina) require stormwater programs to achieve specific nutrient (nitrogen or phosphorus) stormwater standards. The construction, maintenance, and land costs of reducing nitrogen discharge from residential developments using bioretention areas, wet ponds, constructed wetlands, or sand filters range from $60 to $2,500 per pound (Aultman, 2007). These control costs can be an order of magnitude higher than nitrogen control costs from point sources or agricultural nonpoint sources. The high per-pound removal costs are due in part to the relatively low mass load of nutrients carried in stormwater runoff. These estimates, however, assume that all costs are allocated exclusively to nitrogen removal. The high per-pound removal costs from the control of single pollutants highlight the importance of achieving ancillary and offsetting benefits associated with stormwater control (e.g., removal of other pollutants of concern, stream-channel protection from volume reduction, and enhancement of neighborhood amenities).

It should also be noted that installing SCMs in an existing built environment tends to be significantly more expensive than new construction. Construction costs for retrofitted extended detention ponds, wet ponds, and constructed wetlands were estimated to be two to seven times more costly than new SCMs (Schueler et al., 2007). Retrofit costs can be higher for a variety of reasons, including the need to upgrade existing infrastructure (culverts, drainage channels, etc.) to meet contemporary engineering and regulatory requirements. Retrofitting a single existing residential city block in Seattle with a new stormwater drainage system that included reduced street widths, biofiltration practices, and enhanced vegetation cost an estimated $850,000 (see Box 5-5; Seattle Public Utilities, 2007). Estimates suggested that the costs might have been even higher using more conventional stormwater piping/drainage systems (Chris May, personal communication, August 2007; EPA, 2007).

As discussed earlier in the chapter, stormwater runoff can be reduced and managed through better site design to reduce impervious cover. Low- to medium-density developments can reduce impervious cover through cluster development patterns that preserve open space and reduce lot sizes. Impervious surfaces and infiltration rates could be altered by any number of site-design characteristics such as reduction in street widths, reduction in the number of cul-de-sacs, and different setback requirements (CWP, 2000a). Finally, impervious surface per capita could be substantially reduced by increasing the population per dwelling unit.

Quantifying the cost of many of these design features is more challenging, and the literature is much less developed or conclusive than the literature on conventional SCM costs. Many design features described above (clustering, reduced setbacks, narrower streets, less curb and gutter) can significantly lower construction and infrastructure costs (CWP, 2001; EPA, 2007). Such features may reduce the capital cost of subdivision development by 10 to 33 percent (CWP, 2000a).

On the other hand, the evidence is unclear whether consumers are willing to pay for these design features. If consumers prefer features typically associated with conventional developments (large suburban lot, for example), then some aspects of alternative development designs/patterns could impose an opportunity cost on builders and buyers alike in the form of reduced housing value. For example, most statistical studies in the U.S. housing market find that consumers prefer homes with larger lots and are willing to pay premiums for homes located on cul-de-sacs, presumably for privacy and safety reasons (Dubin, 1998; Fina and Shabman, 1999; Song and Knapp, 2003). These effects, however, might be partly or completely offset by the higher value consumers might place on the proximity of open space to their homes (Palmquist, 1980; Cheshire and Sheppard, 1995; Qiu et al., 2006). Anecdotal evidence indicates that residents feel that Seattle's Street Edge Alternative program (the natural drainage system retrofit program that combines swales, bioretention and reduced impervious surfaces) increased their property values (City of Seattle, undated). Studies that have attempted to assess the net change in costs are limited, but some evidence suggests that the amenity values of lower-impact designs may match or outweigh the disamenities (Song and Knapp, 2003).

Incentives for Stormwater Management

The dominant policy approach to controlling effluent discharge under the Clean Water Act is through the application of technology-based effluent standards or the requirements to install particular technologies or practices. Some note that this general policy approach may not provide the regulated community with (1) incentives to invest in pollution prevention activities beyond what is required in the standard or with (2) sufficient opportunities or flexibility to lower overall compliance costs (Parikh et al., 2005).

A loosely grouped set of policies, called here "incentive-based,"[1] aim to create financial incentives to manage effluent or volume discharge. Such policies tend to be classified into two groups: price- and quantity-based mechanisms (Stavins, 2000; Parikh et al., 2005). Price-based mechanisms are created when government creates a charge (tax, fee, etc.) or subsidy (payment) on an outcome that government wants to either discourage or encourage. Ideally, the price would be placed on a target outcome (effluents discharged, volume of water released, etc.) and not on the means to achieve that outcome end (such as a tax or subsidy to adopt specific technologies or practices).[2] Quantity-based policies require government to establish some binding limit or cap on an outcome (e.g., mass load of effluent, volume of runoff, etc.) for an identified group of dischargers, but then allow the regulated parties to "trade" responsibilities for meeting that limit or cap. The opportunity to trade creates the financial incentive. The trading concept is discussed in greater detail in Chapter 6, while this section focuses on price-based incentives.

Some stormwater utilities offer reductions in stormwater fees to landowners who voluntarily undertake activities to reduce runoff from their parcels (Doll and Lindsey, 1999; Keller, 2003). The reduction in tax obligations, called credits, can be interpreted as a financial subsidy or payment for implementing on-site runoff controls. Credit payments are typically made based on the volume of water detained. For example, as part of Portland, Oregon's Clean River Rewards program, residents and commercial property owners can reduce their stormwater utility fee by as much as 35 percent by reducing stormwater runoff from existing developed properties (Portland Bureau of Environmental Services, 2008a). Residential and commercial property owners are given a number of ways to reduce runoff to receive this financial benefit. In addition, Portland has a downspout disconnection program that aims to reduce discharge into CSOs in targeted areas in the city. Property owners may be reimbursed up to $53 per eligible downspout (Portland Bureau of Environmental Services, 2008b).

Alternatively, stormwater utilities could (where allowed) also use fee revenue to provide private incentives for stormwater control through a competitive

---

[1] These policies are sometimes called "market-based" policies, but that term will not be used here because many of the incentive-based policies discussed fail to contain features characteristic of a market system.

[2] The literature on what level to set the price (tax or subsidy) is vast, complex, and controversial. Parikh et al. (2005) seem to wander into this debate (perhaps unwittingly) by making a distinction between taxes based on some optimality rule (marginal damage costs equal to marginal control costs) and those based on some other sort of decision rule. Without getting into the specifics of this debate here, this discussion will simply assert more generally that price-based incentive policies structure taxes and subsidies to induce desirable behavioral change (rather than simply to raise revenue).

bidding process. Such a bidding process ("reverse auction") would request proposals for stormwater reduction projects and fund projects that reduce volume at the least cost. Proposed investments that can meet the program objectives at the lowest per unit cost would receive payments. Such a program creates private incentives to search for low-cost stormwater investments by creating a price for runoff volume reduction. The bidding program could also be used to identify cost-effective stormwater investments in areas targeted for enhanced levels of restoration. A bidding program has been proposed as a way to lower overall costs of a stormwater program in Southern California (Cutter et al., 2008). Revenue to fund such a competitive bid program could come from a variety of sources including stormwater utility fees or fees paid into an in lieu fee program.

Finally, impact fees on new developments can be structured in a way to create incentives to reduce stormwater runoff volumes. Charges based on runoff volume (or a surrogate measure like impervious surface) can provide an incentive for developers to reduce the volume of new runoff created.

CHALLENGES TO IMPLEMENTATION OF WATERSHED-BASED MANAGEMENT AND STORMWATER CONTROL MEASURES

The implementation of SCMs has seen variable success. Environmental awareness, threats to potable water sources or to habitat for threatened and endangered species, problems with combined sewer overflows, and other environmental factors have caused cities such as Portland, Oregon; Seattle, Washington; Chicago, Illinois; and Austin, Texas to aggressively pursue widespread implementation of a broad range of SCMs. In contrast, other cities have been slow to implement recommended practices, for many reasons. This is particularly true for nonstructural SCMs, despite their popularity among planners and regulators for the past two decades. A host of real and perceived concerns about individual nonstructural SCMs are often raised regarding development costs, market acceptance, fire safety, emergency access, traffic and parking congestion, basement seepage, pedestrian safety, backyard flooding, nuisance conditions, maintenance, and winter snow removal operations. While most of these concerns are unfounded, they contribute to a culture of inertia when it comes to code change (CWP, 1998a, 2000a). As a result, some nonstructural SCMs are discouraged or even prohibited by local development codes. Very few communities make the consideration of nonstructural practices a required element of stormwater plan review, nor do they require that they be considered early in the site layout and design process when their effectiveness would be maximized. Finally, many engineers and planners feel they can fully comply with existing stormwater criteria without resorting to nonstructural SCMs.

Cost Issues

There are numerous cost issues that have proven to be significant barriers to the use of innovative SCMs. Special construction techniques required for the proper design and function of SCMs, specially formulated manufactured soils, expensive subsurface vaults, and increased land area requirements as a result of increased stormwater storage requirements can significantly increase site development costs. For smaller projects in highly urbanized areas where land costs are high, there can be a disproportionately large expense to comply with stormwater regulations, causing developers to seek, and often receive, exemption from requirements.

Sediment removal and related maintenance activities required to ensure the proper ongoing functioning of SCMs are activities that are not a part of normal building maintenance. Data on maintenance costs of SCMs on privately owned facilities are limited, and management companies responsible for commercial and office building maintenance have yet to provide SCM maintenance as part of their services.

Additional costs are incurred when development review periods by public agencies get extended because of an increased level of design review required to evaluate the compliance of SCMs with city ordinances. Additional review increases development costs and extends the design process. Even with specialized training for city staff to evaluate SCM submittals, deviation from the most basic type of SCM design seems to require extended review and documentation.

Cost concerns are partly responsible for the markedly slow implementation of the stormwater program. The federal deadlines for permit coverage have long passed; in fact more than 14 years have lapsed for medium and large municipalities. A good part of the delay can be explained by the resistance of states and local governments to the unknown cost burden. Cities contend that the permit requirements are unreasonable, expensive, and unrealistic to achieve. Many local government officials view some permit provisions such as LID or better site design as intrusion into the land-use authority of local governments.

As discussed in Chapter 2, the U.S. Congress provided no start-up or upgrade financial assistance, unlike what it did for municipally owned and operated wastewater treatment plants after the promulgation of the NPDES permit program under the Clean Water Act in 1972. Local governments have been reluctant to tax residents or create stormwater utilities. States like California and Michigan even have laws that require voter approval in order for local governments to assess new fees. Thus, to implement the NPDES stormwater program, states have had to largely rely on stormwater permit fees collected to support a skeletal to modest staff for program oversight. In Denver, and presumably in other cities, there is no reduction in stormwater fees when impervious area is reduced because of construction of on-site SCMs. This amounts to a disincentive to do the "right thing." Meanwhile, the overall federal budget for the NPDES program, including stormwater, has been declining.

Long-Term Maintenance of Stormwater Control Measures

One of the weakest parts of most stormwater management programs is the lack of information about, and funding to support, the long-term maintenance of SCMs. If SCMs are not inspected and maintained on a regular basis, the stormwater management program is likely to fail. This also negatively impacts the design process—if there is no inspection program oand no accountability for maintenance, the designer has no incentive to build better, more maintenance-friendly SCMs. Finally, without an accurate assessment of the maintenance needs of an SCM, land owners and other responsible parties cannot anticipate their total costs over the lifetime of the device.

Almost all SCMs require active long-term maintenance in order to continue to provide volume and water quality benefits (Hoyt and Brown, 2005; Hunt and Lord, 2006b). Furthermore, a typical municipality may contain hundreds or thousands of individual SCMs within its jurisdiction. Thus, the long-term obligations for maintenance are considerable. For example, the annual maintenance cost of 100 medium-sized wet ponds (one-half acre to 2 acres) is estimated to be a quarter of a million dollars (Hunt and Lord, 2006c). Currently, the majority of municipal stormwater programs do not have adequate plans or resources in place for the long-term maintenance of SCMs (GAO, 2007).

A number of issues confront the long-term maintenance of SCMs. First, legal and financial responsibility for maintenance must be assigned. Historically stormwater ownership and responsibility have been poorly defined and implemented (Reese and Presler, 2005). If a party is an industrial facility that is required to obtain a permit, then responsibility for maintaining SCMs rests with the permittee. Other instances are more ambiguous. For residential developments, the responsibility for long-term maintenance could be assigned to the developer (e.g., establishing long-term financial accounts for maintenance), individual landowners, homeowners associations, or the municipality itself. Some cities, like Austin and Seattle, assume responsibility for long-term maintenance of SCMs in residential areas. Concerns over assigning responsibility to individual residential landowners or homeowners associations include insufficient technical and financial resources to conduct consistent maintenance and a lack of inspection to require maintenance. A recent survey of municipal stormwater programs found that less than one-third perform regular maintenance on stormwater detention ponds or water quality SCMs in general residential areas (Reese and Presler, 2005). To ensure that adequate maintenance will occur, municipalities can require performance securities (performance bonds, escrow accounts, letter of credit, etc.) that ensure adequate funds are available for maintenance and repair in the event of failure to maintain the SCM by the responsible party.

An effective maintenance program also requires a system to inventory and track SCMs, inspection/monitoring, and enforcement against noncompliance. The large number of SCMs to track and manage creates management challenges. Municipal stormwater programs must administer their regulatory programs, perform inspection and enforcement activities, and maintain SCMs in public lands/rights-of-way and sometimes in residential areas. Municipal programs often do not have adequate staff to ensure that these maintenance responsibilities are adequately carried out. The lack of adequate staff for inspection and an inadequate system for prioritizing inspections have been repeatedly pointed out (Duke and Beswick, 1997; Duke, 2007; GAO, 2007).

Tracking and monitoring costs may also create disincentives for municipalities to adopt smaller-scale SCMs. Residential-scale rain gardens, porous driveways, rain barrels, and grass swales all have the potential to increase the cost and complexity of compliance monitoring because of the multitude of small infiltration devices that are located on private property as opposed to having fewer SCMs located in public rights-of-way or public lands. Small-scale distributed SCMs located on private property raise concerns of municipal willingness to inspect and enforce against noncompliance. Indeed, some municipalities have banned innovative SCMs like pervious pavement because the municipalities have no means to ensure their maintenance and continued operation.

Finally, there is concern that there is inadequate funding to maintain the growing number of SCMs on the landscape. The long-term funding obligation for maintenance has been difficult to assess (GAO, 2007), partly because many stormwater programs frequently do not have adequate accounting practices to define capital value and depreciation, maintenance, operation, or management programs (Reese and Presler, 2005). The problem is compounded because the long-term maintenance cost associated with various types of SCMs is not well understood. Additional research and information are needed on the costs of maintaining the performance of SCMs as experienced in the field (rather than ex ante estimates based on design plans). Research into long-term maintenance costs should include not only routine operation and maintenance costs but also costs for inspection and enforcement and remediation costs associated with SCM performance failures. Such research is critical to understanding the long-term cost obligation that is being assumed by municipal stormwater programs that are responsible for managing a growing number of SCMs.

At the present time, the maintenance schedule for many of the proprietary and non-proprietary SCMs is poorly defined. It will vary with the type of drainage area and the activities that are occurring within it and with the efficiency of the SCM. (For example, the city of Austin, Texas, has determined that the average lifespan of their sand filters ranges from 5 to 15 years, but can be as little as one year if there is construction in the drainage area.) In order to establish a maintenance schedule, an assessment protocol needs to be adopted by municipalities. The protocol, which is specific to the type of SCM, could consist of the following: each year municipalities would be required to collect data from a subset of their SCMs on public and private property, and then over a period of years these data could be used to determine maintenance schedules, predict performance based on age and sediment loading, and identify failed systems. A measurement of the depth of deposited sediment might be the only test needed for settling devices, such as hydrodynamic devices and wet detention ponds. Two levels of analysis could be performed for infiltration devices—one based

on simple visual observations and the other using an instrument to check infiltration rates. These assessment methods for infiltration devices have been tested at
the University of Minnesota (Gulliver and Anderson, 2007). Without an assessment protocol for SCMs, the chances for poor maintenance and outright
failure are greatly increased, it is difficult if not impossible to determine the actual performance of an SCM, and there will be insufficient data to reduce the
uncertainty in future SCM design.

Lack of Design Guidance on Important SCMs and Lack of Training

Progress in implementing SCMs is often handicapped by the lack of local or
national design guidance on important SCMs, and by the lack of training among
the many players in the land development community (planners, designers, plan
reviewers, public works staff, regulators, and contractors) on how to properly
implement them on the ground. For example, design guidance is lacking or just
emerging for many of the non-traditional SCMs, such as conservation of natural
areas, earthwork minimization, product substitution, reforestation, soil restoration, impervious cover reduction, municipal housekeeping, stormwater education, and residential stewardship. Some LID techniques are better covered, such
as the standards for pervious concrete from the American Concrete Institute and
the National Ready Mixed Concrete Association. Design guidance for traditional SCMs such as erosion and sediment control may exist but is often incomplete, outdated, or lacking key implementation details to ensure proper on-the-
ground implementation. In other cases, design guidance is available, but has not
been disseminated to the full population of Phase II MS4 communities. For
example, in an unpublished survey of state manuals used to develop national
post-construction stormwater guidance, Hirschman and Kosco (2008) found that
less than 25 percent provided sizing criteria, detailed engineering design specifications, or maintenance criteria. Nationwide guidance on SCM design and implementation may not be advisable or applicable to all physiographic, climatic,
and ecoregions of the country. Rather, EPA and the states should encourage the
development of regional design guidance that can be readily adapted and
adopted by municipal and industrial permittees. Improvement of SCM design
guidance should incorporate more direct consideration of the parameters of concern, how they move across the landscape, and the issues in receiving waters—a
strategy both espoused in this report (page 351) and in recent publications on
this topic (Strecker et al., 2005, 2007).

The second key issue relates to how to train and possibly certify the hundreds of thousands of individuals that are responsible for land development and
stormwater infrastructure at the local and state level. New stormwater methods
and practices cannot be effectively implemented until local planners, engineers,
and landscape architects fully understand them and are confident on how to apply them to real-world sites. Currently, stormwater design is not a major com-

ponent of the already crowded curriculum of undergraduate or graduate planning engineering or landscape architecture programs. Most stormwater professionals acquire their skills on the job. Given the rapid development of new stormwater technologies, there is a critical need for implementation of regional or statewide training programs to ensure that stormwater professionals are equipped with the latest knowledge and skills. The training programs should ultimately lead to formal certification for stormwater designers, inspectors, and plan reviewers.

Different Standards in Different Jurisdictions That Are Within the Same Watershed

Governmental and watershed boundaries rarely coincide, with the result that most watersheds are made up of many municipal bodies regulating stormwater management. Unfortunately in most cases there is no overarching stormwater regulatory structure that is based upon a watershed analysis. This can result in many unfortunate conflicts, where approval of a stormwater facility does not affect the community issuing the permit. It is often said that the most effective stormwater management for an area high in the watershed is to speed the water downstream, thus saving the upstream community but severely damaging the downstream rivers. While this may be an exaggeration, the problems downstream are less of a concern to the upper watershed communities, and downstream communities may not be able to solve their water issues without help from the upstream communities.

Often neighboring communities' plans or the methods or data used do not coincide. For example, often out-of-date rainfall distributions, methods, or standards are required in the code that do not apply to the newer focus on smaller storms and volume reduction. If methods that include Modified Rational or TR-55 are used, it is difficult if not impossible to show the benefits in peak flow reduction gained through volume reduction devices. Also, some municipalities may require curb and piping and not allow swales, impending the implementation of a cost-effective design. Finally, it is difficult to observe a measureable impact of SCMs when they are guided by a patchwork of regulations. One community may require removal of the first inch of runoff, and another may require the reduction of the 25-year, post-construction peak to the 10-year pre-construction level.

Water Rights that Conflict with Stormwater Management

In the West, water is considered real property, governed by state law and regional water compacts. Landowners in urban areas rarely own surface water rights and are typically prohibited from "beneficial use" of that water, which affects how SCMs are chosen. For example, current practices in Colorado typically allow stormwater to be infiltrated within a short period of time on-site without violation of water laws. However, storage of and/or pumping this water for broader distribution is considered to be a beneficial use and is therefore prohibited. Moreover, as discussed in Chapter 2, SCMs that manage stormwater by driving the water underground with a bored, drilled, or driven shaft or a hole dug deeper than its widest surface dimension are typically considered to be "injection wells," requiring a federal permit and regular monitoring under the Safe Drinking Water Act.

Some states prohibit infiltration because of concerns over long-term groundwater pollution. In California, which does not have a uniform policy for groundwater management and groundwater rights, authority over groundwater quality management falls to several regional and local agencies. For example, the Upper Los Angeles River Area (ULARA) has a court-appointed Watermaster to manage the complex appropriation of its groundwater to user cities and agencies. The ULARA has clashed with the City of Los Angeles regarding rights to all of the water that normally recharges the Los Angeles River via runoff from precipitation. In 2000, the ULARA Watermaster expressed a concern with certain permit provisions of the Los Angeles County MS4 Permit for New Development/ Redevelopment that promoted infiltration, stating that the MS4 permit interfered with the adjudicated right of the City of Los Angeles to manage groundwater.

Urban Development and Sprawl

The continued expansion of urban areas is inevitable given population increases worldwide and the transition from agricultural to industrial economies. Given that urbanization of almost any magnitude—even less than 10 percent impervious area—has been demonstrated to have an impact on in-stream water quality, a central question to be addressed is how water quality can be maintained as cities grow, without having negative impacts on social and economic systems. Ideally, SCMs would perform their water quality function, contribute to the livability of cities, and enhance their economic and social potentials.

Low-density, auto-oriented urban development, commonly known as sprawl, has been the predominant pattern of development in the United States, and increasingly worldwide, since World War II. It has been widely criticized for its inefficient use of land, its high use of natural resources, and its high energy costs—all of which are associated with the required auto-oriented travel. Additionally, ongoing economic costs related to the provision of widely dispersed services and social impacts of a breakdown in community life have been identified (Bruegmann, 2005). Sprawl and the impacts on in-stream water quality that result from urbanization have been an inevitable consequence of improved economic conditions. In the United States, sprawl constitutes the vast majority of development occurring today because a majority of the population is attracted to the benefits of a suburban lifestyle, government has subsidized roads and highways at the expense of public transit, and local zoning often limits development density.

There has been a great deal of innovation in city planning and design in the past decade that encourages greater density and a return to urban living. New types of zoning, New Urbanism, Smart Growth, and related innovations in urban planning and design have been developed in parallel with environmental regulations at local to national levels (see Chapter 2). They acknowledge the importance of protecting natural resources to maintain quality of life and have established water quality as an important consideration in city building.

It is not clear that current stormwater regulations can be effectively implemented over the broad range of development patterns that characterize contemporary cities or if they inadvertently favor one type of development over another. For example, on-site SMCs are often recommended as the preferred means of stormwater management, although they tend to encourage lower-density development patterns. And while they are easily implemented and regulated given the incremental, site-by-site development that is typical of most urban growth, monitoring and maintenance can be expensive and difficult for both the individual property owner and the regulating authority. In highly urbanized areas, they are often relegated to subsurface systems that are expensive and that, to be effective, require high levels of maintenance.

In newly developing areas, cluster development should be encouraged whenever possible, according to the Smart Growth principles of narrower streets, reduced setbacks, and related approaches to reduce the amount of impervious area required and land consumed. Furthermore, an interconnected series of on-site and consolidated SCMs can reduce subsurface stormwater piping requirements. Most planned communities have dedicated park and open-space areas that can constitute 25 percent or more of a development's total land area, making it feasible to easily accommodate consolidated SCMs (typically 8 to 10 percent of impervious area) within multi-functional open space and park lands. Cost efficiencies such as a 30 percent reduction in infrastructure costs (Duaney Plater-Zyberk & Company, 2006) can be realized through Smart Growth development techniques. Clustered housing surrounded by open space, laced with trails, has appreciated in value at a higher rate than conventionally designed subdivisions (Crompton, 2007).

In order to encourage infill or redevelopment over sprawl patterns of development, innovative zoning and other practices will be needed to prevent stormwater management from becoming onerous. For example, incentive zoning or performance zoning could be used to allow for greater densities on a site, freeing other portions of the site for SCMs. Innovations in governance and finance can also be used to incorporate consolidated SCMs into urban environments. For example, the City of Denver, in updating its Comprehensive Plan, designated certain underdeveloped corridors and districts in the city as "areas of change" where it hoped to encourage large-scale infill redevelopment. Given the scale of redevelopment, it would be feasible to establish special maintenance districts, allowing the development of consolidated SCMs that have multiple functions. To fund land purchase and facility design and construction, cash in lieu of payments could be made.

Safety and Aesthetic Concerns

Vector-borne diseases, especially West Nile virus, are a concern when SCMs such as extended detention basins, constructed wetlands, and rain barrels are proposed. Furthermore, other SCMs that are poorly designed, improperly constructed, or inadequately maintained may retain water and provide an ideal breeding ground for mosquitoes, increasing the potential for disease transmission to humans and wildlife. Kwan et al. (2005) found that water-retaining SCMs increase the availability of breeding habitats for disease vectors and provide opportunistic species an extended breeding season. State Health Departments generally recommend that SCMs be designed to drain fully in 72 hours, which is the minimum time required for a mosquito to complete its life cycle under optimum conditions. In SCMs where there is permanent standing water, such as stormwater wetlands, there is the possibility of introducing biota that might prey on mosquitoes. Municipalities may have to consider the added cost of vector control and public health when implementing stormwater quality management programs.

With larger consolidated and regional extended detention facilities, concerns about the safety of children who may be attracted to such SCMs and ensuing liability must be considered. These SCMs need to be fenced off or otherwise designed appropriately to reduce the risk of drowning.

One aspect of stormwater management that is infrequently considered is the aesthetic appeal, or lack thereof, of SCMs. The visual qualities of SCMs are important because they are a growing part of the urban landscape setting. Although it can be assumed that landscapes that are carefully tended are often preferred over other types of landscapes, it depends substantially on one's point of view. For example, an engineer may consider a particular SCM that is functioning as expected to be beautiful in the sense that its engineering function has been realized, even though there is sediment buildup, algae, or other products of a properly functioning SCM visible. Similarly, a biologist or ecologist evaluating an ecologically healthy SCM in an urban context might find it to be beautiful because of its biological or ecological diversity, whereas another individual who evaluates the same SCM finds it to be "weedy." SCMs can be viewed as a means of restoring a degraded landscape to a state that might have existed before urban development. The desire to "return to nature" is a seductive idea that suggests naturalistic SCMs that may have very little to do with an original landscape, given the dramatic changes in hydrology that are inevitable with urban streams. Each of these widely varied views of SCMs may be appropriate depending on the context and the viewer.

A goal of stormwater management should be to make SCMs desirable and attractive to a broader audience, thereby increasing their potential for long-term effectiveness. For example, the Portland convention center rain gardens demonstrate how native and non-native wetland plantings can be carefully composed as a landscape composition and also provide for stormwater treatment. If context and aesthetics of a chosen SCM are poorly matched, there is a high probability that the SCM will be eliminated or its function compromised because of modifications that make its landscape qualities more appropriate for its context.

CONCLUSIONS AND RECOMMENDATIONS

SCMs, when designed, constructed, and maintained correctly, have demonstrated the ability to reduce runoff volume and peak flows and to remove pollutants. However, in very few cases has the performance of SCMs been mechanistically linked to the guaranteed sustainment at the watershed level of receiving water quality, in-stream habitat, or stream geomorphology. Many studies demonstrate that degradation in rivers is directly related to impervious surfaces in the contributing watershed, and it is clear that SCMs, particularly combinations of SMCs, can reduce the runoff volume, erosive flows, and pollutant loadings coming from such surfaces. However, none of these measures perfectly mimic natural conditions, such that the accumulation of these SCMs in a watershed may not protect the most sensitive beneficial aquatic life uses in a state. Furthermore, the implementation of SCMs at the watershed scale has been too inconsistent and too recent to observe an actual cause-and-effect relationship between SCMs and receiving waters. The following specific conclusions and recommendations about stormwater control measures are made.

Individual controls on stormwater discharges are inadequate as the sole solution to stormwater in urban watersheds. SCM implementation needs to be designed as a system, integrating structural and nonstructural SCMs and incorporating watershed goals, site characteristics, development land use, construction erosion and sedimentation controls, aesthetics, monitoring, and maintenance. Stormwater cannot be adequately managed on a piecemeal basis due to the complexity of both the hydrologic and pollutant processes and their effect on habitat and stream quality. Past practices of designing detention basins on a site-by-site basis have been ineffective at protecting water quality in receiving waters and only partially effective in meeting flood control requirements.

Nonstructural SCMs such as product substitution, better site design, downspout disconnection, conservation of natural areas, and watershed and land-use planning can dramatically reduce the volume of runoff and pollutant load from a new development. Such SCMs should be considered first before structural practices. For example, lead concentrations in stormwater have been reduced by at least a factor of 4 after the removal of lead from gasoline. Not creating impervious surfaces or removing a contaminant from the runoff stream simplifies and reduces the reliance on structural SCMs.

SCMs that harvest, infiltrate, and evapotranspirate stormwater are critical to reducing the volume and pollutant loading of small storms. Urban municipal separate stormwater conveyance systems have been designed for flood control to protect life and property from extreme rainfall events, but they have generally failed to address the more frequent rain events (<2.5 cm) that are key to recharge and baseflow in most areas. These small storms may only generate runoff from paved areas and transport the "first flush" of contaminants. SCMs designed to remove this class of storms from surface runoff (runoff-volume-reduction SCMs—rainwater harvesting, vegetated, and subsurface) can also address larger watershed flooding issues.

Performance characteristics are starting to be established for most structural and some nonstructural SCMs, but additional research is needed on the relevant hydrologic and water quality processes within SCMs across different climates and soil conditions. Typical data such as long-term load reduction efficiencies and pollutant effluent concentrations can be found in the International Stormwater BMP Database. However, understanding the processes involved in each SCM is in its infancy, making modeling of these SCMs difficult. Seasonal differences, the time between storms, and other factors all affect pollutant loadings emanating from SCMs. Research is needed that moves away from the use of percent removal and toward better simulation of SCM performance. Hydrologic models of SCMs that incorporate soil physics (moisture, wetting fronts) and groundwater processes are only now becoming available. Research is particularly important for nonstructural SCMs, which in many cases are more effective, have longer life spans, and require less maintenance than structural SCMs. EPA should be a leader in SCM research, both directly by improving its internal modeling efforts and by funding state efforts to monitor and report back on the success of SCMs in the field.

Research is needed to determine the effectiveness of suites of SCMs at the watershed scale. In parallel with learning more about how to quantify the unit processes of both structural and nonstructural practices, research is needed to develop surrogates or guidelines for modeling SCMs in lumped watershed models. Design formulas and criteria for the most commonly used SCMs, such as wet ponds and grass swales, are based on extensive laboratory and/or field testing. There are limited data for other SCMs, such as bioretention and proprietary filters. Whereas it is important to continue to do rigorous evaluations of individual SCMs, there is also a role for more simple methods to gain an approximate idea about how SCMs are performing. The scale factor is a problem for watershed managers and modelers, and there is a need to provide guidance on how to simulate a watershed of SCMs, without modeling thousands of individual sites.

Improved guidance for the design and selection of SMCs is needed to improve their implementation. Progress in implementing SCMs is often handicapped by the lack of design guidance, particularly for many of the non-traditional SCMs. Existing design guidance is often incomplete, outdated, or lacking key details to ensure proper on-the-ground implementation. In other cases, SCM design guidance has not been disseminated to the full population of MS4 communities. Nationwide guidance on SCM design and implementation may not be advisable or applicable to all physiographic, climatic, and ecoregions of the country. Rather, EPA and the states should encourage the development of regional design guidance that can be readily adapted and adopted by municipal and industrial permittees. As our understanding of the relevant hydrologic, environmental, and biological processes increases, SCM design guidance should be improved to incorporate more direct consideration of the parameters of concern, how they move across the landscape, and the issues in receiving waters.

The retrofitting of urban areas presents both unique opportunities and challenges. Promoting growth in these areas is desirable because it takes pressure off the suburban fringes, thereby preventing sprawl, and it minimizes the creation of new impervious surfaces. However, it is more complex than Greenfields development because of the need to upgrade existing infrastructure, the limited availability and affordability of land, and the complications caused by rezoning. These sites may be contaminated, requiring cleanup before redevelopment can occur. Both innovative zoning and development incentives, along with the selection of SCMs that work well in the urban setting, are needed to achieve fair and effective stormwater management in these areas. For example, incentive or performance zoning could be used to allow for greater densities on a site, freeing other portions of the site for SCMs. Publicly owned, consolidated SCMs should be strongly considered as there may be insufficient land to have small, on-site systems. The performance and maintenance of the former can be overseen more effectively by a local government entity. The types of SCMs that are used in consolidated facilities—particularly detention basins, wet/dry ponds, and stormwater wetlands—perform multiple functions, such as prevention of streambank erosion, flood control, and large-scale habitat provision.

REFERENCES

Alexander, D., and J. Heaney. 2002. Comparison of Conventional and Low Impact Development Drainage Designs. Final Report to the Sustainable Futures Society. University of Colorado, Boulder.

Andrews, E. D. 1984. Bed-material entrainment and hydraulic geometry of gravel-bed rivers in Colorado. Geological Society of America Bulletin 95:371-378.

Angus, R., K. Marion, and M. Lalor. 2002. Continuation of Studies to Evaluate the Effectiveness of Current BMPs in Controlling Stormwater Discharges from Small Construction Sites: Pilot Studies of Methods to Improve their

Effectiveness, and Assessment of the Effects of Discharge on Stream Communities. Alabama Water Resources Research Institute.

Arendt, R. 1996. Conservation Design for Subdivisions. Covelo, CA: Island Press.

ASCE (American Society of Civil Engineers). 2007. 2007 Data Analysis Report. International Stormwater BMP Database. U.S. Environmental Protection Agency and the Water Environment Research Foundation. Available at www.bmpdatabase.orgr.

Athayde, D. N., P. E. Shelly, E. D. Driscoll, D. Gaboury, and G. Boyd. 1983. Results of the Nationwide Urban Runoff Program—Vol. 1, Final Report. EPA WH-554. Washington, DC: EPA.

Aultman, S. 2007. Analyzing Cost Implications of Water Quality Trading Provisions: Lessons from the Virginia Nutrient Credit Exchange Act. M.S. Thesis, Virginia Polytechnic Institute and State University, Blacksburg.

Balusek, J. 2003. Quantifying Decreases in Stormwater Runoff from Deep-Tilling, Chisel-Planting and Compost Amendments. Dane County Land Conservation Department. Madison, WI.

Bannerman, R., J. Konrad, D. Becker, G. V. Simsiman, G. Chesters, J. Goodrich-Mahoney, and B. Abrams. 1979. The IJC Menomonee River Watershed Study—Surface Water Monitoring Data, EPA-905/4-79-029. Chicago: U.S. Environmental Protection Agency.

Bannerman, R., K. Baun, M. Bohn, P. E. Hughes, and D. A. Graczyk. 1983. Evaluation of Urban Nonpoint Sources Pollution Management in Milwaukee County, Wisconsin—Vol. I, Urban Stormwater Characteristics, Constituent Sources, and Management by Street Sweeping: Chicago, U.S. PB 84-114164. Springfield, VA: NTIS.

Barr Engineering Co. 2001. Minnesota Small Urban Sites BMP Manual. St. Paul, MN: Metropolitan Council Environmental Services.

Barrett, M. E., J. F. Malina, Jr., and R. J. Charbeneau. 1998. An evaluation of the performance of geotextiles for temporary sediment control. Water Environment Research 70(3):283-290.

Barten, J., and J. Johnson. 2007. Nutrient management with the Minnesota phosphorus fertilizer law. Lakeline (Summer):23-28.

Bean, E. Z., W. F. Hunt, and D. A. Bidelspach. 2007. Evaluation of Four Permeable Pavement Sites in Eastern North Carolina for Runoff Reduction and Water Quality Impacts. ASCE Journal of Irrigation and Drainage Engineering 133(6):583-592.

Bender, G. M., and M. L. Terstriep. 1984. Effectiveness of street sweeping in urban runoff pollution control. The Science of the Total Environment 33:185-192.

Bernhardt, E., and M. Palmer. 2007. Restoring streams in an urbanizing landscape. Freshwater Biology 52:731-751.

Bernhardt, E., M. Palmer, J. Allen, G. Alexander, K. Barnas, S. Brooks, J. Carr, S. Clayton, C. Dahm, J. Follstad-Shah, D. Galat, S. Gloss, P. Goodwin, D. Hart, B. Hassett, R. Jenkinson, S. Katz, G. M. Kondolf, P. S. Lake, R. Lave,

J. L. Meyer, T. K. O'Donnell, L. Pagano, B. Powell, E. Sudduth. 2005. Ecology: Synthesizing U.S. river restoration efforts. Science: 308:636-637.

Bernot, M., and W. Dodds. 200 5. Nitrogen retention, removal and saturation in lotic ecosystems. Ecosystems 8:442-453.

Black and Veatch. 2005. Stormwater Utility Survey. Overland, KS: Black and Veatch.

Braga, A., M. Horst, and R. Traver. 2007. Temperature effects on the infiltration rate through an infiltration basin BMP. Journal of Irrigation and Drainage Engineering 133(6):593-601.

Braskerud, B. C. 2001. The Influence of vegetation on sedimentation and re-suspension of soil particles in small constructed wetlands. Journal of Environmental Quality 30:1447-1457.

Brinkman, R., and G. A. Tobin. 2001. Urban Sediment Removal: The Science, Policy, and Management of Street Sweeping. Boston, MA: Kluwer Academic.

Brown, E., D. Carac, and R. Pitt. 2004. Illicit discharge detection and elimination: a guidance manual for program development and technical assessments. Ellicott City, MD: Center for Watershed Protection.

Brown, W., and D. Caraco. 1997. Muddy water in, muddy water out. Watershed Protection Techniques 2(3):393-404.

Brown, W., and T. Schueler. 1997. The Economics of Stormwater BMPs in the Mid-Atlantic Region. Ellicott City, MD: Center for Watershed Protection.

Bruegmann, R. 2005. Sprawl: A Compact History. The University of Chicago Press.

Bukaveckas, P. 2007. Effects of channel restoration on water velocity, transient storage and nutrient uptake in a channelized stream. Environmental Science and Technology 41:1570-1576.

Burton, G., and R. Pitt. 2002. Stormwater Effects Handbook: A Toolbox for Watershed Managers, Scientists and Engineers. Boca Raton, FL: CRC/Lewis.

Caltrans. 2004. BMP Retrofit Pilot Program Final Report CTSW-RT-01-050. Sacramento, CA: California Department of Transportation.

Cappiella, K., and K. Brown. 2000. Derivation of impervious cover for suburban land uses in the Chesapeake Bay. Final Report. Chesapeake Research Consortium. Ellicott City, MD: Center for Watershed Protection.

Cappiella, K., T. Schueler, and T. Wright. 2006. Urban Watershed Forestry Manual. Part 2: Conserving and Planting Trees at Development Sites. Newtown Square, PA: USDA Forest Service.

Carling, P. 1988. The concept of dominant discharge applied to two gravel-bed streams in relation to channel stability thresholds. Earth Surface Processes and Landforms 13:355-367.

CASQA (California Stormwater Quality Association). 2007. Municipal Stormwater Program Effectiveness Assessment Guidance. Sacramento, CA: California Association of Stormwater Quality Agencies.

Chang, M. 2006. Forest Hydrology: An Introduction to Water and Forests, 2nd

Ed. New York: CRC Press.

Chang, Y., Chou, C., Su, K., and C. Tseng. 2005. Effectiveness of street sweeping and washing for controlling ambient TSP. Atmospheric Environment 39:1891-1902.

Charles River Watershed Association. 2007. CRWA municipal stormwater financing: Survey results. Charles River Watershed Association. Weston, MA.

Cheng, M., L. Coffman, Y. Zhang, and J. Licsko. 2005. Hydrologic responses from low impact development compared to conventional development. Pp 337-357 *In:* Stormwater Management for Smart Growth. New York: Springer.

Cheshire, P., and S. Sheppard. 1995. On the price of land and the value of amenities. Economica 62:247-267.

Chollak, T., and P. Rosenfeld. 1998. Guidelines for Landscaping with Compost-Amended Soils. Prepared for City of Redmond Public Works, Redmond, WA. Available at http://www.ci.redmond.wa.us/insidecityhall/ publicworks/environment/pdfs/compostamendedsoils.pdf. Last accessed August 26, 2008.

City of Seattle. No date. Seattle's Natural Drainage Systems: A low-impact development approach to stormwater management. Brochure, Seattle, WA.

Clark, S., M. Lalor, R. Pitt, and R. Field. 2005. Wet-weather pollution from commonly used building materials. Paper presented at the 10th International Conference on Urban Drainage, August 21-26, Copenhagen.

Clausen, J. C. 2007. Jordan Cover Watershed Project Final Report. Storrs, CT: University of Connecticut.

Coombes, P. 2004. Water sensitive design in the Sydney Region—Practice Note 4. Rainwater Tanks. Published by the Water Sensitive Design in the Sydney Region Project.

Coombes, P., J. Argue, and G. Kuczera. 2000. Figtree Place: a case study in water sensitive urban development (WSUD). Urban Water Journal 4(1):335-343.

Crompton, J. 2007. The impact of parks and open spaces on property taxes. Chapter 1 *In:* The Economic Benefits of Land Conservation. T. F. de Brun (ed.). Trust for Public Lands.

Cross, L., and L. Duke. 2008. Regulating industrial stormwater: state permits, municipal implementation, and a protocol for prioritization. Journal of the American Water Resources Association 44(1):86-106.

Currier, B., J. Jones, and G. Moeller. 2005. NPDES Stormwater Cost Survey. Report prepared for the California State Water Resources Control Board. Sacramento, CA: California State Water Resources Control Board.

Cutter, W. B., K. A. Baerenklau, A. DeWoody, R. Sharma, and J. G. Lee. 2008. Costs and benefits of capturing urban runoff with competitive bidding for decentralized best management practices. Water Resources Research, doi:10.1029/2007WR006343.

CWP (Center for Watershed Protection). 1998a. Better Site Design: A Handbook for Changing Development Rules in Your Community. Ellicott City, MD: Center for Watershed Protection.

CWP. 1998b. The benefits of better site design in residential subdivisions. Watershed Protection Techniques 3(2):633-646.

CWP. 1998c. The benefits of better site design in commercial developments. Watershed Protection Techniques 3(2):647-656.

CWP. 2000a. Introduction to better site design. Watershed Protection Techniques 3(2):623-632.

CWP. 2000b. National Pollutant Removal Performance Database, 2nd Ed. Ellicott City, MD: Center for Watershed Protection.

CWP. 2001. The Economic Benefits of Better Site Design in Virginia. Document prepared for Virginia Department of Conservation and Recreation.

CWP. 2004. Stormwater pond and wetland maintenance guidebook. Ellicott City, MD: Center for Watershed Protection.

CWP. 2005. Pollution Source Control Practices. Manual 8, Urban Subwatershed Restoration Manual Series. Ellicott City, MD: Center for Watershed Protection.

CWP. 2007a. National Pollutant Removal Performance Database, 3rd Ed. Ellicott City, MD: Center for Watershed Protection.

CWP. 2007b. Urban stormwater retrofit practices. Manual 3, Urban Small Watershed Restoration Manual Series. Ellicott City, MD: Center for Watershed Protection.

CWP. 2008. Municipal good housekeeping practices. Manual 9, Urban Small Watershed Restoration Manual Series. Ellicott City, MD: Center for Watershed Protection.

Davis, A. P., W. F. Hunt, R. G. Traver, and M. E. Clar. 2008. Bioretention technology: an overview of current practice and future needs. ASCE Journal of Environmental Engineering (in press).

Debo, T. N., and A. J. Reese. 2002. Municipal Stormwater Management, Second Edition. Location: Lewis Publishers, CRC Press.

Deschesne, M., S. Barraud, and J. P. Bardin. 2005. Experimental Assessment of Stormwater Infiltration Basin Evolution. J. Env. Engineering 131(7):1090–1098.

DeWoody, A. E. 2007. Determining Net Social Benefits from Optimal Parcel-Level Infiltration of Urban Runoff: A Los Angles Analysis. Master's Thesis, University of California, Riverside.

Dietz, M. E., and Clausen, J. C. 2008. Stormwater Runoff and Export Changes with Development in a Traditional and Low Impact Subdivision. Journal of Environmental Management 87(4):560-566.

Dietz, M., and J. Clausen. 2006. Saturation to improve pollutant retention in a rain garden. Environmental Science and Technology 40(4):1335-1340.

Doll, A., and G. Lindsey. 1999. Credits bring economic incentives for onsite stormwater management. Watershed and Wet Weather Technical Bulletin 4(1):12-15.

Doll, R., and G. Jennings. 2003. Stream restoration: a natural channel design handbook. North Carolina State University Extension, Raleigh.

Duaney Plater-Zyberk & Company. 2006. Light Imprint New Urbanism: A Case Study Comparison. Chicago: Congress for the New Urbanism. Available at http://www.cnu.org/sites/files/Light%20 Imprint%20NU%20Report-web.pdf. Last accessed August 26, 2008.

Dubin, R. A. 1998. Predicting house prices using multiple listings data. Journal of Real Estate Finance and Economics 17:35-59.

Duke, L. D. 2007. Industrial stormwater runoff pollution prevention regulations and implementation. Presentation to NRC Stormwater Committee, August 22, Seattle, WA.

Duke, L. D., and P. G. Beswick. 1997. Industry compliance with storm water pollution prevention regulations: the case of transportation industry facilities in California and the Los Angles Region. Journal of the American Water Resources Association 33(4):825-838.

Duke, L., and C. Augustenberg. 2006. Effectiveness of self regulation and self-reported environmental regulations for industry: the case of stormwater runoff in the U.S. Journal of Environmental Planning and Management 49(3):385-411

Duncan, R. 2004. Selecting the "right" stormwater utility rate model — an adventure in political and contextual sensitivity. In: World Water Congress 2001 Bridging the Gap: Meeting the World's Water and Environmental Resources Challenges. ASCE Conference Proceedings, May 20–24, 2001, Orlando, FL.

Emerson, C. H., C. Welty, and R. Traver. 2005. Watershed-scale evaluation of a system of storm water detention basins. Journal of Hydrologic Engineering 10(3):237-242.

Emerson, C., and R. Traver. 2008. Multiyear and Seasonal Variation of Infiltration from Storm-Water Best Management Practices. J. Irrig. Drain. Engrg. 134(5):598-605.

EPA (U.S. Environmental Protection Agency). 1983. Results of the Nationwide Urban Runoff Program. PB 84-185552. Washington, DC: Water Planning Division.

EPA. 1999. Preliminary Data Summary of Urban Stormwater Best Management Practices. EPA-821-R-99-012. Washington, DC: EPA Office of Water.

EPA. 2000. Social Costs in Guidelines for Preparing Economic Analysis. Publ. 240 R-00-003. Washington DC: EPA.

EPA. 2005a. Environmental Technology Verification Report, Stormwater Source Area Treatment Device—BaySaver Technologies, Inc., BaySaver Separation System, Model 10K. September 2005. EPA/600/R-05/113. Washington DC: EPA.

EPA. 2005b. Environmental Technology Verification Report; Stormwater Source Area Treatment Device--The Stormwater Management StormFilter® using Perlite Filter Media. August 2005. EPA/600/R-05/137. Washington DC: EPA.

EPA. 2005c. Environmental Technology Verification Report: Stormwater Source Area Treatment Device: Vortechnics, Inc. Vortechs System, Model 1000 September 2005 EPA/600/R-05/140. Washington DC: EPA.

EPA. 2005d. National Management Measures to Control Nonpoint Source Pollution from Urban Areas. Washington, DC: U.S. Government Printing Office.

EPA. 2007. Reducing Stormwater Costs Through Low Impact Development (LID) Strategies and Practices. EPA 841-F-07-006. Washington DC: EPA.

Ermilio, J., and R. Traver. 2006. Hydrologic and pollutant removal performance of a bio-infiltration BMP. EWRI 2006, National Symposium.

Ferguson, B. K. 1991. The Failure of Detention and the Future of Stormwater Design. Landscape Architecture 81(12):76-79.

Ferguson, B. K. 2002. Stormwater Management and Stormwater Restoration. Chapter I.1 of Handbook of Water Sensitive Planning and Design. R. L. France (ed.). Lewis Publishers.

Ferguson, B. K. 1994. Stormwater Infiltration. Boca Raton, FL: CRC Press.

Ferguson, B. K. 1998. Introduction to Stormwater. New York: John Wiley & Sons.

Fina, M., and L. Shabman. 1999. Some unconventional thoughts on sprawl. William and Mary Environmental Law Review 23(3):739-775.

FISRWG (Federal Interagency Stream Restoration Working Group). 2000. Stream Corridor Restoration: Principles, Processes and Practices. Washington, DC: USDA Natural Resource Conservation Service.

Florida Stormwater Association. 2003. Stormwater Utilities Survey. Available online at http://www.florida-stormwater.org/surveys/2003/menu.asp. Last accessed July 2006.

GAO. 2007. Further Implementation and Better Cost Data Needed to Determine Impact of EPA's Storm Water Program on Communities. GAO-07-479. Washington, DC: GAO.

Goetz, S., R. Wright, A. Smith, E. Zinecker, and E. Schaub. 2003. IKONOS imagery for resource management: tree cover, impervious surfaces, and riparian buffer analyses in the mid-Atlantic region. Remote Sensing in the Environment 88:195-208.

Gomi, T., R. Sidle, and J. Richardson. 2002. Understanding processes and downstream linkages of headwater systems. BioScience 53(10):905-915.

Greenway, M., P. Dale, and H. Chapman. 2003. An assessment of mosquito breeding and control in 4 surface flow wetlands in tropical–subtropical Australia. Water Science and Technology 48(5):249–256.

Gregory, J., M. Duke, D. Jones, and G. Miller. 2006. Effect of urban soil compaction on infiltration rates. Journal of Soil and Water Conservation 61(3):117-133.

Groffman, P., A. Dorset, and P. Mayer. 2005. N processing within geomorphic structures in urban streams. Journal North American Benthological Society 24(3):613-625.

Gulliver, J. S., and J. L. Anderson. 2007. Assessment of Stormwater Best Management Practices. Regents of the University of Minnesota. Available at http://wrc.umn.edu/outreach/stormwater/ Last accessed December 4, 2007.

Hager, M. C. 2003. Low impact development: lot-level approaches to stormwater management are gaining ground. Stormwater. Available at http://www.lowimpactdevelopment.org/lid%20articles /stormwater_feb2003.pdf. Last accessed August 26, 2008.

Hardy, M, P. Coombes, and G. Kuczera. 2004. An investigation of estate level impacts of spatially distributed rainwater tanks. Proceedings of the 2004 International Conference on Water Sensitive Urban Design—Cities as Catchments, November 21–25, 2004, Adelaide.

Hathaway, J., and W. Hunt. 2006. Level spreaders: Overview, design and maintenance. Urban Waterways. North Carolina State University and Cooperative Extension. Raleigh.

Heaney, J. P., D. Sample, and L. Wright. 2002. Costs of Urban Stormwater Control. EPA-600-02/021. Washington, DC: EPA.

Heasom, W., R. G. Traver, and A. Welker. 2006. Hydrologic modeling of a bioinfiltration best management practice. Journal of the American Water Resources Association 42(5):1329-1347.

Hillel, D. 1998. Environmental Soil Physics. San Diego, CA: Academic Press.

Hirschman, D., and J. Kosco. 2008. Managing stormwater in your community: a guide for building an effective post-construction problem. EPA 833-R-08-001. Tetra-tech, Inc. and Center for Watershed Protection. Ellicott City, MD.

Holman-Dodds, J., A. Bradley, and K. Potter. 2003. Evaluation of hydrologic benefits of infiltration based urban stormwater management. Journal of the American Water Resources Association 39(1):205-215.

Horner, R. R., and C. Chapman. 2007. NW 110th Street Natural Drainage System Performance Monitoring, with Summary of Viewlands and 2nd Avenue NW SEA Streets Monitoring. Report to Seattle Public Utilities by Department of Civil and Environmental Engineering, University of Washington, Seattle.

Horner, R., H. Lim, and S. Burges. 2003. Hydrologic Monitoring of the Seattle Ultra-Urban Stormwater Management Project. Water Resources Series. Technical Report 170. Seattle, WA: University of Washington Department of Civil and Environmental Engineering.

Hoyt, S., and T. Brown. 2005. Stormwater pond and wetland maintenance concerns and solutions. Paper presented at the EWRI 2005: Impacts of Global Climate Change Conference, American Society of Civil Engineers, Anchorage, Alaska, May 15-19.

Huber, W. L. Cannon and M. Stouder. 2006. BMP Modeling Concepts and Simulation. EPA/600/R-06/033. Corvallis, OR: U.S. Environmental Protection Agency.

Hunt, W., and W. Lord. 2006a. Bioretention Performance, Design, Construction, and Maintenance. AG-588-05. North Carolina Cooperative Extension Service. Urban Waterways.

Hunt, W. F., and W. G. Lord. 2006b. Determining Inspection and Maintenance Costs for Structural BMPs in North Carolina. Submitted to University of North Carolina, Water Resources Research Institute. (November).

Hunt, W. F., and B. Lord. 2006c. Stormwater wetlands and wet pond maintenance. AGW-588-07. North Carolina Cooperative Extension Service.

Hunt, W. F., J. T. Smith, S. J. Jadlocki, J. M. Hathaway, and P. R. Eubanks. 2008. Pollutant removal and peak flow mitigation by a bioretention cell in urban Charlotte, NC. ASCE Journal of Environmental Engineering 134(5):403-408.

Jefferies, C. 2004. Sustainable Drainage Systems in Scotland: The Monitoring Programme. Scottish Universities SUDS Monitoring Project. Dundee, Scotland.

Johnson, C., T. Driscoll, T. Siccama and G. Likens. 2000. Elemental fluxes and landscape position in a northern hardwood forest ecosystem. Ecosystems 3:159-184.

Johnston, D. M., J. B. Braden, and T. H. Price. 2006. Downstream economic benefits of conservation development. Journal of Water Resources Planning and Management 132(1):35-43.

Kaspersen, J. 2000. The stormwater utility: will it work in your community? Stormwater 1(1). Available at http://www.forester.net/sw_0011_utility.html. Last accessed August 26, 2008.

Kaushal, S., P. Groffman, P. Meyer, E. Striz, and A. Gold. 2008. Effects of stream restoration on denitrification in an urbanizing watershed. Ecological Applications 18(3):789-804.

Keller, B. 2003. Buddy can you spare a dime? What is stormwater funding? Stormwater 4:7.

Kitchell, A. 2002. Managing for a pure water supply. Watershed Protection Techniques 3(4):800-812.

Konrad, C. 2003. Opportunities and constraints for urban stream rehabilitation. *In:* Restoration of Puget Sound Rivers. D. Montgomery, S. Bolton, D. Booth, and L. Wall (eds.). Seattle, WA: University of Washington Press.

Kwan, J., M. Metzger, M. Shindelbower, and C. Fritz. 2005. Mosquito production in stormwater treatment devices in South Lake Tahoe, California. Proceedings and Papers of the Seventy-Third Annual Conference of the Mosquito and Vector Control Association of California 73:113-119.

Kwiatkowski, M., A. L. Welker, R. G. Traver, M. Vanacore, and T. Ladd. 2007. Evaluation of an infiltration best management practice (BMP) utilizing pervious concrete. Journal of the American Water Resources Association 43(5):1208-1222.

Lager, J. A., W. G. Smith, and G. Tchobanoglous. 1977. Catchbasin Technology Overview and Assessment. EPA-600/2-77-051. Cincinatti, OH: EPA.

Lambe, L., M. Barrett, B. Woods-Ballard, R. Kellagher, P. Martin, C. Jefferies, and M. Hollon. 2005. Performance and Whole Life Costs of Best Management Practices and Sustainable Urban Drainage Systems. Publ. 01-

CTS-21T. Alexandria VA: Water Environment Research Foundation.

Larson, M. L., D. B. Booth, and S. M. Morley. 2001. Effectiveness of large woody debris in stream rehabilitation projects in urban basins. Ecological Engineering 18(2):211-226.

Law, N. 2006. Research in support of an interim pollutant removal rate for street sweeping and storm drain cleanout. Technical Memo No. 2. Prepared for the EPA Chesapeake Bay Program and Urban Stormwater Working Group. Ellicott City, MD: Center for Watershed Protection.

Law, N., K. Diblasi, and U. Ghosh. 2008. Deriving Reliable Pollutant Removal Rates for Municipal Street Sweeping and Storm Drain Cleanout Programs in the Chesapeake Bay Basin. Ellicott City, MD: Center for Watershed Protection.

Legg, A., R. Bannerman, and J. Panuska. 1996. Variation in the relation of runoff from residential lawns in Madison, Wisconsin. USGS Water Resources Investigations Report 96-4194. U.S. Geological Survey.

Lehner, P., G. P. Aponte Clark, D. M. Cameron, and A. G. Frank. 1999. Stormwater Strategies: Community Strategies to Runoff Pollution. Natural Resources Defense Council. Available at http://www.nrdc.org/water/pollution/storm/stoinx.asp. Last accessed October 19, 2008.

Leopold, L. B. 1968. Hydrology for Urban Planning—A Guidebook on the Hydrologic Effects of Urban Land Use. USGS Circular 554. Washington, DC: U.S. Geological Survey.

Lichter, J., and P. Lindsey. 1994. Soil compaction and site construction: assessment and case studies. The Landscape Below Ground. G. W. Watson, and D. Neely (eds.). Savoy, IL: International Society of Arboriculture.

Line, D., and N. White. 2007. Effect of development on runoff and pollutant export. Water Environment Research 75(2):184-194.

Lloyd, S., T. Wong and C. Chesterfield. 2002. Water sensitive urban design: a stormwater management perspective. Cooperative Research Centre for Catchment. Monash University, Victoria 3800 Australia. Industry Report 02/10

Loucks, D. P., E. van Beek, J. R. Stedinger, J. P. M. Dijkman, and M. T. Villars. 2005. Water Resources Systems Planning and Management: An Introduction to Methods, Models, and Applications. Paris: UNESCO.

Low Impact Development Center. 2007. Introduction to low impact development. Available at http://www.lid-stormwater.net/intro/background.htm. Last accessed December 4, 2007.

Lowrance, R., and J. Sheridan. 2005. Surface runoff quality in a managed three zone riparian buffer. Journal of Environmental Quality 34:1851-1859.

MacMullan, E., and S. Reich. 2007. The Economics of Low-Impact Development: A Literature Review. Eugene, OR: ECONorthwest. Available at http://www.econw.com/reports/ECONorthwest_Low-Impact-Development-Economics-Literature-Review.pdf. Last accessed August 26, 2007.

MacCrae, C. R. 1997. Experience from morphological research on Canadian streams: Is control of the two-year frequency runoff event the best basis for stream channel protection? Proceedings of the Effects of Watershed Developments and Management on Aquatic Ecosystems Conference, Snowbird, UT, August 4–9, 1996. L. A. Roesner (ed.). New York: American Society of Civil Engineers.

Manor, A. W. 1966. Public works in ancient Mesopotamia. Civil Engineering 36(7):50-51.

MD DNR. 2005. A Users Guide to Watershed Planning in Maryland. Annapolis, MD: DNR Watershed Services.

McBride, M., and D. Booth. 2005. Urban impacts on physical stream condition: effects on spatial scale, connectivity, and longitudinal trends. Journal of the American Water Resources Association 6:565-580.

McCuen, R. H. 1979. Downstream effects of stormwater management basins. Journal of the Hydraulics Division 105(11):1343-1356.

McHarg, I. L., and F. R. Steiner. 1998. To Heal the Earth, Selected Writings of Ian McHarg. Washington, DC: Island Press.

McHarg, I. L., and J. Sutton. 1975. Ecological Plumbing for the Texas Coastal Plain. Landscape Architecture 65:78-89.

MD DNR (Maryland Department of Natural Resources). 2003. Critical Area 10% Rule Guidance Manual: Maryland Chesapeake and Atlantic Coastal Bays. Annapolis, MD: Critical Area Commission.

Meyers, J. 2003. Where Rivers Are Born: The Scientific Imperative for Defending Small Streams and Wetlands. Washington, D.C.: American Rivers.

Mineart, P., and S. Singh. 1994. Storm Inlet Pilot Study. Performed by Woodward Clyde Consultants for Alameda County Urban Runoff Clean Water Program.

Montalto, F., C. Behr, K. Alfredo, M. Wolf, M. Arye, and M. Walsh. 2007. Rapid assessment of the cost-effectiveness of low impact development for CSO control. Landscape and Urban Planning 82(3):117-131.

Moore, A. and M. Palmer. 2005. Invertebrate diversity in agricultural and urban headwater streams: Implications for conservation and management. Ecological Applications 15(4):1169-1177.

Morgan, R. A., F. G. Edwards, K. R. Brye, and S. J. Burian. 2005. An evaluation of the urban stormwater pollutant removal efficiency of catch basin inserts. Water Environment Research 77(5):500-510.

MSSC (Minnesota Stormwater Steering Committee). 2005. Minnesota Stormwater Manual. St. Paul MN: Emmons & Oliver Resources, Inc., and the Minnesota Pollution Control Agency.

NAFSMA (National Association of Flood and Stormwater Management Agencies). 2006. Guidance for Municipal Stormwater Funding. Washington, DC: NAFSMA. Available at http://www.nafsma.org/Guidance%20Manual%20Version%202X.pdf. Last accessed October 19, 2008.

Narayanan, A., and R. Pitt. 2006. Costs of urban stormwater control practices. Tuscaloosa, AL: Department of Civil, Construction and Environmental Engineering, University of Alabama.

Nichols, D., Akers, M.A., Ferguson, B., Weinberg, S., Cathey, S., Spooner, D., and Mikalsen, T. 1997. Land development provisions to protect Georgia water quality. The School of Environmental Design, University of Georgia. Athens, GA. 35pp.

NRC (National Research Council). 2000. Watershed Management for a Potable Water Supply. Washington, DC: National Academy Press.

PaDEP (Pennsylvania Department of Environmental Protection). 2006. Pennsylvania Stormwater Best Management Practices Manual. Harrisburg, PA: Bureau of Stormwater Management, Division of Waterways, Wetlands and Erosion Control.

Palmquist, R. 1980. Alternative techniques for developing real estate price indexes. Review of Economics and Statistics 62(3):442-448.

Parikh, P., M. A. Taylor, T. Hoagland, H. Thurston, and W. Shuster. 2005. Application of market mechanism and incentives to reduce stormwater runoff: an integrated hydrologic, economic, and legal approach. Environmental Science and Policy 8:133-144.

Perez-Pedini, C., J. Limbruneer, and R. Vogel. 2004. Optimal location of infiltration-based Best management practices for stormwater management. ASCE Journal of Water Resources Planning and Management 131(6):441-448.

Philips, R., C. Clausen, J. Alexpoulus, B. Morton, S. Zaremba, and M. Cote. 2003. BMP research in a low-impact development environment: The Jordan Cove Project. Stormwater 6(1):1-11.

Pickup, G., and R. F. Warner. 1976. Effects of hydrologic regime on magnitude and frequency of dominant discharge. Journal of Hydrology 29:51-75.

Pitt, R. 1979. Demonstration of Nonpoint Pollution Abatement Through Improved Street Cleaning Practices. EPA-600/2-79-161. Cincinnati, OH: EPA.

Pitt, R., with contributions from S. Clark, R. Field, and K. Parmer. 1996. Groundwater Contamination from Stormwater. ISBN 1-57504-015-8. Chelsea, MI: Ann Arbor Press, Inc.

Pitt, R., T. Brown, and R. Morchque. 2004a. National Stormwater Quality Database. Version 2.0. University of Alabama and Center for Watershed Protection Final Report to U.S. Environmental Protection Agency.

Pitt, R., Maestre, A., and Morquecho, R. 2004b. National Stormwater Quality Database. Version 1.1. Available at http://rpitt.eng.ua.edu/Research/ms4/Paper/Mainms4paper.html.

Pitt, R., S. Chen, S. Clark, and J. Lantrip. 2005. Soil structure effects associated with urbanization and the benefits of soil amendments. *In:* World Water and Environmental Resources Congress. Conference Proceedings. American Society of Civil Engineers. Anchorage, AK.

Pouyat, R., I. Yesilonis, J. Russell-Anelli, and N. Neerchal. 2007. Soil chemical and physical properties that differentiate urban land use and cover types. Soil Science Society of America Journal 71(3):1010-1019.

Pouyat, R., M. McDonnel, and S. Pickett. 1995. Soil characteristics of oak stands along an urban-rural land use gradient. Journal of Environmental Quality 24:516-526.

Prince George's County, Maryland. 2000. Low-Impact Development Design Strategies. EPA 841-B-00-003. Washington, DC: EPA.

Qiu, Z., T. Prato, and G. Boehm. 2006. Economic valuation of riparian buffer and open space in a suburban watershed. Journal of the American Water Resources Association 42(6):1583-1596.

Randolph, J., A. C. Nelson, J. M. Schilling, and M. Nowak. 2006. Impact of environmental regulations on the cost of housing. Paper presented at the Association of Collegiate School of Planning, Fort Worth, TX. November 9.

Rea, M., and R. Traver. 2005. Performance monitoring of a stormwater wetland best management practice. National Conference, World Water & Environmental Resources Congress 2005 (EWRI/ASCE).

Reed, S. C., R. W. Crites, and E. J. Middlebrooks. 1998. Natural systems for waste management and treatment. McGraw-Hill Professional. ISBN 0071346627, 9780071346627.

Reese, A. J., and H. H. Presler. 2005. Municipal stormwater system maintenance. Stormwater. Available at http://www.stormh2o.com/sw_0509_municipal.html. Last accessed December 4, 2007.

Richman, T. 1997. Start at the Source: Design Guidance for Storm Water Quality Protection. Oakland, CA: Bay Area Stormwater Management Agencies Association.

Roy, A., C. Faust, M. Freeman, and J. Meyer. 2005. Reach-scale effects of riparian forest cover on urban stream ecosystems. Canadian Journal of Fisheries and Aquatic Science 62:2312-2329.

Roy, A., M. Freeman, B. Freeman, S. Wenger, J. Meyer, and W. Ensign. 2006. Importance of riparian forests in urban subwatersheds contingent on sediment and hydrologic regimes. Environmental Management 37(4):523-539.

Rushton, B. 2002. Low impact parking lot design infiltrates stormwater. *In:* Sixth Biennial Stormwater Research & Watershed Management Conference, September 14-17, 1999, Tampa FL. Brooksville, FL: Southwest Florida Water Management District.

Sample, D. J., J. P. Heaney, L. T. Wright, C.-Y. Fan, F.-H. Lai, and R. Field. 2003. Costs of best management practices and associated land for urban stormwater controls. Journal of Water Resources Planning and Management 129(1):59-68.

Schollen, M., T. Schmidt, and D. Maunder. 2006. Markham Small Streams Study—Policy Update and Implementing Guidelines for the Protection and Management of Small Drainage Courses. Town of Markham, Ontario.

Schueler, T. 1998. Irreducible pollutant concentration discharged from stormwater practices. Watershed Protection Techniques 2(2):369-372. Ellicott City, MD: Center for Watershed Protection.

Schueler, T. 2001a. The compaction of urban soils. Watershed Protection

Techniques 3(2):661-665. Ellicott City, MD: Center for Watershed Protection.

Schueler, T. 2001b. Can urban soil compaction be reversed? Watershed Protection Techniques 3(2):666-669. Ellicott City, MD: Center for Watershed Protection.

Schueler, T. 2001c. On watershed education. Watershed Protection Techniques 3(3):680-689. Ellicott City, MD: Center for Watershed Protection.

Schueler, T. 2002. On watershed behavior and resident education. Watershed Protection Techniques 3(3):671-686. Ellicott City, MD: Center for Watershed Protection.

Schueler, T., and K. Brown. 2004. Urban Stream Repair Practices: Manual 4. Urban Subwatershed Restoration Manual Series. Ellicott City, MD: Center for Watershed Protection.

Schueler, T., D. Hirschman, M. Novotney, and J. Zielinski. 2007. Urban Stormwater Retrofit Practices: Urban Stormwater Restoration Manual 3. Ellicott City, MD: Center for Watershed Protection.

Seattle Public Utilities. 2007. Street Edge Alternatives: Community Cost and Benefits. Available at http://www.seattle.gov/util/About_SPU/Drainage_&_Sewer_System/Natura l_Drainage_Systems/Street_Edge_Alternatives/COMMUNITY_200406180 902084.asp. Last accessed August 2007.

Selbig, W., and R. Bannerman. 2008. A Comparison of Runoff Quality and Quantity from Two Small Basins Undergoing Implementation of Conventional and Low Impact Development Strategies : Cross Plains, WI, Water Years 1999-2005. USGS Scientific Investigations Report 2008-5008. U.S. Geological Survey.

Sharkey, L. J. 2006. The Performance of Bioretention Areas in North Carolina: A Study of Water Quality, Water Quantity, and Soil Media. Thesis: North Carolina State University, Raleigh.

Shaver, E., R. Horner, J. Skupien, C. May, and G. Ridley. 2007. Fundamentals of Urban Runoff Management—Technical and Institutional Issues. Madison, WI: North American Lake Management Society.

Sidle, R. C. 1988. Bed load transport regime of a small forest stream. Water Resources Research 24: 207-218.

Singer, M., and R. Rust. 1975. Phosphorus in surface runoff from a deciduous forest. Journal of Environmental Quality 4:302-311.

Smith, R. A., and W. F. Hunt. 2007. Pollutant removal in bioretention cells with grass cover. Pp. 1-11 In: Proceedings of the World Environmental and Water Resources Congress 2007.

Song, Y., and G. Knaap. 2003. New urbanism and housing values: a disaggregate assessment. Journal of Urban Economics 54(2):218-238

Stagge, J. 2006. Field Evaluation of Hydrologic and Water Quality Benefits of Grass Swales for Managing Highway Runoff. Master's Thesis, University of Maryland.

Stavins, R. N. 2000. Market-based environmental policies. Public Policies for

Environmental Protection. P. R. Portney and R. N. Stavins (eds.). Washington, DC: Resources for the Future.

Stephens, K., P. Graham, and D. Reid. 2002. Stormwater Planning: A Guidebook for British Columbia. Vancouver, BC: Environment Canada.

Strecker, E. W., W. C Huber, J. P. Heaney, D. Bodine, J. J. Sansalone, M. M. Quigley, D. Pankani, M. Leisenring, and P. Thayumanavan. 2005. Critical assessment of Stormwater Treatment and Control Selection Issues. Water Environment Research Federation, Report No. 02-SW-1. ISBN 1-84339-741-2. 290pp. and NCHRP Report 565.

Strecker, E., M. Quigley, and M. Leisenring. 2007. Critical Assessment of Stormwater Treatment and Control Selection Issues- Implications and Recommendations for Design Standards. 6th International Conference on Sustainable Techniques and Strategies in Urban Watershed Management. June 24-28, 2007, Lyon, France. Novatech 2007.

Strecker, E., M. Quigley, B. Urbonas, and J. Jones. 2004. Stormwater management: state-of-the-art in comprehensive approaches to stormwater. The Water Report 6:1-10.

Sudduth, E., J. Meyer, and E. Bernhardt. 2007. Stream restoration practices in the southeastern U.S. Restoration Ecology 15:516-523.

Swamikannu, X. 1994. Auto Recycler and Dismantler Facilities: Environmental Analysis of the Industry with a Focus on Stormwater Pollution. Ph.D. Dissertation, University of California, Los Angeles.

SWRPC (Southeastern Wisconsin Regional Planning Commission). 1991. Costs of Urban Nonpoint Source Water Pollution Control Measures. Technical Report No. 31. Waukesha, WI: Southeastern Regional Planning Commission.

Traver, R. G., and R. A. Chadderton. 1992. Accumulation Effects of Stormwater Management Detention Basins. Hydraulic Engineering: Saving a Threatened Resource—In Search of Solutions. Baltimore, MD: American Society of Civil Engineers.

Turner, M. 2005. Leachate, Soil and Turf Concentrations from Fertilizer-Results from the Stillhouse Neighborhood Fertilzer Leachate Study. Austin: City of Austin Watershed Protection and Development Review Department.

USGS (United States Geological Survey). 2005. Evaluation of Street Sweeping as a Water-Quality Management Tool in Residential Basins in Madison. Scientific Investigations Report. September. Reston, VA: USGS.

Van Metre, P. C., B. J. Mahler, M. Scoggins, and P. A. Hamilton. 2006. Parking Lot Sealcoat: A Major Source of Polycyclic Aromatic Hydrocarbons (PAHs) in Urban and Suburban Environmental. USGS Fact Sheet 2005-3147.

Van Seters, T., D. Smith and G. MacMillan. 2006. Performance evaluation of permeable pavement and a bioretentions swale. Proceedings 8th International Conference on Concrete Block Paving. November 6-8, 2006, San Francisco, CA.

Walsh, C, K. Waller, J. Gehling, and R. MacNally. 2007. Riverine invertebrate assemblages are degraded more by catchment urbanisation than riparian deforestation. Freshwater Biology 52(3):574-587.

Wang, L., J. Lyons, P, Rasmussen, P. Simons, T. Wiley, and P. Stewart. 2003a. Watershed, reach, and riparian influences on stream fish assemblages in the Northern Lakes and Forest Ecoregion. Canadian Journal of Fisheries and Aquatic Science 60:491-505.

Wang, L., J. Lyons, and P. Kanehl. 2003b. Impacts of urban land cover on trout streams in Wisconsin and Minnesota. Transactions of the American Fisheries Society 132:825-839.

WERF (Water Environment Research Foundation). 2006. Performance and Whole-Life Costs of BMPs and SUDs. Alexandria, VA: IWA.

WERF. 2008. Analysis of Treatment System Performance: International Stormwater Best Management Practices (BMP) Database [1999-2008]. Alexandria, VA: WERF.

Whisnant, R. 2007. Universal Stormwater Model Ordinance for North Carolina. Chapel Hill, NC: University of North Carolina Environmental Finance Center.

Wiegand, C., T. Schueler, W. Chittenden, and D. Jellick. 1986. Cost of Urban Stormwater Runoff Controls. Pp. 366-380 *In:* Proceedings of an Engineering Foundation Conference. Urban Water Resource. Henniker, NH: American Society of Civil Engineers.

Winter, T. 2007. The role of groundwater in generating streamflow in headwater areas in maintaining baseflow. Journal of American Water Resources Association 43(1):15-25.

Wisconsin DNR (Department of Natural Resources). 2002. Wisconsin Administrative Code: Environmental Protection General: chaps. TRANS 401.03. Madison, WI: Wisconsin DNR.

Wossink, A., and B. Hunt. 2003. The Economics of Structural Stormwater BMPs in North Carolina. Research Report Number 344. Raleigh, NC: Water Resources Research Institute.

Zarriello, P., R. Breault, and P. Weiskel. 2002. Potential effects of structural controls and street sweeping on stormwater loads to the Lower Charles River, Massachusetts. USGS Water Resources Investigations Report 02-4220. U.S. Geological Survey.

6

Innovative Stormwater Management and Regulatory Permitting

There are numerous innovative regulatory strategies that could be used to improve EPA's stormwater program. This chapter first outlines a substantial departure from the status quo, namely, basing all stormwater and other wastewater discharge permits on watershed boundaries instead of political boundaries. Watershed-based permitting is not a new concept, but it has been attempted in only a few communities. Development of the new permitting paradigm is followed by more modest and easily implemented recommendations for improving the stormwater program, from a new plan for monitoring industrial sites to encouraging greater use of quantitative measures of the maximum extent practicable requirement. The recommendations in the latter half of the chapter do not preclude adoption of watershed-based permitting at some future date, and indeed they lay the groundwork in the near term for an eventual shift to watershed-based permitting.

WATERSHED PERMITTING FRAMEWORK FOR MANAGING STORMWATER

At its initial meeting in January 2007, the committee heard opinions that collectively pointed in a new direction for managing and regulating stormwater that would differ from the end-of-pipe approach traditionally applied by regulatory agencies under the National Pollutant Discharge Elimination System (NPDES) permits and be based instead on a watershed framework. Indeed, the U.S. Environmental Protection Agency (EPA) has already given substantial thought to watershed permitting and issued a Watershed-Based NPDES Permitting Policy Statement (EPA, 2003a) that defined watershed-based permitting as an approach that produces NPDES permits that are issued to point sources on a geographic or watershed basis. It went on to declare that, "The utility of this tool relies heavily on a detailed, integrated, and inclusive watershed planning process. Watershed planning includes monitoring and assessment activities that generate the data necessary for clear watershed goals to be established and permits to be designed to specifically address the goals."

In the statement, EPA listed a number of important benefits of watershed permitting:

- More environmentally effective results;
- Ability to emphasize measuring the effectiveness of targeted actions on improvements in water quality;

- Greater opportunities for trading and other market-based approaches;
- Reduced cost of improving the quality of the nation's waters;
- More effective implementation of watershed plans, including total maximum daily loads (TMDLs); and
- Other ancillary benefits beyond those that have been achieved under the Clean Water Act (e.g., integrating CWA and Safe Drinking Water Act [SDWA] programs).

Subsequent to the policy statement, EPA published two guidance documents that lay out a general process for a designated state that wishes to set up any type of permit or permits under CWA auspices on a watershed basis (EPA, 2003b, 2007a). It also outlined a number of case studies illustrating various kinds of permits that contain some watershed-based elements. Box 6-1 describes in greater detail the more recent report (EPA, 2007a) and its 11 "options" for watershed-based permitting. Unfortunately, the EPA guidance is lacking in its description of what constitutes watershed-based permitting, who would be covered under such a permit, and how it would replace the current program for municipalities and industries discharging stormwater under an individual or general NPDES permit. Few examples are given, some of which are not even watershed-based, with most of the examples involving grouping municipal wastewater treatment works under a single permit with no reference to stormwater. Most of the 11 options are removed from the fundamental concept of watershed-based permitting. Finally, the guidance fails to elaborate on the policy statement goal to make water quality standards watershed-based. The committee concluded that, although the EPA documents lay some groundwork for watershed-based permitting—especially the ideas of integrated municipal permits, water quality trading, and monitoring consortia—the sum total of EPA's analysis does not define a framework for moving toward true watershed-based permitting. The guidance attends to few of the details associated with such a program and it has made no attempt to envision how such a system could be extended to the states and the municipal and industrial stormwater permittees. This chapter attempts to overcome these shortcomings by presenting a more comprehensive description of watershed-based permitting for stormwater dischargers.

The approach proposed in this chapter fits within the general framework outlined by EPA but goes much further. First, it is intended to replace the present structure, instead of being an adjunct to it, and to be uniformly applied nationwide. The proposal adopts the goal orientation of the policy statement and then extends it to root watershed management and permitting in comprehensive objectives representing the ability of waters to actually support designated beneficial uses. The proposal builds primarily around the integrated municipal permit concept in the policy statement and technical guidance. Like EPA's outline, the committee emphasizes measuring the effectiveness of actions in bringing improvements, but goes on from there to recommend a set of monitoring activi-

---

BOX 6-1
EPA's Current Guidance on Watershed-Based Permitting

Rather than explicitly define watershed based permitting, the EPA's recent guidance (EPA, 2007a) groups a large number of activities as having elements of watershed-based permitting, and defines how each might be utilized by a community. They are

- NPDES permitting development on a watershed basis,
- Water quality trading,
- Wet weather integration,
- Indicator development for watershed-based stormwater management,
- TMDL development and implementation,
- Monitoring consortium,
- Permit synchronization,
- Statewide rotating basin planning,
- State-approved watershed management plan development,
- Section 319 planning, and
- Source water protection planning.

Taking these topics in order, the first option is generally similar to that in EPA (2003a,b), but with some more detail on possible permitting forms. "Coordinated individual permits" implies that individual permits would be made similar and set with respect to one another and to a holistic watershed goal. The nature of such permits is not fully described, and there are no examples given. An "integrated municipal permit," also presented in the earlier policy statement, would place the disparate individual NPDES permits in a municipality (e.g., wastewater plants, combined sewer overflows, municipal separate storm sewer systems [MS4s]) under one permit. However, such a permit is not necessarily watershed-based. Finally, the "multi-source permit" could go in numerous directions, none of which are described in detail. In one concept, all current individual permittees who discharge a common pollutant into a watershed would come under one new individual permit that regulates that pollutant, while keeping the existing individual permits intact for other purposes. The Neuse River Consortium is given as an example. Alternatively, a multi-source permit could cover all dischargers of a particular type now falling under one individual permit that regulates all of their pollutants (no examples are given). In yet another application, this permit could be a general permit, and it would be identical to the existing general permits, except that it would be organized along watershed boundaries. As above, it could be refined on the basis of pollutant or discharger type.

The other ten options are more distant from the fundamental concept of watershed-based permitting. The water quality trading description is minimal, though it does mention a new EPA document that gives guidance to permittees for trading. Wet weather integration, the third topic, can mean any number of things, from creating a single permit to cover all discharges of pollutants during wet weather in a municipality, as described above for "coordinated individual permits," to just having all the managers of the systems get together and strategize. Although a stated goal is to reduce the amount of water in the sewer system after a storm, this integration is not particularly well defined in the document, nor is it well differentiated from other activities that would normally occur under an MS4 permit.

continues next page

BOX 6-1 Continued

Indicator development for watershed-based stormwater management refers to identifying indicators that are better than one or a few pollutants at characterizing the degree of impairment wrought by stormwater. Stormwater runoff volume is one indicator being developed by Vermont, and percent impervious surface is another. As discussed in Chapter 2, some states have long used biological indicators that integrate the effects of many pollutants as well as physical stresses such as elevated flow velocities. Indicators can be used as TMDL targets or as goals in NPDES permits. Identifying and adopting indicators is, essentially, a prerequisite to implementing some of the other options listed above.

Regarding the next topic on the list, the option of TMDL development is obvious, since the TMDL program is by definition watershed based. If it can be made the highest priority, and if stormwater is a contributor, then the implementation plan can be an excellent way to combat stormwater pollution on a watershed basis. Reducing the contribution of the pollutant from a stormwater source can involve water quality trading, better enforcement of existing permits, or creating new watershed-based permits. Hence, again, there is considerable overlap with the previously discussed options.

Developing a monitoring consortium is an option that works when sufficient data are not available to do much else. The concept mainly refers to monitoring of ambient waters. The activity is shared among partners (e.g., all wastewater plants in a region), with the goal of collecting and analyzing enough data to improve management decisions on a watershed basis, instead of for a single plant.

The following topic, permit synchronization, refers to having all permits within a watershed expire and be renewed simultaneously. This approach could be helpful for streamlining administrative, monitoring, and management tasks associated with maintaining the permits. Some states have operated in this way, whereas others have decided not to. It is one way to coordinate permits in cases where other types of watershed-based permitting would not work. Similarly, the statewide rotating basin approach, used by many states, relies on a five-year cycle. The state is divided into major watersheds, and each watershed is in a different stage of the cycle every year. It is a way to distribute the workload such that there is never a year when, for example, every watershed would require monitoring. Since it is a statewide program, how it relates to a watershed-based permitting situation is not at all clear.

---

ties designed to support active adaptive management to achieve objectives, aswell as to assess compliance. Credit trading, indicator development, the rotating basin approach, and monitoring should be part of management and permitting programs within watersheds, and ideas are advanced to develop these and other elements.

In addition to building on the work of EPA, the proposed approach tackles many of the impediments to effective watershed management identified in the National Research Council (NRC) treatise on watershed management (NRC, 1999). That report noted that watershed approaches are easiest to implement at the local level; thus, the approach developed in this chapter is a bottom-up process in which programmatic responsibility lies mainly with municipalities. Because the natural boundaries of watersheds rarely coincide with political jurisdictions, watersheds as geographic areas are less useful for political, institutional, and funding purposes, such that initiatives and organizations directed at watershed management should be flexible. The proposed approach recognizes this reality and makes numerous suggestions for pilot testing, funding, and institutional arrangements that will facilitate success. Finally, NRC (1999) notes the

With regard to the next topic, there has been a great deal of watershed planning around the nation and tremendous variety in form and comprehensiveness. Plans generally contain some information on the state of the watershed, goals for the watershed, and activities to meet those goals. Development of such plans in areas that do not have them could facilitate watershed-based permitting by providing much needed information about conditions, sources of pollutants, and methods to reduce pollution. According to EPA, a watershed plan may or may not indicate the need for watershed-based permitting.

The Section 319 Program refers to voluntary efforts to reduce pollution from nonpoint sources. The program in and of itself is not relevant to NPDES permits, since it deals strictly with activities that are not regulated. However, these activities could be traded with more traditional stormwater practices as part of a watershed-based effort to reduce overall pollution reaching waterbodies. Many watershed plans must consider guidance for the 319 program in order to get funding for their management activities.

If the watershed in question contains a drinking water source (either surface water or groundwater), then a good source water protection plan can have a significant impact on NPDES permitting in a watershed. Information collected during the assessment phase of source water protection could be used to help inform watershed-based permitting. Also, NPDES permits could be rewritten taking into account the proximity of discharges to source water intakes.

Following its coverage of the 11 options, EPA (2007a) gives a hypothetical example of picking six of the options to develop permitting for a watershed. It discusses how the options might be prioritized, but in a very qualitative manner, according to considerations such as availability of funding and personnel, stakeholder desires, environmental impacts, and sequencing of events. Chapter 1 of the report ends with a list of performance goals that might apply to the 11 options.

Chapter 2 further explains the multi-source watershed-based permit, discussing, for example, who would be covered by it, who would administer it, and how credit trading fits in. The chapter has a lot of practical, although quite intuitive, information about how to write such a permit. Much of the decision making is left to the permit writer. There are discussions of effluent limitations, monitoring requirements, reporting and record keeping, special conditions, and public notice. Chapter 3 follows by presenting case studies, although fewer than appeared in 2003 and not all truly watershed based.

need to "develop practical procedures for considering risk and uncertainty in real world decision-making in order to advance watershed management." The proposed revised monitoring system presented later in this chapter is designed to provide information in the face of ongoing uncertainty, i.e., adaptive management in a permitting context.

Watershed Management and Permitting Issues

There are many implications of redirecting the stormwater management and regulatory system from a site-by-site, SCM-by-SCM approach to an emphasis on attainment of beneficial uses throughout a watershed. Most fundamentally, the program's focus would shift to a primary concentration on broad goals in terms of, for example, achieving a targeted condition in a biological indicator associated with aquatic ecosystem beneficial uses or no net increase in elevated flow duration. Application of site-specific stormwater control measures (SCMs) would no longer constitute presumptive evidence of permit compliance, as is often the case in permits now, although it would still be an essential means to meeting goals. Achieving those goals, however, would form the compliance criteria.

In recognition of the demonstrated negative effects of watershed hydrologic modification on the attainment of beneficial uses, the proposal steps beyond the generally prevailing practice by embracing water quantity as a concern along with water quality. The inclusion of hydrology is consistent with the CWA on several grounds. First, elevated runoff peak flow rates and volumes increase erosive shear stress on stream beds and banks and directly contribute particulate pollutants to the flow (such as suspended and settleable solids, as well as nutrients and other contaminants bound to the soil material). Conversely, reduced dry-weather flows often occur in urban streams as a result of lost groundwater recharge and tend to concentrate pollutants and, hence, worsen their biological effects. Moreover, pollutant mass loading is the product of concentration and flow volume, and thus increased wet-weather surface runoff directly augments the cumulative burden on receiving waters. Finally, regulatory precedent for incorporating hydrology exists, as demonstrated by Vermont's stormwater program (LaFlamme, 2007).

At this time, stormwater management and regulation are divorced from the management and regulation of municipal and industrial wastewater. A true watershed-based approach would incorporate the full range of municipal and industrial sources, including (1) public streets and highways; (2) municipal stormwater drainage systems; (3) municipal separate and combined wastewater collection, conveyance, and treatment systems; (4) industrial stormwater and process wastewater discharges; (5) private residential and commercial property; and (6) construction sites. These many sources represent an array of uncoordinated permits under the current system and a strong challenge to developing a watershed-based approach. As pointed out in Chapter 2, multi-source considerations are an implicit facet of TMDL assessments, wherein states must consider both point and nonpoint sources. EPA (2003b) identified, among other possible permit types, an Integrated Municipal NPDES Permit, which would bundle all requirements for a municipality (e.g., stormwater, combined sewer overflows, biosolids, pretreatment) into a single permit. The Tualatin River watershed in Oregon has faced this challenge, at least in part, through an innovative watershed permit that combines both wastewater treatment and stormwater, brings in management of agricultural contributions to thermal pollution, and allows for pollutant trading among sources (see Box 6-2). It appears that the various participating parties did not use their energies in trying to allocate blame but instead determined the most effective and efficient ways of improving conditions. For example, the municipal permittees willingly offered incentives to agricultural landowners to plant riparian shade trees as an alternative to more expensive means of reducing stream temperatures under their direct control. Indeed, with agriculture not being regulated by the Clean Water Act, watershed permitting

BOX 6-2
Watershed-Based Permitting in Oregon

Clean Water Services is a wastewater and stormwater utility that covers a special ser-vice district of 12 cities and unincorporated areas in urban Washington County, Oregon. It was originally chartered in the 1970s as the Unified Sewerage Agency to consolidate the management of 26 "package" wastewater treatment facilities. Its responsibilities expanded to stormwater management in the early 1990s and it now serves nearly 500,000 customers. There are four wastewater treatment plants (WWTPs) in the district, with a dry weather capacity of 71 million gallons per day (MGD). During low-flow months, the discharge from these plants can account for 50 percent of the water in the Tualatin River. The district also own rights to one-quarter of the stored water in Hagg Lake. The land use in the watershed is about one-third urban, one-third agriculture, and one-third forest.

In 2001, the region was faced with TMDLs on the Tualatin River or its tributaries for to-tal phosphorus, ammonia, temperature, bacteria, and dissolved oxygen. By 2002, the area was also dealing with four expired NPDES permits and one expired MS4 permit (all of which had been administratively extended), approval of a second TMDL, and an Endan-gered Species Act (ESA) listing. The region decided that it wanted to try to integrate all of these programs using a watershed-based regulatory framework. This would include a TMDL implementation mechanism, an ESA response plan, and integrated water resources management (meaning that water quantity, water quality, and habitat considerations would be made at the same time). Prior to integration, water quality was covered by the TMDL and NPDES programs, but these programs did not cover water quantity and habitat issues. The ESA listing addressed the habitat issues, but it was done totally independently of the TMDLs and NPDES permits.

Thus, the region applied for an integrated municipal NPDES permit that bundles all NPDES permit requirements for a municipality into a single permit, including publicly owned treatment works (POTWs), pretreatment, stormwater, sanitary sewer overflows, and biosol-ids. Initially, it encompassed the four WWTP permits, the one MS4 permit, and the indus-trial and construction stormwater permits. The hope was that this would streamline multiple permits and capture administrative and programmatic efficiencies; provide a mechanism for implementing more cost-effective technologies and management practices including water quality credit trading; integrate watershed management across federal statutes such as the CWA, SDWA, and ESA; and encourage early and meaningful collaboration and coopera-tion among key stakeholders.

This case study was successful because a single entity—Clean Water Services—was already in charge of what would have otherwise been a group of individual permittees. Furthermore, all the NPDES permits had expired and the TMDL had just been issued, pro-viding a window of opportunity. The state regulatory agency was very willing, and EPA provided a $75,000 grant. Finally, there was a robust water quality database and modeling performed for the area because of the previous TMDL work. The watershed-based permit, the first in the nation, was issued February 26, 2004. Among its unique elements are an intergovernmental agreement companion document signed by the Oregon Department of Environmental Quality (DEQ), water quality credit trading, and consolidation of reporting requirements. The water quality trading is one of the most interesting elements, and sev-eral variations have been attempted. Biological oxygen demand (BOD) and NH_3 have been traded both intra-facility and inter-facility.

The temperature TMDL on the Tualatin River is a particularly interesting example of trading because it helped to bring agriculture into the process, where it would otherwise not have been involved. Along the length of the river, there are portions that exceed the tem-perature standard. A TMDL allocation was calculated that would lower temperatures by the

continued next page

BOX 6-2 Continued

same amount everywhere, such that there would be no point along the river that would be in exceedance. Options for reducing temperature include reducing the influent wastewater temperature (which is hard to do), reducing the total WWTP discharge to the Tualatin River (which is not practical), mechanically cooling or refrigerating WWTP discharge (which would require more energy), or trading the heat load via flow augmentation and increased shading (which is what was attempted).

Clean Water Services choose to utilize a market-based, watershed approach to meet the Tualatin temperature TMDL. It was market-based because it had financial incentives for certain groups to participate, it was cost-effective, and it provided ancillary ecosystem services. It was a watershed-based approach because it capitalized on the total assimilative capacity of the basin. What was done was to (1) provide cooling and in-stream flow augmentation by releasing water from Hagg Lake Reservoir, and (2) trade riparian stream surface shading improvement credits. They also reused WWTP effluent in lieu of irrigation withdrawals. For the riparian shading, they developed an "enhanced" CREP program to increase the financial incentives to rural landowners (with Clean Water Services paying the difference over existing federal and state programs). Clean Water Services also made incentive payments to the Soil and Water Conservation District to hire people to act as agents of Clean Water Services. Oregon DEQ's Shadalator model was used to quantify thermal credits for riparian planting projects, which required that information be collected at 100-foot increments along the stream on elevation, aspect, wetted width, Nordfjord-Sogn Detachment Zone, channel incision, and plant type and planting corridor width. To summarize, over the five-year term of the permit, Clean Water Services will release 30 cfs/d of stored water from Hagg Lake each July and August and shade roughly 35 miles of tributary riparian area (they have already planted 34 miles of riparian buffer). This plan involved an element of risk taking, since the actions of unregulated parties (such as farmers) have suddenly become the responsibility of Clean Water Services.

---

and initiatives of this type represent the best, and perhaps only, mechanism for ameliorating negative effects of agricultural runoff that, left unattended, would undo gains in managing urban runoff. The Neuse River case study, discussed later in this chapter, is another example of bringing agricultural contributions to aquatic degradation under control, along with urban sources, through a watershed-based approach.

Significant disadvantages of the current system of separate permits for municipal, construction, and industrial activities are (1) the permits attack the problem on a piecemeal basis, (2) they are hard to coordinate because they expire at different times, (3) they are not designed to allow for long-term operation of SCMs, and (4) they do not cover all discharges. A solution to these problems would be to integrate all discharge permitting under municipal authority, as is proposed here. The lead permittee and co-permittees would bear ultimate responsibility for meeting watershed goals and would regulate all public and private discharges within their jurisdictions to attain them. Municipalities are the natural focus for this role because they are the center of land-use decisions throughout the nation.

Municipalities must be provided with substantially greater resources than they have now to take on this increased responsibility. Beyond funding, regulatory responsibilities must be realigned to some degree. The norm now is for states to administer industrial permits directly and generally attend to all aspects of permit management. However, states, more often than not, are unable because of resource limitations to give permittees much attention in the form of inspection and feedback to ensure compliance. At the same time, some states, explicitly or implicitly, expect municipal permittees to set up programs to meet water quality standards in the waters to which all land uses under their jurisdictions discharge.[1] It only makes sense in this situation to have designated states (or EPA for the others) specify criteria for industrial and construction permits but revise regulations to empower and support municipal co-permittees in compliance-related activities. This paradigm is not unprecedented in environmental permitting, as under the Clean Air Act, states develop state implementation plans for implementation by local entities. For this new arrangement to work, states would have to be comfortable that municipalities could handle the responsibility and be able to exercise the added authority granted. The committee's opinion is that municipalities generally do have the capability, working together as co-permittees with a large-jurisdiction lead permittee and with guidance and support from states.

It bears noting at the outset that the proposed new program would not reduce the present system's reliance on general permits. Whereas a general permit now can be issued to a group of municipalities having differing circumstances, under the new system a permit could just as well be formulated in the same way for a group of varying watersheds. General industrial and construction permits would be just as prevalent too.

Toward Watershed-Based Permitting

Watershed-based permitting is taken in this report to mean regulated allowance of discharges of water and wastes borne by those discharges to waters of the United States, with due consideration of (1) the implications of those discharges for preservation or improvement of prevailing ecological conditions in the watershed's aquatic systems, (2) cooperation among political jurisdictions sharing a watershed, and (3) coordinated regulation and management of all discharges having the potential to modify the hydrology and water quality of the watershed's receiving waters.

---

[1] For example, the second Draft Ventura County [California] Municipal Separate Storm Sewer System Permit states (under Findings D. Permit Coverage), "Provisions of this Order apply to the urbanized areas of the municipalities, areas undergoing urbanization and areas which the Regional Water Board Executive Officer determines are discharging storm water that causes or contributes to a violation of a water quality standard ... ." The permit further states (under Part 2—Receiving Water Limitations), "1. Discharges from the MS4 that cause or contribute to a violation of water quality standards are prohibited. ... 3. ... This Order shall be implemented to achieve compliance with receiving water limitations. If exceedence(s) of water quality objectives or water quality standards persist ... the Permittee shall assure compliance with discharge prohibitions and receiving water limitations ... ."

Determining Watershed Scale for Permitting

A fundamental question that must be answered at the outset of any move to watershed permitting is, What is a watershed? Hydrologically, a watershed is the rain catchment area draining to a point of interest. Hence, the question comes down to, Where should the point of interest be located to define watersheds for permitting purposes? If placed close to the initial sources of surface runoff (e.g., on each first-order stream just above its confluence with another first-order stream), attention would be very specifically directed. However, there would be little flexibility to devise solutions for the greatest good. For example, trading of the commodities runoff quantity and quality would be very restricted. If on the other hand the point of interest is placed far downstream, thus defining a very large watershed, a welter of issues, and probably also of involved jurisdictions, would overly confuse the management and regulatory task.

The U.S. Geological Survey (USGS) delineates watersheds in the United States using a nationwide system based on surface hydrologic features. This system divides the country into 21 regions, 222 subregions, 352 accounting units, and 2,262 cataloging units. These hydrologic units are arranged within each other, from the smallest (cataloging units) to the largest (regions). USGS identifies each hydrologic unit by a unique hydrologic unit code (HUC) consisting of 2 to 16 digits based on the four levels of classification in the hydrologic unit system. Watersheds thus delineated are typically of the order a few square kilometers in area. This system is now being linked to the National Hydrography Dataset (NHD) and the National Land Cover Dataset to produce NHDPlus, an integrated suite of application-ready geospatial datasets.

The USGS system provides a starting point. Ultimately, though, what constitutes a watershed will best be answered with reference to specific biogeophysical conditions and problems and by personnel at relatively close hand (i.e., state or regional oversight agency staff). A general guideline might be the catchment area of a waterbody influenced by a set of similar subwatersheds. Similar subbasins would presumably be amenable to similar solutions and trading off reduced efforts in some places for compensating additional efforts elsewhere, as well as to analysis and monitoring on a representative basis, instead of exhaustively throughout. Often, a watershed defined in this way would flow into another watershed and influence it. Thus, there would have to be coordination among managers and regulators of interacting watersheds. It would be common for several watersheds ranging from relatively small to large in scale to be nested. Each would have its management team, and a committee drawn from those teams should be formed to coordinate goals and actions.

A prerequisite to moving toward watershed permitting, then, is for states or regions within states to delineate watersheds. California took this step early in the NPDES stormwater permitting process and offers a model in this respect, as well as in encompassing all jurisdictions coordinated by a lead permittee. First, the state organized its California EPA regional water boards on a watershed basis. Furthermore, since 1992 it has been common in California to establish one jurisdiction as the lead permittee (e.g., Los Angeles County in the Los Angeles region, Orange County in the Santa Ana Region, and San Diego County in the San Diego Region) and all of the politically separate cities as co-permittees. The lead permittee has typically been the jurisdiction most widely distributed geographically in the region and large enough to develop compliance mechanisms and coordinate their implementation among all participants. Box 6-3 describes the approach taken to delineating management units within the Chesapeake Bay watershed, which comprises parts of Pennsylvania, Maryland, Virginia, and the District of Columbia. The case study illustrates well the approach advocated here of focusing on the outcome in the receiving water and considering all aspects of land and water resources management that determine that outcome.

Steps Toward Watershed-Based Permitting

Once a watershed is defined, a further question arises regarding how much and what part of its territory to cover formally under permit conditions. Under the present system substantial development occurring outside Phase I or Phase II municipal jurisdictions is escaping coverage. Failing to control relatively high levels of development both outside a permitted jurisdiction and upstream of more lightly developed areas within a permitted area is particularly contrary to the watershed approach. Areas having a more urban than rural character are already essentially treated as urban in water supply and sewer planning, and the same should occur in the area of stormwater management. Accordingly, the permit should extend to any area in the watershed, even if outside Phase I or II jurisdictions, zoned or otherwise projected for development at an urban scale (e.g., more than one dwelling per acre). States do have authority under the CWA to designate any area for Phase II coverage based on projected growth or the presence of impact sources. They should be required to do so for nationwide uniformity and best protection of water resources.

It is essential to clarify that watershed-based permitting as formulated in this chapter differs sharply from what has been termed watershed (or basin) planning. According to EPA, watershed planning "identifies broad goals and objectives, describes environmental problems, outlines specific alternatives for restoration and protection, and documents where, how, and by whom these action alternatives will be evaluated, selected, and implemented" (*http://www.epa. gov/watertrain/planning/planning7.htm*). Drawing up such a plan is a time-consuming process, which has often become an end in itself, instead of a means to an end. Completing a full watershed plan, as usually construed, should not be a prerequisite to watershed-based permitting. Rather, the anticipated process would spring much more from comprehensive, advanced scientific and technical analysis of the water resources to be managed and their contributing catchment areas than from a planning framework.

BOX 6-3
Watershed Delineation for the Chesapeake Bay

The "Tributary Strategy Team" approach of the Chesapeake Bay Watershed provides a specific example of a watershed-scale approach to implementation of water quality control measures. Some background on this longstanding program is first provided, before turning to how watersheds were delineated. In 1983, the states of Virginia, Maryland, and Pennsylvania; the District of Columbia; and EPA signed an agreement to form the Chesapeake Bay Program with a goal to restore and protect the bay, which was suffering from nutrient overenrichment, severely reduced submerged aquatic vegetation, and contamination by toxics. In 1987 the program established a target of a 40 percent reduction in the amount of nutrients entering the Bay by 2000. In 1992 the bay program partners agreed to continue the 40 percent reduction goal beyond 2000 by allocating nutrient reduction targets to the bay's tributaries. In Chesapeake 2000, the most recent version of the Chesapeake Bay agreement, the nutrient reduction goals were reaffirmed, and an additional goal of sediment reduction was established. New York, Delaware, and West Virginia, locations of the bay's headwaters, also became involved in nutrient and sediment reduction. Cap load allocations for nutrients (nitrogen and phosphorus) and sediment to be reached by 2010 were agreed upon by the states. The states began developing 36 voluntary watershed-based tributary strategies to meet the state cap load allocations covering the entire 64,000-square-mile Chesapeake Bay watershed.

Watershed-based tributary strategies are developed in cooperation with local watershed stakeholders. For rural areas, where stakeholders include farmers, nutrient strategies include promotion of management practices such as maintaining cover crops on recently harvested cropland to reduce soil erosion, reduction in nitrogen applications, conservation tillage, and establishment of riparian buffers. For urban-area stakeholders such as homeowners and municipalities, tributary strategies include practices such as enhanced nutrient removal at WWTPs, low-impact development (LID) practices, erosion and sediment control practices, and septic system upgrades.

The first cut at delineating the watershed, which was based on hydrography and topography, defined the eight major areas draining to the Chesapeake Bay: six major basins (Susquehanna, Potomac, York, James, Rappahannock, and Patuxent) plus smaller areas not draining to a major river on the Eastern and Western Shores of the bay in Maryland. These subdivisions are disparate with respect to size (the Susquehanna can engulf almost the entire other seven), but direct drainage to the bay was the criterion at this level.

The next cut was made at state borders. For example, the Susquehanna traverses three states and was subdivided at the New York–Pennsylvania and Pennsylvania–Maryland political boundaries. Further cuts were subsequently made within some states. The criteria for these cuts varied from state to state, but generally involved a combination of smaller political jurisdictions (e.g., county, township), subwatershed basin borders, and other local considerations, such as local interest and investment (e.g., watershed associations).

The resulting delineations are highly variable in size but apparently satisfactory to the local parties who decided on the areas. They represent individual "tributary strategy areas" but are also nested within the larger eight designations and involve interjurisdictional and interstate coordination where a subbasin is divided by a political boundary. Although the example of the Chesapeake Bay is at a very large scale, the principles of watershed delineation it illuminates apply at all scales.

Effective watershed-based permitting as outlined in this report is composed of:

- Centralizing responsibility and authority for implementation with a municipal lead permittee working in partnership with other municipalities in the watershed as co-permittees;
- Adopting a minimum goal in every watershed to avoid any further loss or degradation of designated beneficial uses within the watershed's component waterbodies;
- Assessing waterbodies that are not providing designated beneficial uses in order to set goals aimed at recovering these uses;
- Defining careful, complete, and clear specific objectives to be achieved through management and permitting;
- Comprehensive impact source analysis as a foundation for targeting solutions;
- Determining the most effective ways to isolate, to the extent possible, receiving waterbodies from exposure to those impact sources;
- Developing and appropriately allocating funding sources to enable the lead permittee and partners to implement effectively;
- Developing a monitoring program composed of direct measures to assess compliance and progress toward achieving objectives and diagnosing reasons for the ability or failure to meet objectives, in support of active adaptive management; and
- Developing a market system of trading credits as a tool available to municipal co-permittees to achieve watershed objectives, even if solutions cannot be uniformly applied.

The system proposed herein is a significant departure from the road traveled in the 20 years since CWA amendments began to bring stormwater under direct regulation. This reorganization is necessary because of the failure of the present system to achieve widespread and relatively uniform compliance (see Chapter 2) and, ultimately, to protect the nation's water resources from degradation by municipal, industrial, and construction runoff. The workload associated with adopting this approach will be considerable and will take some time to complete. The structure of the new program should be fully in place within five years, which is considered to be a reasonable period to complete the work. It could be fully implemented throughout the nation within ten years. However, interim measures toward its fulfillment should occur sooner, within one to two years. Such measures should be applied to each land-use and impact-source category (i.e., existing residential and commercial development, existing industry, new development, redevelopment, construction sites). For example, measures such as an effective impervious area limit or a requirement to maintain pre-development recharge to the subsurface zone could make early progress in managing new development, and lead toward the ultimate, objective-based management and permitting strategy for that category. Advanced source control performance standards would be appropriate interim measures for existing development.

One innovative approach to watershed-based management that can ease the burden of the proposed new system is the rotating basin approach. As described by EPA (2007a), this option entails delineating state watershed boundaries and grouping the watersheds into basin management units, usually by the state water pollution control agency. Next, states implement a watershed management process on a rotating schedule, which is usually composed of five activities: (1) data collection and monitoring, (2) assessment, (3) strategy development, (4) basin plan review, and (5) implementation. Over time, different waterbodies are intensively studied as part of the rotation. Data collected can be used to support a number of different reporting and planning requirements, including a finding of attainment of water quality standards, a determination of impairment, or possible delisting if the waterbody is found not to be impaired. Florida offers a good example of the rotating basin approach. The Florida Department of Environmental Protection has defined five levels of intensity, or phases, each taking about one year to complete, and it has divided the state into 30 areas based on HUCs. At any one time six areas are in each phase before rotating to a subsequent phase. This division of effort would help alleviate the burden of moving to a new system of watershed-based permitting by programming the work over a period of years. It could certainly be organized on a priority basis, in which the watersheds of greatest interest for whatever reason (e.g., having the highest resource values, being most subject to new impacts) would get attention first.

An Objective-Based Framework

The proposed framework for watershed-based management and regulation of stormwater relies on broad goals to retain and recover aquatic resource beneficial uses, backed by specific objectives (e.g., water quality criteria) that must be achieved if the goals are to be fulfilled. Meeting the objectives and overarching goals is intended to become the basis for determining permit compliance, instead of the current reliance on implementation of SCMs as presumptive evidence of compliance.

The broad goals of retaining and recovering beneficial uses are entirely consistent with the antidegradation clause of the CWA. Antidegradation means that the current level of water quality shall be maintained and protected, unless waters exceed levels necessary for maintaining their beneficial uses *and* the state finds that allowing lower water quality is necessary to accommodate important economic or social development. In accordance with the antidegradation clause, a major pillar of the proposed concept is the goal of preventing degradation from the existing state of biological health, whatever it may be, to a lower state. Thus, fully and nearly pristine watersheds are to remain so and, at a minimum, partially or highly impaired ones are to suffer no further impairment. Beyond this minimum, impaired waters should be assessed to determine if feasible actions can be taken to recover lost designated beneficial uses or at least improve degraded uses.

As discussed in Chapter 2, beneficial uses relate to the social and ecological services offered, or intended to be offered, by waterbodies. For example, California has 20 categories of beneficial uses embracing water supply for various domestic, agricultural, and industrial purposes; provision of public recreation; and support of aquatic life and terrestrial wildlife (CalEPA, Central Coast Regional Water Board Basin Plan). That beneficial uses are usually assigned at the state level by waterbody classes or specific waterbodies would not change under the proposed permitting program revision. Most waters have several beneficial uses encompassing some water supply and ecological functions and, perhaps, some form of recreation. Unlike most current stormwater programs where attainment of beneficial uses is only implicit, these goals would become explicit in the altered system and officially promulgated by the authority operating the permit program (a designated state, in most cases, or EPA). The permitting authority would then partner with municipal permittees to determine the conditions that must be brought to bear to attain beneficial uses, set objectives or criteria to establish those conditions, and follow through with the tasks to accomplish objectives.

The proposed framework's reliance on achieving objectives that reflect the cumulative aquatic resource effects of contributing watershed conditions suggests the following related concepts:

- In whatever manner watershed boundaries are set, the full extent of the watershed from headwaters onward should be considered in defining objectives. This is important even where watershed scale and boundaries are based on local and/or regional hydrogeomorphic circumstances and their associated management and regulatory needs. Watersheds can and often will be defined and nested at different scales (e.g., streams tributary to a lake, a river flowing into an estuary or marine bay).

- The scale of objectives must be consistent with the scale and recognized beneficial uses of the watershed(s) in question; for example, sustaining salmonid fish spawning could be the basis for a stream objective, while retaining an oligotrophic state could be the essential objective for a lake to which the stream is tributary.

- Whenever beneficial uses pertain to living organisms (aquatic life or humans), representing the vast majority of all cases, objectives should be largely in biological terms. That is not to say that supplementary objectives cannot be stated otherwise (e.g., in terms of flow characteristics, chemical water quality constituents, or habitat attributes), but the ultimate direct thrust of the program should be toward the biota.

•   Objectives must be carefully chosen to represent attributes of importance from a resource standpoint, limited in number for feasibility of tracking achievement, and defined in a way that achievement can be measured. For example, nitrogen is generally the nutrient limiting algal growth in saline systems and in excess it stimulates growth that can reduce dissolve oxygen, killing fish and other aerobic organisms. In this case the most productive objectives would probably target reduction of nitrogen concentration and mass flux and maintenance of dissolved oxygen. For waterbodies designated for contact recreation, fecal coliform indicators (although not directly pathogenic when waterborne) have proven to be an effective means of assessing condition and should continue to form the basis for objectives to protect contact recreation until research produces superior measures. If drinking water supply is a designated beneficial use of a lake, it will better serve that function in a lower than a higher state of eutrophication, which can be managed, according to a long limnological research record, by restricting water column chlorophyll a as an objective. Where the beneficial use is fish protection and propagation, biological criteria might include (1) maintenance of a specific population size of a resident fish species when that species' population can be assayed conveniently; (2) maintenance of a numerical index (e.g., benthic index of biotic integrity) when a fish species of ultimate interest cannot be assessed so conveniently but is known or reasonably hypothesized to be associated with the index; or (3) a related parameter, such as eelgrass beds, which are important fish nursery areas in estuarine waters, such that areal coverage by these beds would be an appropriate objective to track over time. An intermittent waterbody could have biological criteria related to, for example, fish migration or amphibian reproduction.

•   The achievement of objectives, or lack thereof, is the basis for followup and prescription of remedies in an active adaptive management mode; that is, falling short of objectives would trigger a search for reasons throughout the watershed, followed by identification of actions necessary and sufficient to remedy the shortfall, assessment of their ability to reach objectives, and the cost of doing so. In the course of this assessment it may be concluded that the objective itself is faulty and should be restated, replaced, or discarded.

Basing the watershed framework principally on biological objectives grows out of the CWA's fundamental charge to protect the biological (as well as physical and chemical) integrity of the nation's waters. The tie between specific physical and chemical conditions and the sustenance of aquatic biological communities is not well established through an extensive, well-verified body of research. Moreover, living organisms consuming or living in water are subject to a vast multitude of simultaneous physical and chemical agents having the potential to harm them individually and interactively. There are no realistic prospects for research to determine the levels of these numerous agents that must be maintained to support beneficial uses. Therefore, their integrative effects must be determined using measures of biological populations or communities of interest.

By and large, state water quality standards as now promulgated would not serve the proposed objective-based system well. They are usually not phrased in biological terms or with respect to hydrologic variables now known to have instrumental negative effects on aquatic organisms, but instead mostly as concentrations of selected chemical elements or compounds. However, there is no prohibition of biological or hydrologic standards in the law. The recommended emphasis is consistent with and informed by the tiered aquatic life uses system applied by some states and illustrated for Ohio in Box 2-1. The use of such systems must expand greatly to support the recommended framework. An opportunity to do so exists through the triennial review already required for each state's water quality standards.

Certain special considerations affect the development and use of objectives as the device to carry forward watershed-based stormwater management and regulation. First, other elements of the CWA beyond the stormwater program and other laws may very well be involved in a watershed (see Chapter 2). Municipal and industrial wastewater discharges will often be contributors along with stormwater. Aquatic organisms may be listed as threatened or endangered under the federal ESA or state authority. Both objectives and the management and regulatory program designed to achieve objectives should reflect any such circumstances.

Instituting the proposed permitting program will require converting the TMDL program to one more suitable for its purposes and structure. The TMDL program is watershed based and hence offers some precedent and experience applicable to the new system. However, for the most part, it has operated only on waters declared to be impaired for specific pollutants, and it relies on management of specific physical and chemical water quality variables. Furthermore, in its current mode it takes no account of potential future impact sources. The TMDL program should be replaced with one adapted to the objective-based framework proposed here. This new program should apply to all waters assigned objectives, "impaired" or not, and formulate limits in whatever terms are best to achieve objectives. Hence, although the program would expand in coverage area, the efficient tailoring of objectives directly to beneficial uses could compensate for the expansion by targeting fewer variables. Finally, the new program should look to the future as well as the present by encompassing the anticipated impacts of prospective landscape changes.

The nature of a program to replace TMDLs can be glimpsed from a few attempts to move in the anticipated direction even under the existing structure. For example, Connecticut collected data directly linking impervious cover to poor stream health in Eagleville Brook (Connecticut Department of Environmental Protection, 2007). The stream's TMDL was developed using watershed impervious cover as a surrogate parameter for a mix of pollutants conveyed by stormwater. The intention is to reduce effective imperviousness by disconnecting impervious areas, installing unspecified SCMs, minimizing additional disturbance, and enhancing in-stream and riparian habitat. Flow was used as a surrogate for stormwater pollution in the Potash Brook, Vermont TMDL (Vermont DEC, 2006). In this waterbody, the impairment was based on biological indices that were then related to a hydrologic condition believed to be necessary to achieve the Vermont criteria for aquatic life. The TMDL will be implemented via the use of runoff-volume-reduction SCMs throughout the watershed.

Impact Sources

The CWA provides for regulating, as specific land-use types, only designated industrial categories, with construction sites disturbing one acre or more considered to be one of those categories. Otherwise, it gives authority to regulate municipal jurisdictions operating separate storm sewer systems. Generally speaking, these jurisdictions encompass, in addition to the industrial categories, the full range of urban land-use types, such as single- and multiple-family residential, various kinds and scales of commercial activity, institutional, and parks and other open space. All of these land uses and the activities conducted on them are, to one degree or another, sources of the agents that physically and chemically modify aquatic systems to the detriment of their biological health. Hence, most of the impact sources to which these aquatic systems are subject are not directly regulated under CWA authority as are industrial sources, but instead are indirectly regulated through the municipal program. Also, as already discussed, the situation is further complicated by the presence of municipal and industrial wastewater sources along with landscape sources contributing flow and pollutants to receiving waters via stormwater discharges.

The watershed-based framework envisioned here relies on municipalities led by a principal permittee. Thus, a fundamental task that municipal permittees charged with operating under a watershed-based permit must do is to find industries and construction sites in the watershed that have not filed for permit coverage and bring them under regulation. Furthermore, municipal co-permittees, with leadership by a watershed lead permittee, must classify industries and construction sites within their borders according to risk and accordingly prioritize them for inspection and monitoring (methods for doing this are discussed later in the chapter). Municipal permittees must have better tools than they have had in the past to assess the various impact sources and formulate strategies to manage them that have a reasonably high probability of fulfilling objectives. The present state of practice and research findings offers some directions for choosing or more completely developing these tools. However, by no means are all the necessary elements available, and substantial new basic and applied research must be performed.

From the literature come several possibilities to improve source analysis in the complex urban environment. Some examples of apparent promise, drawn from Clark et al. (2006) include the following:

- Nirel and Revaclier (1999) used the ratio of dissolved rubidium (Rb) to strontium (Sr) to identify and quantify the impact of sewage effluents on river quality in Switzerland. Rubidium was present in larger quantities than strontium in feces and urine, making the ratio of these two elements an effective tracer that does not vary with river flow for a given water quality condition. Using the ratio alone produced the same conclusions regarding impact as measuring a host of physicochemical water quality variables. The researchers estimated that the Rb:Sr ratio must be lower than 0.007 if biological diversity is to be maintained, which could be the basis of an objective to manage river water quality. Although this case pertains to municipal wastewater and the technique works best in waters with a naturally low Rb:Sr ratio (e.g., calcareous regions), it success points out a potential avenue of research to simplify stormwater management on the basis of quantitative objectives related to biological integrity.

- Cosgrove (2002) described the approach used in New Jersey to characterize the relative contribution of point and nonpoint sources of pollutants in the Raritan River Basin. Twenty-one surface water sampling locations within the watershed were monitored four to five times per year from 1991 to 1997. These data were evaluated by comparing the median concentration at each sampling location with land-use statistics. Cumulative probability curves were also developed for each pollutant to demonstrate the probability that the concentration at a given location would be below a certain level (e.g., a stream standard). These probability curves were useful in determining the risk that a given location would violate a particular standard. The concentration data, coupled with continuous flow monitoring records, were utilized to determine the total load for each constituent. Regression analysis was used to develop a relationship between the total in-stream loads and flow. Such an analysis provided an indication of municipal or industrial discharge versus diffuse-source-dominated locations. Pollutant loads could then be converted to yield (load per unit area) to normalize the results for comparison from one station to another. The "screening level" methodology uses only existing data and, not requiring advanced modeling techniques, can be used to understand where to focus more rigorous modeling techniques.

- Maimone (2002) presented the overall approach that was used to screen and evaluate potential pollutant sources within the Schuylkill River watershed as part of the Schuylkill River Source Water Assessment Partnership. The partnership performed source water assessments of 42 public water supply intakes for the Pennsylvania Department of Environmental Protection. The watershed encompasses over 1,900 square miles with more than 3,000 potential point sources of contamination. In addition, runoff from diverse land uses such as urban and agriculture had to be characterized using the Stormwater Management Model. For all 42 surface water intakes, potential point sources were identified using existing databases. The list was first passed through a series of Geographic Information System-based "screening" sieves to limit the sources to only those considered to be high priority (including proximity and travel time from source to intake). Ten categories were identified that cover the range of the most important contaminants that might be found within the watershed, and a representative or surrogate chemical was identified whose properties were used to stand in for the category. Beyond the geographic screening, a more sophisticated screening was needed to limit the number of sites, using a decision support computer software program called EVAMIX. The greatest benefit of EVAMIX, compared to other software, is that it allows mixed criteria evaluation, qualitative and quantitative, to be considered concurrently. EVAMIX produced source rankings representing an organized and consistent use of both the objective data and the subjective priorities of decision makers.

- Hetling et al. (2003) investigated the effect of water quality management efforts on wastewater discharges to the Hudson River (from Troy, New York to the New York City Harbor) from 1900 to 2000. The paper demonstrated a methodology for estimating historic loadings where data are not available. Under these circumstances, estimated historic sewered and treated populations and per capita values were used to calculate wastewater flow and loadings for 5-day biochemical oxygen demand (BOD_5), total suspended solids (TSS), total nitrogen, and total phosphorus. The analysis showed that dispersed landscape sources have become the most significant contributors of the first two contaminants to the river, while municipal wastewater plants remain the largest sources of nutrients. The methodology presented in this paper could be used by co-permittees to estimate present-day sources of various types and contribute to moving toward a comprehensive permit incorporating multiple sources.

- Zeng and Rasmussen (2005) used multivariate statistics to characterize water quality in a lake and its tributaries. Tributary water was composed of three components. Factor analysis demonstrated that stormwater runoff was the predominant cause of elevation of a group of water quality variables in a factor including TSS, the measurement of which is a convenient surrogate for all variables in the factor. Similarly, municipal and industrial discharges could be characterized by total dissolved solids, and groundwater by alkalinity plus soluble reactive phosphorus. These sources can thus be distinguished through measurement of just four common water quality variables. Reducing the number of analytes reduces laboratory costs and allows resources to be freed up for other purposes. Cluster analyses performed on the data indicated that further savings could be realized by sampling just one among several stations in a cluster and sampling at just one point in time over a period of relatively stable water quality (e.g., a relatively dry period).

A key research need associated with applying the proposed framework is assessment of these and other mechanisms for sorting out the contributions of the variety of impact sources in the urban environment. Leading this effort would be a natural role for EPA.

Impact Reduction Strategies

The philosophical basis for impact reduction under a modified permitting system centered on a lead municipal permittee and associated co-permittees is to avoid, as far as possible, exposing receiving waters to impact sources or to otherwise minimize that exposure. The concept embraces both water quantity and quality impact sources and specifically raises the former category to the same level of scrutiny as traditionally applied to water quality sources. Furthermore, the endpoints upon which success and compliance would be judged are directly related to achievement of beneficial uses. This approach to impact reduction, where the direct focus is on reducing the loss of aquatic ecosystem functioning supportive of beneficial uses, fundamentally contrasts with the currently prevailing system. What are primary concerns in the existing system (e.g., discharge concentrations of certain chemical and physical substances, technological strategies from a menu of practices) are still prospectively important, but only as a means toward realizing functional objectives, not as endpoints themselves. To be sure, attaining beneficial uses will require wise choices among tools to decrease discharges and contaminant emissions. However, the ultimate proof will always be in biological outcomes.

As made clear in Chapters 3 and 4, linkages among myriad stressing agents, impact receptors, and specific mitigating abilities of technological fixes are poorly understood and not easily understandable. The proposed new paradigm acknowledges that the linkages are not established among the voluminous elements in an exceptionally complex system ranging from impact sources, through environmental transport and fate mechanisms, to ecosystem health. However, it is intuitively and theoretically clear that minimizing the generation of impacts in the first place and slowing their progression into aquatic environments can break the chain of landscape alteration that leads to increased runoff and pollutant production, modifies aquatic habitat, and ultimately causes deterioration of the biological community. Landscapes can be managed in a preventive, integrated fashion that deals with the many undifferentiated agents of impact and avoids, or at least reduces, the damage. Although the application of these theories may not automatically and quickly stem biological losses, the powerful mechanism of adaptive management, if correctly applied, can be used to make course corrections toward meeting the defined objectives.

An earlier National Research Council (NRC) committee examined the scientific basis of EPA's TMDL program and recommended "adaptive implementation" (AI) to water quality standards (NRC, 2001a). That committee drew AI directly from the concept of adaptive management for decision making under uncertainty, introduced by Holling and Chambers (1973) and Holling (1978) and described it as an iterative process in which TMDL objectives and the implementation plans to meet those objectives are regularly reassessed during the ongoing implementation of controls. Shabman et al. (2007) and Freedman et al. (2008) subsequently extended and refined the applicability of AI for promoting water quality improvement both within and outside of the TMDL program. In that broader context, AI fits well with the framework put forward here. Indeed, the proposed revised monitoring system presented later in this chapter is designed to provide information to support adaptive management in a permitting context.

The Stages of Urbanization and Their Effects on Strategy

In waterbodies that are not in attainment of designated uses, it is likely that the physical stresses and pollutants responsible for the loss of beneficial uses will have to be decreased, especially as human occupancy of watersheds increases. Reducing stresses, in turn, entails mitigative management actions at every life stage of urban development: (1) during construction when disturbing soils and introducing other contaminants associated with building; (2) after new developments on Greenfields are established and through all the years of their existence; (3) when any already developed property is redeveloped; and (4) through retrofitting static existing development. Most management heretofore has concentrated on the first two of those life stages.

The proposed approach recognizes three broad stages of urban development requiring different strategies: new development, redevelopment, and existing development. *New development* means building on land either never before covered with human structures or in prior agricultural or silvicultural use relatively lightly developed with structures and pavements (i.e., Greenfields development). *Redevelopment* refers to fully or partially rebuilding on a site already in urban land use; there are significant opportunities for bringing protective measures to these areas where none previously existed. The term *existing development* means built urban land not changing through redevelopment; retrofitting these areas will require that permittees operate creatively.

What is meant by redevelopment requires some elaboration. Regulations already in force typically provide some threshold above which stormwater management requirements are specified for the redeveloped site. For example, the third Draft Ventura County Municipal Separate Storm Sewer System Permit defines "significant redevelopment" as land-disturbing activity that results in the creation or addition or replacement of 5,000 square feet or more of impervious surface area on an already developed site. The permit goes on to state that where redevelopment results in an alteration to more than 50 percent of the impervious surfaces of a previously existing development, and the existing development was not subject to postdevelopment stormwater quality control requirements, the entire site becomes subject to application of the same controls required for new development. Where the alteration affects 50 percent or less of the impervious surfaces, only the modified portion is subject to these controls.

All urban areas are redeveloped at some rate, generally slowly (e.g., roughly one or at most a few percent *per annum*) but still providing an opportunity to ameliorate aquatic resource problems over time. Extending stormwater requirements to redeveloping property also gradually "levels the playing field" with new developments subject to the requirements. As pointed out in Chapter 2, some jurisdictions offer exemptions from stormwater management requirements to stimulate desired economic activities or realize social benefits. Such exemptions should be considered very carefully with respect to firm criteria designed to weigh the relative socioeconomic and environmental benefits, to prevent abuses, to gauge just how instrumental the exemption is to gaining the socioeconomic benefits, and to compensate through a trading mechanism as necessary to achieve set aquatic resource objectives.

It is important to mention that not only residential and commercial properties are redeveloped, but also streets and highways are periodically rebuilt. Highways have been documented to have stormwater runoff higher than other urban land uses in the concentrations and mass loadings of solids, metals, and some forms of nutrients (Burton and Pitt, 2002; Pitt et al., 2004; Shaver et al., 2007). Redevelopment of transportation corridors must be taken as an opportunity to install SCMs effective in reducing these pollutants.

Opportunities to apply SCMs are obviously greatest at the new development stage, somewhat less but still present in redevelopment, but most limited when land use is not changing (i.e., existing development). Still, it is extremely important to utilize all readily available opportunities and develop others in static urban areas, because compromised beneficial uses are a function of the development in place, not what has yet to occur. Often, possibly even most of the time, to meet watershed objectives it will be necessary to retrofit a substantial amount of the existing development with SCMs. To further progress in this overlooked but crucial area, the Center for Watershed Protection issued a practical Urban Stormwater Retrofit Practices manual (Schueler et al., 2007).

Practices for Impact Reduction

As described in Chapter 5, in the past 15 to 20 years stormwater management has passed through several stages. First, it was thought that the key to success was to match postdevelopment with predevelopment peak flow rates, while also reducing a few common pollutants (usually TSS) by a set percentage. Finding this to require large ponds but still not forestalling impacts, stormwater managers next deduced that runoff volumes and high discharge durations would also have to decrease. Almost simultaneously, although not necessarily in concert, the idea of LID arose to offer a way to achieve actual avoidance or at least minimization of discharge quantity and pollutant increases reaching far above predevelopment levels. For purposes of this discussion, the SCMs associated with LID along with others are named Aquatic Resources Conservation Design (ARCD). First, this term signifies that the principles and many of the methods apply not only to building on previously undeveloped sites, but also to redeveloping and retrofitting existing development. Second, incorporating aquatic resources conservation in the title is a direct reminder of the central reason for improving stormwater regulation and management. ARCD goes beyond LID to encompass many of the SCMs discussed in Chapter 5, in particular those that decrease surface runoff peak flow rates, volumes, and elevated flow durations caused by urbanization, and those that avoid or at least minimize the introduction of pollutants to any surface runoff produced. This concentration reduction, together with runoff volume decrease, cuts the cumulative mass loadings (mass per unit time) of pollutants entering receiving waters over time. The SCM categories from Table 5-1 that qualify as ARCD include:

- Product Substitution,
- Watershed and Land-Use Planning,
- Conservation of Natural Areas,
- Impervious Cover Minimization,
- Earthwork Minimization,
- Reforestation and Soil Conservation,
- Runoff Volume Reduction—Rainwater Harvesting, Vegetated, and Subsurface,
- Aquatic Buffers and Managed Floodplains, and
- Illicit Discharge Detection and Elimination.

The menu of ARCD practices begins with conserving, as much as possible, existing trees, other vegetation, and soils, as well as natural drainage features (e.g., depressions, dispersed sheet flows, swales). Clustering development to affect less land is a fundamental practice advancing this goal. Conserving natural features would further entail performing construction in such a way that vegetation and soils are not needlessly disturbed and soils are not compacted by heavy equipment. Using less of polluting materials, isolating contaminating materials and activities from contacting rainfall or runoff, and reducing the introduction of irrigation and other non-stormwater flows into storm drain systems are essential. Many ARCD practices fall into the category of minimizing impervious areas through decreasing building footprints and restricting the widths of streets and other pavements to the minimums necessary. Water can be harvested from impervious surfaces, especially roofs, and put to use for irrigation and gray water system supply. Harvesting is feasible at the small scale using rain barrels and at larger scales using larger collection cisterns and piping systems. Relatively low traffic areas can be constructed with permeable surfaces such as porous asphalt, open-graded Portland cement concrete, coarse granular materials, concrete or plastic unit pavers, or plastic grid systems. Another important category of ARCD practices involves draining runoff from roofs and pavements onto pervious areas, where all or much can infiltrate or evaporate in many situations.

If these practices are used, but excess runoff still discharges from a site, ARCD offers an array of techniques to reduce the quantity through infiltration and evapotranspiration and improve the quality of any remaining runoff. These practices include (1) bioretention cells, which provide short-term ponded and soil storage until all or much of the water goes into the deeper soil or the atmosphere; (2) swales, in which water flows at some depth and velocity; (3) filter strips, broad surfaces receiving sheet flows; (4) infiltration trenches, where temporary storage is in below-ground gravel or rock media; and (5) vegetated ("green") roofs, which offer energy as well stormwater management benefits. Natural soils sometimes do not provide sufficient short-term storage and hydraulic conductivity for effective surface runoff reduction because of their composition but, unless they are very coarse sands or fine clays, can usually be amended with organic compost to serve well.

ARCD practices should be selected and applied as close to sources as possible to stem runoff and pollutant production near the point of potential generation. However, these practices must also work well together and, in many cases, must be supplemented with strategies operating farther downstream. For example, the City of Seattle, in its "natural drainage system" retrofit initiative, built serial bioretention cells flanking relatively flat streets that subsequently drain to "cascades" of vegetated stepped pools created by weirs, along more sloping streets. The upstream components are highly effective in attenuating most or even all runoff. Flowing at higher velocities, the cascades do not perform at such a high level, although under favorable conditions they can still infiltrate or evapotranspire the majority of the incoming runoff (Horner et al., 2001, 2002, 2004; Chapman, 2006; Horner and Chapman, 2007). Their role is to reduce runoff from sources not served by bioretention systems as well as capture pollutants through mechanisms mediated by the vegetation and soils. The success of Seattle's natural drainage systems demonstrates that well-designed SCMs can mimic natural landscapes hydrologically, and thereby avoid raising discharge quantities above predevelopment levels.

In some situations ARCD practices will not be feasible, at least not entirely, and the SCMs conventionally used now and in the recent past (e.g., retention/detention basins, biofiltration without soil enhancement, and sand filters) should be integrated into the overall system to realize the highest management potential.

The proposed watershed-based program emphasizing ARCD practices would convey significant benefits beyond greatly improved stormwater management. ARCD techniques overall would advance water conservation, and infiltrative practices would increase recharge of the groundwater resource. ARCD practices can be made attractive and thereby improve neighborhood aesthetics and property values. Retention of more natural vegetation would both save wildlife habitat and provide recreational opportunities. Municipalities could use the program in their general urban improvement initiatives, giving incentives to property owners to contribute to goals in that area while also complying with their stormwater permit.

Municipal Permittee Roles in Implementing Strategies

Municipal permittees sharing a watershed will have key roles in promoting ARCD under the proposed new system. First, the lead permittee and its partners would be called upon to perform detailed scientifically and technically based watershed analysis as the program's foundation. The City of San Diego (2007) offers a model by which permittees could operate with its Strategic Plan for Watershed Activity Implementation. The plan consists of:

- Activity location prioritization—locations prioritized for action based on pollutant loading potential;

- Implementation strategy and activity prioritization—tiered approach identifying activities directed at meeting watershed goals over a five-year period;

- Potential watershed activities—general list of activities required and potentially required to meet goals as guidance for planning and budgeting;

- Watershed activity maps—specified locations for activities; and

- Framework for assessment monitoring—a plan for development of the monitoring and reporting program.

Municipal permittees would be required under general state regulations to make ARCD techniques top priorities for implementation in approving new developments and redevelopments, to be used unless they are formally and convincingly demonstrated to be infeasible. In that situation permit approval would still require full water quantity and quality management using conventional practices. Beyond regulation, municipalities would be called upon to give private property owners attractive incentives to select ARCD methods and support to implement them. Furthermore, they should supplement on-site ARCD installations with municipally created, more centralized facilities in subwatersheds.

Other municipal roles in the proposed program revolve around the prominence of soil infiltration as a mechanism in ARCD. Successful use of infiltration requires achieving soil hydraulic conductivity sufficient to drain the runoff collector quickly enough to provide capacity for subsequent storms and avoid nuisance conditions, while not so rapid that contaminants would reach groundwater. One important task for municipal co-permittees will be defining watershed soils and hydrogeological conditions to permit proper siting and design of infiltrative facilities. A great deal of soils information already exists in any community but must be assembled and interpreted to assist stormwater managers. U.S. Department of Agriculture soil surveys, while a start, are often insufficiently site-specific to characterize the subsurface accurately at a point on the landscape. More localized data available to municipalities come from years of recorded well logs, soil borings, and percolation test results. Municipalities should tap these records to define, to their best ability, soil types, hydraulic conductivities, and seasonal groundwater positions. Although abundant and valuable, these data are unlikely to be sufficient to define subsurface attributes across a watershed. Thus, municipalities should collect additional data (soil borings, soils analyses, and percolation tests) to obtain a good level of assurance of the prospects for infiltrative ARCD.

Part of the task for municipalities will be overcoming opposition to infiltration if it is unjustified. Some opponents discourage infiltration based on coarse soil survey data that may not apply at all at a locality, or they fail to take into account that the well-established ARCD practice of soil amendment, generally with organic compost, can improve the characteristics of somewhat marginal soils sufficiently to function well during infiltration. While such amendment cannot increase hydraulic conductivity sufficiently in restrictive clay soils, the technique has proven to effectuate substantial infiltration and attendant reduction in runoff volumes and peak flow rates in Seattle's natural drainage systems, discussed above. These systems lie on variable soils, including formations categorized by the Natural Resources Conservation Service (2007) as being in hydrologic group C. This group generally has somewhat restricted saturated hydraulic conductivity in the least transmissive layer between the surface and 50 centimeters (20 inches) of between 1.0 micrometers per second (0.14 inches per hour) and 10.0 micrometers per second (1.42 inches per hour). Furthermore, additional runoff reduction often occurs through evapotranspiration, which is enhanced by the vegetation in ARCD systems.

Another objection sometimes raised to infiltrating stormwater is its perceived potential to compromise groundwater quality. Whether or not that potential is very great depends upon a number of variables: rate of infiltration, ability of the soil type to extract and retain contaminants, distance of travel to groundwater, and any contaminated layers through which the water passes. It is unlikely that urban stormwater, with its prevailing pollutant concentrations, will threaten groundwater if it travels at a moderate rate, through soils of medium or fine textures without contaminant deposits, to groundwater at least several meters below the surface. To ensure that groundwater is not compromised when surface water is routed through infiltrative practices, municipalities must establish where appropriate conditions do and do not exist and spot infiltration opportunities accordingly. Records of past waste disposal, leaks, and spills must be consulted to clean up or stay away from contaminated zones. There are alternatives even if documented soils or groundwater limitations rule out infiltrative practices. Much can be accomplished to reduce the quantities of contaminated urban runoff discharged to receiving waters through impervious surface reduction, water harvesting, and green roofs.

One additional problem to infiltrating stormwater runoff exists in some relatively dry areas and must be countered by municipalities. Overirrigation of lawns and landscape plantings has already increased infiltration well over the predevelopment amount and raised groundwater tables, sometimes to problematic levels. This unnecessary use of irrigation not only wastes potable water, often scarce in such areas, but reduces capacity to infiltrate stormwater without further water table rise. Municipalities should set up effective programs to conserve water and simultaneously increase stormwater infiltration capacity.

A final element of an integrated management and permitting program under municipal control is use of capacity in the sanitary sewer and municipal waste-water treatment systems to treat some stormwater. This initiative must be pursued very carefully. One reason for care is that municipal treatment works have historically been overburdened with stormwater flows in combined sewers and have not yet broken free of that burden through sewer separation programs. A second reason is that municipal sewage treatment plants are generally designed to remove particulates and decompose organic wastes and not to capture the array of pollutants in stormwater, many dissolved or associated with the finest and most difficult to capture particles. Toxic contaminants can damage microbes and upset biological treatment plants. Nonetheless, capacity exists in many WWTPs to treat stormwater. The delivery of pollutants the plant was not designed to handle can be managed by pretreatment requirements, applied to industrial stormwater dischargers particularly. Dry weather flows, consisting mostly of excess irrigation runoff, can be diverted to treatment plants to prevent at least some of the nutrient and pesticide contamination that otherwise would flow to receiving waters. Additional capacity to treat stormwater can be gained by repairing defective municipal wastewater pipes that allow groundwater entry.

Special Considerations for Construction and Industrial Land Uses

All of the principles discussed above apply to industrial and construction sites as well: minimize the quantity of surface runoff and pollutants generated in the first place, or act to minimize what is exported off the site. Unfortunately, construction site stormwater now is managed all too often using sediment barriers (e.g., silt fences and gravel bags) and sedimentation ponds, none of which are very effective in preventing sediment transport. Much better procedures would involve improved construction site planning and management, backed up by effective erosion controls, preventing soil loss in the first place, which might be thought of as ARCD for the construction phase of development. Just as ARCD for the finished site would seek to avoid discharge volume and pollutant mass loading increase above predevelopment levels, the goal of improved construction would be to avoid or severely limit the release of eroded sediments and other pollutants from the construction site. Chapter 5 discusses construction-phase stormwater management in more detail.

Other industrial sites are faced with some additional challenges. First, industrial sites usually have less landscaping potentially available for land-based treatments. Their discharges are often more contaminated and carry greater risk to groundwater. On the other hand, industrial operations are amenable to a variety of source control options that can completely break the contact between pollutants and rainfall and runoff. Moving operations indoors or roofing outdoor material handling and processing areas can transform a high-risk situation to a no-risk one. It is recommended that industrial permits strongly emphasize source control (e.g., pollution prevention) as the first priority and the remaining ARCD measures as secondary options (as outlined in Table 5-9). Together these measures would attempt to avoid, or minimize to the extent possible, any discharge of stormwater that has contacted industrial sources.

It is likely that the remaining discharges that emanate from an industrial site will often require treatment and, if relatively highly contaminated, very efficient treatment to meet watershed objectives. Some industrial stormwater runoff carries pollutant concentrations that are orders of magnitude higher than now prevailing water quality standards. In these cases meeting watershed objectives may require providing active treatment, which refers to applying specifically engineered physicochemical mechanisms to reduce pollutant concentrations to reliably low levels (as opposed to the passive forms of treatment usually given stormwater, such as ponds, biofiltration, and sand filters). Examples now in the early stages of application to stormwater include chemical coagulation and precipitation, ion exchange, electrocoagulation, and filtration enhanced in various ways. These practices are undeniably more expensive than source controls and other ARCD options and traditional passive treatments. If they must be used at all, it is to the advantage of all parties that costs be lowered by decreasing contaminated waste stream throughput rates to the absolute minimum.

Administrative and Funding Arrangements

A number of practical, logistical considerations pertain to converting to the permitting and regulatory system discussed above. These considerations include:

- What design and performance standards should be placed on the management systems?
- What administrative vehicles offer the best prospects for success?
- What funding arrangements are necessary to support the revised permitting and management system?

Design and Performance Standards

It has already been asserted under the discussion of objectives above that ultimate performance standards should be based on results in the aquatic systems under protection. The report further advocates promulgating these standards primarily in terms of biological health (for protection of human health, aquatic life, or both), supplemented by measures of conditions well known to influence biological health quite directly, such as hydrologic variables. It was further proposed that active adaptive management be applied in relation to the degree of achievement of water resource objectives. However, it would not be wise to standardize entirely on this level and leave all questions of the means to the end to individual permittees. Certain design-level standards would also be appropriate. An example is provided by the recently issued draft municipal permit for Ventura County, California. In that permit, application of low-impact methods to new development and redevelopment is specified to hold the effective impervious area to 5 percent of the total contributing catchment. While technical experts may disagree on the precise number, the point is that adopting such a standard gives a straightforward design requirement on an evidentiary basis. Results in the receiving waters would still be tracked and used in active adaptive management if necessary, but effective application of the design standard would provide some level of initial assurance that the aquatic health standards can be met.

Forging Institutional Partnerships

At the heart of the proposal for a new system of regulating discharges to the nation's waters is issuing permits to groups of municipalities in a watershed operating as co-permittees under a lead permittee. Furthermore, the proposal envisions these municipal permittees assuming responsibility for and implementing the permits for all public and private dischargers in their jurisdictions. These admittedly sweeping changes in the way waters have been managed almost everywhere in the nation raise serious issues of acquiescence to the new arrangements, compatibility, and devising a sufficient and stable funding base. This section draws from the small number of examples where arrangements like those proposed here have been attempted.

The Los Angeles County Municipal Storm Water Permit offers a case study in how to aggregate municipalities in a co-permittee system while still allowing prospective members latitude should they perceive their own interests to deviate, even considering the advantages of group action. The permit, first issued in 1990, presently covers five watersheds and 86 municipal permittees. During the process of reissuing the 1996 permit, the City of Long Beach challenged the provisions of the Los Angeles County MS4 permit. The city was given the option of applying for its own individual permit, which it did. Long Beach was issued its own individual MS4 permit in 1999 with provisions similar to the Los Angeles County MS4 permit. As another example, a small coastal municipality (Hermosa Beach) covered by the Los Angeles County Municipal Storm Water Permit investigated the possibility of withdrawing from the county permit in 2000 to be reclassified as a Phase II municipality. Just as with Long Beach, Hermosa Beach was given the option of applying for an individual permit as a Phase I MS4, but in the end Hermosa Beach elected to remain within the areawide permit. Although this report strongly encourages cooperative participation of municipalities as co-permittees, it does not mandate it. Rather, the flexibility illustrated above should be retained in the proposed new permitting program. What matters for compliance with the CWA is that a municipality manage discharges in a manner at least equivalent to other permittees in the watershed.

Stephenson and Shabman (2005) gave thought to the dilemma of entities who may not naturally work well together being asked to cooperatively solve a problem that all have had a share in creating. They argued that new organizational forms that consolidate multiple regulated entities under a single organizational umbrella could be used to coordinate and manage jointly the collective obligations of a group of regulated parties at lower costs to members. Private and public regulated entities alike could benefit from participation in these new organizations. Such cooperative organizations could offer participating parties financial incentives and decision-making flexibility through credit trading programs.

Two larger-scale compliance associations exist in the Neuse and Tar-Pamlico river basins in North Carolina (Stephenson and Shabman, 2005). In both programs the state was concerned about nutrient enrichment of estuary waters and imposed an aggregate cap on industrial and municipal wastewater dischargers equivalent to a 30 percent reduction in nitrogen loads. In both programs, the state granted individual point source dischargers a choice: (1) accept new requirements to control nitrogen through individual NPDES permits or (2) form and join a discharger association. The rigidities associated with individual NPDES permits provided enough incentive for most point source dischargers to opt for the second choice. Compliance associations were then created and issued permits.

The Neuse River rules cover nonpoint agricultural sources as well as point discharges. Counties are responsible for reducing nutrient loads, and farmers must either join county associations that apply different strategies or individually contribute to meeting objectives by setting aside 50- to 100-foot buffers along all streams.

North Carolina requires compliance associations to meet a single mass load cap. In the Tar-Pamlico case, the legal requirement to meet the cap was established by an enforceable contractual agreement signed by the association and the state. In the Neuse program, a single "group compliance permit" was issued to the association. Both legal mechanisms established financial penalties for the two associations if aggregate discharges of the group exceed the association cap. A key advantage of the association is similar to that of a formal effluent trading program—granting dischargers flexibility to decide how best to meet the aggregate load cap. To date, the associations have managed to keep nitrogen loads considerably below their respective caps. Compliance costs have also fallen below original projections. Further, there is some evidence that the association concept is producing incentives for strong cooperative behavior that did not exist prior to implementation.

The case studies presented here illustrate ways in which both public and private entities subject to regulation can exercise options for operating autonomously should they not wish to incorporate with a group, while still contributing to the achievement of watershed objectives. The case studies suggest that most dischargers conclude in the end that group membership offers considerable advantages.

Funding Considerations

The existing stormwater permit program is characterized, in most of the nation, by municipal Phase I and now Phase II permittees operating mostly alone. In contrast the new system envisions coalitions of permittees that share a watershed operating in concert, under the coordination and leadership of a principal permittee. The present structure tends to bring about duplication in effort and staff, whereas cooperation should stimulate efficiencies that could defray at least part or even much of the extra local costs associated with new responsibilities for municipal permittees.

As explored in the preceding section, municipalities may not necessarily wish to join in co-permittee arrangements; and mechanisms are proposed to allow them to operate individually, as long as watershed objectives are met. However, the state could encourage participation through financial inducements, for example, by estimating the resources needed to meet the requirements of each watershed permit and pointing out to permittees how shared resources can save each contributor money. The state should also set preferences and better terms for grants in the favor of municipalities who join together.

To the questions of administrative vehicles and funding arrangements, stormwater utilities are the preferred mechanism, and regulations should support creating stormwater utilities. It should be added that, with watershed-based permitting as proposed here, utilities should also be regionalized on a watershed basis. A utility draws funds from the entities served in direct relation to the cost of providing the services, here management of the quantity and quality of stormwater discharged to natural waterbodies. These funds must be dedicated to that purpose and that purpose only, and cannot be redirected to general agency coffers or for any unrelated use.

Not only are more funds from more reliable sources needed, but monies should be redirected in ways differing from their allocation under the current system. It was proposed earlier that a lead municipal permittee, working with other municipal co-permittees, be given responsibility for coordinating permitting and management of municipal, industrial, and construction stormwater permits, and even permits involving other sources, such as industrial process and municipal wastewaters. Those entities would hence be doing work now devolving to individual private developers and industrial plants and other public authorities. They would need to attract the revenue from those other bodies in proportion to the added work taken on. A utility structure would provide a well-tested means of carrying out this reallocation.

Stormwater utility fees are generally assessed according to a simple formula, such as a flat rate for all single-unit dwellings and in proportion to impervious area for commercial property. Some municipalities have investigated charging more directly according to the estimated quantity and quality of stormwater discharged into the public drainage system. Municipal permittees may choose to formulate such a system, but the development process itself is not a trivial task and, being based on general (and usually quite simple) hydrologic and water quality models, can generate considerable arguments from rate payers. Going through this process is probably not necessary or even advisable for most municipal permittees, who will have many new functions should the proposed system be adopted. Instead, they should concentrate on implementing a fee structure based on a simple formula like the one above and then capture additional revenues for special functions that they will take over from industrial and construction permittees.

As discussed previously, in the proposed program municipal co-permittees, with leadership by a watershed lead permittee, will be asked to classify industries and construction sites within their borders according to risk and accordingly prioritize them for inspection and monitoring. It is proposed in the section on Measures of Achievement, below, that inspection include reviewing and approving industrial and construction site stormwater pollution prevention plans (SWPPPs). While many municipalities now inspect construction sites for stormwater compliance and some inspect industries, this work will increase significantly in the new system, and SWPPP review and approval will be a completely new element. Moreover, municipalities would perform some industrial monitoring now conducted by the industries themselves and may monitor high-risk construction sites. These special functions would require different institutional arrangements and substantial new revenue that could not be fairly charged to all rate payers. There are several possible sources for these funds. One way would be to increase industrial and construction permit fees and direct large proportions to municipalities to support inspection and monitoring. The permitting authority (designated state or EPA) would still hold ultimate authority, and municipalities could refer industrial and construction permittees found during inspection to be out of compliance to the permitting authority for enforcement. Another means would be to form consortia of industries of similar type and assess fees directly applicable to inspection and monitoring. For example, scrapyards under the jurisdiction of the California EPA Los Angeles Regional Water Board formed a monitoring consortium under which sample collection by a qualified contractor rotates among the members, with funding by all. While the members operate this system, it could be adapted to operation by municipal co-permittees.

A second-level funding concern is, once revenues are generated, how should they be put to use? It is very important that funds largely be devoted directly to the tasks at hand regarding the achievement of objectives instead of into excessive administrative and bureaucratic structure. These tasks are scientific and technical and are highly oriented toward what is actually going on in the drainage systems and their receiving waters. Thus, the majority of funds should be directed to making scientific and technical judgments based on observations and monitoring results obtained in the field (see the discussion below).

Measures of Achievement

Critique of the Current Monitoring System

No area exemplifies the differences between the present and proposed new stormwater permitting and monitoring systems more than the measures used to gauge achievement. The current monitoring system is characterized by scattered and uncoordinated measurements of discharges from Phase I MS4s and some industries, and some visual observations of construction sites. The system proposed to take its place would emphasize monitoring of receiving water biological conditions as a data source for prescribing management adaptations to meet specified biological objectives. The discussion here first critiques the prevailing system to construct part of the rationale for changing it. It then proceeds to outline a recommended monitoring structure to replace it.

To expand very briefly on the point that the present system is scattered and uncoordinated, monitoring under all three stormwater permits is according to minimum requirements not founded in any particular objective or question. It therefore produces data that cannot be applied to any question that may be of importance to guide management programs, and it is entirely unrelated to the effects being produced in the receiving waters. Phase I municipal permit holders are generally required to monitor some storms at some discharges for no stated purposes but to report periodically to the permitting agency (Phase II municipalities have no monitoring requirements, although they may represent the major or even only impact sources in a given watershed). The usual model for industries across the nation is to collect a few discharge grab samples a year and send the results to the permitting authority, plus occasionally to make observations for obvious signs of pollution (e.g., oil sheen, odor). Construction site monitoring is less standardized and often involves no water quality monitoring at all. Again, no permittee under any of the three programs is obligated according to national standards to check the effects of its discharges on receiving waters. Since the individual effects of any discharger are often not distinguishable from any other, the scattershot system would usually not be able to discern responsibility for negative effects in the receiving water ecosystem.

Input to the committee conveyed the strong sense that monitoring as it is being done is nearly useless, burdensome, and producing data that are not being utilized. For example, the City of Philadelphia conducts substantial amounts of wet weather monitoring, which is very expensive, but it can barely monitor for TSS in many of its heavily impacted streams (Crockett, 2007). The resources to monitor for the more exotic pollutants do not exist. Smaller municipal permittees without the resources and sophistication of a big-city program have difficulty performing even the most basic monitoring. City water managers believe that the traditional stormwater program places too much emphasis on monitoring of individual chemicals rather than looking at ecological results (Crockett, 2007).

Industry representatives have also described several problems they see in industrial stormwater monitoring as it is performed now (Bromberg, 2007; Longsworth, 2007; Smith, 2007). One concerns the high degree of variability, from the methods used to what is actually measured (Stenstrom and Lee, 2005; Lee et al., 2007). Opponents have been quite critical of the benchmarks to which industrial monitoring data are compared, believing that the benchmarks have no basis in direct measurements associating stormwater with impacts. Some have suggested replacing monitoring with an annual stormwater documentation report to the permitting authority. It seems that industry personnel disrespect the current monitoring framework for some good reasons and feel it conveys a burden for little purpose. There was some implication that industry would be receptive to measures offering more meaningful information in place of poorly conceived monitoring requirements (Bromberg, 2007; Longsworth, 2007; Smith, 2007).

Proposed Revised Monitoring System

A structure in several tiers is proposed as a monitoring system to serve the watershed-based permitting and management framework.

Progress Evaluation Tier. This tier would represent the ultimate basis for judgment on whether the objectives adopted for the watershed are being met. Because these objectives would mainly be expressed in terms related to direct support of beneficial uses, so too would monitoring in the Progress Evaluation Tier principally emphasize direct measurements of ecological health. The preferred model for this evaluation would be the paired watershed approach, which is based on the classic method of scientific experimentation and was developed for water resource management investigations by EPA (Clausen and Spooner, 1993). Ideally, conditions in the waterbody under evaluation would be compared to conditions in the same waterbody before imposition of a permit and management scheme (before versus after comparison), as well as to conditions in a similar waterbody not subject to human-induced changes (affected system versus reference system comparison). At least one of these comparisons must be made if both cannot. If the objectives involve improving conditions, and not just avoiding more degradation, the reference should represent that state to which the objective points.

This function has traditionally been the province of the permitting authority (i.e., the designated state or EPA). In the new program, the function is assigned to municipal permittees, guided by the lead permittee, to conduct or contract, but with a substantial contribution by the permitting authority in the form of material support and guidance. The primary vehicle envisioned to perform the progress assessment is a well-qualified monitoring consortium serving the watershed, and perhaps other watersheds in the vicinity. Case studies below present examples of successful joint ventures in monitoring that can serve as models. The proposal is based on the belief that monitoring should be more manageable and effective at the watershed compared to the state level and, furthermore, that utilizing a consortium approach should make it feasible for a coalition of municipal co-permittee partners to commission monitoring.

Findings of objective shortfall would trigger development of active adaptive management strategies. Generally, an assessment should be conducted to determine what additional measures should be put in place in regulating new development and redevelopment, as well as increasing coverage of existing developments with retrofits.

Diagnostic Tier. The second tier would be designed to provide the municipal permittees with the necessary information to formulate active adaptive management strategies, and they would be responsible for this second tier as well as the first. The Diagnostic Tier would be composed of assessment of information from the Compliance Reporting Tier, plus some specific field monitoring to determine the main reasons for ability or failure to meet objectives. Some highly directed monitoring of receiving water conditions could determine the need to improve management of water quantity, water quality, or both. A tool like the Vermont flow-duration curves is an example of a potentially useful device for diagnostic purposes. To allow the use of such a tool, it is important that continuous flow recorders be installed on key streams in the watershed. The techniques described in the Impact Sources section above, once they are further developed, would also be useful in Diagnostic Tier monitoring.

An important dimension of this tier would be prioritized inspection and monitoring of potentially high-risk industrial and construction sites. In addition, data submitted by the industrial and construction permittees according to the Compliance Reporting Tier would assist in targeting dischargers to bring about the necessary improvements in water quantity and/or quality management.

Compliance Reporting Tier. It is proposed that the first step in compliance reporting be submission of SWPPPs by all construction and industrial permittees (plus municipal corporation yards as an industrial-like activity) to the jurisdictional municipal permittee for review and approval. It is further proposed that the industrial permittees and municipal corporation yards be relieved of sample collection, *if* they develop SWPPPs making maximum possible use of ARCD practices, supplemented by active treatment as necessary, and the municipal permittee approves the SWPPP. Construction sites would be given a similar sampling dispensation if they develop an approved SWPPP along the lines of Box 5-3.

Otherwise, the permittees would be required to perform scientifically valid sampling and analysis and report results to the watershed co-permittees. This more comprehensive and meaningful monitoring would increase the burden already felt by permittees and create a strong incentive to apply excellent SCMs. This burden could be relieved to a degree through participation with other similar dischargers in the watershed in a monitoring coalition. As an example, in North Carolina coalitions of wastewater dischargers are working with the state Division of Water Quality (DWQ) to create and manage coalition-led watershed monitoring programs that operate in conjunction with DWQ's ambient chemistry and biological programs (Atkins et al., 2007). Lee et al. (2007), after an assessment of industrial stormwater and other monitoring data, concluded that selecting a subset of permittees from each monitored category would yield better results at lower overall cost compared to monitoring at every location. This strategy would permit the use of more advanced sampling techniques, such as flow-weighted composite samplers instead of grab sampling, to estimate representative loads from each category with improved accuracy and reduced variability.

All permittees would still make observations of the SCMs and discharges and keep records. The final proposed step in compliance reporting is an annual report covering observations, SCM operation and maintenance, SWPPP modifications, and monitoring results (if any), to be sworn as to correctness, notarized, and submitted to the lead municipal permittee. The Massachusetts Environmental Results Program (April and Greiner, 2000) offers a possible model for compliance reporting and verification. This program uses annual self-certification to shift the compliance assurance burden onto facilities. Senior-level company officials certify annually that they are, and will continue to be, in compliance with all applicable air, water, and hazardous waste management performance standards. The state regulatory agency reviews the certifications, conducts both random and targeted inspections, and performs enforcement when necessary.

Research Tier. The final tier would be outside the permit system and exist to develop broad mechanistic understanding of stormwater impacts and SCM functioning important to assist permittees in reaching their objectives. EPA and state agencies designated to operate the permit system would have charge of this tier. These agencies would develop projects and contract with universities and other qualified research organizations on a competitive basis to carry out the research.

Instructive Case Studies for the Proposed Revised Monitoring System

Many municipalities, even large ones, would be challenged and burdened by taking on comprehensive watershed monitoring. The Southern California Coastal Water Research Project Authority (SCCWRP, http://www.sccwrp.org) offers an excellent model of how co-permittees in a watershed or an even broader area could organize to diffuse these challenges and burdens. SCCWRP is a joint-powers agency, one that is formed when several government bodies have a common mission that can be better addressed by pooling resources and knowledge. In SCCWRP's case, the common mission is to gather the necessary scientific information so that member agencies can effectively and cost-efficiently protect the Southern California marine environment. Key goals adopted by SCCWRP are defining the mechanisms by which aquatic biota are potentially affected by anthropogenic inputs and fostering communication among scientists and managers. Comprised of a multidisciplinary staff, SCCWRP encompasses units specializing in analytical chemistry, benthic ecology, fish biology, watershed conditions, toxicology, and emerging research.

SCCWRP's current mission stems from the results of a 1990 NRC review of marine environmental monitoring programs in the Southern California Bight (NRC, 1990). It was determined that although $17 million was being spent annually on marine monitoring, it was not possible to provide an integrated assessment of the status of the Southern California coastal marine environment. Most monitoring was associated with NPDES permit requirements and directed toward addressing questions about site-specific discharge sources. As a result, most monitoring in the bight was restricted to an area covering less than 5 percent of the bight's overall watershed, making it difficult to draw conclusions about the system as a whole. The limited spatial extent of monitoring was also found to limit the quality of local-scale assessments, since the boundaries of most monitoring programs did not match the spatial and temporal boundaries of the important physical and biological processes in the bight.

NRC (1990) further found that there was a lack of coordination among existing programs, with substantial differences in the parameters measured among programs, preventing integration of data. Even when the same parameters were examined, they were often measured with different methodologies or with different (or unknown) levels of quality assurance. Moreover, the NRC found that even when the same parameters were measured in the same way, substantial differences in data storage systems among monitoring programs limited access to the data for more comprehensive assessment. To avoid repetition of these shortcomings, the SCCWRP example should be given very thorough consideration as a template for the Progress Evaluation, Diagnostic, and Research Tiers in the proposed revised monitoring program.

The San Gabriel River Regional Monitoring Program (SGRRMP, *http://www.lasgrwc.org/SGRRMP.html*) is a watershed-scale counterpart to the larger-scale regional monitoring efforts in Southern California. The SGRRMP incorporates local and site-specific issues within a broader watershed-scale perspective. The program exists to improve overall monitoring cost effectiveness, reduce redundancies within and between existing monitoring programs, target monitoring efforts to contaminants of concern, and adjust monitoring locations and sampling frequencies to better respond to management priorities in the San Gabriel River watershed. Five core questions provide the structure for the regional program:

- What is the environmental health of streams in the overall watershed?

- Are the conditions at areas of unique importance getting better or worse?
- Are receiving waters near discharges meeting water quality objectives?
- Are local fish safe to eat?
- Is body-contact recreation safe?

The workgroup convened to establish the program recommended monitoring designs to answer the core questions effectively and efficiently. The resulting program is a multilevel monitoring framework that combines probabilistic and targeted sampling for water quality, toxicity, and bioassessment and habitat condition.

The City of Austin, Texas, has more than 20 years of stormwater monitoring experience and offers additional guidance on designing and implementing watershed monitoring programs (City of Austin, 2006). Austin performs detailed periodic synoptic sampling in the watersheds it manages to track trends in stormwater quantity and quality. The city uses the results to evaluate the impacts of land development on stormwater quantity and pollution, establishing statistical relationships between measures of these conditions and the amount of impervious cover. Trend assessment over time leads to recommended changes to the City of Austin Environmental Criteria Manual as needed.

Creating Flexibility and Incentives
Within a Watershed Approach

A watershed-based permitting approach to stormwater management focuses attention on watershed objectives and endpoints. To be able to achieve these goals, observable performance measures beyond the success of an individual SCM need to be identified that are consistent and necessary to meet designated uses. These might include watershed-level numeric limits on the amount of a particular pollutant allowed to enter a waterbody (e.g., pounds of phosphorus) or various measures of allowable volume of discharge. A watershed focus shifts attention away from specific SCM performance and site-specific technological requirements to achieving a larger watershed goal. As a consequence, there is considerable management flexibility in deciding how these goals will be achieved. Indeed, this flexibility was cited by the NRC (1999) as a prerequisite to successful watershed management.

One way of exercising this flexibility is to create an "incentive-based" or "market-based" approach to choose how watershed goals are met. It is recognized throughout the environmental management field that entities subject to regulation do not necessarily have equal opportunities and qualifications to comply sufficiently to sustain resources. To compensate for this, the market-based approach allows individual discretion to select how effluent (or runoff volume) will be controlled (choice of technology, processes, or practices) and where they will be controlled (on site or off site). That is, any discharger legitimately unable to meet discharge quantity and quality allocations would be able to finance offsets elsewhere to achieve the watershed goals. An important element and challenge is to couple this decision-making flexibility with personal (typically financial) incentives so that people willingly make choices supportive of the watershed objectives. Broadly stated, the idea is to create financial reasons and decision-making opportunities to lower compliance costs and create or implement new effluent/volume control options (Shabman and Stephenson, 2007).

Because incentive-based policies require a shift in emphasis from technologies and practices to outcomes (e.g., volume or quantity of effluents), the municipal manager would not be responsible for deciding what SCM will be implemented in specific areas or hand picking specific practices to promote. Rather the stormwater program manager's responsibilities shift to establishing watershed goals, developing metrics to measure outcomes and performance, and performing necessary inspection and enforcement activities.

Effluent trading, sometimes called "water-quality trading," is one type of incentive-based policy. In an ideal form, effluent trading requires government to establish a binding aggregate limit or cap on an outcome (e.g., mass load of effluent, volume of runoff) for an identified group of dischargers. The cap or aggregate allowable discharge is set to support and achieve a socially determined environmental goal. Because it is fixed, the cap provides the public assurances that environmental objectives will be achieved in the face of a growing and changing economy. The total allowable discharge is then divided into discrete and transferable units, called allowances, and either distributed or auctioned to existing dischargers. All dischargers must own sufficient allowances to cover their discharges. For instance, any new or expanding source must first purchase allowances (and hence effluent or volume reductions) from another source before legally discharging. The requirement to hold allowances on the condition to discharge and the positive allowance price creates financial incentives for pollution prevention. Dischargers holding allowances rather than reducing discharge face forgone revenues that could have been achieved from the sale of allowances. Conversely, expanding dischargers have incentives to invest in pollution prevention in order to avoid the cost of purchasing additional allowances.

In the context of the revised permit system advocated here, achievement of objectives (generally of a biological nature) will require some combination of strategies such as no net increases in hydrologic parameters (e.g., peak flow rates, durations, volumes), water pollutants, forest cover loss, and effective impervious area. If one entity is unable to contribute adequately to meeting its share of compliance, then it must obtain the necessary credit by buying it from another similar entity that is able to contribute more than its designated share. Ideally, all sources of a waterbody's problems, not only stormwater, would come under the trading system.

Implementing the market system requires development of a resource-based currency, a nontrivial exercise but one for which models are available in other fields, especially air emissions. For example, emission trading has been a critical element of the nation's strategy to limit sulfur dioxide and nitrogen oxide emissions (Ellerman et al., 2000). Carbon trading is a cornerstone policy in the European Union effort to limit greenhouse gas emissions. The EPA promotes the use of trading to help achieve the goals of the CWA and has issued several policy statements and recently published guidance on how trading programs can be grafted within existing NPDES permitting programs (EPA, 2003a, 2007b).

However, compared to the air program, experience and success with trading in the water program have been limited (Shabman et al., 2002). Furthermore, programs labeled trading have been implemented in a multitude of ways in the nation's water quality program (Woodward et al., 2002; Stephenson et al., 2005; Shabman and Stephenson, 2007). In many instances, trading programs are case-specific and isolated "trades" that do not fundamentally change the choice and incentives facing dischargers in a conventional permitting system. The extent to which trading policies can be effectively employed on a watershed scale is limited not only by the physical differences between air and water mediums, but also by the unique legal structure of the CWA (Stephenson et al., 1999). For example, the CWA is oriented around imposing technology-based performance requirements on specific subset of discharge sources. Individual NPDES permits require sources to achieve these agency-identified levels of performance and may specify how performance is achieved. The statute also places limits and disincentives on the degree to which permit agencies can deviate from these limits (e.g., "antibacksliding").

Thus, the focus of the NPDES permitting system has been on individual source control and technologies, unlike the air program, which has a stronger statutory orientation around achieving broader air quality goals (ambient air quality standards). The orientation of the NPDES program limits the flexibility and incentives for regulated parties that might make market-oriented trading possible. It turns out that some of the more successful applications of trading in the water program have occurred because of permitting innovations that effectively avoid some of these rigidities (see discussion of North Carolina point source control program on the Neuse River, above).

Trading programs of various types have been proposed or suggested for stormwater (Thurston et al., 2003; Parikh et al., 2006). Although conceptual models of a comprehensive trading program based on the total volume of allowable water to be discharged have been proposed, no working examples have yet to be implemented. More limited versions of trading programs, however, have been developed. These programs provide compliance flexibility for new sources of stormwater runoff. In some locations, new developments face a requirement to provide a specific level of volume or effluent control from the parcel to be developed. The regulated entity is typically obligated to meet this requirement with the applications of on-site SCMs. Trading programs create opportunities for regulated entities to meet their regulatory requirement off site (off the parcel to be developed), called here an offset. In some trading programs, the off-site controls can be accomplished by the creation of an in lieu fee program. Such programs typically occur for dischargers that are not required to hold or obtain individual NPDES permits.

In lieu fee programs offer some opportunity for regulated parties to make a financial payment (fee) to a local government entity in lieu of implementing on-site controls. The fees are collected and used to implement stormwater controls in other areas of the watershed. Controlling runoff at a regional level rather than through the construction of many small on-site controls may be more cost-effective given the economies of scale associated with some SCMs (see Chapter 5 pages 362–363). The option for off-site controls also allows the stormwater program to direct investments in stormwater control to specifically targeted areas of the watershed.

Examples of in lieu fee programs include Santa Monica, California, the Neuse River Basin in North Carolina, and Williamsburg, Virginia. Santa Monica's program requires new and redevelopment projects to treat a specific volume of runoff. The program first requires the regulated entity to take all feasible steps to meet the requirement through the implementation of on-site infiltration practices. If the regulated party can demonstrate why it is economically and physically infeasible to install any type of infiltration or treatment SCM, the regulated party can pay a fee based on the volume of water that needs to be controlled (the total mitigation volume is the volume that would have been attenuated via an SCM). The fee set by Santa Monica is $18/gallon of total required mitigation volume. The $18 reflects the cost of constructing an SCM and maintaining it over 40 years (DeWoody, 2007). Presumably these fees are used to construct infiltration measures elsewhere.

The Neuse River Program requires all new land development to meet a nitrogen export standard of 3.6 pounds per acre per year (North Carolina Division of Water Quality, 1999). The water quality goal for the Neuse basin is to reduce mass nitrogen loads by 30 percent in order to improve water quality in the estuary. The export standard was set to achieve a 30 percent reduction from the average nitrogen load from lands prior to development. Developers have the option to meet this export standard either through the application of on-site SCMs or by paying a fee into a state-administered Riparian Buffer Restoration Fund (see 15A North Carolina Administrative Code 02B .0240), which would be used to reduce nitrogen loads elsewhere in the basin. Developer discretion, however, is not unlimited. Under no circumstances may developers discharge more than an estimated 6.0 pounds per acre per year from a residential site.

The Williamsburg program has an in lieu fee program for total phosphorus loads created by new development (Frie et al., 1996; Stephenson et al., 1998). For every new development, the increase in total phosphorus load from stormwater runoff from impervious surfaces is estimated. Developers have the choice to meet the phosphorus load reduction requirement through the application of on-site controls or by paying a fee to the city. The fee is set at $5,000/lb of phosphorus, with the fees earmarked to the construction of regional stormwater facilities or for the preservation of open space within the city. The presence of a fee option could also provide incentives for developers to implement source reduction practices.

The above programs differ in some important ways. For example, the Santa Monica program requires regulated entities to undergo a "sequencing" process that places regulatory preference on on-site controls before being able to use the fee option. The Williamsburg program allows regulated entities the option to select between constructing on-site controls and paying the fee without a regulatory preference for on-site controls. Sequencing rules tend to limit control options and thus the cost-effectiveness of these types of programs.

In lieu fee programs are distinguished from other offset programs in that it is the responsibility of the local government (or more generally, any designated fee service provider such as a nongovernmental organization) to provide the off-site SCMs. In lieu fee programs, common in the U.S. wetlands program, face a number of implementation and design challenges (Shabman and Scodari, 2004). For example, enforcement sometimes becomes a concern because the local stormwater management agency responsible for constructing and maintaining the SCMs is also responsible for monitoring and enforcement. These dual responsibilities create potential conflicts of interest; if an off-site mitigation project fails, there maybe no apparent overseeing agency to enforce corrective actions. The lack of transparency in accounting to determine whether the offset projects provide enough compensation is also sometimes a challenge. Finally, the ability to fully offset the volume of effluent discharge from a new development is contingent on collecting enough revenue from the fee to pay for the construction and maintenance of offsite SCMs. The delay between impacts and compensation and lack of full public cost accounting complicate the challenges of setting an appropriate fee.

Ensuring that in lieu fee programs provide the necessary mitigation could be accomplished in a number of ways. For example, an oversight agency may be designated to establish tracking and reporting requirements and monitor in lieu fee program performance. Or, the potential conflicts of interest inherent in the lieu fee program design could be avoided by separating the provision of the off-site mitigation service from the monitoring and enforcement. It is possible to imagine that the private sector, rather than an in lieu fee administrator, could provide off-site stormwater reduction services to those subject to the stormwater control requirements. In this case, the private sector would provide stormwater detention/retention services above and beyond what is required by law. These private service providers would receive stormwater runoff credits for these investments ("above baseline") that could be sold to developers who might wish to meet their control obligations in ways other than on-site controls. In essence, the role of searching, designing, and constructing offsite SCMs would be transferred to the private-sector stormwater credit providers. The local stormwater managers, however, would retain full authority to monitor, verify, and enforce to ensure that these offsets are successfully implemented.

The flexibility provided by in lieu fee and trading programs requires that pollutant loads or runoff volume created at one site be reduced at another site. Thus, a design issue confronting these types of programs is the consideration of the spatial extent in which offsetting activities can occur. The extent of the spatial range of offsetting activities in turn will depend partly on the nature and type of service being offset. For example, in the Neuse example nitrogen is a regional, basinwide concern with minimal localized effects. In such cases, the offsetting activities might be allowed basinwide (after adjusting for nitrogen attenuation through the basin). In other situations where localized concerns maybe a greater concern (say from localized flooding), the flexibility offered by such programs may be more limited. However, such spatial flexibility might also be a way to implement and achieve watershed planning objectives. For example, development may be encouraged in high-impact areas, and offsetting fees could be used to protect and enhance water quality objectives in other areas.

This last point deserves further explanation. Although this chapter advocates that biological conditions in waterbodies should be maintained or improved, there are many urban areas where local waterbodies cannot achieve the same designated uses as less developed areas. If a goal-setting entity chose to do so, beneficial uses for waters in these areas could be set at levels that acknowledge this highly altered condition, such that these streams would not be expected to achieve the same biological condition as streams outside the urban core (see Chapter 5 pages 364-366). This might be done to encourage development in high impact areas; San Jose, CA, provides an example (see Chapter 2). In that city's stormwater program, in urban areas where on-site control is either technically impossible (due to soil or space constraints) or prohibitively costly, the developers can meet the post-construction treatment standard by providing volume control either through participation in a regional stormwater project or by providing equivalent projects off site (e.g., stream restoration).

It is also possible to design a stormwater offset program that allows the different functions of stormwater management to be separated to achieve watershed objectives. For example, management of peak flow serves mostly to prevent localized flooding while more stringent volume control maybe required to protect stream channels and aquatic life. Control of peak flow might be required on site or within a narrow geographic region. In areas targeted for development, however, the volume control needed for channel protection might be transferred off site and into areas where watershed planning has identified the need for higher levels of stream channel protection or enhancement (more stringent water quality standards). A similar watershed approach based on functional assessment was recommended for wetland compensation (NRC, 2001b).

Regulatory and Legal Implications of Proposed Watershed-Based Permitting Framework for Managing Stormwater

EPA, the states, and municipal permittees would all have tasks to perform to transform the framework set forth in this report to a fully developed and functioning program. These efforts would be rewarded with a program that is rooted in science, transparent in its aims, fairer for all than the current program, and better for the aquatic environment. This section of the report outlines the tasks necessary to carry the proposal forward to full development.

EPA should seek significant congressional funding to support the states and municipalities in undertaking this new program, in the nature of the support distributed to upgrade municipal WWTPs after the 1972 passage of the Federal Water Pollution Control Act. Beyond financial support, EPA's tasks emphasize broad policy formulation, regulatory modifications and adaptations necessary to initiate the new program, and guidance to the states and permittees. The principal adaptation needed in the regulatory arena involves converting the current TMDL program to a form suitable for the new system. Guidance would be needed in a number of crucial areas, and it is EPA's natural role to develop it.

States (or EPA for states without delegated authority) would have broad responsibilities to translate policies and federal regulations into their own regulatory and management systems. A key task in this regard would be to recast water quality standards into objectives most directly supporting sustenance and improvement of beneficial uses. States already have considerable background for performing this task through their present definitions of beneficial uses, the Section 303(d) process for assessing waterbody compliance with water quality standards, and the triennial review of those standards. However, the added prominence of biological aspects of beneficial uses and associated objectives will require additional analysis. Other prominent state tasks will involve defining the watersheds subject to permits, forming bodies of co-permittees associated with the watersheds, and appointing the lead permittee. Many other state tasks entail cooperative work with the permittees to support and assist them in funding and conducting their activities.

Many aspects of the municipal permittees' roles in implementing strategies were explored above in a section titled accordingly. That section especially focused on activities to advance the use of ARCD methods. More broadly, the permittees will be coordinators of all permits pertaining to the watershed's aquatic resources, collectively pointed toward meeting objectives that the permittees adopt under state oversight. Other categories of tasks assigned to the municipalities under the proposed system include monitoring, in the contexts of both inspections and sampling performed through a consortium, and enforcement actions and program adaptations to promote progress toward achieving objectives. Box 6-4 provides a listing of anticipated tasks for the municipal permittees as well as the states and EPA.

A Pilot Program as a Stepping Stone

The shift of responsibility for stormwater regulation to municipalities under the watershed-based approach may lead to some surprises in implementation and enforcement. Primarily because of this, EPA is well advised to institute a pilot

BOX 6-4
Government Agencies Roles during the Operation of a
Watershed-Based Permitting System

EPA

1. Petition Congress for significant funding support for states and municipal permittees, and develop a program of fairly distributing funds based on environmental and financial needs at the watershed level.
2. Initiate regulatory modifications and clarifications necessary to establish the system.
3. Set policies for watershed permitting based on this report's recommendations.
4. Adapt TMDL program for use in the new program.
5. Produce guidance to assist the states and municipal permittees in the areas of:
   a. Developing a rotating basin approach;
   b. Developing an integrated municipal NPDES permit incorporating the full range of sources;
   c. Developing stormwater utilities and other funding mechanisms;
   d. Using impact source analysis (e.g., using reasonable potential analysis and new research results, industrial and construction site risk assessment);
   e. Using ARCD techniques for new development, redevelopment, and retrofitting;
   f. Developing monitoring consortia;
   g. Developing a credit trading system;
   h. Developing an active adaptive management program

Designated States (or EPA otherwise)

1. Define watersheds for which permits will be issued and set up a rotating basin approach to govern watershed analysis in support of subsequent steps.
2. Formulate and formally adopt goals relative to avoiding any further loss or degradation of designated beneficial uses in each watershed's component waterbodies and recovering lost beneficial uses.
3. Use the results of the existing Section 303(d) process and supplementary work to assess the extent of designated beneficial use achievement in each watershed and set goals for protection and recovery.
4. Match municipal permittees to watersheds and designate a lead permittee for each watershed.
5. Estimate resource needs to fulfill permit requirements in each watershed.
6. Develop a grant program, drawing on EPA and state funds, to support municipal permittees, with incentives for joining co-permittee associations.
7. Identify areas outside the jurisdictions of permitted municipalities that should be brought into the program because of projected development or the existence of problem sources that would compromise the protection and recovery of beneficial uses.
8. Use the triennial review process to modify water quality standards to the objective basis, emphasizing biological outcomes recommended in this report.
9. Revise the TMDL program in accord with the needs of the new program.
10. Set requirements for credit trading systems.
11. Set up an integrated municipal NPDES permit incorporating the full range of sources.
12. Work with municipal permittees to establish specific objectives as the basis for progress assessment.
13. Work with municipalities to develop adaptive management programs responding to progress assessment results.

14. Write municipal permits incorporating the above elements.

15. Write industrial and construction general or individual permits incorporating the recommendations in this report.

16. Allocate a substantial portion of industrial and construction permit fees to municipal permittees to oversee those sectors.

17. Set requirements for municipalities and private properties to opt out of the defined program without compromising the achievement of objectives.

18. Provide consultation, support, and guidance (adapted from EPA materials or originally produced) to municipal permittees in the areas of:
   a.   Developing stormwater utilities and other funding mechanisms;
   b.   Using impact source analysis (e.g., industrial and construction site risk assessment);
   c.   Using ARCD techniques for new development, redevelopment, and retrofitting;
   d.   Developing monitoring consortia;
   e.   Developing a credit trading system

19. Perform enforcement actions on non-complying dischargers referred by municipal permittees.

20. Assess performance of municipal permittees and specify corrections, rewards, and penalties accordingly.

Municipal Co-permittees (led by Lead Permittee)

1.   Adopt specific objectives as the basis for program progress assessment.

2.   Convert ordinances and regulations as needed to implement the modified program.

3.   Supplement and reorganize staffing to emphasize progress and compliance assessment as the principal functions of the program.

4.   Perform or contract detailed scientifically and technically based watershed analysis as a foundation for permit compliance.

5.   Assemble existing data on soils and hydrogeologic properties and supplement with additional data collection as necessary to assess infiltration prospects across the municipality.

6.   Create incentives for private property owners to maximize the use of ARCD methods in new development and redevelopment.

7.   Build subwatershed-scale, publicly owned ARCD works to supplement on-site management measures and as retrofits.

8.   Develop capacity for stormwater management in municipal WWTPs by reducing groundwater inflows to sanitary sewer lines.

9.   In areas experiencing excessive infiltration and groundwater table rise resulting from non-stormwater flows, develop capacity for stormwater management through infiltration by formulating water conservation programs.

10. Identify industries and construction sites that are required to apply for permits but have not done so and compel their filing.

11. Establish or enhance existing programs to inspect and oversee industries and construction sites; report non-complying dischargers to the state for enforcement actions.

12. Set up or join a monitoring consortium structured to implement the progress evaluation and diagnostic tiers of the proposed monitoring program.

13. Annually report monitoring results to the permitting authority; submit a comprehensive progress assessment triennially.

program that provides some experience in municipality-based stormwater regulation before instituting a nationwide program. This pilot program will also allow EPA to work through more predictable impediments to this watershed-based approach. The most obvious impediment arises from the inevitable limits of an urban municipality's responsibility within a larger watershed: substantial growth and accompanying stormwater loading may occur on the outside periphery of a municipality's designated boundaries. If an urban authority lacks legal authority over this future growth, and if this growth contributes significantly to water quality degradation, then a considerable share of the urban stormwater problem could remain poorly addressed. A pilot program should help identify the extent of this jurisdictional slippage and help identify ways to overcome it. Second, it is possible that some municipalities will balk at the added responsibility involved with the watershed-based approach, even with adequate funding. Unless the objective performance standards are rigid, the monitoring requirements substantial, and the rewards for compliance compelling for municipalities that meet the standards, it is quite possible that noncompliance or bare minimal compliance will be the norm. A pilot program provides a less politically charged atmosphere to experiment with the benefits of watershed-based regulation at the local level and to generate local government support for the approach. Finally, because the watershed-based approach necessitates legislative amendments to the CWA, instituting a pilot program in the interim—both to improve the design of a watershed-based program as well as to generate enthusiasm for it—seems a sensible course.

The pilot program should target those local governments that are most eager to redress water quality degradation in their watersheds, but feel stymied by what they perceive as inadequate legal authority and flexibility to make the necessary improvements. Willing municipalities or regional governments would thus opt-in to the program. The pilot program entices these more progressive municipalities to participate by allowing them to serve as the lead authority and providing them with much greater flexibility to determine how to meet their performance-based water quality goals with fewer legal constraints.

Under the pilot program, a municipal government or similar legal authority would apply to EPA or a delegated state to be designated as the lead agency for that portion of the watershed within its legal jurisdiction. In the application itself the municipality would establish—using modeling and ambient data—how it plans at a general level to maintain or exceed its water quality goals (objective performance standards). These goals must be at or above the state water quality goals, or if they are different (i.e., use biological criteria when the state adopts chemical criteria), the municipality must demonstrate how its performance standards will attain the equivalent of the state water quality goals at the downstream edge of the municipality's border. The municipality would also be required to provide assurance of sufficient infrastructure and funding to allow it to develop a water quality plan, implement that plan, issue permits, and enforce the requirements within its boundaries. Finally, municipal plans, once finalized, would need to meet minimum federal procedural requirements. For example, the plans must be transparent and provide opportunities for public comment; they must be enforceable; and they must establish monitoring programs that will track whether they in fact meet the objective performance standards. If a municipality fails to meet any of its performance standards by the requisite deadline, the state and EPA would have the option of revoking the municipality's program, and reinstituting federal requirements. Ideally, federal guidance would also be available to municipalities to provide direction on how they might institute a watershed-based plan within their boundaries, while still reserving considerable flexibility to allow them to develop creative and progressive stormwater solutions. For example, municipalities would be encouraged to form stormwater utilities that are financed from point and even nonpoint sources that assist them in establishing rigorous permitting and enforcement of their water quality plan.

Municipalities that voluntarily take on this role as lead authority will be rewarded with few legal constraints on how they meet their performance-based objectives. NPDES permits for major sources will still be required and must meet federal minima (technology-based controls) to avoid possible hot spots surrounding large dischargers, and states would remain listed as the lead permittee for these permits, but the lead municipality or other regional government would be able to propose new, more stringent limits that are presumptively favored in revised NPDES permits. Stormwater permits would also be mandatory, but their substantive requirements would be left wholly within the discretion of the lead municipality. Finally, states and municipalities would *not* be required to comply with all of the federal regulations governing TMDLs (they would make a basic load calculation for pollutants contributing to degraded conditions, 33 U.S.C. § 1313(d), but would not be required to do more). Instead, the watershed-based program would be considered the functional equivalent of TMDLs for at least the municipality's portion of the watershed since the program ensures that water quality objectives are met. Municipalities could even be allowed to set interim goals over a period of a decade or more so that TMDLs need not be achieved in a single permit cycle.

Other than federal minimum standards for major NPDES sources, municipalities would have primary if not exclusive authority to decide what types of sources (including nonpoint) require permits, whether certain land uses might be taxed for stormwater management fees, and whether and how to create trading programs among the contributors to water quality impairments within their watershed. Municipalities would also have legal authority to petition EPA to restrict upstream sources that contribute significantly to water quality degradation in ways that make it difficult for them to reach their goals. Upstream governments or sources could be subject to more rigorous federal or state TMDLs and could be vulnerable to tort and related claims from downstream municipalities.

This added flexibility and authority for municipalities to control water quality problems within their legal jurisdiction—coupled with objective performance standards—should lead to more creative approaches to stormwater management that create significant benefits to the municipality (i.e., more green-space buffers along waterways for recreation) and stronger planning and taxation of new developments that otherwise might be uncontrolled. Municipal green space, parks, and a variety of other public goods that both reduce stormwater and enhance the public enjoyment of the surface waters could result from allowing a municipality the freedom to determine how best to regulate sources within its local boundaries. For example, rather than automatically allowing federally approved SCMs that have little aesthetic or recreational qualities, alternative approaches to SCMs that retain their effectiveness but provide other qualities (particularly qualities that draw the public outdoors for recreation or relaxation) are more likely to be encouraged or even required by a municipality that serves as lead over implementation of its water quality program.

Although a national watershed-based approach to stormwater regulation is likely to require legislative amendments, the pilot program may not necessitate additional legislative authorization. It is possible that through regulation, EPA may be able to develop "in lieu of" or "functional equivalent" requirements that allow a rigorous watershed plan to substitute for the bare federal requirements governing stormwater regulation, general permits, and TMDL planning laid out in the CWA. This type of intricate legal analysis, however, is beyond the scope of this document.

Final Thoughts

The watershed-based stormwater permitting program outlined above is ultimately essential if the nation is to be successful in arresting aquatic resource depletion stemming from sources dispersed across the landscape. EPA is called upon to adopt the framework now and set in motion a process to move it toward implementation over the next five to, at most, ten years. This chapter deals with some but not the entire realm of political, legal, regulatory, and logistical issues raised by converting to a fundamentally different system of management and permitting. Ideas are contributed regarding piloting and transitioning toward the new program, altering institutional arrangements to accommodate it, and incentives for effective participation. For watershed-based permitting to take hold, specific actions will have to be undertaken by EPA, state permitting authorities, and municipal permittees during the adoption and transition process.

The proposed program could be implemented by EPA in a number of ways, ranging from making it mandatory without any exception in all states and jurisdictions to leaving it entirely voluntary. The committee recommends neither extreme and believes the best course would be: (1) pilot test and refine the program as described in the report section titled "A Pilot Program as a Stepping Stone;" (2) make the refined program the default to be followed by all designated states (and EPA in others) and all municipal, industrial, and construction permittees, unless a state permitting authority convincingly demonstrates to EPA's satisfaction than an alternative approach will accomplish the program's overall goal of retaining and recovering aquatic resource beneficial uses; (3) develop very significant incentives for states and permittees to participate; and

(4) require objective demonstration by any state opting for an alternative that it is broadly achieving the goal to at least the same extent as states within the program, with appropriate sanctions for noncompliance.

ENHANCEMENT OF EXISTING PERMITTING BASIS

The current federal stormwater regulatory framework has been in place since 1990, and the point source NPDES program under which it is being implemented has existed since 1972. The U.S. Congress deliberately acted in 1987 to amend the federal CWA with the goal of addressing stormwater pollution because it had been identified as a leading cause of surface water impairments, and regulations were inadequate to address it effectively. The total rethinking of the current framework of regulating stormwater pollution described above may require changes in statute and take a long time to implement. Thus, in addition to the longer-term approach that integrates a watershed-wide planning and permitting strategy into the program, several near-term solutions are also offered, with the objective of improving the current regulatory implementation and which at most might require changes in regulation.

Problems Complying with Both Municipal and General Industrial Permits

The NPDES permitting authority issues (1) separate individual permits or general permits to impose discharge requirements on small, medium, and large MS4s; (2) general permits that require construction activity operators who discharge stormwater to waters of the United States, including those who discharge via MS4s, to implement SCMs; and (3) general permits for operators of stormwater discharges associated with industrial activity who discharge to waters of the United States, including those who discharge via MS4s, to implement SCMs. The MS4 operators in turn are also required under the terms of their MS4 permits to require industries and construction site operators who discharge stormwater via the MS4 to implement controls to reduce pollutants in stormwater discharges to the maximum extent practicable, including those covered under the permitting authority's NPDES general permits. This dual-coverage scheme appears intended to recognize the separation of governmental authorities. Unfortunately, in practice it is duplicative, inefficient, and ineffective in controlling stormwater pollution that enters the MS4 from diffuse and dispersed sources. Particularly in the area of monitoring of water quality, the dual approach seems to have resulted in a lack of prioritization of high-risk industrial sources and the purposeless collection of industrial stormwater monitoring data or the poor use of it to strategically reduce the discharge of stormwater pollutants to the MS4.

The preference of EPA to use general NPDES permits to alleviate the administrative burden associated with permitting more than a 100,000 point sources discharging stormwater is understandable. It would have been prudent to have some form of prioritization to select some subset of the whole as high-risk or have a strategy for identifying a subset for individual NPDES permits to better achieve the objective of ensuring compliance with water quality standards on the basis of potential risk. As discussed in Chapter 2, there are no federal guidelines for prioritization (determining what industries are high-risk for stormwater discharges), and the state permitting authorities have largely not prioritized because of the overwhelming burden of administering a very expansive stormwater permitting program.

In the existing permitting scheme, the MS4 operator cannot be faulted for having a reasonable expectation that the permitting authority's general NPDES permits that regulate industrial activities and construction that discharge to the MS4 would require, at a minimum, a sufficient level of identification and implementation of SCMs to facilitate the MS4 operator's compliance with the MS4 permit. However, such controls are not identified by the NPDES permitting authority and rather are left to the choice of the industrial facility and construction site operators. Furthermore, the NPDES permitting authority imposes weak to no discharge sampling requirements on industrial facility and construction activity operators, which greatly impairs the MS4's ability to determine and control the worst regulated stormwater discharges to the MS4. Similarly, the NPDES permitting authority's general permit for construction activity encourages construction facility operators to consider post-construction stormwater controls, but it does not require them, even though the MS4 permit's programmatic measures mandate new development planning and post-construction controls as essential elements of the MS4 program. The lack of integration among stormwater permits and the absence of objective measures of compliance that are quantifiable is a glaring shortcoming in current stormwater permits and renders them difficult to enforce for water quality protection.

The California EPA State Water Board asked an expert panel to evaluate the extent of implementation success of the stormwater program in California and the feasibility of numeric effluent limits in stormwater permits. In its report (CA SWB, 2006), the panel concluded that the flexible approach of allowing a permittee to self-select SCMs for the purpose of controlling stormwater pollution was largely ineffective. The reasons stated were: (1) the SCMs were selected without proper consideration of design, performance, hydraulics, and function; (2) the MS4 permittees were not accountable for the performance of the SCMs; (3) the industrial and construction permittees were not responsible for the performance of the SCMs; and (4) the SCMs were seldom maintained properly except for aesthetic purposes. In other words, the flexibility provided by self-determination, self-evaluation, and self-reporting did not assure that SCMs were being implemented to effectively reduce stormwater pollutants to the MEP. Rather, the flexibility resulted in a lack of coordination of purpose and accountability between the MS4 permittees who owned or operate the MS4 and the industry and construction permittees who discharge to the MS4. Although typically enforcement by the permitting authority would have restored the integrity of the stormwater program, that remedy is likely to be ineffective here because the choice of SCMs is left too much to discretion and there are no quantifiable performance or design criteria for water quality purposes.

Integration and Dissemination of Authority

This section offers a near-term alternative solution to the problem cited above that utilizes the existing framework of the NPDES stormwater program. The strategy builds on the authority of MS4s over industry and construction sites to implement an integrated permitting scheme to reduce stormwater pollution into the waters of the United States. Unlike the first section of this chapter, it does not take a watershed approach to protecting water quality, even though the municipal stormwater programs may be more cost-effective if implemented on a watershed scale. It also addresses a significant shortcoming of the current scheme, that is, failure to recognize the enormous staff resources that it would take at the federal and state level for successful implementation in the absence of the leadership of local governments. Further, federal and state NPDES permitting authorities do not presently have, and can never reasonably expect to have, sufficient personnel under the principles of democratic governance, such as in the United States, to inspect and enforce stormwater regulations on more than 100,000 discrete point source facilities discharging stormwater. A better structure would be one where the NPDES permitting authority empowers the MS4 permittees, who are local governments working for the public good, to act as the first tier of entities exercising control on stormwater discharges to the MS4 to protect water quality—an approach here called "integration."

The central concept of integration is to give the MS4s controlling jurisdiction and responsibility over discharges from construction and industry to the MS4 in addition to their responsibility to implement the programmatic minimum measures identified in regulation. This approach would be similar to the current NPDES permitting scheme for publicly owned WWTPs, where a WWTP operator controls the quality of wastewater inputs (industrial waste streams) to make sure that the total output will not exceed water quality standards (see Box 6-5 on the National Pretreatment Program). The WWTP operators establish additional criteria such as local limits, require discharge monitoring of industrial wastes, and conduct inspections to make sure industrial discharges implement adequate wastewater treatment technologies, so that treated effluent from the wastewater treatment can comply with water quality standards to protect receiving waters. The same could be done for stormwater, except here the WWTP is replaced by the MS4, and the other inputs in this case are all industrial and construction discharges of stormwater into the MS4. The criteria by which the outputs of the industries are judged could be either water quality- or technology-based criteria. This arrangement puts the burden on the MS4 to identify high-risk industries because the MS4 is now responsible for the overall output (which could be, for example, the concentration of pollutants in stormwater monitored during

BOX 6-5
National Pretreatment Program

EPA's NPDES Permitting Program requires that all point source discharges to waters of the United States (i.e., "direct discharges") must be permitted. To address "indirect discharges" from industries to Publicly Owned Treatment Works (POTWs), EPA, through CWA authorities, established the National Pretreatment Program as a component of the NPDES Permitting Program. The National Pretreatment Program requires industrial and commercial dischargers to treat or control pollutants in their wastewater prior to discharge to POTWs.

In 1986, more than one-third of all toxic pollutants entered the nation's waters from POTWs through industrial discharges to public sewers. Certain industrial discharges, such as slug loads, can interfere with the operation of POTWs, leading to the discharge of untreated or inadequately treated wastewater into rivers, lakes, etc. Some pollutants are not compatible with biological wastewater treatment at POTWs and may pass through the treatment plant untreated. This "pass through" of pollutants impacts the surrounding environment, occasionally causing fish kills or other detrimental alterations of the receiving waters. Even when POTWs have the capability to remove toxic pollutants from wastewater, these toxics can end up in the POTW's sewage sludge, which in many places is land-applied to food crops, parks, or golf courses as fertilizer or soil conditioner.

The National Pretreatment Program is unique in that the general pretreatment regulations require all large POTWs (i.e., those designed to treat flows of more than 5 MGD) and smaller POTWs with significant industrial discharges to establish local pretreatment programs. These local programs must enforce all national pretreatment standards (effluent limitations) and requirements, in addition to any more stringent local requirements necessary to protect site-specific conditions at the POTW. More than 1,500 POTWs have developed and are implementing local pretreatment programs designed to control discharges from approximately 30,000 significant industrial users.

EPA has supported the pretreatment program through development of more than 30 manuals that provide guidance to EPA, states, POTWs, and industry on various pretreatment program requirements and policy determinations. Through this guidance, the pretreatment program has maintained national consistency in interpretation of the regulations.

The general pretreatment regulations establish responsibilities of federal, state, and local government, industry, and the public to implement pretreatment standards to control pollutants that pass through or interfere with POTW treatment processes or that may contaminate sewage sludge. The general pretreatment regulations apply to all non-domestic sources that introduce pollutants into a POTW. These sources of "indirect discharge" are more commonly referred to as industrial users (IUs). Since IUs can be as simple as an unmanned coin-operated car wash to as complex as an automobile manufacturing plant or a synthetic organic chemical producer, EPA developed four criteria that define a significant industrial user (SIU). Many of the general pretreatment regulations apply to SIUs as opposed to IUs, based on the fact that control of SIUs should provide adequate protection of the POTW.

Unlike other environmental programs that rely on federal or state governments to implement and enforce specific requirements, the Pretreatment Program places the majority of the responsibility on local municipalities. Specifically, Section 403.8(a) of the general pretreatment regulations states that any POTW (or combination of treatment plants operated by the same authority) with a total design flow greater than 5 million MGD and smaller POTWs with SIUs must establish a local pretreatment program. As of early 1998, 1,578 POTWs were required to have local programs. Although this represents only about 15 percent of the total treatment plants nationwide, these POTWs account for more than 80 percent (i.e., approximately 30 billion gallons a day) of the national wastewater flow.

Consistent with Section 403.8(f), POTW pretreatment programs must contain the six minimum elements described below (EPA, 1999):

1. Legal Authority

The POTW must operate pursuant to legal authority enforceable in federal, state, or local courts, which authorizes or enables the POTW to apply and enforce any pretreatment regulations developed pursuant to the CWA. At a minimum, the legal authority must enable the POTW to:

i. deny or condition discharges to the POTW,

ii. require compliance with pretreatment standards and requirements,

iii. control IU discharges through permits, orders, or similar means,

iv. require IU compliance schedules when necessary to meet applicable pretreatment standards and/or requirements and the submission of reports to demonstrate compliance,

v. inspect and monitor IUs,

vi. obtain remedies for IU noncompliance, and

vii. comply with confidentiality requirements.

2. Procedures

The POTW must develop and implement procedures to ensure compliance with pretreatment requirements, including:

i. identify and locate IUs subject to the pretreatment program,

ii. identify the character and volume of pollutants contributed by such users,

iii. notify users of applicable pretreatment standards and requirements,

iv. receive and analyze reports from IUs,

v. sample and analyze IU discharges and evaluate the need for IU slug control plans,

vi. investigate instances of noncompliance, and

vii. comply with public participation requirements.

3. Funding

The POTW must have sufficient resources and qualified personnel to carry out the authorities and procedures specified in its approved pretreatment programs.

4. Local Limits

The POTW must develop local limits or document why those limits are not necessary.

5. Enforcement Response Plan (ERP)

The POTW must develop and implement an ERP that contains detailed procedures indicating how the POTW will investigate and respond to instances of IU noncompliance.

6. List of SIUs

The POTW must prepare, update, and submit to the approval authority a list of all significant industrial users (SIUs).

In addition to the six specific elements, pretreatment program submissions must include:

- A statement from the city solicitor (or the like) declaring the POTW has adequate authority to carry out program requirements;
- Copies of statutes, ordinances, regulations, agreements, or other authorities the POTW relies upon to administer the pretreatment program, including a statement reflecting the endorsement or approval of the bodies responsible for supervising and/or funding the program;

continues next page

BOX 6-5 Continued

- A brief description and organizational chart of the organization administering the program; and
- A description of funding levels and manpower available to implement the program.

The objectives of the National Pretreatment Program are achieved by applying and enforcing three types of discharge standards: (1) prohibited discharge standards, (2) categorical standards, and (3) local limits.

Prohibited Discharge Standards

All IUs, whether or not subject to any other national, state, or local pretreatment requirements, are subject to the general and specific prohibitions identified in 40 C.F.R. §§403.5(a) and (b), respectively. General prohibitions forbid the discharge of any pollutant(s) to a POTW that cause pass-through or interference. These prohibited discharge standards are intended to provide general protection for POTWs. Examples of these include prohibitions on discharges of pollutants that can create fire or explosion hazards, cause corrosive structural damage, obstruct flow within the POTW, and interfere with the POTW's biological treatment activity. However, their lack of specific pollutant limitations creates the need for additional controls, namely categorical pretreatment standards and local limits.

Categorical Standards

Categorical pretreatment standards (i.e., categorical standards) are national, uniform, technology-based standards that apply to discharges to POTWs from specific industrial categories (i.e., indirect dischargers) and limit the discharge of specific pollutants. Categorical pretreatment standards for both existing and new sources are promulgated by EPA pursuant to Section 307(b) and (c) of the CWA. Limitations developed for indirect discharges are designed to prevent the discharge of pollutants that could pass through, interfere with, or otherwise be incompatible with POTW operations. The categorical pretreatment standards can be concentration based or mass based. For example, the pretreatment standard for the electrical and electronic component manufacturing industry (40 C.F.R. Part 469, Subparts A-D) are concentration-based daily maximum and monthly average limits that vary by subpart and pollutant parameter.

Local Limits

Prohibited discharge standards are designed to protect against pass-through and interference generally. Categorical pretreatment standards, on the other hand, are designed to ensure that IUs implement technology-based controls to limit the discharge of pollutants. Local limits, however, address the specific needs and concerns of a POTW and its receiving waters. Federal regulations at 40 CFR §§403.8(f)(4) and 122.21(j)(4) require control authorities to evaluate the need for local limits and, if necessary, implement and enforce specific limits as part of pretreatment program activities. Local limits are developed for pollutants (e.g., metals, cyanide, BOD_5, TSS, oil and grease, organics) that may cause interference, pass-through, sludge contamination, and/or worker health and safety problems if discharged in excess of the receiving POTW treatment plant's capabilities and/or receiving water quality standards.

events). If put in this position, municipalities will make intelligent choices and adopt effective strategies to identify which industries and sources to focus upon. Each of these issues is discussed in greater detail below.

Determination of High-Risk Dischargers

At present, the federal stormwater regulations do not specifically identify which sources would be considered high risk given the common pollutants in MS4 stormwater discharges. With the exception of the category of municipal landfills and hazardous waste treatment, storage, and disposal facilities, it does not even state that the other nine categories of industry singled out in the regulations for permitting under the multi-sector industrial stormwater general permit (MSGP) are really high risk. The devolution of this responsibility to the municipality is sensible because the municipality, as the land-use authority, already conducts development review and issues industrial conditional-use permits. The permitting authority would still be responsible for inspecting high-risk state, federal, and other facilities over which the MS4 permittee has no jurisdiction. In addition, the permitting authority would inspect municipal facilities such as airports, ports, landfills, and waste storage facilities to avoid the situation of self-inspection. Methods for ranking industries according to risk are discussed in a subsequent section.

It is likely that some of the designated high-risk facilities would be better regulated by individual stormwater NPDES permits. In particular, good candidates for individual NPDES permits include international ports, airports, and multiphase construction land developments, which are similar (in the potential risk they pose to water quality) to traditional major wastewater facilities such as petroleum refineries and large POTWs.

SCM Design Parameters, Numerical SCM Performance Criteria, and Monitoring

For the integration approach to work, the permitting authority and the MS4 permittee must better delineate SCM design parameters, numerical performance criteria, and default SCMs based on best available technology or water quality standards for the discharge of industrial and construction stormwater. Both the ASCE International Storm Water Database (which is now called the WERF International Storm Water Database because it is maintained by the Water Environment Research Foundation) and the National Stormwater Quality Database (NSQD), which were developed with EPA funding, are comprehensive datasets that can be used to develop numeric technology-based effluent criteria or limits for industrial and construction stormwater discharges. The MS4 can then determine the compliance of industry and construction activity with its requirements by using either some numeric criteria or a suite of SCMs that have been presumptively determined as capable of achieving the performance criteria. The EPA MSGP includes a general list of sector-specific SCMs, but these presently have no performance criteria associated with them. It is important that the EPA continue to support both the WERF and the NSQD databases as the repositories of SCM performance and MS4 monitoring data, so that MS4s can use them to establish local limits and update the performance criteria periodically to fully effectuate the iterative approach to ensuring that MS4 discharges eventually will meet water quality standards.

The proposed integration scheme will also facilitate the MS4 permittee's implementation of a purpose-oriented stormwater monitoring program directed toward identifying problematic industrial or construction stormwater discharges or high-risk industrial facility sectors. The current benchmark monitoring conducted by MSGP facilities would be eliminated. Instead, MSGP facilities would have the option of performing scientifically valid stormwater discharge sampling to demonstrate their compliance with performance criteria or to participate in an MS4-led monitoring program by paying in lieu fees to support the cost of the purpose-oriented MS4 monitoring program. The net effect of this alternative is to pool the resources to come up with an optimal sampling strategy to replace what is now a stormwater monitoring strategy that is haphazard and not useful.

MS4 Responsibilities

Under integration, the MS4 permittee would be primarily responsible for the quality of stormwater discharges that exit the MS4 to the waters of the United States. The MS4 permittee would not be responsible for stormwater discharges from federal and state facilities or for facilities that have been issued an individual NPDES permit for stormwater discharges. The MS4 permittee would be responsible for implementing the six minimum program measures, assisting in the oversight and inspection of facilities covered under the MSGP and the construction general permit (CGP), and implementing a strategic water quality monitoring program to identify and control pollutant discharges from high-risk sites. The permitting authority would share any fees collected under the MSGP and CGP with the MS4, and facilities covered by them would have the option to opt-out of self-monitoring and contribute equivalent funds to an MS4-led monitoring program. Similarly, the permitting authority would be expected to support research and special studies that address issues of regional or national significance through partnerships with the MS4 permittees.

Some MS4s may balk at taking on more responsibility for the control of stormwater pollution, as required for integration to succeed. However, there are already several case examples that exist. The State of Oregon requires facilities that discharge industrial stormwater to file a Notice of Intent (NOI) for coverage under the MSGP with both the state and the local MS4 (Campbell, 2007). The state has an agreement with the local MS4s for the inspection of the facilities covered under the MSGP and the sharing of NOI fees. The State of Tennessee has a statewide pilot program to partner with local MS4s for the inspection of construction sites that are covered under the CGP.

Analogy to the WWTP Pretreatment Program

It is certainly true that the MS4s are a more challenging point source to regulate for the discharge of pollutants than WWTPs. WWTPs have fewer out-falls discharging to waters of the United States than MS4s, and inputs into them are through discrete rather than diffuse sources as in the case of MS4s. It is thus expected to be more difficult to identify problem stormwater sources and to hold them accountable for discharges in excess of standards. This problem is not insurmountable, however. Watershed and land-use hydrologic models can be developed and refined by strategic sampling of pollutant sources for use by MS4 permittees and regulatory agencies. If EPA and state permitting authorities es-tablish measurable outcomes as expected endpoints of progress, MS4 permittees will make intelligent choices about which measures to implement in order to meet these endpoints. In large part, the lack of progress nationally towards con-trolling pollutants in stormwater discharges from the MS4s has been due to the absence of national SCM design standards, MS4 discharge performance criteria, and stormwater effluent guidelines. Presently, the MS4 permittees as owners and operators of the MS4 affirmatively approve connections to the conveyance system for rainfall runoff. Historically the issuance of the MS4 connection per-mit has been based on the sizing of the pipes for the conveyance of flood waters. There are few barriers to including water quality considerations in reauthorizing these connections and adding new ones.

Note that EPA did initially consider using the WWTP pretreatment ap-proach for stormwater discharges by requiring MS4 permittees to be primarily responsible for discharges of stormwater associated with industrial activity through the MS4 (53 Fed. Reg. 49428; December 7, 1988). However, EPA de-viated from this approach in issuing its Final Storm Water Rule (55 Fed. Reg. 48006; November 16, 1990). In the absence of regulations that specifically con-fer authority on MS4 permittees to establish local limits for stormwater dis-charges to the MS4 from industry and businesses, the EPA should promulgate specific SCMs and performance guidelines with rigorous requirements for self-monitoring and compliance in order to support the integrated framework for controlling stormwater pollution from MS4s.

Potential Legal Barriers

A revised stormwater program that requires MS4s to play a more significant role in enforcement and oversight and that provides greater specificity in permit requirements is not only contemplated, but arguably demanded by Congress in the CWA. Specifically, Congress directs that MS4 permits be conditioned on the requirement that the MS4s "shall require controls to reduce the discharge of pollutants to the maximum extent practicable" 42 U.S.C. § 1342(p)(3)(B)(iii). EPA has already conditioned Phase I MS4 permits on the requirement that the municipality establish that it has the legal authority to inspect discharges into the system and take regulatory and enforcement action against excessive or violating sources [40 C.F.R. § 122.26(d)(2)(i)]. Nevertheless, to ensure that MS4s play an even more active role, EPA should include several additional requirements in its implementing regulations. In addition to promulgating more detailed and specific SCM requirements as discussed above, EPA should also require that the Phase I MS4s establish that they possess sufficient funding and staff to effectuate their responsibilities [see, e.g., 40 C.F.R. § 403.8(f)(2) and (3) requiring this showing for the POTW program]. Like the POTW program, states should also be authorized as MS4 permittees when the local governments are unable or unwilling to carry out their mandatory stormwater permit responsibilities [see, e.g., 40 C.F.R. § 403.10(e) providing this authority for the POTW program].

Industrial Program

The industrial stormwater permit program presently incorporates a menu of SCMs that are to be selected by the facility operator, a rudimentary monitoring program that includes visual observations, some water quality sampling for selected parameters for certain types of industries subject to numerical effluent limitations (see Table 2-6) or a set of pollutant-level benchmarks that are to be used as a measure to appropriately revise the SWPPP (see Table 2-5), and annual reporting. Neither SCM performance criteria nor the characteristics of a design storm for water quality purposes have been established. Given the broad discretion that facility operators enjoy as a result, it has been difficult to gauge compliance with the MSGP and initiate enforcement for non-compliance even though industrial stormwater discharges are required to meet effluent limitations (technology- or water quality-based) that reflect water quality standards (Duke and Beswick, 1997; Duke and Augustenborg, 2006; Wagner, 2006). Several ideas to address some of the shortcomings in the implementation of the permitting program for industrial stormwater discharges are offered as *additions* to the concept of MS4 regulatory integration discussed previously. They would substantively improve the current industrial stormwater permitting program even if the integration recommendations were not acted upon.

Criteria for a Water Quality Design Storm and Subsequent SCM Selection

To improve the quality of stormwater discharges from industry, provide for better accountability, and advance the objectives of the CWA, it is important first to identify the criteria for a water quality design storm as opposed to one for flood control design, where the objective is to protect human life and real property. It is important that the permitting authority designate the basis for the determination of the water quality design storm, and explicitly state that it would form the criteria for evaluation of compliance with technology-based standards or water quality-based standards. This is essential because the engineering design decisions that determine how much stormwater is to be treated to remove toxic pollutants that pose a risk to human health or aquatic life is more a policy matter than a scientific one (Schiff et al., 2007). While modeling exercises using continuous simulation methods in theory could be performed for every project or subwatershed or region to support planning decisions on how much stormwater needs to be treated for optimum water quality benefits, such a detailed analysis will be too cumbersome and cost-prohibitive for routine planning and implementation purposes. Thus it is recommended that the EPA establish guidelines for the selection of water quality design storms for controlling pollution from MS4 and industrial stormwater discharges. This would not be a new practice for EPA because the agency has previously established design storms for certain industrial sectors when promulgating effluent guidelines (Table 2-6). Conceivably, unlike the technology limiting design storms that are set on rainfall recurrence intervals, the design storm to protect surface water quality and beneficial uses could be different for different eco-regions of the United States.

The water quality design storm, which may be expressed as total rainfall depth, runoff volume, or rainfall intensity, incorporates the concept that extreme rainfall events are rare, and that a few times each year the runoff volume or flow rate from a storm will exceed the design volume or rate capacity of an SCM. Therefore, for the purpose of best available technology and cost-effectiveness, industrial facility operators should not be held accountable for pollutant removal from storms beyond the size for which an SCM is designed.

For MS4 operators, the concept of designing MS4s for both flood control conveyance (capital flood design) and for water quality protection (water quality design) involves a fundamental shift. Whereas flood control engineers design conveyance systems with return frequencies of two years (streets), ten years (detention basins), 50 years, and 100 years (channels), the water quality design storm event is for a return frequency of six months to a year. The water quality design implicitly focuses on treating the first flush of runoff, which contains the highest load and concentration of pollutants and which occurs in the first half to one inch of runoff. In contrast, flood control designs are built to convey tens of inches of runoff.

In addition to issuing the guidelines to support the setting of stormwater criteria for water quality design, it is important that the EPA establish SCM performance criteria based on best technologies and identify the "presumptive technologies" that have been demonstrated to achieve the performance criteria. The water quality design storm and the best available technologies with their associated criteria can then form a basis for technology-based effluent limitations to be included in industrial stormwater permits. If the facility operator elects the identified presumptive technology, then compliance monitoring requirements can be scaled down to a minimum to ensure that the treatment systems are being properly maintained. On the other hand, if the operator elects to go with a suite of alternative SCMs, then the monitoring requirements sufficient to demonstrate that the suite of alternative SCMs are in fact achieving the effluent quality of the selected technology can be prescribed. In such a scheme, visual monitoring will serve to ensure that the treatment systems are being properly maintained, and compliance can be reported using the same procedures as required presently for the industrial wastewater permits.

How to Identify a High-Risk Industry

Both the watershed-based permitting approach described previously in this chapter and the integration approach call for municipal permittees, as part of their responsibilities, to identify high-risk industrial stormwater dischargers. This involves identifying the potential sources of concern, evaluating the extent of their potential impacts, and then prioritizing them for attention—a classic risk assessment. Municipalities would generally not be able to give equal and full attention to all sources, nor should they. Unfortunately, what constitutes high risk or any level of risk for industries covered by NPDES stormwater permits has not been defined by EPA, although the states have developed various interpretations (see Appendix C).

Two methodologies for identifying industrial and commercial facilities that are considered high-risk for discharging pollutants in stormwater are presented below. Box 6-6 describes the "intensity of industrial activity" method devised for the City of Jacksonville (Duke, 2007). This method uses telephone queries and a point scale system to visually score each facility based on the intensity of the industrial activities exposed to stormwater, and groups the results into categories A, B, C, or D in increasing order of intensity (Cross and Duke, 2008). The categories are designed to distinguish high-risk facilities from low-risk facilities, and not to make fine distinctions among facilities with similar characteristics. This typology is sufficient to distinguish facilities with little or no potential for discharging pollutants associated with stormwater from facilities that might discharge those pollutants. More than half of the facilities that were subject to Florida's MSGP were determined to be low-risk (Cross and Duke, 2008).

Box 6-7 outlines an empirical methodology used by the County of Los Angeles to rank the risk of industrial facilities for stormwater pollution on the basis of pollution potential P. The pollution potential P was computed as a product of the number of on-site sources, percent imperviousness, pollutant toxicity, degree of exposure, and the number of facilities (Los Angeles County, 2001). Based on this ranking scheme, five top high-risk industries were selected: (1) automobile dismantlers, (2) automobile repair, (3) metal fabrication, (4) motor freight, and (5) automobile dealers. Stormwater discharges from six facilities in each category were characterized over a two-year period, and the effectiveness of SCMs

BOX 6-6
Risk Assessment for Industrial Dischargers of Stormwater

The City of Jacksonville has had very good success in determining what industries pose the highest stormwater risks by starting with businesses having the Standard Industrial Classification (SIC) codes designated for permit coverage but using multiple lists of potential sources and cross checking them to target inspections and other interventions where they will have the best effect. Other clues to sources of interest include other environmental permits (e.g., wastewater NPDES permits, permits for discharge to sanitary sewer), tax records, records of fire code inspections, building permit filings, planning agency proceedings, contacts with business associations, marketing information put out by companies, Resource Conservation and Recovery Act hazardous waste reports, and telephone and field surveys.

Duke (2007) proposed a 0- to 8-point scoring scheme (shown below) to rate the intensity of industrial activities exposed to stormwater. The system is based on the relative amount of exposure to precipitation and runoff by industrial materials, processes, wastes, and vehicles. Once municipalities gather the data and then classify their industries accordingly, they would have a very useful tool to program inspections and monitoring emphasizing the industries most risking their success in achieving established objectives. A similar system could and should be developed for construction sites.

<u>0 points</u>
Small bulk waste, e.g., covered dumpster: area <100 m^2
Hazardous waste: containers not exposed to precipitation

<u>1 point</u>
Outdoor vehicle use: 1-2 vehicles, outdoors occasionally/never, not used in precipitation
Vehicle washing outdoors, 1-2 vehicles, rarely or occasionally done

<u>2 points</u>
Outdoor vehicles, e.g., forklifts: 1-2, outdoors occasionally/never, used in precipitation
Outdoor vehicles, e.g., forklifts: 1-2, outdoors every day, not used in precipitation
Outdoor vehicles, e.g., forklifts: 3-4, outdoors occasionally/never, not used in precipitation
Vehicle maintenance or re-fueling, 1-2 vehicles, rarely or occasionally done, outside
Vehicle washing outdoors, 1-2 vehicles, regularly done
Vehicles washing outdoors, 3 vehicles, rarely or occasionally done

<u>4 points</u>
Storage of materials or products: area < 100m^2 and/or < five 55-gallon drums
Fixed outdoor equipment: 1-2 small or large item(s)
Outdoor vehicles, e.g., forklifts: 1-2, outdoors every day, used in precipitation
Outdoor vehicles, e.g., forklifts: 3-4, outdoors occasionally/never, used in precipitation
Outdoor vehicles, e.g., forklifts: 3-4, outdoors every day, not used in precipitation
Uncovered shipping/receiving area: 1-2 docks
Vehicle maintenance or re-fueling outdoors, 1-2 vehicles, regularly done
Vehicle maintenance or re-fueling outdoors, vehicles, rarely or occasionally done
Plant yard, rail lines, access roads: 1,000 ft^2
Small process equipment, e.g., compressors, generators: exposed to precipitation

continues next page

BOX 6-6 Continued

6 points
   Outdoor vehicles, e.g., forklifts: 3-4, outdoors every day, used in precipitation
   Outdoor vehicles, e.g., forklifts: > 5 or heavy, outdoors occasionally, used in precipitation
   Outdoor vehicles, e.g., forklifts: > 5 or heavy, outdoors every day, not used in precipitation
   Vehicle maintenance or re-fueling outdoors, 3 vehicles, regularly done
   Plant yard, rail lines, access roads: 1,000 ft^2

8 points
   Storage of materials or products: area 100^2 and/or five 55-gallon drums
   Boneyard of scrap, disused equipment, similar
   Hazardous waste: containers exposed to precipitation
   Fixed outdoor equipment: small or 2 large items
   Outdoor vehicles, e.g., forklifts: > 5 or heavy, outdoors every day, used in precipitation
   Uncovered shipping/receiving area: 3 docks
   Plant yard, rail lines, access roads: 5,000 ft^2
   Manufacturing activities, e.g., cutting, painting, coating materials: exposed to precipitation

SOURCE: Duke (2007).

---

was assessed at a subset of them. However, the monitoring was minimal, and so much of the prioritization was based on best professional judgment about pollutant discharges.

Industrial Stormwater Discharge Monitoring

Monitoring data from Phase I MS4s have been compiled in the NSQD for several years, making possible a number of important findings about the quality of municipal stormwater (see Chapter 3). Although industry that occurs within MS4s is technically included in the NSQD, the data are lumped together and not sector specific. There is no comparable, reliable source of data specifically on industrial discharges, even though EPA requires benchmark monitoring for MSGP industrial permittees. The intent was that industrial facility operators would use benchmark exceedances as action levels to improve SCMs, but this self-directed approach has been largely a failure. Many industrial facilities reported repeated exceedances of benchmark values without action, and others have failed to report any monitoring data at all. In addition, the representativeness of single grab samples taken to characterize the discharge and less-than-rigorous sample collection and quality assurance procedures have resulted in monitoring data that are not very useful. One of the only analyses of benchmark monitoring data ever done evaluated California's program between 1992 and 2001 (see Box 4-2; Stenstrom and Lee, 2005; Lee et al., 2007). The study showed no relationship between facility type and stormwater discharge quality. The cited reasons for the poor relationship included variability in sampling parameters, sampling time, and sampling strategy—that is, poor data.

BOX 6-7
Los Angeles County Critical Facilities Monitoring Data

One of the few sources of data on industrial stormwater discharges comes from the County of Los Angeles. A stepwise process was used to identify the highest-risk industrial/commercial facilities, which were then monitored to measure the quality of their stormwater discharges and to evaluate the effectiveness of SCMs. The initial list of candidate facilities was identified from their relative numbers and the extent of their outdoor activities. This list was then refined using an empirical equation for pollutant potential P:

$P = Q \times R \times T \times E \times N$

where

Loading (Q) is the number of sources at a site and the likelihood of release;
Imperviousness (R) of a site is the percent of paved area;
Pollutant toxicity (T) denotes the number of toxic pollutants and the inherent toxicity of the mix;
An exposure factor (E) signifies if activities are exposed to rainfall; and
The Number (N) represents the total number of sites in the county.

Each variable was assigned a qualitative number from 1 to 10, with 10 representing the worst condition.

Based on this equation, five top "critical source" industries were determined: (1) automobile dismantlers; (2) automobile repair; (3) metal fabrication; (4) motor freight; and (5) automobile dealers. Six facilities from each of these categories were monitored during five storms a year for two years. The stormwater discharge samples were analyzed for general conventional pollutants, heavy metals, bacteria, and semi-volatile organic compounds. Half of the facilities were then fitted with SCMs, which were monitored to evaluate their effectiveness.

The highest median values were observed for total zinc (approx. 450 µg/L), dissolved zinc (approx. 360 µg/L), total copper (approx. 240 µg/L), and dissolved copper (approx. 110 µg/L) in stormwater discharges from fabricated metal sites. However, levels for total and dissolved zinc did not appear to be significantly different among the industry types. SCMs in the form of good housekeeping and spill containment measures were installed at half of the sites. For total and dissolved zinc, the median concentration lowered or stayed nearly the same with the implementation of SCMs at the auto dismantling, auto repair, and fabricated metals industries (i.e., in none of the circumstances was the difference significant). For total and dissolved copper, however, where the fabricated metal industry had displayed the highest median concentrations, levels were significantly reduced with the implementation of SCMs. The auto dismantling and auto repair businesses showed no significant differences in copper after the implementation of SCMs.

SOURCE: Los Angeles County (2001).

In the past, it has been proposed to EPA that it fund a project that would systematically collect the benchmark monitoring data across the nation, as has been done for MS4s, but these suggestions have been rejected. To get better data from specific industrial sectors, it is recommended that a small subset of industrial users and sectors be selected for composite sampling in a program directed by the MS4. Alternatively, making a trained team responsible for monitoring of small-business industrial dischargers would reduce, if not eliminate, current problems with quality assurance.

Monitoring of industrial stormwater discharges could be streamlined by considering the adoption of a Reasonable Potential Analysis (RPA), which is already part of the existing practice in developing limits for NPDES wastewater permits (EPA, 1991). The RPA is a procedure that uses statistical distribution assumptions in association with a limited number of wastewater discharge quality measurements to determine the likelihood that a receiving water quality standard would be violated, which assists the permitting authority in determining what permit limitations should be set to protect receiving water quality. The effluent data from any treatment system may be described using standard descriptive statistics such as the mean concentration and the coefficient of variation. Using a statistical distribution such as the lognormal, an entire distribution of values can be projected from limited data; limits on pollutant concentrations in discharge can then be set at a specified probability of occurrence so that the receiving water is protected. An RPA for stormwater pollutants may be particularly relevant in developing performance criteria for SCMs for facilities discharging stormwater within the integrated framework of MS4 permitting. Also, MS4 permittees could use the method to reduce the number of pollutants that high-risk industries would be required to monitor in order to demonstrate to the municipality that they are not the source of pollutants in MS4 discharges that are impairing surface waters.

Construction Program

The recommendations for stormwater discharges associated with construction activity are very similar to those offered for stormwater discharges associated with industrial activity. The integration with the MS4 program is less of a challenge because municipalities have always had primacy on land development planning and construction activity. Most municipalities have had requirements for soil erosion and sediment control plans on construction sites that precede the federal stormwater regulations. EPA regulations already allow permitting authorities to approve Phase I and Phase II MS4 permittee oversight of CGP construction sites under the qualifying local program provision (40 C.F.R. 122.44(s)) (Grumbles, 2006). The weakness in the implementation of this provision currently is the absence of rigorous SCM performance criteria guidelines for MS4s permittees to meet in order to be deemed as qualifying.

The construction stormwater general permit program requires the development and implementation of an SWPPP. The SWPPP, which must be prepared before construction begins, focuses on two major requirements: (1) describing the site adequately and identifying the sources of pollution to stormwater discharges associated with construction activity on site and (2) identifying and implementing appropriate measures to reduce pollutants in stormwater discharges to ensure compliance with the terms and conditions of this permit. The SWPPP must describe the sequence of major stormwater control activities and the kinds of SCMs that will be in place, and it must identify interim and permanent stabilization practices, including a schedule of their implementation. There is an expectation that the construction site operator will use good site planning, preserve mature vegetation, and properly stage major earth-disturbing activities to avoid sediment loss and prevent erosion. Post-construction stormwater controls need to be considered, but are not required. Construction site operators are required to visually inspect the construction site weekly and perform a walk through before predicted storm events. No annual reports are required, but records must be kept for a period of three years after permit coverage has been terminated. There are no SCM performance criteria, other than a suggestion that most SCMs should be able to achieve 80 percent TSS removal. As with industry, it is difficult to gauge compliance with the CGP except when inadequate SCMs result in a massive discharge of sediment from a construction site.

The pollutant parameters that are of concern in stormwater discharges from construction activity are TSS, settleable solids, turbidity, and nutrients from erosion; pH from concrete and stucco; and a wide range of metallic and organic pollutants from construction materials, processes, wastes, and vehicles and other motorized equipment. The permitting authority, in addition to guidelines for the water quality design storm, must establish SCM performance criteria for stormwater discharges associated with construction activity. The construction site operator should be given the option of implementing SCMs that are the presumptive technology, or equivalent SCMs that can achieve the performance criteria. For example, the recommended SCMs in Box 5-3 could serve as the presumptive construction SCMs on a typical construction site that is less than 50 acres in size. If the operator elects to go with a suite of alternative SCMs, then adequate monitoring must be performed to demonstrate that the alternative SCMs are in fact achieving the performance criteria. In addition, the CGP presently does not mandate or require that post-construction SCMs be integrated with the MS4 permittee requirements under its New Development/Redevelopment Program requirements. The proper planning for and implementation of SCMs that will help mitigate stormwater pollution from planned future use of the site will be critical to protecting water quality. Thus the post-construction requirements of the CGP should be strengthened and better integrated with the new development/redevelopment requirements of the MS4 permits.

Municipal Program

Several key enhancements to the MS4 permitting program are needed to ensure that resources are targeted to achieve the greatest on-the-ground implementation of SCMs to make incremental progress in meeting water quality standards. Six specific issues are discussed below; their implementation will require greater collaboration and flexibility among regulators and permitted parties. These recommendations are suggested for communities that are not ready for the integrated watershed approach proposed in the prior section, and represent a bridge toward building internal capacity to implement them.

Numeric Expression of "Maximum Extent Practicable"

The ambiguity of the term "maximum extent practicable" (MEP) has been a major impediment to achieving meaningful water quality results in the MS4 program. The EPA should develop numerical expressions of MEP in the next round of permit renewals that can be measured and tracked. A national numeric benchmark should be avoided; states should focus on regional benchmarks that are tied to their water quality problems. Four examples of methods to define MEP in a numeric manner are provided below: the first three are applied at a regional or state level, whereas the last (impervious cover-based TMDLs) offers more flexibility to be applied at individual sites.

Establish Municipal Action Levels. This approach relies on the use of a national database of stormwater runoff quality to establish reasonable expectations for outfall monitoring in highly developed watersheds. The NSQD (Pitt et al., 2004) allows users to statistically establish action levels based on regional or national event mean concentrations developed for pollutants of concern. The action level would be set to define unacceptable levels of stormwater quality (e.g., two standard deviations from the median statistic, for simplicity). Municipalities would then routinely monitor runoff quality from major outfalls. Where an MS4 outfall to surface waters consistently exceeds the action level, municipalities would need to demonstrate that they have been implementing the stormwater program measures to reduce the discharge of pollutants to the maximum extent practicable. The MS4 permittees can demonstrate the rigor of their efforts by documenting the level of implementation through measures of program effectiveness, failure of which will lead to an inference of noncompliance and potential enforcement by the permitting authority.

Site-Based Runoff and/or Pollutant Load Limits. This approach is primarily used for watersheds that are experiencing rapid development; it establishes numeric targets or performance standards for pollutant or runoff reduction that must be met on individual development sites. The numeric targets may involve specific pollutant load limits or runoff reduction volumes. For example,

Virginia DCR (2007) and Hirschman et al. (2008) established a statewide computational method to ensure that SCMs are sized, designed, and sequenced to comply with specific nutrient-based load and runoff reduction limits. The nutrient load limits of 0.28 lb/acre/yr for total phosphorus and 2.68 lb/acre/yr for total nitrogen were computed using the Chesapeake Bay Model for Virginia tributaries to the bay. The design process also requires the computation of runoff reduction volumes achieved to promote the use of nonstructural SCMs. The basic concept is that new development on non-urban land must not exceed the average annual nutrient load and runoff volume for non-urban land using effective SCMs in the watershed. This blended site-based runoff and load limit approach has been advocated by the Office of Inspector General (2007) and Schueler (2008a) and is under active consideration by several other Chesapeake Bay states.

Wenger et al. (2008) reports on a no-net-hydrologic-increase strategy to protect endangered fish species in the northern Georgia Piedmont that sets specific on-site runoff reduction requirements for a range of land uses and design storm events. A similar approach has been incorporated into the recently enacted Energy Independence and Security Act of 2007 that contains provisions that require that the "sponsor of any development or redevelopment project involving a Federal facility with a footprint that exceeds 5,000 square feet shall use site planning, design, construction, and maintenance strategies for the property to maintain or restore, to the maximum extent technically feasible, the predevelopment hydrology of the property with regard to the temperature, rate, volume, and duration of flow."

The challenge of defining MEP as a runoff reduction or pollutant load limit is that considerable scientific and engineering analysis is needed to establish the performance standards, evaluate SCM capability to meet them, and devise a workable computational approach that links them together at both the site and watershed levels. In addition, care must be taken to define an appropriate baseline to represent predevelopment conditions that does not unduly penalize redevelopment projects or make it impossible to comply with limits at new development sites after maximum effort to apply multiple SCMs is made.

Turbidity Limits for Construction Sites. Numeric enforcement criteria can be used to define what constitutes an egregious water quality violation at construction sites and provide a technical criterion to measure the effectiveness of erosion and sediment control practices. Currently, most states and localities do not specify either numeric enforcement criteria or a monitoring requirement within their CGP (see the survey data contained in Appendix C).

A maximum turbidity limit would establish definitive criteria as to what constitutes a direct sediment control violation and trigger an assessment for remediation and prevention actions. For example, local erosion and sediment control ordinances could establish a numeric turbidity limit of 75 Nephelometric Turbidity Units (NTU) as an instantaneous maximum for rainfall events less than an inch (or a 25 NTU monthly average) and would prohibit visible sediment in water discharged from upland construction sites. While the exact turbidity limit would need to be derived on a regional basis to reflect geology, soils, and receiving water sensitivity, research conducted in the Puget Sound of Washington indicates that turbidity limits in the 25 to 75 NTU can be consistently achieved at most highway construction sites using current erosion and sediment control technology that is properly maintained (Horner et al., 1990). If turbidity limits are exceeded, a detailed assessment of site conditions and follow-up remediation actions would be required. If turbidity limits continue to be exceeded, penalties and enforcement actions would be imposed. Enforcement of turbidity limits could be performed either by state, local, or third party erosion and sediment control inspectors, or—under appropriate protocols, training, and documentation—by citizens or watershed groups.

Impervious Cover Limits and IC-based TMDLs. MS4s that discharge into TMDL watersheds also require more quantitative expression of how MEP will be defined to reduce pollutant loads to meet water quality standards. Maine, Vermont, and Connecticut have recently issued TMDLs that are based on impervious cover rather than individual pollutants of concern (Bellucci, 2007). In such a TMDL, impervious cover is used as a surrogate for increased runoff and pollutant loads as a way to simplify the urban TMDL implementation process. Impervious cover-based TMDLs have been issued for small subwatersheds that have biological stream impairments associated with stormwater runoff but no specific pollutant listed as causing the impairment (in most cases, these subwatersheds are classified as impacted according to the Impervious Cover Model [ICM]—see Box 3-10). A specific subwatershed threshold is set for effective impervious cover, which means impervious cover reductions are required through removal of impervious cover, greater stormwater treatment for new development, offsets through stormwater retrofits, or other means.

Traditional pollutant-based TMDLs would continue to be appropriate for "non-supporting" and "urban drainage" subwatersheds, although they could be modified to focus compliance monitoring on priority urban source areas or subwatersheds that produce the greatest pollutant loads. Although EPA (2002) indicates that this analysis does not extend to demonstrating that changes will occur in receiving waters, it does outline a rigorous process for evaluating pollutant discharges and SCM performance. More recent EPA guidance (2007c) recommends that MS4s conduct a four-step analysis, which is distilled to its essence below:

Step 1: Estimate loads for pollutant of concern for the watershed.

Step 2: Provide a specific list of SCMs that will be applied in the listed watershed.

Step 3: Estimate the pollutant removal capability of the individual SCMs applied.

Step 4: Compute aggregate watershed pollutant reduction achieved by the MS4.

Although this is not a particularly new interpretation of addressing stormwater loads in watersheds listed as impaired and/or having written TMDLs, it is exceptionally uncommon for individual MS4s to document the link between their stormwater discharges and water quality standard exceedances, as modified by the system of SCMs that they used to reduce these pollutants. As of 2007, EPA could only document 17 TMDLs that addressed stormwater discharges using this sequential analysis. EPA and states need to provide more specific guidance for MS4s to comply with TMDLs in their permit applications and annual reports.

Focus MS4 Permit Implementation at the Subwatershed Level

Chapter 5 noted the importance of the watershed context for making better local stormwater decisions. This context can be formally incorporated into local MS4 permits by focusing implementation on a subwatershed basis, using the ICM, as described in Box 3-10 and outlined in Table 6-1. When urban streams are classified by the ICM, this basic subwatershed planning process can be used to establish realistic water quality and biodiversity goals for individual classes of subwatersheds, as shown in Table 6-2. As can be seen, goals for water and habitat quality become less stringent as impervious cover increases within the subwatershed. This subwatershed approach provides stormwater managers with more specific, measurable, and attainable implementation strategies than the one-size-fits-all approach that is still enshrined in current wet-weather management regulations.

Some examples of how to customize stormwater strategies for different subwatersheds are described in Table 6-3. This approach enables MS4s to utilize the full range of watershed planning, engineering, economic, and regulatory tools that can manage the intensity, location, and impact of impervious cover on receiving waters. In addition, the application of multiple tools in a given subwatershed class helps provide the maximum level of protection or restoration for an individual subwatershed when impervious cover is forecast to increase due to future growth and development. The conceptual management approach shown in Table 6-3 is meant to show how urban stream classification can be used to guide stormwater decisions on a subwatershed basis. The first column of the table lists some key stormwater management issues that lend themselves to a subwatershed approach and are explained in greater detail below.

Linkage with Local Land-Use Planning and Zoning. Given the critical relation between land use and the generation of stormwater, communities should ensure that their planning tools (e.g., comprehensive plans, zoning, and watershed planning) are appropriately aligned with the intended management classification for each subwatershed. For example, it is reasonable to encourage redevelopment, infill, and other forms of development intensification within non-

TABLE 6-1 Components of Subwatershed-Based Stormwater Management

1.   Define interim water quality and stormwater goals (i.e., pollutants of concern, biodiversity targets) and the primary stormwater source areas and hotspots that cause them.

2.   Delineate subwatersheds within community boundaries.

3.   Measure current and future impervious cover within individual subwatersheds.

4.   Establish the initial subwatershed management classification using the ICM.

5.   Undertake field monitoring to confirm or modify individual subwatershed classifications.

6.   Develop specific stormwater strategies within each subwatershed classification that will guide or shape how individual practices and SCMs are generally assembled at each individual site.

7.   Undertakes restoration investigations to verify restoration potential in priority subwatersheds.

8.   Agree on the specific implementation measures that will be completed within the permit cycle. Evaluate the extent to which each of the six minimum management practices can be applied in each subwatershed to meet municipal objectives.

9.   Agree on the maintenance model that will be used to operate or maintain the stormwater infrastructure, assign legal and financial responsibilities to the owners of each element of the system, and develop a tracking and enforcement system to ensure compliance.

10. Define the trading or offset system that will be used to achieve objectives elsewhere in the local watershed objectives in the event that full compliance cannot be achieved due to physical constraints (e.g., indexed fee-in-lieu to finance municipal retrofits).

11. Establish sentinel monitoring stations in subwatersheds to measure progress towards goals.

12. Revise subwatershed management plans in the subsequent NPDES permitting cycle based on monitoring data.

supporting or urban drainage subwatersheds, whereas down-zoning, site-based IC caps, and other density-limiting planning measures are best applied to sensitive subwatersheds.

Stormwater Treatment and Runoff Reduction MEP. Subwatershed classification allows managers to define achievable numerical benchmarks to define treatment in terms of the maximum extent practicable. Thus, a greater level of treatment is required for less-developed subwatersheds and a reduced level of treatment is applied for more intensely developed subwatersheds. This is most frequently expressed in terms of a rainfall depth associated with a given design storm. Designers are required to treat and/or reduce runoff for all storm events up to the designated storm event. This flexibility recognizes the greater difficulty and cost involved in providing the same level of treatment in an in-

TABLE 6-2 Expectations for Different Urban Subwatershed Classes

Lightly Impacted Subwatersheds (1 to 5% IC)	• Consistently attain scores for specific indicators for hydrology, biodiversity, and geomorphology that are comparable to streams whose entire subwatersheds are fully protected in a natural state (e.g., national parks). Should provide for healthy reproduction of trout, salmon, or other keystone fish species.
Moderately Impacted Subwatersheds (6 to 10% IC)	• Consistently attain scores for specific stream indicators that are comparable to the highest 10 percent of streams in a population of rural watersheds in order to maintain or restore ecological structure, function, and diversity of the streams. The "good to excellent" indicator scores for this category of subwatersheds will be the benchmark against which the relative quality of more developed subwatersheds will be measured.
Heavily Impacted Subwatersheds (11 to 25% IC)	• Consistently attain good stream quality indicator scores to ensure enough stream function to adequately protect downstream receiving waters from degradation. • Function is defined in terms of flood storage, in-stream nutrient processing, biological corridors, stable stream channels, and other factors.
Non-Supporting Subwatersheds (26 to 60% IC)	• Consistently attain "fair to good" stream quality indicator scores. • Meet bacteria standards during dry weather and trash limits during wet weather. • Maintain existing stream corridor to allow for safe passage of fish and floodwaters.
Urban Drainage Subwatersheds (61 to 100% IC)	• Maintain "good" water quality conditions in downstream receiving waters. • Consistently attain "fair" water quality scores during wet weather and "good" water scores during dry weather. • Provide clean "plumbing" in upland land uses such that discharges of sewage and toxics do not occur.

Note: the objectives presume some portion of the subwatershed has already been developed, thereby limiting attainment of objectives. If a subwatershed is not yet developed, managers should shift expectations up one category (e.g., urban drainage should behave like non-supporting). Also, the specific ranges of IC that define each management category should always be derived from local or regional monitoring data. Note that the ranges in IC shown to define a subwatershed management category are illustrative and will vary regionally.

tensely developed subwatershed, as well as the fact that less treatment is needed to maintain stream condition in a highly urban subwatershed.

The other key element of defining MEP is to specify how much of the treatment volume must be achieved through runoff reduction. The runoff reduction volume has emerged as the primary performance benchmark to maintain predevelopment runoff conditions at a site after it is developed. In its simplest terms, this means achieving the same predevelopment runoff coefficient for each storm up to a defined storm event through a combination of canopy interception, soil infiltration, evaporation, rainfall harvesting, engineered infiltration, extended filtration, or evapotranspiration (Schueler, 2008b). Once again, the physical feasibility and need to provide treatment through runoff reduction becomes progressively harder as subwatershed impervious cover increases.

TABLE 6-3 Examples of Customizing Stormwater Strategies on a Subwatershed Basis

Stormwater Management Issue	Lightly Impacted Subwatershed (1 to 5% IC)	Moderately Impacted Subwatershed (6 to 10% IC)	Impacted (IC 11 to 25%)	Non-Supporting (IC 26 to 60%)	Urban Drainage (61% + IC)
Linkage with Local Land-Use Planning and Zoning	Utilize extensive land conservation and acquisition to preserve natural land cover	Implement site-based or watershed-based IC caps and maximize conservation of natural areas	Reduce the IC created for each zoning category by changing local codes and ordinances	Encourage redevelopment, development intensification and mass transit to decrease per-capita IC utilization in the urban landscape.	Develop watershed restoration plans to maintain or enhance existing aquatic resources.
Site-based Stormwater Reduction and Treatment Limits	Allow no net increase in runoff volume, velocity and duration up to the five-year design storm	Treat runoff from two-year design storm, using SCMs to achieve 100% runoff reduction		Treat runoff from the one-year design storm, using SCMs to achieve at least 75% runoff reduction	
Site-Based IC Fees	None	Establish Excess IC fee for projects that exceed IC for zoning category		Allow IC mitigation fee	
Subwatershed Trading	Receiving Area for Conservation Easements	Receiving Area for Restoration Projects and/or Retrofit		Receiving or Sending Area for Retrofit	Sending Area for Restoration Projects
Stormwater Monitoring Approach	Measure in-stream metrics of biotic integrity		Track subwatershed IC and measure SCM performance	Check outfalls and measure SCM performance	Check stormwater quality against municipal actions levels at outfalls
TMDL Approach	Protect using antidegradation provisions of the CWA	Use IC-based TMDLs that use flow or IC as a surrogate for traditional pollutants		Use pollutant TMDLs to identify problem subwatersheds	Use pollutant TMDLs to identify priority source areas
Dry Weather Water Quality	Perform in-stream grab sampling of water quality at sentinel stations	Check for failing septic systems	Screen outfalls for illicit discharges	Perform dry weather sampling in streams and outfall screening	Perform dry weather sampling in receiving waters
Addressing Existing Development	Protect or conserve natural areas, enhance riparian cover, assess road crossings, and ensure farm, forest, and pasture best practices are used		Perform stream repairs, riparian reforestation, and residential stewardship	Perform storage retrofits and stream repairs	Use pollution source controls and municipal housekeeping

Site-Based IC Fees. Several economic strategies can be used to promote equity and efficiency when it comes to managing stormwater in different kinds of subwatersheds. In lower-density subwatersheds, an excess impervious cover fee can be charged to individual sites that exceed a maximum threshold for impervious cover for their zoning category. Similarly, an impervious cover mitigation fee can be levied at individual development sites in more intensely developed subwatersheds when on-site compliance is not possible or it is more cost-effective to provide an equivalent amount of treatment elsewhere in the watershed. The type of fee and the frequency that is used is expected to be closely related to the subwatershed classification.

Subwatershed Trading. The degree of impervious cover in a subwatershed also has a strong influence on the feasibility, cost, and appropriateness of restoration projects. Consequently, any revenues collected from various site IC fees can be traded among subwatersheds to arrive at the least-cost, effective solutions. In general, the most intensely developed subwatersheds are sending areas and the more lightly developed subwatersheds are used as receiving areas for such projects.

Stormwater Monitoring Approach. Subwatershed classification can also be used to define the type and objectives for stormwater monitoring to track compliance over time. For example, in sensitive subwatersheds, it may be advisable to routinely measure in-stream metrics of biological integrity to ensure stream quality is being maintained or enhanced. As impervious cover increases, stormwater managers may want to shift toward tracking of subwatershed impervious cover and actual performance monitoring of select SCMs to establish their effectiveness (e.g., impacted subwatersheds). At even higher levels of impervious cover, streams are transformed into urban drainage, and monitoring becomes more focused on identifying individual stormwater outfalls with the worst quality during storm conditions.

TMDL Approach. Subwatershed classification may also serve as a useful tool to decide how to apply TMDLs to impaired waters, or how to ensure that healthy waters are not degraded by future land development. For example, most lightly developed subwatersheds will seldom be subject to a TMDL, or if so, urban stormwater is often only a minor component in the final waste load allocation. Antidegradation provisions of the CWA are often the best means to protect the quality of these healthy waters before they are degraded by future land development. By contrast, impaired watersheds appear to be the best candidates to apply impervious cover-based TMDLs, as described earlier in this section. As subwatershed impervious cover increases, more traditional pollutant-based TMDLs are warranted, with a focus on problem subwatersheds for non-supporting streams and priority source areas for urban drainage.

Dry Weather Water Quality. The type, severity, and sources of illicit discharges often differ among different subwatershed classifications, which can have a strong influence on the kind of dry weather detective work needed to isolate them. For example, in lightly developed subwatersheds, failing septic systems are often the most illicit discharges, which prompts assessments at the lot or ditch level. The storm-drain network and potential discharge source areas becomes progressively more complex as subwatershed impervious cover increases. Consequently, illicit-discharge assessments shift toward outfall screening, catchment analysis, and individual source analysis.

Addressing Existing Development. The need for, type of, and feasibility for restoration efforts shift as subwatershed impervious cover increases. In general, lightly developed watersheds have the greatest land area available for retrofits and restoration projects in the stream corridor. Consequently, unique restoration strategies are developed for different subwatershed classifications (Schueler, 2004).

Require More Quantitative Evaluation of MS4 Programs

The next round of permit renewals should contain explicit conditions to define and measure outcomes from the six minimum management measures that constitute a Phase II MS4 program. Measurable program evaluation is critical to develop, implement, and adapt effective local stormwater programs, and has been consistently requested in permits and application guidance. To date, however, only a small fraction of MS4 communities have provided measurable outcomes with regard to aggregate pollutant reduction achieved by their municipal stormwater programs.

CASQA (2007) defines a six-level pyramid to assess program effectiveness, beginning with documenting activities, raising awareness, changing behaviors, reducing loads from sources, improving runoff quality, and ultimately leading to protection of receiving water quality (see Figure 6-1).

At the current time, most MS4s are struggling simply to organize or document their program activities (i.e., the first level), and few have moved up the pyramid to provide a quantitative link between program activities and water quality improvements. The framework and methods to evaluate program effectiveness for each of the six minimum management measures has been outlined by CASQA (2007). Regulators are encouraged to work with permitted municipalities to define increasingly more specific quantitative measures of program performance in each succeeding permit cycle.

FIGURE 6-1 Pyramid of Assessment Outcome Levels for an MS4. SOURCE: CASQA (2007).

Shift Monitoring Requirements to Measure the Performance of Stormwater Control Measures

The lack of monitoring requirements in the Phase II stormwater program makes it virtually impossible to measure or track actual pollutant load or runoff volume reductions achieved. While the existing Phase I outfall monitoring requirements have improved our understanding of urban stormwater runoff quality, they are also insufficient to link program effort to receiving water quality. It is recommended that both Phase I and II MS4s shift to a more collaborative monitoring effort to link management efforts to receiving water quality, as described below:

• If a review of past Phase 1 MS4s stormwater outfall monitoring indicates no violations of the Municipal Action Limits, then their current outfall monitoring efforts can be replaced by pooled annual financial contributions to a regional stormwater monitoring collaborative or authority to conduct basic research on the performance and longevity of range of SCMs employed in the community.

• If some subwatersheds exceed Municipal Action Levels, outfall monitoring should be continued at these locations, as well as additional source area sampling in the problem subwatershed to define the sources of the stormwater pollutant of concern.

• Phase II MS4s should be encouraged to make incremental financial contributions to a state or regional stormwater monitoring research collaborative to conduct basic research on SCM performance and longevity. Although the committee knows of no examples where this has been accomplished, this pooling of financial resources by multiple MS4s should produce more useful scientific data to support municipal programs than could be produced by individual MS4s alone. Phase II communities that do not participate in the research collaborative would be required to perform their own outfall and/or SCM performance monitoring, at the discretion of the state or federal permitting authority.

• All MS4s should be required to indicate in their annual reports and permit renewal applications how they incorporated research findings into their existing stormwater programs, ordinances, and design manuals.

CONCLUSIONS AND RECOMMENDATIONS

The watershed-based permitting program outlined in the first part of this chapter is ultimately essential if the nation is to be successful in arresting aquatic resource depletion stemming from sources dispersed across the landscape. Smaller-scale changes to the EPA stormwater program are also possible. These include integration of industrial and construction permittees into municipal permits ("integration"), as well as a number of individual changes to the current industrial, construction, and municipal programs.

Improvements to the stormwater permitting program can be made in a tiered manner. Thus, individual recommendations specific to advancing one part of the municipal, industrial, or construction stormwater programs could be implemented immediately and with limited additional funds. "Integration" will need additional funding to provide incentives and to establish partnerships between municipal permittees and their associated industries. Finally, the watershed-based permitting approach will likely take up to ten years to implement. The following conclusions and recommendations about these options are made:

The greatest improvement to the EPA's Stormwater Program would be to convert the current piecemeal system into a watershed-based permitting system. The proposed system would encompass coordinated regulation and management of all discharges (wastewater, stormwater, and other diffuse sources), existing and anticipated from future growth, having the potential to modify the hydrology and water quality of the watershed's receiving waters.

The committee proposes centralizing responsibility and authority for implementation of watershed-based permits with a municipal lead permittee working in partnership with other municipalities in the watershed as co-permittees, with enhanced authority and funding commensurate with increased responsibility. Permitting authorities would adopt a minimum goal in every watershed to avoid any further loss or degradation of designated beneficial uses in the watershed's component waterbodies and additional goals in some cases aimed at recovering lost beneficial uses. The framework envisions the permitting authorities and municipal co-permittees working cooperatively to define careful, complete, and clear specific objectives aimed at meeting goals.

Permittees, with support from the permitting authority, would then move to comprehensive scientific and technically based watershed analysis as a foundation for targeting solutions. The most effective solutions are expected to lie in isolating, to the extent possible, receiving waterbodies from exposure to those impact sources. In particular, low-impact design methods, termed Aquatic Resources Conservation Design in this report, should be employed to the full extent feasible and backed by conventional SCMs when necessary. This report also outlines a monitoring program structured to assess progress toward meeting objectives and the overlying goals, diagnosing reasons for any lack of progress, and determining compliance by dischargers. The new concept further includes market-based trading of credits among dischargers to achieve overall compliance in the most efficient manner and adaptive management to program additional actions if monitoring demonstrates failure to achieve objectives.

Integration of the three permitting types, such that construction and industrial sites come under the jurisdiction of their associated municipalities, would greatly improve many deficient aspects of the stormwater program. Federal and state NPDES permitting authorities do not presently have, and can never reasonably expect to have, sufficient personnel to inspect and enforce stormwater regulations on more than 100,000 discrete point source facilities discharging stormwater. A better structure would be one where the NPDES permitting authority empowers the MS4 permittees to act as the first tier of entities exercising control on stormwater discharges to the MS4 to protect water quality. The National Pretreatment Program, EPA's successful treatment program for municipal and industrial wastewater sources, could serve as a model for integration.

Short of adopting watershed-based permitting or integration, a variety of other smaller-scale changes to the EPA stormwater program could be made now, as outlined below.

EPA should issue guidance for MS4, MSGP, and CGP permittees on what constitutes a design storm for water quality purposes. Precipitation events occur across a spectrum from small, more frequent storms to larger and more extreme storms, with the latter being a more typical focus of guidance manuals to date. Permittees need guidance from regional EPA offices on what water quality considerations to design SCMs for beyond issues such as safety of human life and property. In creating the guidance there should be a good faith effort to integrate water quality requirements with existing stormwater quantity requirements.

EPA should issue guidance for MS4 permittees on methods to identify high-risk industrial facilities for program prioritization such as inspections. Two visual methods for establishing rankings that have been field tested are provided in the chapter. Some of these high-risk industrial facilities and construction sites may be better covered by individual NPDES stormwater permits rather than the MSGP or the CGP, and if so would fall directly under the permitting authority and not be part of MS4 integration.

EPA should support the compilation and collection of quality *industrial* stormwater effluent data and SCM effluent quality data in a national database. This database can then serve as a source for the agency to develop technology-based effluent guidelines for stormwater discharges from industrial sectors and high-risk facilities.

EPA should develop numerical expressions to represent the MS4 standard of Maximum Extent Practicable. This could involve establishing municipal action levels based on expected outfall pollutant concentrations from the National Stormwater Quality Database, developing site-based runoff and pollutant load limits, and setting turbidity limits for construction sites. Such numerical expressions would create improved accountability, bring about consistency, and result in implementation actions that will lead to measurable reductions in stormwater pollutants in MS4 discharges.

Communities should use an urban stream classification system, such as a regionally adapted version of the Impervious Cover Model, to establish realistic water quality and biodiversity goals for individual classes of subwatersheds. The goals for water and habitat quality should become less stringent as impervious cover increases within the subwatershed. This should not become an excuse to work less diligently to improve the most degraded waterways—only to recognize that equivalent, or even greater, efforts to improve water quality conditions will achieve progressively less ambitious results in more highly urbanized watersheds. This approach would provide stormwater managers with more specific, measurable, and attainable implementation strategies than the one-size-fits-all approach that is promoted in current wet weather management regulations.

Better monitoring of MS4s to determine outcomes is needed. Only a small fraction of MS4 communities have provided measurable outcomes with regard to aggregate flow and pollutant reduction achieved by their municipal stormwater programs. A framework and methods to evaluate program effectiveness for each of the six minimum management measures have been outlined by CASQA (2007) and should be adopted. In addition, the lack of monitoring requirements in the Phase II stormwater program makes it virtually impossible to measure or track actual pollutant load or runoff volume reductions achieved. It is recommended that both Phase I and II MS4s shift to a more collaborative monitoring paradigm to link management efforts to receiving water quality.

***

Watershed-based permitting will require additional resources and regulatory program support. Such an approach shifts more attention to ambient outcomes as well as expanded permitting coverage. Additional resources for program implementation could come from shifting existing programmatic resources. For example, some state permitting resources may be shifted away from existing point source programs toward stormwater permitting. Strategic planning and prioritization could shift the distribution of federal and state grant and loan programs to encourage and support more watershed-based stormwater permitting programs. However, securing new levels of public funds will likely be required. All levels of government must recognize that additional resources may be required from citizens and businesses (in the form of taxes, fees, etc.) in order to operate a more comprehensive and effective stormwater permitting program.

REFERENCES

April, S., and T. Greiner. 2000. Evaluation of the Massachusetts Environmental Results Program. Washington, DC: National Academy of Public Administration..

Atkins, J. R., C. Hollenkamp, and J. Sauber. 2007. Testing the watershed: North Carolina's NPDES Discharge Coalition Program enables basinwide monitoring and analysis. Water Environment & Technology 19(6).

Bellucci, C. 2007. Stormwater and Aquatic Life: Making the Connection Between Impervious Cover and Aquatic Life Impairments for TMDL Development in Connecticut Streams. Pp. 1003-1018 *In:* TMDL 2007. Alexandria, VA: Water Environment Federation.

Bromberg, K. 2007. Comments to the NRC Committee on Stormwater Discharge Contributions to Water Pollution, January 22, 2007, Washington, DC.

Burton, G. A., and R. E. Pitt. 2002. Stormwater Effects Handbook. Boca Raton, FL: Lewis/CRC Press.

California EPA, State Water Board. 2006. Storm Water Panel Recommendations—The Feasibility of Numeric Effluent Limits Applicable to Discharges of Storm Water Associated with Municipal, Industrial, and Construction Activities. Available at http://www.cacoastkeeper.org/assets/pdf/StormWaterPanelReport_06.pdf.

Campbell, R. M. 2007. Achieving a Successful Storm Water Permit Program in Oregon. Natural Resources & Environment 21(4):39-44.

CASQA (California Stormwater Quality Association). 2007. Municipal Stormwater Program Effectiveness Assessment Guidance. Los Angeles. Available at info.casqa@org.

Chapman, C. 2006. Performance Monitoring of an Urban Stormwater Treatment System. Master's Thesis, University of Washington, Seattle.

City of Austin. 2006. Stormwater Runoff Quality and Quantity from Small Watersheds in Austin, TX. Austin, TX: Watershed Protection Department, Environmental Resources Management Division.

City of San Diego. 2007. Strategic Plan for Watershed Activity Implementation. San Diego, CA: Stormwater Pollution Prevention Division.

Clark, S., R. Pitt, S. Burian, R. Field, E. Fan, J. Heaney, and L. Wright. 2006. The Annotated Bibliography of Urban Wet Weather Flow Literature from 1996 through 2005. *Note: Publisher not shown.*

Clausen, J. C., and J. Spooner. 1993. Paired Watershed Study Design, 841-F-93-009. Washington, DC: EPA Office of Water.

Connecticut Department of Environmental Protection. 2007. A Total Maximum Daily Load Analysis for Eagleville Brook, Mansfield, CT. Hartford: State of Connecticut Department of Environmental Protection. Available at http://www.ct.gov/dep/lib/dep/water/tmdl/tmdl_final/eaglevillefinal.pdf.

Cosgrove, J. F. 2002. TMDLs: A simplified approach to pollutant load determination. WEFTEC 2002 Conference Proceedings September 2002. Alexandria, VA: Water Environment Federation.

Crockett, C. 2007. The regulated perspective of stormwater management. Presentation to the NRC Committee on Stormwater Discharge Contributions to Water Pollution. January 22, 2007. Washington, DC.

Cross, L. M., and L. D. Duke. 2008. Regulating industrial stormwater: state permits, municipal implementation, and a protocol for prioritization. Journal of the American Water Resources Association 44(1):86-106.

Cutter, W. B., K. A. Baerenklau, A. DeWoody, R. Sharma, and J. G. Lee. 2008. Costs and benefits of capturing urban runoff with competitive bidding for decentralized best management practices. Water Resources Research, doi:10.1029/2007WR006343.

DeWoody, A. E. 2007. Determining Net Social Benefits from Optimal Parcel-Level Infiltration of Urban Runoff: A Los Angles Analysis. M.S. Thesis. University of California, Riverside.

Doll, A., and G. Lindsey. 1999. Credits bring economic incentives for onsite stormwater management. Watershed and Wet Weather Technical Bulletin 4(1):12-15.

Duke, L. D. 2007. Industrial stormwater runoff pollution prevention regulations and implementation. Presentation to the National Research Council Committee on Reducing Stormwater Discharge Contributions to Water Pollution, Seattle, WA, August 22, 2007.

Duke, L. D., and C. A. Augustenborg. 2006. Effectiveness of self identified and self-reported environmental regulations for industry: the case of storm water runoff in the U.S. Journal of Environmental Planning and Management 49:385-411.

Duke, L. D., and P. Beswick. 1997. Industry compliance with storm water pollution prevention regulations: the case of transportation industry facilities in California and the Los Angeles region. Journal of the American Water Resources Association 33:825-838.

Ellerman, A. D., P. L. Joskow, R. Schmalensee, J. P. Montero, and E. M. Bailey. 2000. Markets for Clean Air: the U.S. Acid Rain Program. New York: Cambridge University Press.

EPA (U. S. Environmental Protection Agency). 1991. Technical Support Document for Water Quality-Based Toxics Control. EPA-505/2-90-001. Washington, DC: EPA Office of Water Enforcement and Permits.

EPA. 1999. Introduction to the National Pretreatment Program. EPA-833-B-98-002. Washington, DC: EPA Office of Wastewater Management.

EPA. 2002. Establishing total maximum daily load (TMDL) Wasteload allocations (WLAs) for storm water sources and NPDES permit requirements based on those WLAs. Memorandum from Robert Wayland, Director, Office of Wetlands, Oceans, and Watersheds to Jim Hanlon, Director, Office of Water, November 22, 2002. Available at www.epa.gov/npdes/pubs/final-wwtmdl.pdf.

EPA. 2003a. Watershed-Based NPDES Permitting Policy Statement. In Watershed-Based National Pollutant Discharge Elimination System (NPDES) Permitting Implementation Guidance. EPA, Washington, DC.

EPA. 2003b. Watershed-Based National Pollutant Discharge Elimination System (NPDES) Permitting Implementation Guidance. EPA, Washington, DC.

EPA. 2007a. Watershed-Based NPDES Permitting Technical Guidance (draft). EPA, Washington, DC.

EPA. 2007b. Water Quality Trading Toolkit for Permit Writers. EPA 833-R-07-004. Washington, DC: EPA Office of Wastewater Management, Water Permits Division.

EPA. 2007c. Understanding Impaired Waters and Total Maximum Daily Load (TMDL Requirements for Municipal Stormwater Programs. EPA 883-F-07-009. Philadelphia, PA: EPA Region 3.

Freedman, P., L. Shabman, and K. Reckhow. 2008. Don't Debate; Adaptive implementation can help water quality professionals achieve TMDL goals. WE&T Magazine August 2008:66-71.

Frie, S., L. Curtis, and S. Martin. 1996. Financing regional stormwater facilities. *In:* Managing Virginia's Watersheds in the 21st Century: Workable Solutions. Proceedings from the 9th Annual Virginia Water Conference, Staunton.

Grumbles, B. 2006. Qualifying Local Programs for Construction Site Storm Water Runoff. Memorandum from EPA Assistant Administrator Ben Grumbles to James Mac Indoe, Alabama Dept. of Environmental Management. May 8.

Hetling, L .J., A Stoddard, and T. N. Brosnan. 2003. Effect of water quality management efforts on wastewater loadings during the past century. Water Environment Research 75(1):30.

Hirschman, D., K. Collins, and T. Schueler. 2008. Draft Virginia Stormwater Management Nutrient Design System. Prepared for Technical Advisory Committee and Virginia Department of Conservation and Recreation. Ellicott City, MD: Center for Watershed Protection.

Holling, C. S., ed. 1978. Adaptive Environmental Assessment and Management. Caldwell, NJ: Blackburn Press.

Holling, C. S., and A. D. Chambers. 1973. Resource science: The nurture of an infant. BioScience 23:13-20.

Horner, R. R., and C. Chapman. 2007. NW 110th Street Natural Drainage System Performance Monitoring, with Summary of Viewlands and 2nd Avenue NW SEA Streets Monitoring. Report to Seattle Public Utilities by Department of Civil and Environmental Engineering, University of Washington, Seattle.

Horner, R. R., H. Lim, and S. J. Burges. 2002. Hydrologic Monitoring of the Seattle Ultra-Urban Stormwater Management Projects, Water Resources Series Technical Report No. 170. Department of Civil and Environmental Engineering, University of Washington, Seattle.

Horner, R. R., H. Lim, and S. J. Burges. 2004. Hydrologic Monitoring of the Seattle Ultra-Urban Stormwater Management Projects: Summary of the 2000-2003 Water Years. Water Resources Series Technical Report 181. Department of Civil and Environmental Engineering, University of Washington, Seattle.

Horner, R., C. May, E. Livingston, D. Blaha, M. Scoggins, J. Tims, and J. Maxted. 2001. Structural and non-structural BMPs for protecting streams. Pp. 60-77 *In:* Linking Stormwater BMP Designs and Performance to Receiving Water Impact Mitigation. Proceedings Engineering Research Foundation Conference. American Society of Civil Engineers.

Horner, R., J. Guedry, and M. Kortenhof. 1990. Improving the Cost-Effectiveness of Highway Construction Site Erosion and Sediment Control. Washington State Department of Transportation. Seattle, WA: Department of Civil Engineering, University of Washington.

Keller, B. 2003. Buddy can you spare a dime? What is stormwater funding. Stormwater 4:7.

LaFlamme, P. 2007. Presentation to the Committee on Stormwater Discharge Contributions to Water Pollution, January 22, 2007, Washington, DC.

Lee, H., X. Swamikannu, D. Radulescu, S. Kim, and M. K. Stenstrom. 2007. Design of stormwater monitoring programs. Journal of Water Research, doi:10.1016/j.watres.2007.05.016.

Longsworth, J. 2007. Comments to the NRC Committee on Stormwater Discharge Contributions to Water Pollution. January 22, 2007, Washington, DC.

Los Angeles County Department of Public Works. 2001. Los Angeles County 1994-2000 Integrated Receiving Water Impacts Report. Available at http://ladpw.org/wmd/NPDES/IntTC.cfm.

Maimone, M. 2002. Prioritization of contaminant sources for the Schuylkill River source water assessment. Presentation at the Watershed 2002 Specialty Conference, Ft. Lauderdale, FL, February 23-27.

NRC (National Research Council). 1990. Monitoring Southern California's Coastal Waters. Washington, DC: National Academy Press.

NRC. 1999. New Strategies for America's Watersheds. Washington, DC: National Academy Press.

NRC. 2001a. Assessing the TMDL Approach to Water Quality Management. Washington, DC: National Academy Press.

NRC. 2001b. Compensating for Wetland Losses Under the Clean Water Act. Washington, DC: National Academy Press.

Natural Resources Conservation Service. 2007. Part 630 Hydrology, National Engineering Handbook, Chapter 7, Hydrologic Soil Groups. Washington, DC: U.S. Department of Agriculture.

Nirel, P. M., and R. Revaclier. 1999. Assessment of sewage treatment plant effluents impact on river water quality using dissolved Rb:Sr ratio. Environmental Science and Technology 33(12):1996.

North Carolina Division of Water Quality. 1999. Neuse River Basin Model Stormwater Program for Nitrogen Control. Available at http://h2o.enr.state.nc.us/su/Neuse_SWProgram_Documents.htm. Last accessed November 2007.

Office of Inspector General. 2007. Development Growth Outpacing Progress in Watershed Efforts to Restore the Chesapeake Bay. Report 2007-P-000031. Washington, DC: U.S. Environmental Protection Agency.

Parikh, P., M. A. Taylor, T. Hoagland, H. Thurston, and W. Shuster. 2005. Application of market mechanism and incentives to reduce stormwater runoff: an integrated hydrologic, economic, and legal approach. Environmental Science and Policy 8:133-144.

Pitt, R., A, Maestre, and R. Morquecho. 2004. National Stormwater Quality Database. Version 1.1. Available at http://rpitt.eng.ua.edu/Research/ms4/Paper/Mainms4paper.html. Last accessed January 28, 2008.

Schiff, K., D. Ackerman, E. Strecker, and M. Leisenring. 2007. Concept development: design storm for water quality in the Los Angeles region. Southern California Coastal Water Research Project. Costa Mesa.

Schueler, T. 2008b. Technical Support for the Baywide Runoff Reduction Method. Baltimore, MD: Chesapeake Stormwater Network. Available at www.chesapeakestormwater.net.

Schueler, T. 2004. An Integrated Framework to Restore Small Urban Watersheds. Manual 1. Urban Subwatershed Restoration Manual Series. Ellicott City, MD: Center for Watershed Protection.

Schueler, T. 2008a. Final Bay-wide Stormwater Action Strategy: Recommendations for Moving Forward in the Chesapeake Bay. Baltimore, MD: Chesapeake Stormwater Network.

Schueler, T., D. Hirschman, M. Novotney, and J. Zielinski. 2007. Urban

Stormwater Retrofit Practices. Ellicott City, MD: Center for Watershed Protection.

Shabman, L., and K. Stephenson. 2007. Achieving nutrient water quality goals: bringing market-like principles to water quality management. Journal of American Water Resources Association 43(4):1076-1089.

Shabman, L., and P. Scodari. 2004. Past, Present, and Future of Wetland Credit Sales. Discussion Paper 04-48. Washington, DC: Resources for the Future.

Shabman, L., K. Reckhow, M. B. Beck, J. Benaman, S. Chapra, P. Freedman, M. Nellor, J. Rudek, D. Schwer, T. Stiles, and C. Stow. 2007. Adaptive Implementation of water quality improvement plans: opportunities and challenges. Durham, NC: Duke University.

Shabman, L., K. Stephenson, and W. Shobe. 2002. Trading programs for environmental management: reflections on the air and water experiences. Environmental Practice 4:153-162.

Shaver, E., R. Horner, J. Skupien, C. May, and G. Ridley. 2007. Fundamentals of Urban Runoff Management: Technical and Institutional Issue, 2nd Ed. Madison, WI: North American Lake Management Society.

Smith, B. 2007. Comments to the NRC Committee on Stormwater Discharge Contributions to Water Pollution. January 22, 2007, Washington, DC.

Stenstrom, M. K., and H. Lee. 2005. Final Report. Industrial Stormwater Monitoring Program. Existing Statewide Permit Utility and Proposed Modifications.

Stephenson, K., and L. Shabman. 2005. The use and opportunity of cooperative organizational forms as an innovative regulatory tool under the Clean Water Act. Paper presented at the Southern Agricultural Economics Association Annual Meetings Little Rock, AK, February 5-9, 2005.

Stephenson, K., L. Shabman, and J. Boyd. 2005. Taxonomy of trading programs: Concepts and applications to TMDLs. Pp. 253-285 *In:* Total Maximum Daily Loads: Approaches and Challenges. Tamim Younos (ed.). Tulsa, OK: Pennwell Press.

Stephenson, K., L. Shabman, and L. Geyer. 1999. Watershed-based effluent allowance trading: Identifying the statutory and regulatory barriers to implementation. Environmental Lawyer 5(3):775-815.

Stephenson, K., P. Norris, and L. Shabman. 1998. Watershed-based effluent trading: the nonpoint source challenge. Contemporary Economic Policy 16:412-421.

Thurston, H. W., H. C. Goddard, D. Szlag, and B. Lemberg. 2003. Controlling storm-water runoff with tradable allowances for impervious surfaces. Journal of Water Resources Planning and Management 129(5):409-418.

Vermont Department of Environmental Conservation. 2006. Total Maximum Daily Load to Address Biological Impairment Potash Brook (VT05-11). Chittenden County, Vermont.

Virginia DCR (Virginia Department of Conservation and Recreation). 2007. Virginia Stormwater Management Program Permit Regulations, Chapter 60.

Wagner, W.E. 2006. Stormy regulations: The problems that result when storm water (and other) regulatory programs neglect to account for limitations in scientific and technical programs. Chapman Law Review 9(2):191-232.

Wenger, S., T. Carter, R. Vick, and L. Fowler. 2008. Runoff limits: an ecologically based stormwater management program. Stormwater April/May. Available at http://www.stormh2o.com/march-april-2008/ecologically-stormwater-management.aspx.

Woodward, R. T., R. A. Kaiser, and A. B. Wicks. 2002. The structure and practice of water quality trading markets. Journal of the American Water Resources Association 38:967-979.

Zeng, X., and T. C. Rasmussen. 2005. Multivariate statistical characterization of water quality in Lake Lanier, Georgia, USA. Journal of Environmental Quality 34(6):1980-1991.

Appendixes

Appendix A
Acronyms

BAC	best attainable conditions
BAT	best available technology
BCG	Biological Condition Gradient
BCT	best control technology
BOD	biochemical oxygen demand
CAFO	concentrated animal feeding operation
CBWM	Chesapeake Bay Watershed Model
CCI	Census of Construction Industries
CERCLA	Comprehensive Environmental Response, Compensation, and Liability Act
CGP	Construction General Permit
CN	Curve Number
COD	chemical oxygen demand
COV	coefficient of variability
CWA	Clean Water Act
DHSVM	Distributed Hydrology, Soil, and Vegetation Model
EIA	effective impervious area
EMC	event mean concentration
ERP	Enforcement Response Plan
ETV	Environmental Technology Verification Program
EWH	exceptional warmwater habitat
FEMA	Federal Emergency Management Agency
FHWA	Federal Highway Administration
FIFRA	Federal Insecticide, Fungicide, and Rodenticide Act
GIS	Geographic Information System
GWLF	General Watershed Loading Function
HRU	Hydrologic Response Unit
HSPF	Hydrologic Simulation Program–Fortran
HUC	hydrologic unit code
ICM	Impervious Cover Model
KCRTS	King County Runoff Time Series
LDC	least disturbed conditions
LEED	Leadership in Energy and Environmental Design
LID	low-impact development
MDC	minimally disturbed conditions
MEP	maximum extent practicable
MGD	million gallons per day
MSGP	multi-sector industrial stormwater general permit
MTBE	methyl tert-butyl ether
NCSI	Normalized Channel Stabilization Index
NOI	Notice of Intent

NPDES	National Pollutant Discharge Elimination System
NRDC	Natural Resources Defense Council
NRI	National Resource Inventory
NSQD	National Stormwater Quality Database
NTU	Nephelometric Turbidity Unit
NURP	National Urban Runoff Program
PAH	polycyclic aromatic hydrocarbons
PCB	polychlorinated biphenyl
POTW	publicly owned treatment works
PUD	planned unit development
RCRA	Resource Conservation and Recovery Act
RPA	Reasonable Potential Analysis
SBUH	Santa Barbara Unit Hydrograph
SCCWRP	Southern California Coastal Water Research Project Authority
SCM	stormwater control measure
SIC	Standard Industrial Classification
SLAMM	Source Loading and Management Model
SMDR	Soil Moisture Distributed and Routing
SWAT	Soil and Water Assessment Tool
SWMM	Stormwater Management Model
SWPPP	stormwater pollution prevention plan
TALU	tiered aquatic life use
TARP	Technology Acceptance and Reciprocity Partnership
TIA	total impervious area
TKN	total Kjedahl nitrogen
TMDL	total maximum daily load
TND	traditional neighborhood development
TOD	transit-oriented development
TSCA	Toxic Substances Control Act
TSS	total suspended solids
UAA	Use Attainability Analysis
UDC	unified development code
ULARA	Upper Los Angeles River Area
USLE	Universal Soil Loss Equation
WERF	Water Environment Research Foundation
WQA	Water Quality Act
WQS	water quality standard
WWH	warmwater habitat
WWHM	Western Washington Hydrologic Model
WWTP	wastewater treatment plant

Appendix B
Glossary

Antidegradation: Policies which ensure protection of water quality from a particular waterbody where the water quality exceeds levels necessary to protect fish and wildlife propagation and recreation on and in the water. This also includes special protection of waters designated as outstanding natural resource waters. Antidegradation plans are adopted by each state to minimize adverse effects on water.

Best Management Practice (BMP): Physical, structural, and/or managerial practices that, when used singly or in combination, reduce the downstream quality and quantity impacts of stormwater. The term is synonymous with Stormwater Control Measure (SCM).

Biofiltration: The simultaneous process of filtration, infiltration, adsorption, and biological uptake of pollutants in stormwater that takes place when runoff flows over and through vegetated areas.

Bioinfiltration: A particular SCM that is like bioretention but has more infiltration, and thus would be categorized as an infiltration process.

Bioretention: A stormwater management practice that utilizes shallow storage, landscaping, and soils to control and treat urban stormwater runoff by collecting it in shallow depressions before filtering through a fabricated planting soil media. This SCM is often categorized under "filtration" although it has additional functions.

Buffer: The zone contiguous with a sensitive area that is required for the continued maintenance, function, and structural stability of the sensitive area. The critical functions of a riparian buffer (those associated with an aquatic system) include shading, input of organic debris and coarse sediments, uptake of nutrients, stabilization of banks, interception of fine sediments, overflow during high-water events, protection from disturbance by humans and domestic animals, maintenance of wildlife habitat, and room for variation of aquatic system boundaries over time due to hydrologic or climatic effects. The critical functions of terrestrial buffers include protection of slope stability, attenuation of surface water flows from stormwater runoff and precipitation, and erosion control.

> **Stream buffers** are zones of variable width that are located along both sides of a stream and are designed to provide a protective natural area along a stream corridor.

Combined Sewer Overflow (CSO): A discharge of untreated wastewater from a combined sewer system at a point prior to the headworks of a publicly owned treatment works. CSOs generally occur during wet weather (rainfall or snowmelt). During periods of wet weather, these systems become overloaded, bypass treatment works, and discharge directly to receiving waters.

Combined Sewer System: A wastewater collection system that conveys sanitary wastewaters (domestic, commercial, and industrial wastewaters) and stormwater through a single pipe to a publicly owned treatment works for treatment prior to discharge to surface waters.

Constructed Wetland: A wetland that is created on a site that previously was not a wetland. This wetland is designed specifically to remove pollutants from stormwater runoff.

Created Wetland: A wetland that is created on a site that previously was not a wetland. This wetland is created to replace wetlands that were unavoidably destroyed during design and construction of a project. This wetland cannot be used for treatment of stormwater runoff.

Detention: The temporary storage of stormwater runoff in an SCM with the goals of controlling peak discharge rates and providing gravity settling of pollutants.

Detention Facility/Structure: An above- or below-ground facility, such as a pond or tank, that temporarily stores stormwater runoff and subsequently releases it at a slower rate than it is collected by the drainage facility system. There is little or no infiltration of stored stormwater, and the facility is designed to not create a permanent pool of water.

Drainage: Refers to the collection, conveyance, containment, and/or discharge of surface and stormwater runoff.

Drainage Area: That area contributing runoff to a single point measured in a horizontal plane, which is enclosed by a ridge line.

Drainage Basin: A geographic and hydrologic subunit of a watershed.

Dry Pond: A facility that provides stormwater quantity control by containing excess runoff in a detention basin, then releasing the runoff at allowable levels. Synonymous with detention basin, it is intended to be dry between storms.

Effluent Limitation: Any restriction imposed by the EPA director on quantities, discharge rates, and concentrations of pollutants that are discharged from point sources into waters of the United States, the waters of the contiguous zone, or the ocean.

Effluent Limitation Guidelines: A regulation published by the EPA Administrator under Section 304(b) of the Clean Water Act that establishes national technology-based effluent requirements for a specific industrial category.

Exfiltration: The downward movement of water through the soil; the downward flow of runoff from the bottom of an infiltration SCM into the soil.

Extended Detention: A stormwater design feature that provides for the gradual release of a volume of water in order to increase settling of pollutants and protect downstream channels from frequent storm events. When combined with a pond, the settling time is increased by 24 hours.

Filter Strip: A strip of permanent vegetation above ponds, diversions, and other structures to retard the flow of runoff, causing deposition of transported material and thereby reducing sedimentation. As an SCM, it refers to riparian buffers, which run adjacent to waterbodies and intercept overland flow and shallow subsurface flow (both of which are usually sheet flow rather than a distinct influent pipe). The term is borrowed from the agricultural world.

Flood Frequency: The frequency with which the flood of interest may be expected to occur at a site in any average interval of years. Frequency analysis defines the n-year flood as being the flood that will, over a long period, be equaled or exceeded on the average once every n years.

Frequency of Storm (Design Storm Frequency): The anticipated period in years that will elapse, based on average probability of storms in the design region, before a storm of a given intensity and/or total volume will recur; thus, a 10-year storm can be expected to occur on the average once every 10 years. Sewers designed to handle flows which occur under such storm conditions would be expected to be surcharged by any storms of greater amount or intensity.

General Permit: A single permit issued to a large number of dischargers of pollutants in stormwater. General permits are issued by the permitting authority, and interested parties then submit a Notice of Intent (NOI) to be covered. The permit must identify the area of coverage, the sources covered, and the process for obtaining coverage. Once the permit is issued, a permittee may submit an NOI and receive coverage within a very short time frame.

Grab Sample: A sample which is taken from a stream on a one-time basis without consideration of the flow rate of the stream and without consideration of time.

Hotspot: An area where land use or activities generate highly contaminated runoff, with concentrations of pollutants in excess of those typically found in stormwater.

Hydrograph: A graph of runoff rate, inflow rate, or discharge rate, past a specific point as a function of time.

Hydroperiod: A seasonal occurrence of flooding and/or soil saturation; it encompasses depth, frequency, duration, and seasonal pattern of inundation.

Hyetograph: A graph of measured precipitation depth (or intensity) at a precipitation gauge as a function of time.

Impervious Surface or Impervious Cover: A hard surface area which either prevents or retards the entry of water into the soil. Common impervious surfaces include roof tops, walkways, patios, driveways, parking lots or storage areas, concrete or asphalt paving, gravel roads, packed earthen materials, and oiled surfaces.

Infiltration: The downward movement of water from the surface to the subsoil.

Infiltration Facility: A drainage facility designed to use the hydrologic process of runoff soaking into the ground, commonly referred to as percolation, to dispose of stormwater.

Infiltration Pond: A facility that provides stormwater quantity control by containing excess runoff in a detention facility, then percolating that runoff into the surrounding soil.

Level Spreader: A temporary SCM used to spread stormwater runoff uniformly over the ground surface as sheet flow. The purpose of level spreaders is to prevent concentrated, erosive flows from occurring. Levels spreaders will commonly be used at the upstream end of wider biofilters to ensure sheet flow into the biofilter.

Municipal Separate Storm Sewer System: A conveyance or system of conveyances (including roads with drainage systems, municipal streets, catch basins, curbs, gutters, ditches, man-made channels, or storm drains) owned by a state, city, town, or other public body that is designed or used for collecting or conveying stormwater, which is not a combined sewer and which is not part of a publicly owned treatment works.

National Pollutant Discharge Elimination System: A provision of the Clean Water Act that prohibits the discharge of pollutants into waters of the United States unless a special permit is issued by EPA, a state, or, where delegated, a tribal government on an Indian reservation. The permit applies to point sources of pollutants to ensure that their pollutant discharges do not exceed specified effluent standards. The effluent standards in most permits are based on the best available pollution technology or the equivalent.

Nonpoint Source: Diffuse pollution source, but with a regulatory connotation; a source without a single point of origin or not introduced into a receiving stream from a specific outlet. The pollutants are generally carried off the land by stormwater. Some common nonpoint sources are agriculture, forestry, mining, dams, channels, land disposal, and saltwater intrusion.

Nonstructural SCM: Stormwater control measure that uses natural measures to reduce pollution levels, does not require extensive construction efforts, and/or promotes pollutant reduction by eliminating the pollutant source.

Peak Discharge Rate: The maximum instantaneous rate of flow during a storm, usually in reference to a specific design storm event.

Point Source: Any discernible, confined, and discrete conveyance, including but not limited to any pipe, ditch, channel, tunnel, conduit, well, discrete fixture, container, rolling stock, concentrated animal feeding operation, landfill leachate collection system, vessel, or other floating craft from which pollutants are or may be discharged.

Pollutant: A contaminant in a concentration or amount that adversely alters the physical, chemical, or biological properties of the natural environment. Dredged soil, solid waste, incinerator residue, filter backwash, sewage, garbage, sewage sludge, munitions, chemical wastes, biological materials, radioactive materials (except those regulated under the Atomic Energy Act of 1954, as amended), heat, wrecked or discarded equipment, rock, sand, cellar dirt and industrial, municipal, and agricultural waste discharged into water (EPA, 2008).

Polutograph: A graph of pollutant loading rate (mass per unit time) as a function of time.

Predevelopment Conditions: Those conditions that existed at a site just prior to the development in question, which are not necessarily pristine conditions.

Pretreatment: The removal of material such as gross solids, grot, grease, and scum from flows prior to physical, biological, and chemical treatment processes to improve treatability. The reduction of the amount of pollutants, the elimination of pollutants, or the alteration of the nature of pollutant properties in wastewater prior to or in lieu of discharging or otherwise introducing such pollutants into a publicly owned treatment works [40 C.F.R. § 403.3(q)]. Pretreatment may include screening, grit removal, stormwater, and oil separators. With respect to stormwater, it refers to techniques employed in stormwater SCMs to help trap coarse materials and other pollutants before they enter the SCM.

Recharge: The flow of groundwater from the infiltration of stormwater runoff.

Recharge Volume: The portion of the water quality volume used to maintain groundwater recharge rates at development sites.

Retention: The process of collecting and holding stormwater runoff with no surface outflow. Also, the amount of precipitation on a drainage area that does not escape as runoff. It is the difference between total precipitation and total runoff.

Retention/Detention Facility: A type of drainage facility designed either to hold water for a considerable length of time and then release it by evaporation, plant transpiration, and/or infiltration into the ground, or to hold stormwater runoff for a short period of time and then release it to the stormwater management system.

Runoff: The term is often used in two senses. For a given precipitation event, direct *storm runoff* refers to the rainfall (minus losses) that is shed by the landscape to a receiving waterbody. In an area of 100 percent imperviousness, the runoff equals the rainfall. Over greater time and space scales, *surface water runoff* refers to streamflow passing through the outlet of a watershed, including base flow from groundwater that has entered the stream channel.

Soil Stabilization: The use of measures such as rock lining, vegetation, or other engineering structure to prevent the movement of soil when loads are applied to the soil.

Source Control: A type of SCM that is intended to prevent pollutants from entering stormwater. A few examples of source control are erosion control practices, maintenance of stormwater facilities, constructing roofs over storage and working areas, and directing wash water and similar discharges to the sanitary sewer or a dead end sump.

Stormwater: That portion of precipitation that does not naturally percolate into the ground or evaporate, but flows via overland flow, interflow, channels, or pipes into a defined surface water channel or a constructed infiltration facility. According to 40 C.F.R. § 122.26(b)(13), this includes stormwater runoff, snow melt runoff, and surface runoff and drainage.

Stormwater Control Measure (SCM): Physical, structural, and/or managerial measures that, when used singly or in combination, reduce the downstream quality and quantity impacts of stormwater. Also, a permit condition used in place of or in conjunction with effluent limitations to prevent or control the discharge of pollutants. This may include a schedule of activities, prohibition of practices, maintenance procedures, or other management practices. SCMs may include, but are not limited to, treatment requirements; operating procedures; practices to control plant site runoff, spillage, leaks, sludge, or waste disposal; or drainage from raw material storage.

Stormwater Drainage System: Constructed and natural features which function together as a system to collect, convey, channel, hold, inhibit, retain, detain, infiltrate, divert, treat, or filter stormwater.

Stormwater Facility: A constructed component of a stormwater drainage system, designed or constructed to perform a particular function or multiple functions. Stormwater facilities include, but are not limited to, pipes, swales, ditches, culverts, street gutters, detention basins, retention basins, constructed wetlands, infiltration devices, catch basins, oil/water separators, sediment basins, and modular pavement.

Structural SCMs: Devices which are constructed to provide temporary storage and treatment of stormwater runoff.

Swale: A shallow drainage conveyance with relatively gentle side slopes, generally with flow depths of less than one foot.

> **Biofilter** (same as a **Biofiltration Swale**): A sloped, vegetated channel or ditch that provides both conveyance and water quality treatment to stormwater runoff. It does not provide stormwater quantity control but can convey runoff to SCMs designed for that purpose.

> **Dry Swale**: An open drainage channel explicitly designed to detain and promote the filtration of stormwater runoff through an underlying fabricated soil media. It has an underdrain.

> **Wet Swale:** An open drainage channel or depression, explicitly designed to retain water or intercept groundwater for water quality treatment.

Technology-Based Effluent Limit: A permit limit for a pollutant that is based on the capability of a treatment method to reduce the pollutant to a certain concentration.

Time of Concentration: The time period necessary for surface runoff to reach the outlet of a subbasin from the hydraulically most remote point in the tributary drainage area.

Total Maximum Daily Load (TMDL): The amount, or load, of a specific pollutant that a waterbody can assimilate and still meet the water quality standard for its designated use. For impaired waters the TMDL reduces the overall load by allocating the load among current pollutant loads (from point and nonpoint sources), background or natural loads, a margin of safety, and sometimes an allocation for future growth.

Volumetric Runoff Coefficient (R_v): The value that is applied to a given rainfall volume to yield a corresponding runoff volume based on the percent impervious cover in a drainage basin.

Water Quality-Based Effluent Limit (WQBEL): A value determined by selecting the most stringent of the effluent limits calculated using all applicable water quality criteria (e.g., aquatic life, human health, and wildlife) for a specific point source to a specific receiving water for a given pollutant.

Water Quality SCM: An SCM specifically designed for pollutant removal.

Water Quantity SCM: An SCM specifically designed to reduce the peak rate of stormwater runoff.

Water Quality Volume (W_{qv}): The volume needed to capture and treat 90 percent of the average annual stormwater runoff volume equal to 1 inch times the volumetric runoff coefficient (R_v) times the site area.

Wetlands: Those areas that are inundated or saturated by surface water or groundwater at a frequency and duration sufficient to support, and that under normal circumstances do support, a prevalence of vegetation typically adapted for life in saturated soil conditions. Wetlands generally include swamps, marshes, bogs, and similar areas. This includes wetlands created, restored, or enhanced as part of a mitigation procedure. This does not include constructed wetlands or the following surface waters of the state intentionally constructed from sites that are not wetlands: irrigation and drainage ditches, grass-lined swales, canals, agricultural detention facilities, farm ponds, and landscape amenities.

Wet Pond: A facility that treats stormwater for water quality by utilizing a permanent pool of water to remove conventional pollutants from runoff through sedimentation, biological uptake, and plant filtration. Synonymous with a retention basin.

SOURCES: Most of the definitions are from EPA (2003), "BMP Design Considerations," 600/R-03/103, or EPA (2008), "Handbook for Developing Watershed Plans to Restore and Protect Our Waters," EPA 841-B-08-002.

Appendix C
Summary of Responses from State Stormwater Coordinators

On February 21, 2007, on behalf of the committee, Jenny Molloy of EPA's Office of Wastewater Management sent the following questions to a group of state stormwater program managers and received six responses (found in Tables C-1 and C-2).

1. For industrial and/or construction: do you have information on non-filers, i.e., folks who should have submitted NOIs, but did not? If so, how old are these data, and how do they compare to overall numbers of those with permit coverage? How did you find and/or estimate the number of non-filers?

2. Also for industrial and/or construction: do you have information on compliance rates? Yes, this is a really broad question, but something along the lines of: based on inspections (or monitoring data, or whatever metric you use), have you made any determinations on numbers of facilities out of compliance, or alternatively, in compliance? If so, define what you mean by compliance (paper violations, SWPPP/BMP inadequacies, water quality standards violations, etc.).

TABLE C-1 Nonfilers

State	Information on Industrial Non-Filers	Estimate Percent Non-Filers as of Total	Basis of Estimate	Period of Estimate	Comment
CA	Yes	50 percent of heavy industry statewide	Study—CA Water Board, 1999; Duke and Shaver, 1999.	1995–1998	
		69 percent Of industry within City of Los Angeles	Study— Swamikannu et al., 2001	1998–2000	
MN	No				Study in progress
OH	No				Plan outreach to business
OR	No				Do not compile data
VT	Yes	88–90 percent of industry	Mass mailing	2006	No response from 2,400 of 3,000 mailings
WI	No				

TABLE C-2 Compliance

State	Information on Compliance Rates	Estimate of Covered Facilities Non-Compliant	Basis of Estimate	Period of Estimate	Comment
CA	Yes (Construction)	40 percent deficient in paperwork; 30 percent with inadequate E&S controls	MS4 construction audit in Los Angeles and Ventura counties, and large CGP construction sites	2002, 2004, and 2005	Prioritized large CGP sites for inspection
	Yes (Industrial)	60 percent poor housekeeping practices; 40 percent incomplete SWPPPs	Transportation sector, plastics manufacturing inspections in Los Angeles County	2005 and 2007	
NH	No				Inspect in response to complaints
OH	No				Inspect construction sites as a priority
OR	No				Do not compile data
VT	No				Plan to inspect for compliance
WV	Yes (Industrial)	66 percent failed to submit report	Monitoring report submittal tracking	2007	Mailed deficiency notices
WI	Yes (Construction)	38 percent with minor and 43 percent with major violations	A subsample of 1 percent of CGP sites	2007	Perform inspections annually; no central database tracking

In September 2007, the NRC Committee on Reducing Stormwater Discharge Contributions to Water Pollution sent the following survey to 50 state stormwater program managers. Responses were received from 18 states, including at least one from every EPA region. The blank survey is shown below, and Tables C-3 through C-9 contain the states' responses.

The NRC committee members will greatly appreciate receiving the following information from State Stormwater Coordinators. Please complete both sides of this form and return to Xavier Swamikannu, CalEPA, Los Angeles Regional Water Board, xswamikannu@waterboards.ca.gov or Fax: (213) 576-6625.

State:
Name of information provider:

Please summarize your State's Stormwater Permit Program

	Municipal Permit	Industrial General Permit	Construction General Permit
What are the monitoring requirements?			
How is compliance demonstrated (monitoring or other activity)?			
To whom is the SWPPP submitted?			
Can an MS4 perform an inspection of an industry within its boundary?			
What industries are considered "high-risk"?			
Do BMP manuals exist for implementation guidance?			
No. of dedicated staff or FTEs			

Does your State Storm Water BMP Manual contain the following, and what are they?	
WQ sizing criteria	
Recharge criteria	
Channel protection criteria	
Overbank flood criteria	
Extreme flows	
Acceptable BMP list	
Detailed engineering specs for BMPs	
Soil and erosion control requirements (unless this is left to the local government)	

TABLE C-3 Monitoring Requirements

State	Municipal	Industrial	Construction
Alabama	Monitoring requirements are specific to the Phase I MS4. MS4 Phase II permit does not require monitoring.	Monitoring is specific to the General Permit type and associated discharge. Alabama has 18 NPDES Industrial Stormwater General Permits. http://www.adem.state.al.us/genpermits.htm	Monitoring is required under specific conditions, but in general compliance with the permit does not require monitoring. ADEM Admin. Code Chapter 335-6-12 is attached.
California	Monitoring requirements are specific to the Phase 1 MS4 permits. MS4 Phase II permit monitoring is discretionary.	2 wet weather sampling events per year – 4 basic parameters and other pollutants known to be on site. Quarterly visual monitoring.	Visual monitoring before, during, and after rain events. Analytical monitoring for discharges to sediment-impaired waterbodies.
Connecticut	Sample six outfalls once a year. Twelve chemical parameters.	Sample all outfalls once a year. Ten chemical parameters plus aquatic toxicity.	None, yet. Soon to modify permit to sample for turbidity.
Georgia	Dry weather outfall screening.	Standard monitoring from the EPA MSGP. Additional monitoring for the pollutant of concern for industries that may be causing or contributing to stream impairment.	Monitoring is required for a qualifying rain event (0.5 inch) once after clearing and grubbing, and once after mass grading.
Hawaii	Visual and water chemistry sampling.	Visual and water chemistry sampling.	Visual
Maine	None	No benchmark monitoring, only effluent limitations. Additional monitoring upon request based on discharges, complaints, audits, or inspections	None
Minnesota	The Phase I MS4 permits for Minneapolis and St. Paul require monitoring. MS4 Phase II permit does not require monitoring.	The current state MSGP does not have monitoring requirements. The proposed next term draft permit would require at least 4 stormwater monitoring events per year.	The current state CGP does not require monitoring. The proposed next term draft permit is not expected to include monitoring.

continued next page

580

TABLE C-3 Continued

State	Municipal	Industrial	Construction
Nebraska	Stormwater monitoring required on different use sites. BMP monitoring.	None. Monitoring can be required by the director through permit.	None. Monitoring can be required by the director through permit.
Nevada	Required for storm events that produce runoff.	None	None
New York	Ad hoc	Similar to monitoring in the EPA MSGP.	None. Self-inspection.
Ohio	Phase I MS4 permits require some chemical and biological monitoring. Phase II MS4 permit does not require mandatory monitoring, although recommended as part of IDDE program.	Similar to monitoring in the EPA MSGP, except annually. No priority chemical monitoring required.	For the state CGP, no chemical monitoring. For special watershed CGPs associated with TMDLs, TSS monitoring required.
Oklahoma	Phase 1 MS4s permits require dry weather monitoring, floatables monitoring, and watershed characterization monitoring, including biological assessments.	Quarterly visual monitoring and annual analytical monitoring.	None
Oregon	Monitoring requirements are specific to the Phase I MS4. The Phase II MS4 permit does not require monitoring, though some permittees do monitor on their own accord. The average frequency is 2-4 times a year.	Industrial facilities required to sample their stormwater discharge 4 times per year. Also required to conduct visual monitoring of their discharge on a monthly basis when discharge is present. Mining sites in addition are subject to the same requirements as in the state CGP since sediment is the main pollutant of concern.	None. However, permittees discharging stormwater to waters listed specifically for turbidity/sedimentation on the most recent 303(d) list or that have a TMDL for turbidity/sedimentation have the option of either monitoring for turbidity or implementing additional BMPs.

State	Municipal	Industrial	Construction
Vermont	None other than the development of an IDDE program and follow-up until elimination occurs	Benchmark monitoring for individual sectors, quarterly for the first year. Visual inspection 4 times per year. Effluent limitations (if applicable) once per year.	None at present. Turbidity monitoring for moderate-risk projects included in draft CGP.
Virginia	Monitoring requirements are specific to the Phase I MS4 permit. The Phase II MS4 permit does not require monitoring.	Benchmark and effluent limitation (the same as EPA's 2000 MSGP), except we only require one sample per year for benchmark samples.	None
Washington	Monitoring requirements are specific to the Phase I MS4, *Outfall conveyance system monitoring.* Selected outfalls for representative land uses are monitored intensively for a wide range of chemical constituents including toxicity. *BMP effectiveness monitoring.* Selected stormwater BMPs are monitored to determine performance and how effective the designs are. The Phase II MS4 permit does not require monitoring, except as required under the IDDE program or for a TMDL.	Industry required to sample for turbidity, pH, zinc, and petroleum oil and grease. If exceeds zinc benchmark, then also need to monitor for total copper, total lead, and hardness. There are additional monitoring requirements for different industry categories. For discharges to impaired 303(d) waters monitor required for the pollutants for which the waterbody is impaired.	All state CGP sites are required to do weekly monitoring for turbidity and pH. If benchmark exceeded, specific actions/responses are triggered. For sites which discharge to waters impaired by phosphorous, turbidity, fine sediments, or high pH, monitoring required for these parameters additionally.
West Virginia	NA	Benchmark monitoring. Sector specific.	None
Wyoming	None	Benchmark monitoring for timber, metal mining, concrete and gypsum, junkyards and recycling. Effluent limitation monitoring for coal piles, concrete manufacture, and asphalt emulsion.	None

NOTE: NA, not answered

TABLE C-4 How is Compliance Demonstrated?

State	Municipal	Industrial	Construction
Alabama	MS4 Phase I – monitoring and BMPs MS4 Phase II – BMPs	Monitoring reporting and BMP implementation	Inspections. Monitoring; SWPPP implementation during inspection; aerial reconnaissance
California	Annual and monitoring reporting. MS4 audits and inspections.	Annual and monitoring reporting. Inspections.	Annual certifications. Inspections
Connecticut	Annual and monitoring reporting.	Annual and monitoring reporting. Inspections.	Inspections. SWPPP review and implementation for large projects.
Georgia	Annual and monitoring reporting.	Annual and monitoring reporting.	Reporting.
Hawaii	Annual and Monitoring reporting. Inspections.	Annual and monitoring reporting. Inspections.	Inspections. Reporting.
Maine	Annual reporting and municipal audits.	Inspections and audits, at least two per 5-year permit term.	NA
Minnesota		Annual reporting and inspections.	
Nebraska	MS4 audits and annual reporting.	Inspections and SWPPP implementation.	Inspections and SWPPP implementation—complaint only.
Nevada	Annual reporting. MS4 audits, inspections.	Annual reporting, inspections	Inspections.
New York	Annual reporting and MS4 audits.	Annual and monitoring reporting. Inspections.	Inspections and SWPPP implementation.
Ohio	Annual reporting.	SWPPP implementation.	SWPPP implementation.
Oklahoma	Annual reporting. MS4 audits and compliance schedules.	Annual and monitoring reporting. Inspections.	SWPPP implementation and inspections based on complaints received.

State	Municipal	Industrial	Construction
Oregon	Annual and monitoring reporting.	Annual and monitoring reporting. Action Plan approval.	Inspections and SWPPP implementation.
Vermont	Annual reporting and MS4 audits.	Monitoring reporting.	Inspections, recordkeeping.
Virginia	Registration statement BMP implementation.	Monitoring reporting and inspections.	Inspections. SWPPP and E&S plan implementation.
Washington	Implementation of prescriptive stormwater management program.	Monitoring reporting and inspections.	Inspections and monitoring reporting.
West Virginia	NA	SWPPP implementation and monitoring reporting.	Inspections. SWPPP implementation.
Wyoming	Periodic MS4 audits.	Inspections, monitoring reporting.	Inspections.

NOTE: NA, not answered.

TABLE C-5 To Whom Is the SWPPP Submitted?

State	Municipal	Industrial	Construction
Alabama	MS4 Phase I – Storm Water Management Program (SWMP) sent to state. Should be available for review at the time of inspection. (SWPPP information should also be provided to the department.) MS4 Phase 2 – SWMP submitted with the Notice of Intent (NOI).	No submittal to state. The SWPPP must be kept on site and made available for review at the time of inspection.	No submittal to state. The SWPPP must be kept on site and made available for review at the time of inspection. SWPPP required to be submitted under certain circumstance during registration and re-registration.
California	MS4 Phase 1 – SWMP incorporated as prescriptive requirements in the permit. MS4 Phase 2 – SWMP submitted to state with NOI	No submittal to state. The SWPPP must be kept on site and made available for review at the time of inspection.	No submittal to state. The SWPPP must be kept on site and made available for review at the time of inspection.
Connecticut	NA	The SWPPP is submitted to the state only if requested.	The SWPPP is submitted to the state only if requested.
Georgia	The SWMP is submitted to the state.	The SWPPP is submitted to the state only if requested. Otherwise it is kept on-site.	The E&S Control Plan equivalent to the SWPPP is submitted to the Local Issuing Authority. It is also submitted to the state if the project disturbs more than 50 ac, or if there is no LIA.
Hawaii	NA	The SWMP is submitted to the state.	The SWMP is submitted to the state.
Maine	NA	The SWPPP is submitted to the state only if requested.	The E&S Control Plan equivalent to the SWPPP is submitted to the state for review.

State	Municipal	Industrial	Construction
Minnesota	Phase 1 MS4 - The SWMP is submitted to the state for review and public notice.	The SWPPP is not required to be submitted to the state.	The SWPPP must be must be submitted to the state for review for projects disturbing 50 acres or more, and has a discharge point within 2,000 feet of an impaired or special water listed in the state CGP. A SWPPP must also be submitted for projects proposing to use alternative method(s) for the permanent stormwater management system.
Nebraska	NA	The SWPPP is submitted to the state only if requested.	The SWPPP is submitted to the MS4 permittee and to the state when requested.
Nevada	NA	No submittal to state. The SWPPP must be kept on site.	No submittal to state. The SWPPP must be kept on site.
New York	NA	Some SWPPPs submitted to state (very few).	About 1/6 SWPPPs submitted to state.
Ohio	NA	The SWPPP is submitted to the MS4 permittee and to the state when requested.	The SWPPP is submitted to the state.
Oregon	NA	The SWPPP is submitted to the state on first application and when renewing coverage under the state MSGP.	The SWPPP is submitted to the state on first application and when renewing coverage under the state CGP. Projects that are greater than 5 acres are subject to public notice and comment.

continued next page

TABLE C-5 Continued

State	Municipal	Industrial	Construction
Vermont	NA	A copy of the SWPPP is submitted to the state, and the original kept on site.	The E&S Control Plan is submitted to the state. Low-risk projects have a standard assigned E&S Control Plan – "Low Risk Handbook".
Virginia	NA	No submittal to the state. The SWPPP must be kept on-site.	No submittal to the state. The SWPPP must be kept on-site.
Washington	NA	The SWPPP is submitted to the state upon first application only. Otherwise, the SWPPP must be kept on site and must be made available to the state, the MS4 permittee, or the public upon request.	The SWPPP is not submitted to the state. The SWPPP must be kept on site and must be made available to the state, the MS4 permittee or the public upon request.
West Virginia	NA	The SWPPP is submitted to the state upon first application only.	The SWPPP is submitted to the state.
Wyoming	NA	The SWPPP is submitted to the state for facilities >50 ac. Class 1 waters not eligible for coverage under the state MSGP.	The SWPPP is submitted to the state for projects >100 ac or on Class 1 waters.

NOTE: NA, not applicable.

TABLE C-6 Can an MS4 Inspect Industries Within Its Boundary?

Alabama	Yes, if adequate legal authority exists.
California	Yes. Local agencies inspection to ensure compliance with local stormwater or municipal ordinance.
Connecticut	Yes. Nothing specific. State MSGP requires industries to comply with the stormwater management program of the MS4 in which they are located.
Georgia	Yes
Hawaii	Yes
Maine	Yes
Minnesota	Yes. Capability to do this varies with the MS4.
Nebraska	Yes. Phase 1 MS4s only.
Nevada	Yes
New York	Yes. MS4s can inspect for illicit discharge detection and elimination. Industries can be inspected under local authority, but local inspections are infrequently conducted.
Ohio	Yes. Phase I MS4s can check for MSGP coverage and that a SWPPP exists in conjunction with pretreatment inspections.
Oklahoma	Yes
Oregon	Yes, under various authorities. Pretreatment, industrial stormwater, construction stormwater, etc.
Vermont	Yes. The MS4 can request an inspection but can be denied access.
Virginia	No. No state statute for private property access to inspect for stormwater management. Some do use Fire Marshall's authority through the fire code.
Washington	Yes
West Virginia	NA
Wyoming	Yes. If the MS4 has authority.

NOTE: NA, not answered.

TABLE C-7 What Industries Are Considered High Risk?

State	Description
Alabama	Metal foundries.
California	None specified in the state MSGP. Some MS4 permits may specify high-risk industries. Construction activity discharging to sediment-impaired waterbodies are identified as high risk in the state CGP.
Connecticut	None specified in the state MSGP.
Georgia	None specified in the state MSGP. Facilities that may be causing or contributing to stream impairment are high risk.
Hawaii	None specified in the state MSGP.
Maine	Auto salvage, scrap metal recycling, boatyards and marinas, concrete and asphalt, batch plants, vehicle maintenance facilities.
Minnesota	None specified in the state MSGP. Heavy industries are considered higher risk.
Nebraska	Ethanol, scrap metal recycling.
Nevada	Waste oil recyclers, auto salvage, aggregate mines, cement plants.
New York	Auto salvage, scrap recycling.
Ohio	None specified in the state MSGP. Individual stormwater permits required for some airports, landfills, sand and gravel operations, and bulk terminals.
Oklahoma	None specified in the state MSGP.
Oregon	None specified in the state MSGP.
Vermont	None specified in the state MSGP. Gravel pits, salvage yards, scrap recycling facilities are considered high risk.
Virginia	None specified in the state MSGP.
Washington	MS4 permit identifies a list of industries and land uses that the permittee must inspect (See Permit appendix 8).
West Virginia	None specified in the state MSGP. Mills and auto salvage yards are considered high risk.
Wyoming	None specified in the state MSGP. Case by case based on proximity to high class waters and industry type.

TABLE C-8 Do State BMP Manuals Exist for Implementation Guidance?

State	Municipal	Industrial	Construction
Alabama	No. Use EPA materials.	No. Use EPA Materials.	Yes. State E&S Manual. http://swcc.state.al.us/erosion_hand book.htm
California	Yes. CASQA and Caltrans manuals. Not officially adopted.	Yes. CASQA and Caltrans manuals. Not officially adopted	Yes. CASQA and Caltrans manuals. Not officially adopted
Connecticut	No	No. An SWPPP guidance document is available online.	Yes. E&S Guidelines (2002) and CT Stormwater Quality Manual (2004).
Hawaii	No. Use EPA materials.	No. Use EPA materials.	No. Use EPA materials.
Georgia	Yes. Georgia Stormwater Management Manual.	No. Use EPA materials.	Yes. Manual for Erosion and Sediment Control in Georgia.
Maine	Yes	Yes	Yes
Minnesota	Yes. The Minnesota Stormwater Manual at: http://www.pca.state.mn.us/water/st ormwater/stormwater-manual.html Stormwater BMPs – Protecting Water Quality in Urban Areas at: http://www.pca.state.mn.us/water/p ubs/sw-bmpmanual.html	No. Plan to develop one.	Yes. Fact sheets and guidance at: http://www.pca.state.mn.us/water/st ormwater/stormwater-ms4.html#bmp
Nebraska	No	No	No
Nevada	Yes	Yes	Yes
New York	Yes	Yes. A few state materials.	Yes

continued next page

TABLE C-8 Continued

State	Municipal	Industrial	Construction
Ohio	No. Use EPA materials.	No. Use EPA materials.	Yes. http://www.dnr.state.oh.us/water/rainwater/default/tabid/9186/Default.aspx
Oklahoma	No. Use EPA materials.	No. Use EPA materials.	No. Use EPA materials.
Oregon	No	No. Have BMP technical assistance guidance documents.	Yes. Use of Oregon BMP manual is optional.
Vermont	Yes	No	Yes. Standards for designers, a field guide for contractors (2006), and the Low Risk Handbook.
Virginia	Yes. E&S control and stormwater handbooks.	No	Yes. E&S control and stormwater handbooks.
Washington	Yes. Stormwater Management Manual for Western Washington (2005) and Stormwater Management Manual for Eastern Washington (2004)	Yes. http://www.ecy.wa.gov/programs/wq/stormwater/manual.html	Yes. http://www.ecy.wa.gov/programs/wq/stormwater/eastern_manual/index.html
West Virginia	No	No	Yes
Wyoming	No	No. Refer to manuals from other states.	No. Refer to manuals from other states.

TABLE C-9 Full-Time Staff Dedicated to the Stormwater Program

State	Municipal	Industrial	Construction	Total Statewide
Alabama	1.5	7	25–30	33.5–38.5
California				89
Connecticut				5
Georgia	4.5	2.5	46	53
Hawaii	0.5	1	2	3.5
Maine	0.7	2.5	NA	
Minnesota		4.3	14	36
Nebraska				3
Nevada	1	1.5	3	5.5
New York	7	1	11	19
Ohio				18
Oklahoma				7
Oregon	1	4–5 (shared with construction)	4–5 (shared with industrial)	5–6
Vermont	0.5	2	5	7.5
Virginia	3	8 (shared with other programs)	10	13
Washington	10	17	16	43
West Virginia	NA	1	5	
Wyoming				4

NOTE: NA, not answered.

Appendix D
Biographical Information for the Committee on Reducing Stormwater Discharge Contributions to Water Pollution

Claire Welty, *Chair*, is the Director of the Center for Urban Environmental Research and Education and Professor of Civil and Environmental Engineering at University of Maryland, Baltimore County (UMBC). Dr. Welty's work has primarily focused on transport processes in aquifers; her current research interest is in watershed-scale urban hydrology, particularly in urban groundwater. Prior to her appointment at UMBC, Dr. Welty was a faculty member at Drexel University for 15 years, where she taught hydrology and also served as Associate Director of the School of Environmental Science, Engineering, and Policy. Dr. Welty is the chair of the National Research Council's (NRC's) Water Science and Technology Board and has previously served on three NRC study committees. She is the Chair-Elect of the Consortium of Universities for the Advancement of Hydrologic Science Inc. Dr. Welty received a B.A. in environmental sciences from the University of Virginia, an M.S. in environmental engineering from the George Washington University, and a Ph.D. in civil and environmental engineering from the Massachusetts Institute of Technology.

Roger T. Bannerman has been an environmental specialist for the Wisconsin Department of Natural Resources for over 30 years. For most of that time he has directed research projects investigating urban runoff. Topics addressed by his studies over the years include the quality of urban streams, identification of problem pollutants in stormwater, toxicity of stormwater pollutants, effectiveness of different stormwater control practices, sources of stormwater pollutants, selection of cost-effective control practices, and benefits of low-impact development. He has applied these results to management plans developed for most urban areas in Wisconsin. This includes the calibration of the urban runoff model called the Source Loading and Management Model. The results of his research projects have been used to develop Wisconsin's new administrative rules that regulate stormwater management. Mr. Bannerman received his B.S. in chemistry from Humboldt State College and an M.S. from the University of Wisconsin in water chemistry.

Derek B. Booth has joint positions as Senior Geologist at Stillwater Sciences, Inc., and Adjunct Professor at the University of Washington where he is senior editor of the international journal *Quaternary Research* and holds faculty appointments in Civil Engineering and Earth & Space Sciences. Prior to this, he was director of the Center for Urban Water Resources Management (and its successor, the Center for Water and Watershed Studies) at the university. He maintains active research into the causes of stream-channel degradation, the effectiveness of stormwater mitigation strategies, and the physical effects of urban development on aquatic systems, with over a dozen publications and a wide range of national and international invited presentations on the topic. Dr. Booth received a B.A. in literature from Hampshire College, a B.A. in geology from the University of California at Berkeley, an M.S. in geology from Stanford University, and a Ph.D. in geological sciences from the University of Washington.

Richard R. Horner is a professor in the Department of Civil and Environment Engineering at the University of Washington, with adjunct appointments in Landscape Architecture and in the College of Forest Resources' Center for Urban Horticulture. He received his Ph.D. from the University of Washington's Department of Civil and Environmental Engineering and previous engineering degrees from the University of Pennsylvania. Dr. Horner splits his time between university research and private practice. In both cases his work concerns how human occupancy of and activities on the landscape affect natural waters, and how negative effects can be reduced. He has been involved in two extended research projects concerning the ecological response of freshwater resources to urban conditions and the urbanization process. The first studied the effect of human activities on freshwater wetlands of the Puget Sound lowlands and led to a comprehensive set of management guidelines to reduce negative effects. A ten-year study involved the analogous investigation of human effects on Puget Sounds' salmon spawning and rearing streams. In addition, he has broad experience in all aspects of stormwater management, having helped design many stormwater programs in Washington, California, and British Columbia. He previously served on the NRC's Committee on the Comparative Costs of Rock Salt and Calcium Magnesium Acetate for Highway Deicing.

Charles R. O'Melia (NAE) is the Abel Wolman Professor of Environmental Engineering and Chair of the Geography and Environmental Engineering Department at the Johns Hopkins University, where he has served on the faculty for over 25 years. Dr. O'Melia's research areas include aquatic chemistry, environmental colloid chemistry, water and wastewater treatment, modeling of natural surface and subsurface waters, and the behavior of colloidal particles. He has served on the advisory board and review committees for the environmental engineering departments of multiple universities. He has served in a range of advising roles to professional societies including the American Water Works Association and Research Foundation, the Water Pollution Control Federation, the American Chemical Society, and the International Water Supply Association. He has served on several NRC committees, including chairing the Steering Committee, Symposium on Science and Regulation, and the Committee on Watershed Management for New York City. He was also a member of the

NRC Water Science and Technology Board and the Board on Environmental Studies and Toxicology. Dr. O'Melia earned a Ph.D. in Sanitary Engineering from the University of Michigan. In 1989, Dr. O'Melia was elected to the National Academy of Engineering for significant contributions to the theories of coagulation, flocculation, and filtration leading to improved water-treatment practices throughout the world.

Robert E. Pitt is the Cudworth Professor of Urban Water Systems in the Department of Civil, Construction, and Environmental Engineering at the University of Alabama (UA). He is also Director of the UA interdisciplinary Environmental Institute. Dr. Pitt's research concerns the effects, sources, and control of urban runoff, which has resulted in numerous development management plans, stormwater ordinances, and design manuals. Dr. Pitt has also developed and tested procedures to recognize and reduce inappropriate discharges of wastewaters to separate storm drainages. He has investigated the sources and control of stormwater toxicants and examined stormwater effects on groundwater. He has also carried out a number of receiving water impact studies associated with stormwater. These studies have included a variety of field monitoring activities, including water and sediment quality, fish and benthos taxonomic composition, and laboratory toxicity tests. His current research includes developing a nationwide database of national stormwater permit information and conducting comprehensive evaluations of these data. Dr. Pitt received a B.S. in engineering science from Humboldt State University, an M.S. in civil engineering from San Jose State University, and a Ph.D. in civil and environmental engineering from the University of Wisconsin.

Edward T. Rankin is an Environmental Management Associate with Ohio University at the Institute for Local Government Administration and Rural Development (ILGARD) which is the Voinovich School of Leadership and Public Affairs located in Athens, Ohio. He had previously been a Senior Research Associate in the Center for Applied Bioassessment and Biocriteria within the Midwest Biodiversity Institute (MBI). Prior to 2002, he was an aquatic ecologist with Ohio EPA for almost 18 years. Mr. Rankin's research centers around the effects of stormwater and other urban stressors on aquatic life, development and application of stream habitat assessment methodologies, development and application of biological criteria and biological-based chemical criteria for aquatic life, and improving the accuracy of total maximum daily loads for nutrients and sediment. He is particularly interested in the application of research to management of aquatic life issues and has extensive experience with the development of tiered aquatic life uses and use attainability analyses in streams. Mr. Rankin received his B.S. in biology from St. Bonaventure University and his M.S. in zoology from The Ohio State University.

Thomas R. Schueler founded the Center for Watershed Protection in 1992 as a nonprofit organization dedicated to protecting our nation's streams, lakes and wetlands through improved land management. In 2007, he launched the Chesapeake Stormwater Network, whose mission is to improve on-the-ground implementation of more sustainable stormwater management and environmental site design practices in each of 1,300 communities and seven states in the Chesapeake Bay Watershed. He has conducted extensive research on the pollutant removal performance, cost, and longevity of stormwater control measures, and he has developed guidance for both Phase I and Phase II communities to meet minimum management measures to comply with municipal stormwater permits, including development of a national stormwater monitoring database and national guidance on illicit discharge detection and elimination. Mr. Schueler has written several widely referenced manuals that describe how to apply the tools of watershed protection and restoration, and he is working on a wide range of research projects and watershed applications across the United States. Prior to founding the Center, he worked for ten years at the Metropolitan Washington Council of Governments, where he led the Anacostia Watershed Restoration Team, one of first efforts to comprehensively restore an urban watershed. He received his B.S. in environmental science from the George Washington University.

Kurt Stephenson is an associate professor of Environmental and Natural Resource Economics in the Department of Agricultural and Applied Economics at the Virginia Polytechnic Institute and State University. His professional objective is to better integrate economic perspectives and analysis into decision making related to water resource issues. Particular emphasis is placed on the application of economic analysis to interdisciplinary research of policy issues. The design and implementation of market-based policies to secure environmental objectives is a primary area of study within this context. He is currently involved in determining effective strategies for reducing nutrient loads in the Opequon Watershed in Virginia and West Virginia, including evaluating the cost effectiveness and feasibility of using urban nonpoint source controls (including stormwater management) as an offset to growth in point source loads. He is a member of the Virginia Department of Environmental Quality's Nutrient Trading Technical Advisory Committee and the Academic Advisory Committee. Dr. Stephenson received his B.S. in economics from Radford University, his M.S. in agricultural economics from Virginia Tech, and his Ph.D. in economics from the University of Nebraska.

Xavier Swamikannu is Chief of the Stormwater Permitting Program for the Los Angeles Regional Water Board and the California EPA, where he has worked for nearly 20 years. He has extensive experience with the implementation of municipal and industrial stormwater programs in Southern California, including the evaluation of pollutant discharges, determining the effectiveness of stormwater control measures in treating stormwater runoff, developing performance criteria and better understanding of their costs. He has participated on EPA's General Permits and Total Maximum Daily Load Work Groups and he has served on many state and regional technical advisory committees concerned with stormwater regulations. He was recognized by the California Water Boards in 2007 for his national leadership in the stormwater program, and by the California State Senate for his service on the technical advisory committee of the Santa Monica Bay Restoration Commission. Dr. Swamikannu received his B.S. in natural and chemical sciences from St. Joseph's College in Bangalore, India, his M.S. in environmental sciences from Texas Christian University, and his Ph.D. in environmental science and engineering from the University of California, Los Angeles.

Robert G. Traver is a professor of Civil and Environmental Engineering at Villanova University and the Director of the Villanova Urban Stormwater Partnership. He conducts research on topics that include modeling of stream hydraulics, urban hydrology, water quality, and measures to mitigate stormwater effects of urbanization. Most recently he has created a Stormwater Best Management Practice Demonstration and Research Park on the Villanova Campus. Dr. Traver is also involved with the implementation of stormwater policy. He has participated in a team study to review the effects of Pennsylvania's water regulation from a watershed sustainability viewpoint, acted as a reviewer for Pennsylvania's 1995 Best Management Practice Handbook, and has served as Chair for the 1998, 1999, 2001, 2003, and 2005 Pennsylvania Stormwater Management Symposiums held at Villanova. More recently he was selected to serve on the American Society of Civil Engineers' External Review Panel of the Corps investigation of Hurricane Katrina. Dr. Traver is a retired LTC in the Army Reserves and a veteran of Operation Desert Storm. He received his B.S. in civil engineering from the Virginia Military Institute, his M.S. in civil engineering from Villanova, and his Ph.D. in civil engineering from Pennsylvania State University.

Wendy E. Wagner is the Joe A. Worsham Centennial Professor at the University of Texas School of Law. Before joining the UT faculty, she was a professor at Case Western Reserve University School of Law and a visiting professor at Columbia Law School and the Vanderbilt School of Law. Wagner's research focuses on the interface between science and environmental law, and her articles have appeared in numerous journals, including the Columbia, Cornell, Duke, Georgetown, Illinois, Texas, Wisconsin, and Yale Law Reviews. She has published on the practical problems with EPA's current approach to stormwater regulation. She has also written several articles on the challenges of regulating media like stormwater, on restoring polluted waters with public values, on the legal aspects of the regulatory use of environmental modeling, and on technology-based standards. Ms. Wagner received a master's degree in environmental studies from the Yale School of Forestry and

Environmental Studies and a law degree from Yale Law School. She clerked for the Honorable Judge Albert Engel, Chief Judge of the U.S. Court of Appeals for the 6th Circuit.

William E. Wenk is founder and president of Wenk Associates, Inc., a Denver-based landscape architectural firm. He is also an Adjunct Associate Professor of Landscape Architecture at the University of Colorado in Denver. For over 20 years, he has been influential in the restoration and redevelopment of urban river and stream corridors, the transformation of derelict urban land, and the design of public parks and open spaces. Mr. Wenk was the Principal Urban Designer for the Menomonee River Valley Redevelopment, an award-winning "green infra-structure" redevelopment in Milwaukee that integrated a network of parks and open spaces through stormwater infrastructure, regional and local trails, and a restored river corridor into a proposed 130-acre mixed-use and light industrial development. Other projects of his include the Prairie Trail Community Master Plan in Ankeny, Iowa (a surface stormwater system designed to provide flood control and water quality for a new 1000-acre mixed-use community), and the Stapleton Airport Parks and Open Space Redevelopment (a surface stormwater drainage design for the 4,500-acre redevelopment), as well as the Stapleton Water Quality Guidelines book to guide planners and developers on how to inte-grate stormwater best management practices into redevelopment. Mr. Wenk received a B.S.L.A. and M.L.A. from Michigan State University and the Univer-sity of Oregon, respectively.

Laura J. Ehlers is a senior staff officer for the Water Science and Technology Board of the National Research Council. Since joining the NRC in 1997, she has served as the study director for eleven committees, including the Committee to Review the New York City Watershed Management Strategy, the Committee on Bioavailability of Contaminants in Soils and Sediment, the Committee on Assessment of Water Resources Research, and the Committee on Public Water Supply Distribution Systems: Assessing and Reducing Risks. Ehlers has periodically consulted for EPA's Office of Research Development regarding their water quality research programs. She received her B.S. from the California Institute of Technology, majoring in biology and engineering and applied science. She earned both an M.S.E. and a Ph.D. in environmental engineering at the Johns Hopkins University. Her dissertation, entitled RP4 Plasmid Transfer among Strains of *Pseudomonas* in a Biofilm, was awarded the 1998 Parsons Engineering/Association of Environmental Engineering Professors award for best doctoral thesis.